S0-BLA-371

THE SCIENTIFIC PRINCIPLES OF CROP PROTECTION

THE SCIENTIFIC PRINCIPLES OF CROP PROTECTION

By

HUBERT MARTIN

D.Sc. (Lond.), A.R.C.S., F.R.I.C., F.C.I.C.

Director, Science Service Laboratory,
Canada Agriculture,
London, Ontario.

LONDON
EDWARD ARNOLD (PUBLISHERS) LTD.

First published in 1928
Second edition 1936
Third edition 1940
Reprinted 1944, 1948
Fourth edition 1959

Printed in Great Britain by
Metcalfe & Cooper, Ltd. 10-24 Scrutton Street, London, E C 2

PREFACE

Entomology and Mycology are concerned with the organisms responsible for injury and disease among crop plants, but the problems of the control of crop pests—Crop Protection—involve, not only biological, but equally important, chemical and physical studies. Indeed, so extensive is the field that co-operation between the entomologist or mycologist and the chemist, physicist or plant physiologist is recognized as the most economical and successful method of progress. The primary purpose of this book is to assist this co-operation by a survey of the scientific principles underlying modern methods of control of crop pests.

Advisory officers and research workers in Plant Pathology find that the literature of Crop Protection is scattered and, moreover, that the discussion of control measures of any particular fungus or pest is frequently overshadowed by purely biological data. Yet the principles underlying these control measures permit of classification and co-ordinated discussion.

As successful control is based only upon accurate knowledge of the life-history and habits of the organism responsible, the Advisory Officer is rightly recruited from biologists. He is therefore better acquainted with the biological than with the physico-chemical aspects. For this reason, the latter aspects are dealt with in greater detail, though it is hoped that this emphasis will not be misinterpreted as an implication of the greater importance of chemical over biological and cultural methods of control. The object has been to present to the mycologist and entomologist a detailed survey of the physico-chemical factors and to provide the chemist and physicist with a means of approach to the biological side.

During the years that have elapsed since the writing of the third edition (1940) the search for synthetic pesticides by the chemical manufacturer has met astounding success and the range of chemicals available for crop protection has been vastly extended. Fortunately their technical data are available elsewhere and attention can be turned to their biochemistry and toxicology.

The details of actual methods of insecticide and fungicide application used in practice, including also the agricultural engineering problems, have formed the subject of bulletins and leaflets in the different countries of the world. As they must vary with the conditions found in those countries their applicability is limited and, except where examples have been found useful, such practical details have not been mentioned. Nor has the inclusion of the principles underlying the control of non-parasitic diseases been possible. A discussion of these diseases, due in a broad sense to unfavourable soil and climatic conditions, would involve a consideration of nutritional and environmental factors beyond the scope of a single volume.

The present edition follows the plan of the third edition except that the expansion of knowledge of the mechanisms of toxic action now permits a co-ordinated treatment and a discussion of the general principles prior to their specific application to the several groups of pesticides.

In 1928 when this book was first prepared, the lack, in common English usage, of a suitable word to cover the subject led to the choice of the German " Pflanzenschutz ". With the passage of years Crop Protection has become more familiar and the change of title follows this trend.

The author acknowledges with gratitude the advice and help of his colleagues both at the Long Ashton Research Station and at the Science Service Laboratory in London, Ontario. His contacts with the biologists have been so close and numerous that personal acknowledgments are impossible except to mention that certain chapters of the present edition have been much improved by the help of Drs. R. A. Ludwig and E. Y. Spencer of this laboratory and by Dr. E. Somers of the Long Ashton Research Station.

H. Martin

Science Service Laboratory,
Canada Agriculture,
University Sub Post Office,
London, Ontario, Canada.

CONTENTS

CHAP.		PAGE
I	INTRODUCTION	1
II	PLANT RESISTANCE	7
	Production of resistant varieties.—Use of resistant varieties.—Nature of plant resistance : to pathogens ; to insects.	
III	THE INFLUENCE OF EXTERNAL FACTORS ON THE SUSCEPTIBILITY AND LIABILITY OF THE PLANT TO ATTACK	29
	Nutritional factors : fertilizers, soil conditions.—Climatic factors : meteorological forecasting of epidemics, control of temperature and humidity relationships.—Modification of time of sowing.	
IV	BIOLOGICAL CONTROL	44
	General principles.—Use of higher animals. Use of insects against insects, weeds.—Use of fungi against insects, fungi.—Bacterial diseases of insects, higher animals.—Virus diseases of insects, higher animals.	
V	FUNGICIDES AND INSECTICIDES	60
	Principles underlying use of sprays, dusts, fumigants.—Spreaders, mode of action, comparison and classification ; stickers ; protective colloids and dispersing agents ; emulsifying agents, properties and preparation of emulsions.	
VI	THE MEASUREMENT AND MECHANICS OF TOXICITY	85
	The principles of bioassay.—Probit analysis.—Hypotheses on the mechanics of toxic action.	
VII	FUNGICIDES	106
	The sulphur group.—Sulphur : action on fungus, on plant ; Lime sulphur and the polysulphide group : action on fungus, on plant ; The copper group : action on fungus, on plant.—The dithiocarbamate and thiuram derivatives.—Glyodin.—Captan. — The quinone group. — Antibiotics. — Miscellaneous fungicides.	
VIII	INSECTICIDES (I)—INORGANIC COMPOUNDS	147
	The arsenic group : chemistry and action on insect, on plant, supplements.—The fluorine group.—Miscellaneous stomach insecticides.	

CHAP.		PAGE
IX	INSECTICIDES (II)—NATURALLY-OCCURRING CONTACT INSECTICIDES	160
	Nicotine: chemistry, action on insect, supplements, dusts.—Compounds of structure akin to nicotine.—Pyrethrum: toxicity and structure relationships, synergists.—Rotenone and related insecticides, the chemistry of the rotenoids.—Quassia.—The hydrocarbon oils: historical, phytotoxic and insecticidal properties.—The glyceride oils and soaps.—Miscellaneous contact insecticides of plant origin.	
X	INSECTICIDES (III)—SYNTHETIC CONTACT INSECTICIDES . .	198
	The dinitrophenols.—The organic thiocyanates.—The chlorinated hydrocarbons: DDT, relationships between structure and insecticidal properties, action on insect; lindane; chlorinated terpenes; the cyclodiene group.—Organophosphorus compounds, action on insect.—The carbamate group.—The bridged diphenyl group.	
XI	WEEDKILLERS	232
	General.—The 2,4-D group: action on plant.—The carbamate group.—The heterocyclic nitrogen group.—The substituted urea group.—The trichloroacetate group.—Miscellaneous group.	
XII	FUMIGANTS	245
	Principles.—Hydrocyanic acid, action on insects, on plants.—Naphthalene. — Tetrachloroethane. — Nicotine. — Methyl bromide.—Miscellaneous fumigants.	
XIII	SEED TREATMENT	254
	Mechanical methods.—Chemical methods: the copper group; formaldehyde; the sulphur group; the mercury group, action on spore, on seed.—Miscellaneous seed disinfectants.—Physical methods.	
XIV	SOIL TREATMENT	269
	Partial sterilization; use of heat, carbon disulphide, coal tar antiseptics, halogenated hydrocarbons, formaldehyde, miscellany; effects on plant growth.—Soil conditions and the pest.—Mechanical methods.	
XV	TRAPS	282
	Chemotropism: attractants, trap crops, repellents.—Phototropism. — Stereotropism. — Miscellaneous traps. — Poisons for use in traps and baits.	
XVI	THE TREATMENT OF THE CENTRES AND VECTORS OF INFECTION	297
	Elimination of infection centres; general, eradication of host plant, weed hosts, alternate hosts.—Elimination of infection vectors: wind, water and insect transmission: cultural and accessory vectors.	
	AUTHOR INDEX	311
	SUBJECT INDEX	327

THE SCIENTIFIC PRINCIPLES OF CROP PROTECTION

CHAPTER I
INTRODUCTION

DARWIN'S concept of a "struggle for existence" among living organisms has as a corollary the idea of a "natural balance". The perpetuation of a plant species growing under conditions undisturbed by man will depend upon its ability to reach maturity and reproduce despite the many factors which tend towards its suppression. Of these factors, an important group comprises the pests of the plant, which in the course of their existence cause direct injury or prejudice its successful development. If the plant tends towards an increase, conditions will favour the multiplication of the pest; the increase of the pest brings about, in turn, a reduction of the productivity of the plant. These processes result in the establishment of an unstable equilibrium—the "natural balance". By the cultivation of those plants useful to him as crop plants, man has upset this natural balance and, in improving growth conditions for his crops, he has provided improved conditions for the pests and diseases of that crop. Hence the need for crop protection.

If wheat, for example, grew only as isolated plants among other vegetation, the chance of survival of the spore of the fungus to which smut of wheat is due would be meagre. But when wheat is cultivated as a crop, not only is the spore of the smut fungus more likely to find a suitable host, but the conditions for the fungus are rendered so favourable that it may become of primary importance in the determination of the yield and quality of the wheat crop.

Prior to 1850, the small beetle now known as the Colorado beetle was an insignificant insect feeding upon solanaceous plants of the Rocky Mountains. The settler, who brought with him and who planted the cultivated potato, innocently supplied the beetle, not only with an abundance of suitable food, but with the means whereby to extend its activities. From 1850 onwards, the beetle has slowly spread from its original home; by 1859 it had reached Nebraska, by 1870 it had travelled as far north as Ontario, and by 1874, continuing its career, it had reached the Atlantic coast. Prior to 1922, sporadic outbreaks occurred on a few occasions in Europe, but were, with one exception, obliterated by eradicative measures. In 1922 colonies of the beetle were found in the Gironde district of France, brought, it is thought (*1*), in food supplies with American Expeditionary Forces during the 1914–18 war. Although this outbreak was held in check until 1927, the beetle was, by 1946, established throughout western Europe to the Elbe and to Austria.

Drastic action reported annually since 1946 (*2*) has held Great Britain clear and, in Europe, its control is co-ordinated by the European Plant Protection Organization.

The history of agriculture abounds with such examples of the result of man's interference with the original natural balance, yet it is comparatively recent that crop protection has received scientific study. This delay arises from the belated recognition of the causative agents of what were previously considered uncontrollable disorders. Larger organisms, such as higher plants and animals, are automatically controlled to a great extent by the processes of cultivation and of civilization. The smaller, unknown to the early husbandman, have only within the last century been recognized and studied. With the understanding of the nature of plant diseases, the possibilities and importance of crop protection have increased.

The urgency of the need for methods of control of the biological factors responsible for crop losses has increased from yet other aspects. In nature, as a result of the struggle for existence, the tendency has been for the evolution of a variety of plant whose reproductive powers are delicately balanced by its ability to withstand the attack of pests. In cultivated varieties, however, qualities such as high productivity or special market virtues may outweigh in importance the ability of the variety to withstand pests and diseases. Indeed, it may be said that in general the heavier cropping and more specialized varieties of plants are more susceptible to attack by these organisms. Again, present knowledge of the soil and nutrient factors of crop production is such that the grower is able to modify and control them so as to secure the most profitable yield. It becomes all the more important therefore that he should be able to control the biological factors in a similar manner. The study of crop protection may then be granted an importance no less fundamental to the study of crop production than that of soil tillage—agriculture in its primary sense—and that of plant nutrition.

Of the organisms which may be classed as the plant's pests, representatives are to be found from the whole range of the plant and animal kingdoms. Of the plant kingdom, there are first those of the higher plants, classed under the general term of " weed ", whose presence is detrimental. The methods of soil preparation, of sowing and of soil tillage after sowing, are designed, not only to improve the growth conditions of the crop, but also to reduce weeds. In this sense, cultivation is perhaps the most important of the methods of pest control. There are further to be considered those of the higher plants (Spermatophytes), such as the Dodders (*Cuscuta* spp.), which are true parasites in that they derive all or part of their food direct from the living plant. Algae are not important as crop pests ; the fungi are, however, in the front rank. Botanically, the fungi (Eumycetes) are cellular plants differing from the algae in their complete lack of chlorophyll or chromatophores. For this reason, the fungi are unable to build up the food they require from the constituents of soil, air and water as do green plants, but, in order to live, must feed upon organic matter, either in the form of decaying plant or animal remains (saprophytic fungi), or in the form of living plant or animal matter (parasitic fungi). Of the Myxomycetes or " Slime Fungi " certain members, such as *Plasmodiophora brassicae* Woron. causing the club root of cruciferous plants, need consideration. At the lower end of the scale are the Schizomycetes or bacteria, of which a number are the causal agents of disease in plants though, on the whole, bacterial pathogens have not the importance in phytopathology that they have in animal pathology. These three groups, the Eumycetes, the Myxomycetes and the Bacteria, were formerly all classed as fungi. This older classification is often retained, as in the word " Fungicide ", a chemical toxic to these organisms, and a term used for such substances employed for their control.

Among the higher animals, the vertebrates, there are special cases where civilization has not entirely controlled those injurious to crops. The question of whether any species of bird causes, in the long run, any serious damage to crops, is controversial. Rodents, such as voles, rats and rabbits, may sometimes cause injury through their abnormal abundance. The ravages of these pests are also serious in orchards or vineyards, which represent a long-term investment for their owners.

Of the invertebrates, the arthropods form a group of great importance as crop pests for it includes the true insects, animals having three pairs of legs and whose adult body consists of three parts. The arachnids, represented by the red spider and other mites, which differ from insects in having no distinct division of the body into three parts and usually possessing four pairs of legs, also figure prominently. Of the lower orders of animals, certain of the nematodes (thread-like worms, generally with unsegmented bodies), and possibly of the protozoa, must be included as plant pests.

The Arachnida and Insecta are usually classed together in the popular sense of the word " insect ". Therefore, in a manner analogous to the use of " Fungicide ", the term " Insecticide " is taken to embrace those substances toxic to insects in this wider sense.

This list would be incomplete without viruses. Virus diseases are common to both the animal and plant worlds and are a type of infectious disease of which the causative agent cannot be made visible under the light microscope. The electron microscope has revealed that certain viruses, such as bacteriophages, possess a well-defined structure whereas others, such as the tobacco mosaic virus, are simply cylindrical protein rods with a core of ribonucleic acid. All viruses appear to have the remarkable property of building their nucleic acid and protein components from the nucleic acids and amino acids of their hosts, thus acquiring a power of multiplication or reproduction which justifies their inclusion in the catalogue of crop pests.

The damage caused by the individual species of this list may be but slight or it may be serious enough to threaten the cultivation of the attacked plant. The selection of control measures will be guided, in each case, by cost in relation to loss or inconvenience caused by the pest. To give an estimate of the monetary loss due to diseases and pests presents obvious difficulties, but reliable estimates are of staggering dimensions (*3*). It is doubtful whether the labour involved in compiling such estimates is repaid by their trustworthiness for, in their calculation, due weight cannot be given to the factors such as the effects of overproduction had the disease or pest not reduced yield and the possibility that nutritional or other factors might have limited yield had the pest not intervened. A more profitable study is of the type carried out by the California Agricultural Experiment Station (*4*) or by Stevens and Wood (*5*) on the economic effects of crop protection. Probably the greatest benefit of crop protection is the additional powers it confers on growers to control and stabilize production. The plum or hop grower need no longer gamble his livelihood on the weather being unfavourable to the aphids infesting his crop. The potato grower is now able to curb the violent fluctuations of the potato crop (*6*) due to blight, the direct cause of the Irish famines of the 1840's and indirectly responsible for such diverse events as the Repeal of the Corn Laws and, through the Irish immigrations to the New World, the preponderance of Irishmen among the police of New York.

Historically the question of pest control is of great antiquity. The importance of the loss caused to the early land-worker by the ravages of an insect such as the locust, is reflected in the story of the plagues of Egypt, and it is rather surprising to find " blasting and mildew " of corn, the result of fungus attack, recognized in the catalogue of afflictions which, according to the Mosaic law, would befall the evildoer (*7*). The Romans, to preserve their fields from " mildew " (probably the rust of wheat), held an annual festival—the Robigalia—in honour of the deity Robigus or Robigo, a rite the institution of which is considered to be due to Numa, the legendary second king of the Romans. But more mundane remedies were also employed. Pliny, in his *Historia Naturalis*, Books 17 and 18 (A.D. 77), gave a series of measures for the control of this particular group of cereal diseases. Apart from mechanical methods of dealing with the more obvious insect pests, such early control measures were of a rule-of-thumb type ; rational means of control could only arise after the cause of the injury was known.

From this point of view, the history of crop protection is seen to follow closely the history of the scientific study of the organisms which have been described as the plant's enemies. Originally the study of plant diseases—since no responsible organism was recognized—fell to the botanist and was entirely physiological in outlook. Thus, even in 1841, Meyen (*8*) considered that smut spores were the product of an excessive and abnormal nutrition. The recognition of the parasitism of fungi dates from 1853, when Anton de Bary published his classical work upon " die Brandpilze " (*9*). Thus did plant pathology pass slowly to the realm of mycology, and even the study of the non-parasitic diseases became the work of the applied mycologist. The first general treatise embracing this wider view of plant pathology appeared as a series of articles by Berkeley (*10*). Later, as a result of the work of Louis Pasteur on bacteria and of the introduction of the " plate " method of isolating bacteria by Robert Koch (1881), the existence of bacterial diseases of plants became established, to the credit mainly of Thomas

Burrill (1879–81). The proof of the infectious nature of the mosaic disease of tobacco, by Iwanowski in 1892, provided yet a further section for the work of the phytopathologist.

Today, therefore, plant pathology, phytopathology, or applied mycology—the terms are often used as if synonymous—embraces the study of the non-parasitic diseases and those due to faulty cultivation and malnutrition of the plant in addition to that of the diseases due to pathogens. Further the study of the virus diseases, and even of diseases due to animal parasites such as nematodes and protozoa, is part of phytopathology.

So, by the accident of its evolution, plant pathology is frequently considered not to include the study of those diseases due to insects, using the word " insects " in its widest sense. Applied mycology and economic entomology have remained apart, the latter subject embracing not only the control of " insect " pests but also the control of noxious higher animals.

If the development of the study of crop protection were to be traced in a like manner, employing as milestones the prominent events in its history, those events are found to coincide closely with the occurrence of particular epiphytotics. The appearance in Europe of the powdery mildew of the vine (1845) was followed by an extension of the use of sulphur as a fungicide (1848); the severe outbreak, in France, of the downy mildew of the vine (1879) was followed by Millardet's discovery of the fungicidal value of the copper sulphate-lime mixture (1882). The spread of the Colorado beetle (1850–9), which has been mentioned above, was countered by the introduction of Paris Green (1867), an arsenical later displaced by lead arsenate (1892), a material first used in the United States during the Gipsy moth plague of 1889. The introduction of hydrocyanic acid and the " tent " method of fumigation (1886) followed the rapid spread, in California, of certain scale insects (1886); the loss caused by these pests also led to the discovery of the possibilities of biological control (1889).

Tremendous developments in the range of chemicals available for crop protection followed by the 1939–45 War. The need for insecticides for use against insect vectors of human diseases such as malaria and trench fever increased, and the reduction of supplies of the non-poisonous derris and pyrethrum prompted an organized search for alternative materials by the Western Powers. The discovery of the synthetic insecticides DDT and BHC was therefore brought to practical use in short time, though these discoveries as well as those of the weed killers 2,4–D and MCPA, were held secret until the end of the war. By 1951 the domestic annual consumption of DDT in the United States alone had reached 70 million pounds whereas that of 2,4–D amounted to over 20 million pounds (*11*).

The profitable market for the synthetic pesticide revealed by this development attracted chemical industry from whose laboratories have emerged an apparently inexhaustible supply of compounds of potential usefulness in crop protection. Many are already established but the search for others continues even though it has been estimated that the cost of discovery and developing a new pesticide is well over a million dollars (*12*).

It is interesting to follow the history of the control of one or more particular pests, but as this is the method which has been adopted for the discussion of the chemical methods of seed treatment, it is unnecessary to give further examples.

Any attempt to classify, under definite headings, the various methods now employed in crop protection is rendered difficult, not so much by their complexity as by their interdependence one upon the other. By adopting groups sufficiently wide, it is convenient to summarize these methods by the following scheme. The large majority fall into the first group (A) and may be termed—

(A) *Preventive*, under which head may be included those methods concerning more directly:
- (1) The Host Plant:
  - (a) The use of resistant varieties.
  - (b) The treatment of the plant with a chemical capable of protecting it from attack.
  - (c) The use of the influence of external factors on the susceptibility of the plant to attack.
- (2) The Parasite:
  - (a) The modification of external factors to curb the activity of the parasite.
  - (b) The use of biological agencies deleterious to the parasite.
  - (c) The destruction of the parasite either—
    - (i) on the "seed" } prior to planting.
    - (ii) in the soil } prior to planting.
  - (d) Mechanical means whereby the parasite is trapped or infection prevented.
  - (e) The treatment of centres and vectors of infection.

(B) *Curative*: the destruction of the parasite, after its attack on the plant is established, by the application of toxic chemicals.

REFERENCES

(*1*) Sascer, E. R., *J. econ. Ent.*, 1940, **33**, 1.
(*2*) *Agriculture, Lond.*, 1946, **53**, 129; up to 1956, **63**, 92.
(*3*) See Morstatt, H., *Handbuch der Pflanzenkrankheiten, Berlin*, 1952, Vol. 6, p. 9.
(*4*) Smith, H. S. *et al.*, *Bull. Calif. agric. Exp. Sta.* 553, 1933.
(*5*) Stevens, N. E. and Wood, J. I., *Bot. Rev.*, 1937, **3**, 277.
(*6*) Hartley, C. and Rathbun-Gravatt, A., *Phytopathology*, 1937, **27**, 159.
(*7*) Deuteronomy xxviii, 22.
(*8*) Meyen, F. J. F., *Pflanzen-Pathologie*, Berlin, 1841.
(*9*) de Bary, A., *Untersuchungen über die Brandpilze und die durch sie verursachten Krankheiten der Pflanze*, Berlin, 1853.
(*10*) Berkeley, M. J., *Gdnrs.' Chron.*, 1854, p. 4, *et seq.*
(*11*) Shepard, H. H., *J. Agr. Food Chem.*, 1951, **1**, 756.
(*12*) Wellman, R. H., *Agr. Chem.*, 1952, **7**(9), 32.

CHAPTER II

PLANT RESISTANCE

THE PRODUCTION OF RESISTANT VARIETIES

Nature favours varieties of plants possessing a high average resistance to deleterious biological factors. Man intervenes by a purposeful selection for propagation of those plants which he considers better than average in yield or some other quality. This practice, based on the rule " like begets like ", must be as old as civilization, but proof of the hereditary nature of resistance in plants was first clearly established by Biffen (*1*) who showed that the degree of susceptibility of cereals towards yellow rust (*Puccinia glumarum* (Schm.) Erikss. & Henn.) is a property which follows Mendelian laws.

The selection of resistant varieties is a process which may be accelerated in several ways. As the object of the selector is the production of varieties resistant to but one or two diseases or pests, he can maintain conditions of infestation as high as possible. He can control the fertilization of the selected variety preventing dilution by susceptible parents and " fixing " the resistant variety. By maintaining, in plant growth chambers, those conditions of temperature, day length, and nutrition most suitable for rapid growth, he can speed up the development of the plant ; at Ottawa, e.g., it has been possible to grow as many as five generations of barley within a year (*2*).

An example of the production of resistant varieties by selection is afforded by the work of the Empire Cotton Growing Corporation Station at Barberton in South Africa. The cultivation of cotton in South Africa had been severely handicapped by the uncontrollable attack of the cotton jassid *Empoasca facialis* Jac. Following the observation that certain strains of the American upland cotton (*Gossypium hirsutum* L.) and the Cambodia variety resisted jassid attack (*3*), a programme of selection was initiated to discover strains possessing both resistance and the desirable qualities of good staple and high yield. Worrall (*4*) observed that resistance was accompanied by hairiness, but as hairiness is associated in most cases with short staple, he advised that only hairy plants of good staple be used for selection. This work, which later was supplemented by a breeding programme (*4*), led to the introduction of a number of jassid-resistant cotton strains most successful in Rhodesia and South Africa.

In crop plants, with the important exception of those capable of vegetative reproduction, selection can be successful only if the qualities which confer resistance are hereditary. The selector is dependent on the genetic variability within the strains he has available ; the re-discovery of Mendel's law opened up the possibility of a purposeful use of hybridization to create new genetic variability on which to practise selection.

The qualities of the parents are inherited by the offspring through the agency of genes, nucleoprotein units of which most, if not all, are carried by the chromosomes. Each gene is of remarkable stability and but rarely undergoes the change known as mutation which is revealed by an alteration of heritable qualities. During normal growth the cell divides and the nucleus undergoes mitosis, a process in which each chromosome re-duplicates itself so that daughter nuclei have a complement of chromosomes identical to that of the parent nucleus. During the formation of the germ cells or gametes, the nucleus undergoes meiosis, a process which produces germ cells each containing but half the number of chromosomes (i.e., is haploid) of the normal cell. Fertilization results in the union of two germ cells, one from each parent, re-establishing the normal (diploid) complement of chromosomes. In broad terms, half of the heritable characters of each parent are combined in the offspring and will become manifest, to reappear in subsequent generations accordingly as the character is dominant or recessive.

The phenomenon of dominant and recessive characters may be most easily explained by supposing that each heritable character is controlled by two genes. Resistance, for example, may be determined by a particular gene " R " which is accompanied by an allelomorphic gene " r ". The combination " Rr " continues through all normal cell divisions, but at meiosis, " R " and " r " go to different gametes. At mating the genes recombine and new individuals will appear in which the combinations " RR ", " Rr " and " rr " occur in the numerical ratio of 1 : 2 : 1. Since " R " is dominant, the mating of " Rr " individuals will give rise to offspring of which three-quarters are resistant (i.e., RR or Rr) and one-quarter are susceptible (i.e., rr). Mating of " RR " individuals will result only in " RR " offspring, i.e., the offspring breed true to type.

In practice, however, hybridization rarely produces the expectations of the simple situation outlined above. Cytological studies have revealed that the situation is much more complicated. Many cultivated plants are not simple diploids but polyploids, and the genes governing any characteristic are repeated several times over. Among cereals, for instance, the valuable varieties of wheat and oats are either hexaploid or tetraploid; among barleys, however, the cultivated forms are diploid though polyploidy is usual among wild grasses of the *Hordeum* species.

Clearly, before the plant breeder can hope to produce resistant varieties, he must have evidence of the existence of resistant genes and his work will begin in a survey of the available varieties, not only of the cultivated varieties but of the wild species from which they were derived. Vavilov (*6*) and his colleagues were the first to organize extensive expeditions to collect the wild species of crop plants. Hawkes (*7*) described the British expedition to collect wild *Solanum* species to serve as sources of the genes required for potato breeding.

The infrequent and spontaneous mutation is most certainly responsible for genetic variability and hopes have arisen that it may be possible to enlarge the gene material available to the plant breeder by the use of methods for accelerating the rate of mutation. In 1927, Muller (*8*) showed that mutation rates in the fruit fly can be increased by one or more hundred-fold by exposure to X-rays, and in 1942, Auerbach and Robson (*9*) found that certain chemicals, such as $\beta\beta'$-dichloroethyl sulphide, increased mutation rates in the same insect. Since that date, many such mutagens have been discovered and, in 1951, Levan (*10*) was able to cite twenty-five different types of compounds able to induce chromosome change in *Allium cepa* L. and in *Vicia faba* L.

It is still too early to quote an example of the successful use of mutagens for the production of resistant varieties for, obviously the use of these agents must be clumsy until more is known of the biochemistry and cytology of the process.

A second way of intervention was opened up by the observation of Blakeslee and Avery (*11*) that treatment of a growing plant with colchicine will sometimes produce shoots of which the chromosome number is double that of the nuclei of untreated plants. The number of known chemicals able to induce polyploidy is still limited though the result is probably due to an interference in the normal division of the chromosome at mitosis. Under the microscope, the chromosomes appear to adhere to each other and this "stickiness" is reported to be produced to some degree by insecticides such as BHC, and by herbicides such as maleic hydrazide. Although the artificial induction of polyploidy does not create new genes for the breeder to manipulate, it may provide him with a means of restoring fertility to those varieties which are sterile by reason of an unbalanced chromosome content. Many outstanding varieties of apple, for instance, are triploids and hence cannot be used for hybridization because effective germ cells cannot be produced. This reason for sterility vanishes in the hexaploid varieties.

REFERENCES

(*1*) Biffen, R. H., *J. agric. Sci.*, 1905, **1**, 4; 1907, **2**, 109.
(*2*) *Report of the Minister of Agriculture of Canada* for the year ending March 31, 1955, p. 91.
(*3*) Parnell, F. R., *Empire Cotton-Growing Corp., Rep. Exp. Stas.*, 1923–5, p. 5, London, 1925.
(*4*) Worrall, L., *J. Dept. Agric. S.A.*, 1923, **7**, 225; 1925, **10**, 487.
(*5*) Parnell, F. R., King, H. E. and Ruston, D. F., *Bull. ent. Res.*, 1949, **39**, 539.
(*6*) Darlington, C. D. and Janaki Ammal, E. K., *Chromosome Atlas of Cultivated Plants*, London, 1945.
(*7*) Hawkes, J. G., *Emp. J. exp. Agric.*, 1945, **13**, 11.
(*8*) Muller, H. J., *Science*, 1927, **66**, 84.
(*9*) Auerbach, C. and Robson, J. M., *Nature* (*Lond.*), 1946, **157**, 302.
(*10*) Levan, A., *Cold Spr. Harb. Symp. quant. Biol.*, 1951, **16**, 233.
(*11*) Blakeslee, A. F. and Avery, A. G., *J. Hered.*, 1937, **28**, 393.

THE USE OF RESISTANT VARIETIES

The use of resistant varieties was suggested as far back as 1815 by Thomas Andrew Knight (*1*) who advocated the cultivation of those varieties of cereals which appear to be free from attack by fungi. Since that time the method has met with marked success for every type of crop production, but especially in cereal growing and in horticulture (*2*).

The advent of resistant varieties has, in a number of cases, saved the threatened crop. Thus, the wart disease of potato (*Synchytrium endobioticum* (Schilb.) Perc.) seemed an insoluble problem to the applied mycologist until Gough (*3*) observed, in 1908, that certain varieties were unaffected by the disease. The discovery of the cucumber Butcher's Disease Resister, immune from blotch (*Cercospora melonis* Cooke), has been a major factor in the disappearance of this disease from Great Britain (*4*). It could be claimed that the sugar-cane industry of Java would not have survived the competition of sugar-beet and the serious losses caused, in the early 1880's, by the " Sereh " disease but for the chance discovery of resistant strains. Wild varieties of sugar-cane were found which were immune to this disease and to mosaic, another virus disease, and by systematic crossing with cultivated varieties, strains were evolved which have enabled the industry to survive (*5*, *6*).

An important method of utilizing this principle is by grafting, which is of special value in fruit culture where the raising of resistant varieties is made difficult by the slow growth of the trees and by the complicated chromosome complement. The failure of early attempts to produce grape vines resistant to Phylloxera by hydridization was countered by the grafting of the European varieties on American root stocks, whereby the severe injury which the insect causes to the roots was prevented. In the eastern states of the United States, collar blight of apple, due to *Bacterium amylovorum* (Burr.) Chester, is avoided by topworking the desired variety on a resistant stock in such a manner that the collar and trunk of the tree will not take the disease (*7*). The pear is, however, even more susceptible to this disease and a resistant pear stock has now been found. Certain Asiatic species of pear, notably *Pyrus ussuriensis*, of which the fruit is of no commercial value, have been shown by Reimer (*8*) to be resistant. Difficulty was experienced in that this variety intergrafted poorly with the common pear, *P. communis*, but by the use of an intermediate scion the desired pear can be grafted on resistant stock. Serious repercussions of the high infestation of cucumber soils in Holland are avoided by grafting the cucumber to roots of *Cucurbita ficifolia*, a wild species resistant to the pathogenic Fusaria (*9*).

Strictly speaking, the use of resistant varieties is a method of pest and disease avoidance rather than of control. Its main advantage, once resistant varieties of the requisite qualities whether of yield or of market and agronomic characters are available, is that there is little or no annual expenditure by the grower on other control measures against the particular disease or pest to which the variety is resistant. But unfortunately the genetic variability which permitted the production of the resistant variety is not confined to the host plant but is present also in the pathogen or insect. A variety proved to be resistant to one strain of a pathogen may be susceptible to another strain. Among phytopathogenic fungi, for instance, it is a general rule that varieties of the pathogen exist, indistinguishable morphologically, but differing widely in such physiological properties as their ability to establish parasitism. Such strains are referred to as " biologic forms ", " biologic races " and other names, of which " physiologic races " was recommended in 1935 by the International Botanical Congress.

This phenomenon, first observed among pathogenic fungi by Schroeter (*10*), is shown most strikingly by highly specialized fungi such as the rusts (*Uredineae*) and the powdery mildews (*Erysiphaceae*). At the turn of the century Eriksson (*11*) had established the existence of three morphologically-identical sub-species or varieties of stem rust (*Puccinia graminis* Pers.); one f.sp. *secalis* attacking rye, barley and a few grasses; a second f.sp. *tritici*, attacking wheat but rarely found on other cereals; a third, f.sp. *avenae* attacking oats and a number of grasses. Currently six such varieties are recognized (*12*). By the use of differential hosts, a method initiated by Stakman (*13*) and his colleagues at the University of Minnesota, the recognized physiologic races of the sub-species *tritici* now number about 275 (*14*).

The difficulties of the wheat breeder seeking varieties resistant to this rust when faced with the multiplicity of variants of the pathogen are well described by Hanna (*15*). In 1934, the variety Thatcher, deriving its resistance from the winter wheat Kanred and the durum wheat Iumillo, was released by the Minnesota Experimental Station. In 1936, the Rust Research Laboratory at Winnipeg released Renown deriving its resistance from the emmer (*Triticum dicoccum*) which was also used in the Canadian varieties Regent and Redman distributed in 1939 and 1947 respectively. The success of these wheats in countering losses due to stem rust, is strikingly shown by the yields quoted by Neatby (*16*); in field trials at Winnipeg from 1935 to 1941 in only two years did the yield of Marquis, a rust-susceptible wheat, exceed 20 bu./acre with an average of 11.5 for the period, whereas over the same period the average yield of Thatcher was 26.8 bu./acre. In 1950 the entire spring wheat area of the Prairies fell to a new physiologic race 15B. The existence of this race had been recognized in 1939 and in anticipation of its appearance, the Winnipeg Laboratory had sought resistant parental material and found it in the variety McMurachy, which originated from a single rust resistant plant selected in 1930 by a Manitoban farmer. From this source arose Selkirk of which in 1955 about 3 million acres was sown in Manitoba and in Saskatchewan. Selkirk is now threatened by the rust strain 15B–3 and the Winnipeg Laboratory is already seeking to counter this threat by the use of resistant genes in the Kenya wheat K–338AC–2E–2. This Canadian experience has revealed the hazards of reliance on a single type of resistance, but by bringing together the genes for resistance in *Triticum* and closely related genera, and by crossing wheat with grasses such as *Agropyron elongatum*, it may be possible to obtain a higher level of resistance than can be found in a single variety.

The maintenance of resistance is thus dependent upon the non-appearance of physiologic races of the pathogen of new infective capacity. It was suggested by Ward (*17*) that the pathogenicity of a fungus may be modified by gradual adaptation or by the phenomenon of "bridging hosts" (see p. 300). Following Craigie's demonstration (*18*) of sexuality in the rust fungi, Waterhouse (*19*) established that the main source of new physiologic races of stem rust is from the sexual stage of the fungus on barberry, where occur opportunities for both hydridization and recombination of the genes for pathogenicity. Hybridization is probably the more effective method for all heteroecious fungi (see p. 300). In less specialized fungi such as the *Fusaria* or *Botrytis cinerea* Pers. there is evidence of erratic and uncontrollable variance in biological properties (*20*, *21*), evidence which would seem to reduce the trustworthiness of the varieties resistant to such fungi.

The causes of this variation, reviewed by the two Christensens and Stakman (*22*), are now becoming understood. Hansen and Smith (*23*), in 1932, discovered that, in *B. cinerea*, the hyphal cells and spores contain two or more nuclei each genetically different. Methods for the quantitative examination of this heterocaryosis have been devised largely by Pontecorvo and his associates. In a review of the various mechanisms in fungi which permit a recombination of heritable characters other than the sexual method, Pontecorvo (*24*) quoted the results obtained by Buxton with two strains of *Fusarium oxysporum pisi*. One strain was pathogenic to the pea variety " Onward " but non-pathogenic to the varieties " Alaska " and " Delwiche Commando " ; the second strain was pathogenic only to " Onward " and " Alaska ". From these two strains a new strain was synthesized pathogenic to all three pea varieties.

The widespread cultivation of a resistant variety directly favours the spread of physiologic races able to attack it, for not only has the fungus an abundance of host plant, but it meets no competition from other physiologic races. Moreover, the possibility that the physiologic races of one area differ from those of another requires that field trials for resistance be renewed in every area of cultivation. The potato varieties immune to wart disease elsewhere failed when grown in the potato fields of Newfoundland where only the mauve-blossomed strain of Sebago remained highly resistant (*25*).

Evidence of physiologic races among insects is reviewed by Thorpe (*26*) and the repercussions of racial segregation in insect populations on applied entomology are surveyed by Smith (*27*). But examples of the sudden appearance of races adapted to food plants not previously attacked are few. The capsid bug *Plesiocoris rugicollis* Fall was, prior to 1900, a pest of willow but since that time has become a serious apple pest (*28*). Examples of the successful use of insect-immune varieties are few. Resistance and susceptibility are the resultants of the interplay of many factors involving both host and pest and, in most cases, these factors are influenced by environmental conditions. Lees (*29*) assembled fourteen examples and the complexity of the problem was demonstrated by Painter (*30*) in a study of the economic value of insect resistant crop plants.

REFERENCES

(*1*) Knight, T. A., *Pamphleteer*, 1815, **6**, 402.
(*2*) Roemer, T., Fuchs, W. J. and Isenbeck, K., *KühnArchiv.*, 1938, **45**, 427 pp. ; Coons, G. H., *Phytopathology*, 1937, **27**, 622.
(*3*) Gough, G. C., *J. R. hort. Soc.*, 1920, **45**, 301.
(*4*) Moore, W. C., *Ann. appl. Biol.*, 1949, **36**, 295.
(*5*) Barber, C. A., *Trop. Agriculture, Trin.*, 1927, **4**, 15.

(6) Cross, E. W., *Plant. Sug. Mfr.*, 1928, **80**, 321.
(7) Waite, M. B. *et al.*, *Yearb. U.S. Dep. Agric.*, 1925, p, 453.
(8) Reimer, F. C., *Bull. Ore. agric. Exp. Sta.* 214, 1925.
(9) Maan, W. J., *Tuinbouw*, 1946, **9**(6), 9.
(10) Schroeter, J., *Beitr. Biol. Pfl.*, 1879, iii, 69.
(11) Eriksson, J., *Jb. wiss. Bot.*, 1896, **29**, 499 ; *Z. Bakt.*, 1902, **9**, 590.
(12) Fischer, G. W., *Yearb. U.S. Dep. Agric.*, 1953, p, 276.
(13) Stakman, E. C., *Science*, 1947, **105**, 627.
(14) Stakman, E. C., *Ann. appl. Biol.*, 1955, **42**, 22.
(15) Hanna, W. F., *Pl. Prot. Conf.* 1956, Butterworths Scientific Publications (Lond.) 1957, p, 31.
(16) Neatby, K. W., *Emp. J. exp. Agric.*, 1941, **9**, 245.
(17) Ward, H. M., *Ann. mycol., Berl.*, 1903, **1**, 132.
(18) Craigie, J. H., *Nature, Lond.*, 1927, **120**, 116, 765.
(19) Waterhouse, W. L., *Proc. Linn. Soc. N.S.W.*, 1929, **54**, 96.
(20) Leonian, L. H., *Phytopathology*, 1929, **19**, 753.
(21) Brierley, W. B., *Ann. appl. Biol.*, 1931, **18**, 420.
(22) Christensen, C. M., Stakman, E. C. and Christensen, J. J., *Annu. Rev. Microbiol.*, 1947, **1**, 61.
(23) Hansen, H. N. and Smith, R. E., *Phytopathology*, 1932, **22**, 953.
(24) Pontecorvo, G., *Annu. Rev. Microbiol.*, 1956, **10**, 393.
(25) Conners, I. L., *30–34th Annual Report of the Can. Dis. Sur.*, 1950–1954.
(26) Thorpe, W. H., *Biol. Rev.*, 1930, **5**, 177.
(27) Smith, H. S., *J. econ. Ent.*, 1941, **34**, 1.
(28) Fryer, J. C. F., *Trans. IV. Congr. Ent., Ithaca*, 1929, ii, 229.
(29) Lees, A. H., *Ann. appl. Biol.*, 1926, **13**, 506.
(30) Painter, R. H., *Insect Resistance in Crop Plants*, New York, 1951.

THE NATURE OF PLANT RESISTANCE

The problem of the nature of plant resistance, which involves not only the means whereby the plant counters parasitic attack but also the question of how the parasite makes its attack, is an immense field which has attracted much attention. The botanical aspects of this work have been systematically examined by Gäumann (*1*) but a brief review must be given here for, in the practical application of this knowledge, lies one of the most promising fields for development in crop protection.

Although the two appear to be closely related from the standpoint of plant resistance it will be better to separate, for the purpose of discussion, relationships of the host to pathogen from those of plant to pest.

Plant Resistance to Pathogens

Most phytopathogens live within the plant tissue and are known as " endophytes ", in contradistinction to those which live on the surface of the plant, a class—the " ectophytes "—most of which develop haustoria serving as organs whereby the mycelium of the fungus, which remains outside the plant tissue, derives its food.

Infection is generally the result of two processes, firstly the penetration of the pathogen into the plant tissue, secondly, the establishment of the parasitism on the host—the infection proper. The successful accomplishment of the first stage will, in the case of fungal pathogens, result only when the spore remains in contact with the plant surface under conditions necessary for germination. The waxy surface of certain leaves and fruit skins may render that plant resistant by permitting water to flow off preventing the accumulation necessary for the germination of the spore. Those varieties of raspberry which are least damaged by *Coniothyrium* spp. possess a more waxy surface. Again potatoes with small hairy leaves and an open habit of growth permitting the leaf surface to dry rapidly appear to be less liable to infection by blight (*Phytophthora infestans* (Mont.) de Bary).

The germination of fungus spores may be stimulated by the presence of plant tissue. Brown (*2*) first showed that the vigour of germination of spores in the infection drop is influenced by the exosmosis of material, which he assumed to be food substances, from the host tissue. Durrell (*3*) demonstrated that stimulation was not obtained in the presence of barium hydroxide, an indication that the active agent is carbon dioxide, a conclusion confirmed by Platz, Durrell and Howe (*4*). On the other hand, Arens (*5*) found that water sprayed on leaves became alkaline within a few hours due to the exosmosis, in particular, of potassium salts. He showed that, in some species, the mineral content of the leaf tissue may be halved in a day by this exosmosis. The role of nutrient, carbon dioxide and other agencies in the germination of fungus spores has been reviewed by Gottlieb (*6*) and by Kovács and Szeöke (*7*).

Penetration. If conditions favour the germination of the fungal spore, it will proceed with the process of penetration, which may occur by direct passage through the epidermis or via the stomatal pore. In the simpler case—penetration by way of the stomata—it would seem that the growth of the germ tube towards the stomata is controlled by some physical or chemical stimulus, perhaps the greater humidity of the air in the stomatal cavity—an example of positive hydrotropism.* Arens (*8*) demonstrated that the movement of zoospores is an example of positive chemotropism for it is influenced by the attractant properties of substances, presumed to be phosphatides, exuded from the stomata. Upon leaves giving no phosphatide reactions, he found that the zoospores did not collect at the stomata.

Immunity will result if this stimulus is lacking or if the penetration of the germ tube into the stomata is mechanically prevented. The susceptibility of sugar-beet towards leaf spot (*Cercospora beticola* Sacc.) has been considered to be dependent, among other factors, upon the length of the stomatal pore (*10*). Doran (*11*) demonstrated that the resistance of snapdragons to *Puccinia antirrhini* Diet. & Holow. is correlated with the number of stomata per unit area. Resistance of stone fruit to *Sclerotinia laxa* Aderh. & Ruhl. was considered by Curtis (*12*) to be determined by the toughness of the cuticle and the size of the stomata. Further, the growth of the germ tube towards the stomata may be hindered mechanically. The young leaves of the grape vine are not attacked by black rot (*Guignardia bidwellii* (Ellis)

---

*The direction of growth or movement of an organism is, at times, controlled by external influences such as light, gravitation, heat, chemicals, etc. According to the nature of the controlling influence this phenomenon, a tropic response, is known as phototropism, geotropism, etc., positive or negative according as the direction is towards or away from the source of the stimulus. In zoology the expression " taxis " is nowadays more frequently used instead of " tropism " for a directed orientation whereas " kinesis " is used when the intensity of the stimulus affects a characteristic of the reaction such as its speed but not its direction ; this distinction was first proposed by Fraenkel and Gunn (*9*).

V. & R.), partly because of their dense covering of hairs; apple scab (*Venturia inaequalis* (Cooke) Wint.) is less common on the under-surface of the leaves, perhaps because of their hairy nature.

The entry of the fungal germ tube into the host by direct penetration through the outer wall of the epidermis appears to result from a more complex series of reactions on the part of both pathogen and host. Not only has the direction of growth of the germ tube to be controlled but the germ tube has to penetrate the epidermal layer of the plant tissue. Various mechanisms have been suggested controlling the directional growth of the germ tube: the contact of the spore with the solid substratum, an example of stereotropism (thigmotropism or haptotropism); a movement from light, an example of negative phototropism. Miyoshi (*13*) developed the chemotropic idea that the fungus, responding to nutrient material diffusing from the host tissue, grows towards the source of that nutrient. Fulton (*14*) suggested that the directional growth is due to the tendency of the fungus to grow away from its own waste products—a negative chemotropism. Graves (*15*) found support for both hypotheses but concluded that negative chemotropism was the more important.

In these few cases, the germination and growth of the spore will be independent of the host plant and entirely non-specific. But if positive chemotropism is involved, specificity may be introduced if susceptible varieties possess, in their cells, attractive substances absent from resistant plants. This idea, elaborated by Massee (*16*), has received little support from subsequent work. Brown and Harvey (*17*) showed that penetration of epidermal tissue and of artificial membranes can be achieved by *Botrytis cinerea* in the absence of chemotropic stimuli and they suggest that, until found inadequate, the hypothesis of a contact stimulus is sufficient.

Penetration may be effected either by the growth of the germ tube or by special mechanisms. Most, if not all, fungal and bacterial phytopathogens are capable of producing, at least in artificial culture, chemicals toxic to the plant. Higgins (*18*) for example found that oxalic acid produced by the pathogen *Sclerotium rolfsii* Sacc. killed the host tissue permitting the entry of the hyphae. Ludwig and his students (*19*) demonstrated that the infection of barley roots by *Helminthosporium sativum* P. K. & B. is preceded by the production of substances toxic to the host tissue which facilitate the infection of the host not only by the pathogen but by other soil micro-organisms not normally regarded as pathogens. Such cases recall the significance of plant wounds, such as pruning cuts (see p.298), as points of entry for fungal pathogens. Apples injured by codling moth (*Carpocapsa pomonella* L.), for example, are liable to attack by brown rot (*Sclerotinia fructigena* Aderh. & Ruhl.) and an appropriate insecticide treatment will often lead to a reduction in losses due to rot (*20*).

Plant viruses have in general no mechanism for penetration and before infection can proceed must be placed within the host plant either by an insect puncture, by some penetrating parasite such as Dodder, or by such processes as grafting or abrasion.

Much work, springing from the idea that the energy necessary for penetration is furnished by the growth of the fungus, has been done in the correlation of disease resistance and the mechanical strength of the cell wall. De Bary (*21*) regarded the turgidity of the cell membrane as a decisive factor, and it has been shown that the fungus makes an easier entry into leaves rendered flaccid by chloroforming or by wilting. Water congestion in relation to disease was extensively studied by Johnson (*22*) at Wisconsin, who has given several examples of greater susceptibility in water-congested plants and has discussed the practical implications of this work. The waxy leaf and fruit coverings may also resist the penetration of the germ tube just as, as has been mentioned above, they render conditions unfavourable for the germination of the spore. The germ tubes from sporidia of *Puccinia graminis*, for instance, penetrate the cuticle of barberry leaves directly by mechanical pressure. Melander and Craigie (*23*) demonstrated that barberries with leaves having external epidermal cells resistant to puncture were usually resistant to infection. The comparative resistance of the fruit of certain plum varieties to brown rot (*Sclerotinia laxa*) has been attributed to the toughness of their skin and flesh (*24*). Willaman (*25*), by the measurement of these two factors, concluded that the " skin test " alone is an adequate guide to the resistance of plum varieties to this fungus.

Establishment of Infection. Establishment of infection will follow successful penetration of the germ tube when the conditions within the host are suitable. The spore may germinate and invade the tissue but establishment of the infection then fails and the fungus dies off. Thus Marryat (*26*) showed that the hyphae of yellow rust (*Puccinia glumarum*) will penetrate the stomata of resistant wheats but that further progress of the fungus is prevented by some unfavourable condition within the host. Cytological studies, which are well exemplified by the work of Allen (*27*), have established that with highly-specialized fungi such as the rusts, establishment of infection fails through the rapid death of host tissue at the seat of infection. Resistance in such cases is due to the high susceptibility (hypersensitivity) of the host plant. The fungus, after expending its food reserves in penetration, is unable to derive nutrient from the barrier of dead host cells.

In other cases the extension of fungal invasion may be prevented by various mechanical, chemical and physiological factors. Mechanical resistance to the spread of the infection is not easy to differentiate from resistance to penetration (e.g., thickness of cuticle, *28*), but the resistance of wheat seedlings to *Gibberella saubinetii* (Mont.) Sacc. was correlated by Dickson, Erkerson and Link (*29*) with the pectin content of the cell wall. Alternatively, the host plant may be able to protect itself from the extension of fungal invasion by morphological changes induced by the attack. Appel and Schuster (*30*) showed that the varieties of potato resistant to rotting bacteria are those which after twenty-four hours have formed a connected cork tissue across a cut surface. Treatment of the cut potato tuber with copper sulphate for the prevention of fungal attack was indeed shown by Dillon Weston and Taylor (*31*) to lead to greater loss by bacterial rots, presumably because the chemical treatment delayed suberization. Tisdale (*32*) found that varieties of flax resistant to wilt (*Fusarium lini*

Bolley) developed a layer of suberized cells effectively walling off the pathogen. Cunningham (*33*), studying the spread in host tissue of various leaf-spotting fungi, showed that when wound cork was formed the spread of the pathogen was limited. Brookes and Moore (*34*) mentioned plum varieties (e.g., Pershore) resistant to silver leaf (*Stereum purpureum* Pers.) because of their ability to produce a gum barrier preventing extension of fungal growth. Hirst and Jones (*35*) pointed out that protective gums of this type may contain terpene-like substances of known antiseptic value.

Among the chemical factors by which the host plant may resist the spread of the pathogen, that of the acidity of the cell sap was an early suggestion due to Comes (*36*). In general, however, experimental work lends little support to this idea ; in his detailed studies, for instance, on the varietal resistance of wheat to stem rust (*Puccinia graminis*), Hurd (*37*) found that the hydrogen ion concentration of the cell sap varies throughout growth and that resistant wheats pass through the period of low *p*H at the same stage of growth as do susceptible varieties, yet no breakdown of resistance could be detected at this stage. On the other hand, Gardner and Kendrick (*38*) considered that failure of inoculations of ripe tomato fruit with the pathogen of bacterial spot, *Xanthomonas vesicatoria* (Gardner & Kendrick) Dows., and the ready infectivity of young seedlings is due to the acidity of the sap of the mature plant. A recent example is due to Tolba and Saleh (*39*) who traced a correlation between the *p*H of tomato tissue and its susceptibility to attack by *Fusarium culmorum* (Smith) Sacc. and *F. oxysporum* Schlecht.

An hypothesis less amenable to experiment is that the oxidation-reduction potential of the host tissue has an influence on the pathogen. This idea is perhaps visible in the finding of Hughes and Fowler (*40*) that the resistance of cotton to blackarm disease (*Xanthomonas malvacearum* (E. F. Smith) Dows.) is coupled with the greater reducing action on Fehlings solution of the water extract of leaves of the resistant varieties. Resistance is certainly not due to the glucose thus measured but may be associated with some factor which gives rise to the high glucose content.

The ability of the pathogen to thrive after penetration will clearly depend on its ability to utilize the host plant as a source of nutrient. The extreme specificity of certain fungi in their choice of food material was utilized by Pasteur (*41*) in 1860 for the separation of the optical isomers of tartaric acid.* It has since been shown (*42*) that the fungus attacks the dextro-isomer with far greater rapidity than it does the laevo-form. Leach (*43*) suggested that such a specificity of food requirements on the part of the pathogen and a similar specificity of food production on the part of the host may be the cause of obligate parasitism.

An hypothesis, similar in certain respects, was put forward by Coons and Klotz (*44*), who considered that the ability of an organism to thrive in the

---

*Tartaric acid is known in four forms, differing only in the arrangement of the relative positions of the groups of atoms within the molecule. Two of these forms are optically active ; they have, in solution, the power of rotating the plane of polarized light, the one being dextrorotatory, the other laevorotatory.

tissue of a plant will be dependent upon its ability to attack the food substance, especially the protein, of the host. The breaking down of the protein is accomplished by proteolytic enzymes, which may be furnished by the fungus itself, or which may arise from the host plant as a result of the invasion of its tissue. The role of the enzyme pectinase, which assists in the breakdown of pectin, in determining the parasitizing ability of *Botrytis cinerea* has been examined by Brown and his co-workers (*45*) whose results have shown the enzyme hypothesis to be of use in interpreting the behaviour of non-specialized parasites of this type. For example, Vasudeva (*46*) showed that the fact that the fungus *Botrytis allii* Munn which is normally non-parasitic on apples (in storage), but which produces definite attack if supplied with a small quantity of nitrogenous nutrient, is associated with the ability of the fungus to excrete pectinase when supplied with nitrogenous food. In the same way, Horne's (*47*) observation, that there is a high degree of correlation between the nitrogen content of different apples of the same variety and the rate of invasion by parasitic fungi, may be explained by the intensifying influence of the nitrogen status upon the secretion, by the fungus, of pectinase.

The role of the pectolytic enzymes in the host-pathogen relationship has been explored more particularly among the wilt diseases caused by *Fusarium oxysporum*, a species containing physiologic races each able to attack a single or closely-related group of host species. This work has been reviewed by Walker and Stahmann (*48*) who advanced the idea that each race produces a specific metabolite which predisposes the tissue of the susceptible host only to the one race enabling that race to induce the typical wilt syndrome. This hypothesis in principle ascribes the difference in host range to an enzyme specificity.

The presence of material toxic to the invading pathogen is a well-examined reason for host resistance, supported in a number of cases by the isolation and identification of the responsible fungicide. The resistance of the heartwood of many conifers can be ascribed to such toxic compounds, the chemistry of which was reviewed by Erdtman (*49*). The thujaplicins, from the western red cedar, *Thuja plicata* Donn. were shown by Rennerfeldt (*50*) to be as toxic to blueing and decay fungi as sodium pentachlorophenate. He (*51*) also found pinosylvin (3 : 5-dihydroxy-*trans*-stilbene), from the heartwood of many species of pines, to be many times more toxic than phenol to wood-rotting fungi. Another stilbene derivative, 2 : 3′ : 4 : 5′-tetrahydroxystilbene, was isolated by Barnes and Gerber (*52*) from the heartwood of the Osage orange, *Toxylon pomiferum* Raf., and shown to be toxic to cellulolytic fungi.

Evidence that varietal resistance in herbaceous plants is due to the specific fungicidal properties of components of the plant has been more elusive. The colorimetric reactions of phenols and other hydroxyaromatic compounds have revealed their presence in plant tissue. Dufrénoy and his colleagues (*53*) considered that the accumulation of phenolic compounds in plant cells adjacent to those invaded by pathogens was a defence mechanism. Newton and Anderson (*54*) related the resistance of certain wheat varieties to rust to their phenol content. Gäumann (*55*) was sceptical of the general hypothesis for these phenols also occur in susceptible varieties, though he granted the possibility that the phenols may be released from more complex derivatives by the enzymatic reactions of the invading pathogen. This suggestion appears to have been made first by Cook and Taubenhaus (*56*) to explain the greater susceptibility, to rots, of mature over immature fruit; mature fruit is poorer in oxidases which serve to protect the infected fruit by liberating, from the tannins, phenols toxic to the invading fungus.

A well-verified example of the phenol hypothesis is due to Walker and his colleagues (*57*) who showed that the resistance of onions to onion smudge (*Colletotrichum circinans* (Berk.) Vogl.) is a genetic factor linked to scale colour and is due to phenolic substances, such as protocatechuic acid and catechol, associated with pigment metabolism. As the resistant yellow- and red-scaled varieties are attacked by the fungus if the outer scales are removed, it would seem that the fungicidal phenols are either present in the living tissue in a non-toxic form or do not come into contact with the invading pathogen. When, however, the outer scales wither the phenols are liberated and serve to protect the living scales of the resistant variety. The same phenols were considered by Ramsey, Heiberg and Wiant (*58*) to be responsible for the resistance of coloured onions to the rot caused by *Diplodia natalensis* Pole-Evans. The allyl sulphides responsible for the pungency of onions are toxic not only to *C. circinans* but to *Botrytis allii* causing neck rot and were considered by Owen, Walker and Stahmann (*59*) to be of more importance than the phenol factor in determining resistance to the latter disease.

The glycosides provide another type of fungicide thought to be the factor responsible for resistant varieties. Tims (*60*) considered a sulphur-containing glucoside to be present in the tissue of cabbage varieties resistant to cabbage yellows (*Fusarium conglutinans* Wollenw.). Reynolds (*61*) recovered from flax extracts two materials toxic to the pathogen *F. lini* responsible for flax wilt; the first, the glucoside linamarine which appeared to be present in greater amounts in resistant strains of flax; the second, heat-stable toxins which varied in amount according to variety and environmental factors. From tomato leaves Irving, Fontaine and Doolittle (*62*) isolated a substance, which they named lycopersicin, toxic to certain pathogenic fungi including *F. lycopersici* Sacc. responsible for tomato wilt. This substance, recovered in crystalline condition from several species of *Lycopersicum* by Kuhn and his associates (*63*) and by Fontaine *et al.* (*64*), was renamed tomatine and is a tetrasaccharide of the aglycone tomatidine for which a structure was proposed by Kuhn, Löw and Trischmann (*65*). Although tomatine effectively inhibits the growth of *F. lycopersici* in pure culture and has been described by Gäumann (*55*) as a preformed resistance factor, no direct evidence has been obtained that it is responsible for wilt resistance; indeed, Irving, Fontaine and Doolittle (*62*) recovered it from susceptible varieties. Biosynthetically it is of interest because of its close chemical kinship to the glycosides associated with the resistance to the Colorado potato beetle (see p. 26).

The examples of the previous paragraph concern species of *Fusarium* and glycosides; to prevent this coincidence from being misleading the case of rye resistance to *F. nivale* (Fr.) Ces. may be quoted. This resistance was traced by Virtanen and Hietala (*66*) to 2(3)-benzoxazolinone which they recovered from rye seedlings.

The suggestion that some specific chemical substance toxic to the fungus is formed by the interaction of host plant and fungus has been subjected to experiment. On the chance that this chemical is translocatable, Roach (*67*) grafted the root, shoot and tuber systems of varieties of potato susceptible and resistant to wart disease (*Synchytrium endobioticum*) and showed that the " cause " of immunity is not a substance produced in one part of the plant, and translocated thence to other parts of the plant. Nor was evidence of the interchange of the " cause " of resistance across grafts obtained by Leach (*68*) in beans resistant and susceptible to *Colletotrichum lindemuthianum* (Sacc. & Magn.) Bri. & Cav., by May (*69*) examining the response of tomatoes to *Fusarium lycopersici*, or by Salmon and Ware (*70*) working with hops resistant and susceptible to *Sphaerotheca humuli* (DC.) Burr. It would appear that, in these cases, the substances to which resistance or susceptibility is due is inseparably connected with the protoplasm, or possibly with the proteins of the protoplasm. Roach suggested that, in view of the specificity of the proteins shown in immuno-chemical reactions, the most hopeful line of attack in the determination of the cause of immunity is by immuno-chemical methods.

That, however, the scion may, in some cases, have an influence upon the susceptibility of the stock is shown by an example quoted from Wormald and Grubb (*71*). These workers found that the susceptibility of the apple stock to crown gall (*Bacterium tumefaciens* E.F. Sm. & Towns.) is modified by the variety of the scion grafted on it. Lord Derby rendered the stock more susceptible, whereas Bramley's Seedling scions appeared to increase resistance to this disease.

The suggestion that the factor determining resistance may prove to be related to the specificity of the proteins of the plant tissue has been examined by Nelson and Dworak (*72*), in flax plants resistant to wilt (*Fusarium lini*). They separated the globulins from the seed of a resistant and of a susceptible variety into two fractions, both of which proved highly specific in precipitin tests. These workers concluded that their trials indicated, though did not prove, that the globulins from the resistant variety possessed a serological peculiarity which is absent from those extracted from susceptible seed. Resistance to wilt may then be associated with the globulin fraction, though whether it is a result of the globulin structure or whether it is due to some tactor connected with the globulin vehicle, they were unable to determine.

The possibility that immuno-chemistry is concerned in disease resistance in plants has resulted in speculations on the feasibility of utilizing, for crop protection, the immunological methods employed in animal pathology. These methods mainly depend on the principle that the animal responds of the introduction, within its body, of an antigen by the formation of an antibody which reacts specifically with the antigen. By injecting the antibody or by stimulating its production within the animal, that animal acquires resistance to the corresponding antigen or to the organism producing that antigen. The ability of fungal and bacterial pathogens and of viruses to produce antibodies in the animal body is proved. Coons and his co-workers (*73*) employed the complement fixation test, the precipitin and agglutinin tests, to differentiate between strains of Fusaria ; Edgecombe (*74*) adapted a precipitin test to distinguish varieties of wheat of varying resistance to rust ; Link *et al.* (*75*) correlated the serological specificity of certain phytopathogenic fungi with their biological specificity. Serological methods have proved exceptionally useful in the differentiation of virus strains, a subject reviewed by Chester (*76*), and in the diagnosis of virus diseases, particularly in van Slogteren's laboratory (*77*).

If a pathogen could induce antibody formation in the plant, the plant should acquire resistance to that pathogen. The first attempt to prove antibody formation in the plant was by Potter (*78*) who isolated, from turnips infected with white rot (*Bacterium carotovorum* (L. R. Jones) Lehm. & Neum.), a substance which prevented the growth of this bacterium on inoculated turnip. Mallman and Hemstreet (*79*) isolated an inhibiting substance from a rotten cabbage which was active against an organism obtained from the same cabbage. Sardiña (*80*) was unsuccessful in his attempts to isolate specific agglutinins and precipitins from plants inoculated with certain phytopathogenic bacteria. But the difficulty of proving antibody formation may hinge on the possibility, indicated from the results of the grafting technique described above, that the antibody, the " cause " of resistance, cannot leave the plant cell in which it is formed. In the animal, the blood circulatory system provides a channel for antibody translocation but, in plants, it would seem that immunological phenomena must be largely if not entirely cullular. Chester (*81*), who has provided a general survey of plant serology, advanced arguments against this distinction between zoo- and phyto-immunology, but it is maintained by Butler and by Salaman (*82*).

Although partial success has been claimed by Carbone and his colleagues (*83*) in the immunization of plants, cases of acquired immunity have since been established. Wingard (*84*) found that many different varieties of host plant, when infected with tobacco ringspot virus, develop symptoms but that, after about two weeks, new growth shows but faint symptoms, until finally growth is normal. Re-inoculation of the recovered plants or of cuttings taken from such plants failed though the sap from these plants induced the disease in other plants, showing that the virus was still present. The treated plants thus acquired tolerance to the virus but have become " carriers " (see p. 299). The second type of acquired immunity was first reported, in 1931, by Thung (*85*). He observed that tobacco plants suffering from tobacco mosaic failed to develop symptoms of a " white mosaic " when inoculated with this virus. The practical significance of this discovery was developed mainly by Salaman (*86*) who suggested that, by infecting a plant with a mild strain of a particular virus causing minor loss of cropping value, the plant can be protected against more virulent strains of the same virus. As protection is not necessarily obtained against other viruses, the phenomenon exhibits a specificity comparable to that of acquired immunity in animals. There is, however, no evidence that in the plant the process of acquired immunity is similar to that in animals. No antibody has been isolated from the resistant plants which remain carriers of the "benign" virus. Bawden (*87*) suggested that there may be a maximum amount of any virus that a plant can contain and if one strain has already grown to this amount a second is unable to multiply. As a plant fully infected with one virus is still susceptible to other unrelated viruses, it seems that related viruses either occupy the same sites or utilize the same materials whereas unrelated viruses multiply at different sites or use different materials. The difficulties of the practical use of this phenomenon are discussed by Henderson Smith (*88*).

Such is a brief review of some of the many factors which may contribute towards the resistance of plants to fungus attack. The entry of the fungus may be prevented by the inability of the spore to find those conditions necessary for its germination ; the stimulus necessary to secure penetration may be lacking ; the penetration of the germ tube may be prevented by morphological features of the host ; the establishment of infection may be prevented by unsuitable conditions within the host such as hypersensitiveness or the presence of specific chemical substances toxic to the fungus ; the development of the fungus may be inhibited by a defensive response of the host plant which may involve morphological changes or the production of specific toxins.

Finally, it is evident that it is most improbable that any single explanation of plant resistance can be adopted as a generalization covering all types of host-parasite relationship. The nature of plant resistance is clearly complicated but the hypotheses so far advanced are, in general, sufficiently elastic to permit a reasonable explanation of the profound effect which environmental conditions frequently display in determining the degree of attack.

REFERENCES

(*1*) Gäumann, E., *Pflanzliche Infektionslehre*, 2nd Ed., Birkhäuser, Basel, 1951.
(*2*) Brown, W., *Ann. Bot.*, 1922, **36**, 285.
(*3*) Durrell, L. W., *Res. Bull. Iowa agric. Exp. Sta.* 84, 1925.
(*4*) Platz, G. A., Durrell, L. W. and Howe, M. F., *J. agric. Res.*, 1927, **34**, 137.
(*5*) Arens, K., *Jb. wiss. Bot.*, 1934, **80**, 248.
(*6*) Gottlieb, D., *Bot. Rev.*, 1950, **16**, 229.
(*7*) Kovács, A. and Szeöke, E., *Phytopath. Z.*, 1956, **27**, 335.
(*8*) Arens, K., *Jb. wiss. Bot.*, 1929, **70**, 93.
(*9*) Fraenkel, G. S. and Gunn, D. L., *The Orientation of Animals*, Oxford University Press, 1940.
(*10*) Pool, V. W. and McKay, M. B., *J. agric. Res.*, 1916, **5**, 1011.
(*11*) Doran, W. L., *Bull. Mass. agric. Exp. Sta.*, 202, 1921.
(*12*) Curtis, K. M., *Ann. Bot.*, 1928, **42**, 39.
(*13*) Miyoshi, M., *Jb. wiss. Bot.*, 1895, **28**, 269.
(*14*) Fulton, H. R., *Bot. Gaz.*, 1906, **41**, 81.
(*15*) Graves, A. H., *Bot. Gaz.*, 1916, **62**, 337.

(16) Massee, G., *Phil. Trans.*, 1905, **197**(B), 7.
(17) Brown, W. and Harvey, C. C., *Ann. Bot.*, 1927, **41**, 643.
(18) Higgins, B. B., *Phytopathology*, 1927, **17**, 417.
(19) Ludwig, R. A., Clark, R. V., Julien, J. B. and Robinson, D. B., *Can. J. Bot.*, 1956, **34**, 653.
(20) Bennett, S. H., Kearns, H. G. H. and Marsh, R. W., *Rep. agric. hort. Res. Sta. Bristol*, 1944, p. 157.
(21) de Bary, A., *Bot. Z.*, 1886, **44**, 377.
(22) Johnson, J., *Res. Bull. Wis. agric. Exp. Sta.*, 160, 1947.
(23) Melander, L. W. and Craigie, J. H., *Phytopathology*, 1927, **17**, 95.
(24) Valleau, W. D., *J. agric. Res.*, 1915, **5**, 365.
(25) Willaman, J. J., *Proc. Soc. exp. Biol. N.Y.*, 1926, **23**, 680.
(26) Marryat, D. C. E., *J. agric. Sci.*, 1907, **2**, 129.
(27) Allen, R. F., *J. agric. Res.*, 1923, **23**, 131 ; 1924, **26**, 571 ; 1926, **32**, 701 ; **33**, 201 ; 1927, **34**, 697 ; 1928, **36**, 487.
(28) Hawkins, L. A. and Harvey, R. B., *J. agric. Res.*, 1919, **18**, 275.
(29) Dickson, J. G., Erkerson, S. H. and Link, K. P., *Proc. Nat. Acad. Sci. Wash.*, 1923, **9**, 434.
(30) Appel, O. and Schuster, J., *Arb. biol. Abt.* (*Anst. Reichsanst.*), *Berl.*, 1912, **8**, 451.
(31) Dillon Weston, W. A. R. and Taylor, R. E., *J. agric. Sci.*, 1944, **34**, 93.
(32) Tisdale, W. H., *J. agric. Res.*, 1917, **11**, 573.
(33) Cunningham, H. S., *Phytopathology*, 1928, **18**, 539 ; 717.
(34) Brooks, F. T. and Moore, W. C., *J. Pomol.*, 1926, **5**, 61.
(35) Hirst, E. L. and Jones, J. K. N., *Research*, 1951, **4**, 411.
(36) Comes, O., abstr. in *Int. Inst. Agric. Bull. agric. Intell.*, 1913, **4**, 1117.
(37) Hurd, A. M., *J. agric. Sci.*, 1924, **27**, 725.
(38) Gardner, M. W. and Kendrick, J. B., *J. agric. Res.*, 1921, **21**, 123.
(39) Tolba, M. K. and Saleh, A. M., *Nature, Lond.*, 1954, **173**, 87.
(40) Hughes, L. C. and Fowler, H. D., *Nature, Lond.*, 1953, **172**, 316.
(41) Pasteur, L., *C. R. Acad. Sci. Paris*, 1860, **51**, 298.
(42) McKenzie, A. and Harden, A., *J. chem. Soc.*, 1903, **83**, 424.
(43) Leach, J. G., *Phytopathology*, 1919, **9**, 59.
(44) Coons, G. H., and Klotz, L. J., *J. agric. Res.*, 1925, **31**, 287.
(45) Brown, W., *Trans. Brit. mycol. Soc.*, 1934, **19**, 11 ; *Ann. appl. Biol.*, 1955, **43**, 325.
(46) Vasudeva, R. S., *Ann. Bot.*, 1930, **44**, 469.
(47) Horne, A. S., *Rep. Fd. Invest. Bd.*, 1932, p. 292.
(48) Walker, J. C., and Stahmann, M. A., *Annu. Rev. Pl. Physiol.*, 1955, **6**, 351.
(49) Erdtman, H., " Chemistry of some heartwood constituents of conifers and their physiological and taxonomic significance." *In Progress in Organic Chemistry. I.* Academic Press Inc., New York, 1952.
(50) Rennerfeldt, E., *Physiol. Plant.*, 1948, **1**, 245.
(51) Rennerfeldt, E., *Svensk. bot. Tidskr.*, 1943, **37**, 83.
(52) Barnes, R. A. and Gerber, N. N., *J. Amer. chem. Soc.*, 1955, **77**, 3259.
(53) Dufrénoy, J., *Rev. Plant veg. Ent. agric.*, 1928, **15**, 106.
(54) Newton, R. and Anderson, J. A., *Canad. J. Res.*, 1929, **1**, 86.
(55) Gäumann, E., *Pflanzliche Infektionslehre*, 2nd Ed., Birkhäuser, Basel, 1951 ; *Principles of Plant Infection*, English trans. of 1st Ed. by W. B. Brierley. Crosby Lockwood & Son Ltd., London, 1950.
(56) Cook, M. T. and Taubenhaus, J. J., *Bull. Delaware agric. Exp. Sta.* 91, 1911 ; 97, 1912.
(57) Angell, H. R., Walker, J. C. and Link, K. P., *Phytopathology*, 1930, **20**, 431 ; Link, K. P. and Walker, J. C., *J. biol. Chem.*, 1933, **100**, 379 ; Hatfield, W. C., Walker, J. C. and Owen, J. H., *J. agric. Res.*, 1948, **77**, 115.
(58) Ramsey, G. B., Heiberg, B. C. and Wiant, J. S., *Phytopathology*, 1946, **36**, 245.
(59) Owen, J. H., Walker, J. C. and Stahmann, M. A., *Phytopathology*, 1950, **40**, 292.
(60) Tims, E. C., *J. agric. Res.*, 1926, **32**, 183.
(61) Reynolds, E. S., *Ann. Mo. bot. Gdn.*, 1931, **18**, 57.
(62) Irving, G. W., Fontaine, T. D. and Doolittle, S. P., *Science*, 1945, **102**, 9.

(63) Kuhn, R. and Löw, I., *Chem. Ber.*, 1948, **81**, 552 ; Kuhn, R., Löw, I. and Gauhe, A., *Chem. Ber.*, 1950, **83**, 448.
(64) Fontaine, T. D., Irving, G. W., Ma, R., Poole, J. B. and Doolittle, S. P., *Arch. Biochem.*, 1948, **18**, 467.
(65) Kuhn, R., Löw, I. and Trischmann, H., *Angew Chem.*, 1952, **64**, 397 ; *Chem. Ber.*, 1952, **55**, 416.
(66) Virtanen, A. I. and Hietala, P. K., *Acta chem. scand.*, 1955, **9**, 1543.
(67) Roach, W. A., *Ann. appl. Biol.*, 1923, **10**, 142 ; 1927, **14**, 181.
(68) Leach, J. G., *Phytopathology*, 1929, **19**, 875.
(69) May, C., *Phytopathology*, 1930, **20**, 519.
(70) Salmon, E. S. and Ware, W. M., *Ann. appl. Biol.*, 1927, **14**, 276.
(71) Wormald, H. and Grubb, N. H., *Ann. appl. Biol.*, 1924, **11**, 278.
(72) Nelson, C. I. and Dworak, M., *Bull. N. Dakota agric. Exp. Sta.*, 202, 1926.
(73) Coons, G. H., *Plant Pathology and Physiology*, Philadelphia, 1928.
(74) Edgecombe, A. E., *Bot. Gaz.*, 1931, **91**, 1.
(75) Link, G. K. K., Link, A. de S., Cross, G. L. and Wilcox, H., *Proc. Soc. exp. Biol.*, 1932, **29**, 1278.
(76) Chester, K. S., *Phytopathology*, 1937, **27**, 903 ; *Quart. Rev. Biol.*, 1937, **12**, 19, 165, 294.
(77) Van Slogteren, E., *Ann. appl. Biol.*, 1955, **42**, 122 ; with Van Slogteren, D. H. M., *Annu. Rev. Microbiol.*, 1957, **11**, 149.
(78) Potter, M. C., *J. agric. Sci.*, 1908, **3**, 102.
(79) Mallman, W. L. and Hemstreet, C., *J. agric. Res.*, 1924, **28**, 599.
(80) Sardiña, J. R., *Angew Bot.*, 1926, **8**, 289.
(81) Chester, K. S., *Quart. Rev. Biol.*, 1933, **8**, 129, 875 ; 1937, **12**, 19, 165, 294.
(82) Butler, E. J., *Rep. 3rd int. Congr. compar. Path.*, 1936, **1**, 2, pp. 1–16. Salaman, R. N., *ibid.*, pp. 167–168.
(83) Carbone, D. and Arnaudi, C., *L'immunita nelle piante*, Milan, 1930. Carbone, D. and Kalajev, A., *Phytopath. Z.*, 1933, **5**, 91.
(84) Wingard, S. A., *J. agric. Res.*, 1928, **37**, 127.
(85) Thung, T. H., abstr. in *Rev. appl. Mycol.*, 1932, **11**, 750.
(86) Salaman, R. N., *Nature*, 1933, **131**, 468.
(87) Bawden, F. C., *Plant Viruses and Virus Diseases*, 3rd Ed., Waltham, Mass., 1950, p. 122.
(88) Henderson Smith, J., *Ann. appl. Biol.*, 1938, **25**, 227.

Plant Resistance against Insects

The factors contributing to the resistance to pathogens, examples of which have been cited above, are not so apparent in the case of the resistance towards insect injury. The conditions necessary for the germination and subsequent growth of the fungus spore, especially of the endophytes, are more delicate than those which govern the extent of injury caused by insects. The methods of attack of insects are far more varied than are found in fungus diseases.

It is possible to classify the methods by which the insect feeds on plants on the basis of the construction of the mouth parts of the insect. Firstly, there are those insects, possessing horny jaws which chew the plant tissue, known as chewing or mandibulate insects. Secondly, there are those insects whose mouth parts are modified to form needle-like organs by means of which the insect pierces into the plant tissue and sucks sap.

With suctorial insects there is an analogy to fungi in the penetration by the pest into the host tissue ; an analogy which may be extended to those mandibulate insects in which the egg is deposited, by the penetration of the ovipositor, within the tissue of the host. With these insects, one factor to which the plant's resistance is due may be the mechanical resistance of the tissue. Haseman (*1*) showed that the extent of infestation of three varieties of wheat by the Hessian fly (*Phytophaga destructor* Say) varied with the ash content; McColloch and Salmon (*2*) suggested that the varietal resistance of wheat to this pest is associated with the presence of silica. Refai (*3*) found no correlation between ash or silica content of winter wheats and resistance to Hessian fly but, finding one in the resistance of the stem to shear, considered that, in resistant varieties, the strains are tough enough to cramp the activities of the fly larvae. If so the case is similar to that of the wheat stem sawfly, *Cephus cinctus* Nort., against which the spring wheats Rescue and Chinook were developed. This work, described by Platt and Farstad (*4*), arose from the observation of H. J. Kemp that solid-stemmed wheat varieties were less frequently damaged by the sawfly. The larvae develop within the hollow stems of susceptible wheat and are presumably handicapped by the presence of pith in the resistant varieties.

While on the subject of mechanical factors responsible for the plant's freedom from insect pests the example of the long-haired jassid-resistant cotton already mentioned (see p. 7) may be parallelled by those of the hooked hairs which entrap aphids on French bean (*Phaseolus vulgaris* L.) (*5, 6*). The freedom of *Solanum polyadenium* from aphids was ascribed by Stringer (*7*) to the gummy secretion of its leaves which hampers the aphid in its feeding.

With suctorial insects, more particularly the aphids, there is a rough parallel to the more specialized nature of fungi in that the separate species of both are tolerant to but a few species of host plant. For example, the apple varieties Northern Spy and Majetin are resistant to the attack of woolly aphis (*Eriosoma lanigerum* Hausm.) a property which Le Pelley's results (*8*) indicated is inherited on Mendelian lines. Northern Spy has been widely used as a rootstock because its immunity from woolly aphis confers freedom of roots and stems from this pest. But, as it exhibits undesirable rootstock qualities (*9*), a survey has been made (*10*) of the genetical basis of this imumnity in order to combine this property with better pomological qualities. At the same time the causes of immunity were investigated, for the hypothesis suggested by Staniland (*11*), that resistance is associated with the completeness of the sclerenchymatous ring whereby the entry of the insect stylet into the vascular tissue is prevented, has proved inadequate. The work of Massee and Roach (*12*) indicated that resistance to woolly aphis is associated with an alcohol-insoluble but ether-soluble material present in host tissue.*

As in resistance to fungal attack, nutrient conditions within the host may determine the degree of infestation by pests through influence on the rate of reproduction, the rate of growth and longevity. For example, Davidson (*13*) found that on a susceptible variety of bean, the mean infestation figure fourteen days after the plants had been infested with a single

*In a later report (17) Roach modified this description to suggest that the factor responsible for the difference between resistant and susceptible varieties is insoluble both in alcohol and in ether, but the experimental data presented are compatible with those of the earlier publication (12).

female of *Aphis rumicis* L. was, 1,037±51.8 whereas, under the same conditions, the shoots of the more resistant winter bean bore but 286±27.2 aphids. Similarly Evans (*14*) was able to correlate the rate of reproduction of the aphis *Brevicoryne brassicae* L. with the nitrogen content, in particular the protein content, of the host tissue. Nutritional factors seem to determine the degree of attack of pea varieties by the aphid *Acyrthosiphum pisum* Harris, for Maltais (*15*) established a correlation between the total water-soluble nitrogen and amino-nitrogen of the tissue and its susceptibility. Auclair (*16*) narrowed the basis for this correlation ; he found that the amino acids arginine, threonine and valine, known to be essential for aphid growth, are present in higher amounts in susceptible than in resistant varieties whereas non-essential amino acids were present in about equal amounts in resistant and susceptible pea varieties.

The larger topic of the properties of the host plant which determine its selection by the pest was discussed in a symposium at the 9th International Congress of Entomology and there Fraenkel (*18*) reviewed the nutritional value of green plants for insects. He concluded that, as green leaves are excellent sources of all the food materials that insects appear to require, host plant specificity is determined by the presence or absence of odd chemical substances in which the plants differ but which render the plant a suitable or unsuitable food. This view he admitted was extreme and one-sided, but he pointed out that our knowledge of the food requirements of phytophagous insects is meagre.

Three stages in the process by which the insect establishes normal feeding on its host plant were differentiated by Dethier (*19*) ; (i) finding the plant ; (ii) the initiation of biting ; (iii) continued feeding. The first process may be by chance as in the case of the peach aphid *Myzus persicae* Sulz. recorded by Broadbent (*20*), though many examples of chemotropism are available. The earliest of these examples is due to Grevillius (*21*) who showed that larvae of the brown tail moth (*Euproctis chrysorrhoea* L.) taken from chickweed, would feed on other plants smeared with a paste prepared from chickweed. Verschaffelt (*22*) similarly induced larvae of the cabbage butterflies *Pieris brassicae* L. and *P. rapae* L., which normally feed only on leaves containing mustard oil glycosides, to feed on leaves of other plants by treating their leaves with mustard oil preparations. Thorsteinson (*23*) extended this work to larvae of the diamondback moth (*Plutella maculipennis* Curt.) to which he made pea meal an attractive food by the addition of sinigrin, sinalbin or glucocheriolin, three components of the mustard oils of cruciferous plants on which the caterpillar normally feeds.

The potato (*Solanum tuberosum* L.) provides an example of Dethier's second and third stages, for certain species of *Solanum* are not attacked by the Colorado potato beetle (*Leptinotarsa decemlineata* Say). The incorporation of the genes responsible for this resistance into the cultivated potato is being attempted by geneticists in many European countries invaded by the beetle in the last decade. Kuhn and Löw (*24*) have summarized the work which has been carried out at Heidelberg on the chemistry of the compounds associated with this resistance. Leaves of *S. demissium* are eaten by beetles but not by larvae; leaves of *S. chacoense* are refused by both beetles and larvae. From *S. tuberosum*, Kuhn and his colleagues isolated six glycosides, the aglycone of which is solanidine, but as the glycosides of *S. chacoense* also appear to be saccharides of solanidine, the cause of the resistance of the latter is not revealed. The main alkaloidal glycoside of *S. demissium*, however, is demissine, the tetrasaccharide of demissidine which is readily convertible to tomatidine (see p. 19). Potato leaves infiltrated with either demissine or tomatine are so highly toxic to larvae of the Colorado potato beetle that these two glycosides are considered responsible for the natural resistance of *S. demissium* and of tomato leaves to the beetle.

The protective role of the essential oils present in the rind of citrus fruits against the Mediterranean fruit fly (*Ceratitis capitata* Wild.) was established by Back and Pemberton (*25*). They showed that, although the presence of the oil does not prevent oviposition on the fruit, the female inevitably punctures one or more of the oil cells in the process and the oil so freed kills the eggs.

The relative freedom of soybeans from insect infestation, at least in the U.S.A., and of soybean products from attack by stored product pests, was traced by Lipke, Fraenkel and Liener (*26*) to the presence of a heat-labile substance, probably protein in character, that interferes with digestion of proteins by the insect. This anti-proteolytic toxin has not been further characterised, but it is distinct from soyin (*27*) which is held responsible for the poor nutritional value to vertebrates of untreated soybean protein.

A recent example of the association of resistance with a specific chemical product of the host plant is due to Smissman and his colleagues (*28*) who isolated 6-methoxy-2(3)-benzoxazolinone from Indian corn and considered it responsible, in part at least, for the resistance to the European corn borer *Pyrausta nubilalis* Hbn. It will be recalled (p. 20) that 2(3)-benzoxazolinone was thought to be the agent responsible for the resistance of rye seedlings to the fungal pathogen *Fusarium nivale*.

Finally, the extent of injury caused by the attack of the insect pest may be related to the response of the plant to that attack, e.g., by the development of hypertrophic tissue such as galls. The " Big Bud " of black currant is the result of attack by the mite *Eriophyes ribis* Nal., and Lees (*29*) ascribed the apparent immunity of certain varieties of black currant to hypersensitivity. The growing point of the attacked bud in these varieties is killed and not stimulated to the irregular growth which occurs in those varieties which show Big Bud.

REFERENCES

(*1*) Haseman, L., *J. econ. Ent.*, 1916, **9**, 291.

(*2*) McColloch, J. W. and Salmon, S. C., *J. econ. Ent.*, 1923, **16**, 293.

(*3*) Refai, F. J., *Kansas State Coll. Ph.D. Thesis*, 1955.

(*4*) Platt, A. W. and Farstad, C. W. *In* Agricultural Research in Canada, pp. 47–51. *The Agric. Inst. of Canada*, 1951–52.

(*5*) McKinney, K. B., *J. econ. Ent.*, 1938, **31**, 630.

(*6*) Johnson, B., *Bull. ent. Res.*, 1953, **44**, 779.

(*7*) Stringer, A., *Rep. agric. hort. Res. Sta. Bristol*, 1946, p. 88.

(*8*) Le Pelley, R. H., *J. Pomol.*, 1927, **6**, 209.

(*9*) Hatton, R. G., *Ann. appl. Biol.*, 1937, **24**, 173.
(*10*) Crane, M. B., Greenslade, R. M., Massee, A. M. and Tydeman, H. M., *J. Pomol.*, 1936, **14**, 137.
(*11*) Staniland, L. N., *Bull. ent. Res.*, 1924, **15**, 157.
(*12*) Greenslade, R. M., Massee, A. M. and Roach, W. A., *Rep. E. Malling Res. Sta.*, 1933, p. 220.
(*13*) Davidson, J., *Ann. appl. Biol.*, 1922, **9**, 135.
(*14*) Evans, A. C., *Ann. appl. Biol.*, 1938, **25**, 558.
(*15*) Maltais, J. B., *Canad. Ent.*, 1951, **83**, 29.
(*16*) Auclair, J. L., *Canad. Ent.*, 1953, **85**, 63.
(*17*) Roach, W. A., *Ann. appl. Biol.*, 1937, **24**, 206.
(*18*) Fraenkel, G. S., *Trans. IX. int. Congr. Ent.*, 1953, **2**, 90.
(*19*) Dethier, V. G., *Trans. IX. int. Congr. Ent.*, 1953, **2**, 81.
(*20*) Broadbent, L., *Ann. appl. Biol.*, 1949, **36**, 334.
(*21*) Grevillius, A. Y., *Beih. bot. Zbl.*, 1905, **18**, 222.
(*22*) Verschaffelt, E., *Proc. Sect. Sci. K. Akad. Wet.*, 1910, **13**, i, 536.
(*23*) Thorsteinson, A. J., *Canad. J. Zool.*, 1953, **31**, 52.
(*24*) Kuhn, R. and Löw, I., *In Origins of Resistance to Toxic Agents.* Academic Press Inc., New York, 1955, p. 122.
(*25*) Back, E. A. and Pemberton, C. E., *J. agric. Res.*, 1915, **3**, 311.
(*26*) Lipke, H., Fraenkel, G. S. and Liener, I. E., *J. Agr. Food Chem.*, 1954, **2**, 410.
(*27*) Liener, I. E. and Pallansch, M. J., *J. biol. Chem.*, 1952, **197**, 29.
(*28*) Smissman, E. E., LaPidus, J. B. and Beck, S. D., *J. org. Chem.*, 1957, **22**, 220.
(*29*) Lees, A. H., *Ann. appl. Biol.*, 1918, **5**, 11.

CHAPTER III

THE INFLUENCE OF EXTERNAL FACTORS ON THE SUSCEPTIBILITY AND LIABILITY OF THE PLANT TO ATTACK

The factors which determine the resistance of the plant to pest or pathogen, discussed in the preceding chapter, may be classified broadly into two groups : firstly, physiological factors intimately bound up in the protoplasm of the plant ; secondly, morphological factors connected with the physical and chemical properties of its tissue and sap. Both groups contain factors capable of modification by the environment. This chapter deals with the modification of the resistance of the plant by the external factors, and the utilization of such modifications as pest-control measures.

The problem is two-sided for those environmental conditions affecting the host plant will also, directly or indirectly, affect the pest or pathogen. Indeed, in some instances the immunity of the plant is only apparent and results because conditions are unfavourable to the pest. These factors may be divided into two groups : firstly, non-biological, such as nutritional and climatic factors ; secondly, biological. The employment of the biological factors restraining the increase and the virulence of the parasite is termed " Biological Control ", the subject of the next chapter.

The present subject is, therefore, the utilization of the influence of non-biological ecological factors upon the extent of parasitism. This subject may be viewed from two aspects, the effect of external factors, firstly, upon the host plant and, secondly, upon the parasite. For convenience, a differentiation will be made between these two aspects, and effects ascribable to changes in the resistance of the host will be referred to as changes in its susceptibility, whereas effects due to alterations in the numbers or virulence of the pest or pathogen will be regarded as changes in liability to attack.

In many cases it is impossible to make a sharp division between " susceptibility " and " liability ". It has been found, for example, that excessive nitrogenous manuring " predisposes " the plant to attack, a predisposition which has been ascribed to changes in the host plant which render it more susceptible. Alternatively, the effect may be due, not to an inherent change in the plant, but to the greater vegetative growth and delay in maturation by which infection of the plant is rendered more probable. Further, it is conceivable that the greater vegetative growth, by shading and interference with air circulation, renders the conditions around the plant more favourable to the parasite. Thus there may be a transition from extreme cases in which external factors cause a breakdown in the plant's resistance (*1*) to those in which the external factors ensure the absence of the pest. The arbitrary distinction of susceptibility and liability is, however, useful even if it cannot be discerned in intermediate cases.

The external growth factors may be classified under two heads : nutritional and climatic. The former embraces the wide field of soil conditions, fertilizer and water supply ; the latter includes temperature, humidity and rainfall.

REFERENCE

(*1*) Salmon, E. S., *Ann. appl. Biol.*, 1927, **14**, 263.

NUTRITIONAL FACTORS

Of these factors, that which lends itself most easily to study is that of the supply of fertilizers.

The Influence of Fertilizers. The observance of the profound influence of nutrient supply on the health of the plant must have followed rapidly the pioneer work of Lawes and of Liebig on plant nutrition. Liebig (*1*), himself, observed that on two fields differently manured the same potato variety behaved differently towards disease. He quoted the experiments of Nagelli and Zoller on the pot culture of potatoes, who concluded that those factors which promote the normal growth of the plant are simultaneously those factors which protect it from disease.

Subsequently it was found that nitrogenous manuring, especially with the more readily assimilated nitrates, creates a predisposition towards certain diseases. Laurent (*2*) found that the susceptibility of potato to blight (*Phytophthora infestans*) was increased by nitrogenous manuring ; a similar effect induced by liming he attributed to the indirect stimulation of the nitrification in the soil. Spinks (*3*) proved that the susceptibility of wheat grown in nutrient solution towards mildew (*Erysiphe graminis* DC.) and yellow rust (*Puccinia glumarum*) could be enhanced by supplying available nitrogen, a verification of the observation of Little (*4*) that heavy manuring, especially with nitrogen, predisposed wheat to rust. Last (*5*) confirmed that nitrogenous manuring increased the extent of mildew infection in wheat and he concluded that this increase was associated with physiological changes within the leaves rather than a greater liability due to a change in the external environment of the plant.

In support of this view it has been found that lack of available phosphate and potash induces a decline in resistance. At Rothamsted, the potash-starved mangold plots suffer from leaf rust (*Uromyces betae* (Pers.) Tul.) whereas the surrounding plots, equally liable to infection, remain healthy. Also, it has been the usual experience that an increased supply of potash or phosphate renders the plant more resistant to fungal and insect pests. Stakman and Aamodt (*6*) reported that, while the direct effect of fertilizers on the development of stem rust (*Puccinia graminis*) of wheat seems to be slight, the indirect effect is profound. By the avoidance of excessive fertilization with nitrogen and by the use of phosphate and potash, they reduced the damage caused by this pathogen. Gassner and Hassebrauk (*7*) demonstrated a clear correlation between susceptibility to rust and mineral nutrition, a factor being the protein content of the cereal foliage.

An example of the successful use of potash is in the control of " Streak " disease of tomato (originally called " Stripe " and attributed to *Bacterium lathyri* Manns & Taubenh.). The incidence of this disease on plots of the variety Comet receiving complete artificials was 40 whereas, on the plots not receiving potash, it was 78 per 120 plants. On 120 plants of the variety Kondine Red, 13 were diseased on the complete artificial plots but 33 on the plots from which potash had been withheld (*8*).

The manner in which the supply of nutrient affects the resistance of the plant is unknown. Its effect on the more profound physiological factors associated with resistance, is but slight. It is more probable that any alteration of relative susceptibility is due to some modification of the morphological characters of the plant. Nitrogenous manuring induces a " soft " growth, an increased leaf area of a more succulent, sappy nature. Remer (*9*) ascribed the predisposition of cereals to rust by nitrogenous manures to the greater vegetative vigour and the increase of leaf surface; Bewley (*10*) thought that excess of nitrogen induces a rapid-growing, soft and sappy plant susceptible to disease, ascribing the special value of potash to its " hardening " effect. In addition, the view has been advanced that any alteration of the degree of attack dependent on manuring may be due to some change in the plant's mode of growth. Gassner (*11*) concluded that the differences in the attack of stem rust (*Puccinia graminis*) on differently manured plots arise through differences in the stage of development of the plant, for, in the different manure plots, plants of the same stage of growth showed an approximately equal development of rust. Armstrong (*12*) referred to the delay of maturation of wheat manured with nitrate of soda and considered that the opportunity for the development of yellow rust (*Puccinia glumarum*) is thereby increased. Phosphates, in general, have the reverse effect so that the plant is enabled to reach maturity more rapidly. Frew (*13*) explained the relative freedom of summer barley from infection by gout fly (*Chlorops taeniopus* Meig.) when manured with phosphates to the earlier development of the head. The number of leaf laminae arising from a shoot above the level of the apex of the ear at the time of oviposition is thereby lessened and the chance of the downward-crawling larvae, hatching from eggs on these leaves, of reaching the ear is correspondingly decreased.

The generalization that nitrogenous manuring predisposes the plant to attack and that potassic and phosphatic fertilizers have an opposite effect has notable exceptions. Bewley (*10*, p. 48) showed, for example, that the *Verticillium* wilt of tomato is more destructive to hard, underfed plants. Nitrogenous manuring may indeed enable the plant to grow away from or to recover more rapidly from the damage caused by insect attack. The shot-hole borer, *Xyleborus fornicatus* Eich., tunnels in the stem and branches of tea bushes, weakening the branches so that they are easily broken by wind or by coolies when pruning. Jepson and Gadd (*14*) showed that secondary mechanical damage of this kind was reduced on plots receiving nitrogenous manuring because of the more rapid healing-over of the wounds. On these plots healing-over was completed in 2.9 months as compared with 3.75 months on the control plots. In one experiment, however, Gadd (*15*) found that the yield of tea from the manured plot was greater in spite of greater borer infestations as measured by the number of broken branches.

Opinions differ as to whether the effect of the separate fertilizer constituents is individual or whether it depends on the general composition of the total nutrient available to the plant. Spinks (*3*) concluded that though the application of mineral manures decreased susceptibility it could not counteract the application of large quantities of nitrogenous manures. A more general view favours the idea of a " balance " in the manuring, and that the ill effects of nitrogenous manuring result from dressings in which nitrogen is in excess. Thus, Schaffnit and Volk (*16*) found that excessive application of phosphates and potash failed to increase resistance if the supply of nitrogen was correspondingly increased. Garrett (*17*) has included, in Table 9 of his masterly review of the effects of environment on root-infecting fungi, references to the evidence that abundant nitrogen and deficient potassium favours the development of wilts caused by various physiologic races of *Fusarium oxysporum*.

The influence of the lime status of the soil upon attack by parasites is shown by the froghopper *Tomaspis saccharina* Dist., which causes blight of sugar-cane. Turner (*18*) found that, in blighted areas, the ratio of exchangeable calcium to the clay and fine silt fraction averaged 0.28–0.63, whereas in blighted areas it was between 0.11–0.25. Why greater damage should be caused in the more acid soils is not clear, and, in a later paper by Pickles (*19*), the influence of soil factors apart from lime status is discussed.

The influence of the remaining elements (" micronutrients ") concerned in plant nutrition on the relations of plant and parasite is not clearly marked. Spinks (*3*) observed in the Woburn pot-culture experiments that wheat plants receiving small amounts of lithium salts were remarkably free from the attack of yellow rust and mildew, whereas the application of small amounts of lead or zinc nitrates rendered the plants extremely susceptible. The influence of lithium on resistance was more fully investigated by Kent (*20*). Cadmium salts have also been found to confer resistance to mildew to wheat (*21*) growing in sand culture. One rather special phenomenon may be mentioned to show the possibility of the influence of trace elements in the relationship of host and parasite. Brenchley and Thornton (*22*) found that, for the symbiotic development of the nodule bacteria of certain Leguminosae, the presence of boron is necessary. The special case of selenium is outside the subject of nutrition. Similarly, the question of the effect of manuring on soil pests or pathogens must be left until the consideration of Soil Treatment.

REFERENCES

(*1*) Liebig, J. von, *Die Chemie in ihrer Anwendung auf Agricultur und Physiologie*, 9th ed., Brunswick, 1876.

(*2*) Laurent, E., abstr. in *Exp. Sta. Rec.*, 1900, **11**, 550.

(*3*) Spinks, G. T., *J. agric. Sci.*, 1913, **5**, 231.

(*4*) Little, W. C., *J. R. agric. Soc.*, 1883, ser. 2, **19**, 634.

(*5*) Last, F. T., *Ann. appl. Biol.*, 1953, **40**, 312.

(*6*) Stakman, E. C. and Aamodt, O. S., *J. agric. Res.*, 1924, **27**, 341.

(*7*) Gassner, G. and Hassebrauk, K., *Phytopath. Z.*, 1931, **3**, 535.
(*8*) Bewley, W. F. and Paine, S. G., *Rep. exp. Res. Sta. Cheshunt*, 1919, p. 22.
(*9*) Remer, W., *Z. PflKrankh.*, 1904, **14**, 65.
(*10*) Bewley, W. F., *Diseases of Glasshouse Plants*, London, 1923, p. 46.
(*11*) Gassner, G., *Zbl. Bakt.*, 1916, ii, **44**, 512.
(*12*) Armstrong, S. F., *J. agric. Sci.*, 1922, **12**, 57.
(*13*) Frew, J. G. H., *Ann. appl. Biol.*, 1924, **11**, 175.
(*14*) Jepson, F. P. and Gadd, C. H., *Bull. Dep. Agric. Ceylon*, 78, 1926.
(*15*) Gadd, C. H., *Ann. appl. Biol.*, 1944, **31**, 47.
(*16*) Schaffnit, E. and Volk, A., abstr. in *Rev. appl. Mycol.*, 1927, **6**, 570.
(*17*) Garrett, S. D., *Biology of Root-infecting Fungi*. Cambridge Univ. Press, 1956.
(*18*) Turner, P. E., *J. agric. Sci.*, 1929, **19**, 26.
(*19*) Pickles, A., *Trop. Agric.*, 1937, **14**, 5.
(*20*) Kent, N. L., *Ann. appl. Biol.*, 1941, **28**, 189.
(*21*) Meyer, H., *Phytopath. Z.*, 1950, **17**, 63.
(*22*) Brenchley, W. E. and Thornton, H. G., *Proc. roy. Soc.*, 1925, **98**, B., 373.

The Influence of General Soil Conditions

Soil conditions influence the extent of attack not only by their indirect effect on the action of fertilizers but by their direct effect on the growth both of the plant and parasite. The factors involved are interrelated in complex fashion and, although for convenience an attempt is made below to deal separately with soil type, soil moisture and soil temperature, the interrelationships between these various soil conditions must not be forgotten.

Soil Type. Müller and Molz (*23*) concluded, that in Germany, less damage is caused by yellow rust (*Puccinia glumarum*) on heavy, deep and moist soils than on shallow dry soils ; that, whereas this rust is rarely found on marshy soils rich in mineral matter, it is abundant on poor soils. Howard (*24*) believed that the resistance of the plant towards rust may be affected profoundly by soil aeration. He showed that in India several of the most rust-resistant wheats are shallow-rooted, whereas most of the wheats liable to rust are deep-rooted, and cited the observation of Clouston that the red rot of sugar-cane (*Colletotrichum falcatum* Went.) is common on the stiff black soils of India but is rare on the porous " bhata " soils. The reverse effect of soil aeration is noted by Butler (*25*) from Reunion, where the arabica variety of coffee grows well on the drier side of the island where the soil is deep, but in shallow soils it is severely attacked by leaf disease.

The type of soil frequently exerts a profound influence upon crop infestation by soil-inhabiting insects. Williams (*26*) stated that the pea thrips, *Kakothrips robustus* Uzel., is more serious on light soils than on heavy soils. MacGill (*27*), who reviewed the evidence upon the relationship between thrips infestation and soil type, also found that *Thrips tabaci* Lind. was less prevalent upon untilled clay soils than upon light tilled soils and suggested that the thrips were unable to emerge as easily from soils forming a hard surface as from loose soils or from clays which readily form shrinkage cracks on drying. The general experience of French viticulturists that vines growing on loose textured soils suffer less from the attacks of *Phylloxera vitifoliae* Fitch than vines on heavy soils, was critically examined by Nougaret and Lapham (*28*) in California. They confirmed this conclusion

2

and suggested that heavy soils, on drying, form cracks in which the root forms of *Phylloxera* shelter, whereas on loose friable soils, the insects are forced to come to the surface where they are unable to survive the arid conditions.

In such cases, an obvious method of avoiding damage by disease or pest is to select soil types unfavourable to the parasite. For this purpose the data now being collected by systematic soil surveying should be of great use.

Soil Moisture. For the well-being of the crop an optimum range of moisture content of the soil is necessary. If the moisture content is too low, drought may profoundly affect the susceptibility of the crop, or it may have a direct effect upon the activities of the parasite. The effect of soil moisture on the parasitism of cabbage by cabbage yellows (*Fusarium conglutinans*) was studied by Tisdale (*29*), who found that strains of host plant normally resistant became attacked if deprived of water. On the other hand, Glynne (*30*) showed that, in dry soil, infection of potato by *Synchytrium endobioticum* did not occur. It has been suggested (*31*) that a high moisture content of the soil is necessary to secure infection of the potato by this fungus as the zoospores require a film of water for migration.

The case of abnormally high soil moisture is most complex and its effects on parasitism are difficult to analyse. Thus, as badly drained soils are cold and poorly aerated, the factors of soil temperature and oxygen supply come into play, a complexity well illustrated by the work of McKinney and Davis (*32*) on the infection of wheat by take-all (*Ophiobolus graminis* Sacc.). The reduced liability to attack in soils of high moisture content is exemplified by the observation of Jones (*33*) that the germination of the spores of loose smut of oats (*Ustilago avenae* (Pers.) Rostr.) is inhibited when the soil is saturated or, in other words, when the supply of available oxygen is reduced to a minimum.

Where the soil moisture is so small in amount that irrigation is necessary, this may be employed as a pest-control measure. Parker (*34*) found that an early irrigation of sugar beet at the first sign of wilting of the plants reduced infestation by the root aphis (*Pemphigus betae* Doane).

The flooding of soil to asphyxiate the insects living in it was applied practically in French vineyards against *Phylloxera*. The method met with a certain success and its use extended for a while despite its ill after-effects on the soil. Wardlaw (*35*) reported some success in the elimination of *Fusarium oxysporum* f. *cubense* (E.F.S.) Wollenw., the pathogen of the Panama wilt disease of bananas, by flooding infested soil. The effects of flood-fallowing on the fungus flora of a banana soil and the usually successful eradication of *F. oxysporum* f. *cubense* were studied by Stover (*36*) and his colleagues. After 40 days of flooding at least 85 per cent of the indigenous *Fusaria* were eradicated; deprivation of oxygen appeared to be the main factor involved, which in turn is accelerated by microbiological activity in large masses of decomposing plant tissue.

At one time it was usual in North America to flood cranberry bogs in June to check insect pests. Franklin (*37*) advised flooding every third year but considered it a dubious annual practice for it encouraged fungal infection of new growth and sometimes caused crop reduction through the death of the flower buds.

The part played by soil selection and cultivation has been stressed most forcibly by Howard (*38*), who maintained that by proper cultivation of suitable varieties the problems of Crop Protection are completely solved. Howard's faith in this hypothesis rested on his claim of its successful application in India, where, by the selection of varieties suited to local conditions and by attention to soil conditions, in particular, aeration, drainage and humus content, the economic consequences of pests and diseases are rendered negligible. Indeed, he insisted that insects and fungi serve as indicators of imperfect cultivation and therefore that their destruction by other methods, such as the use of insecticides and fungicides, is fundamentally unsound. His extreme views do not survive critical examination (*39*).

Soil Temperature. The effect of soil temperature and moisture upon infection by soil-borne organisms has been extensively studied, in particular, by Jones, Walker and their colleagues at the Wisconsin Agricultural Experiment Station (*40*). Of the fungi which require relatively high soil temperatures in order to establish infection, the *Fusaria* are prominent, and Jones, Walker and Monteith (*41*) showed that at soil temperatures below 17° C, wilt of cabbage (*Fusarium conglutinans*) failed to develop. Scab of potatoes, caused by *Streptomyces scabies* (Thaxt.) Skinner, is likewise more prevalent under warm soil conditions, the optimum temperature being 20–23° C (*42*).

The prevalence of onion smut (*Urocystis cepulae* Frost) in the northern states of the United States and its absence from the southern states was explained by Walker and Jones (*43*) on the basis of temperature relations. Onion seedlings become resistant to this disease at a certain stage of growth and, as increase of temperature promotes rapid growth, the susceptible period is passed more quickly at the higher temperatures. The fungus is also favoured by high temperatures, but Walker and Wellman (*44*) have shown that, above 25° C, the growth of the fungus is retarded and that of the host is stimulated. This preferential effect of temperature, permitting a disease escape, was found by Leach (*45*) to operate with many of the pathogens associated with the damping-off of seedlings.

The influence of soil temperature on insect pests may be illustrated from the work of Cook (*46*), who studied the effect of climatic factors on the relative abundance of cutworms in the soil. The optimum soil moisture content for the development of *Lycophotia margaritosa* Haw., he found to be about 60 per cent of the total water capacity. As a heavy rainfall at a high temperature will result in the same degree of soil moisture as a lighter rainfall at a lower temperature, this optimum condition may be expressed as an equation involving rainfall and temperature. If these conditions approach this optimum over a series of years, there is likelihood of an increase of cutworm during the following year.

The direct application of this principle to crop protection would appear to be limited to circumstances when a control of soil temperature is possible, as in the glasshouse provided with methods of soil heating. The indirect application is, however, of obvious importance and, with increasing knowledge of the effect of environmental conditions upon crop growth and the activity of parasites, successful examples of the application of the principle may become more numerous.

REFERENCES

(*23*) Müller, H. and Molz, E., *Fühlings landw. Z.*, 1917, **66**, 42.
(*24*) Howard, A., *Ann. appl. Biol.*, 1921, **7**, 273.
(*25*) Butler, E. J., *Fungi and Disease in Plants*, Calcutta, 1918, p. 117.
(*26*) Williams, C. B., *Ann. appl. Biol.*, 1915, **1**, 222.
(*27*) MacGill, E. I., *Ann. appl. Biol.*, 1930, **17**, 150.
(*28*) Nougaret, R. L. and Lapham, M. H., *Tech. Bull. U.S. Dep. Agric.*, 20, 1928.
(*29*) Tisdale, W. B., *J. agric. Res.*, 1923, **24**, 55.
(*30*) Glynne, M. D., *Ann. appl. Biol.*, 1925, **12**, 34.
(*31*) Foister, C. E., *Conf. Empire Meteorologists*, 1929, ii, 168.
(*32*) McKinney, H. H. and Davis, R. J., *J. agric. Res.*, 1925, **31**, 827.
(*33*) Jones, E. S., *J. agric. Res.*, 1923, **24**, 577.
(*34*) Parker, J. R., *J. agric. Res.*, 1915, **4**, 241.
(*35*) Wardlaw, C. W., *Nature*, 1947, **160**, 405.
(*36*) Stover, R. H., Thornton, N. C. and Dunlap, V. C., *Soil Sci.*, 1953, **76**, 225; Stover, R. H., *ibid.*, 1954, **77**, 401.
(*37*) Franklin, H. J., *Bull. Mass. agric. Exp. Sta.*, 447, 1948.
(*38*) Howard, A., *Emp. Cott. Gr. Rev.*, 1936, **13**, 186; *An Agricultural Testament*, Oxford, 1940.
(*39*) Hopkins, D. P., *Chemicals, Humus and the Soil*, Faber and Faber Ltd., London, 1957.
(*40*) See Jones, L. R., Johnson, J. and Dickson, J. G., *Res. Bull. Wis. agric. Exp. Sta.* 71, 1926.
(*41*) Jones, L. R., Walker, J. C. and Monteith, J., *J. agric. Res.*, 1925, **30**, 1027.
(*42*) Jones, L. R., McKinney, H. H. and Fellows, H., *Res. Bull. Wis. agric. Exp. Sta.* 53, 1922.
(*43*) Walker, J. C. and Jones, L. R., *J. agric. Res.*, 1921, **22**, 235.
(*44*) Walker, J. C. and Wellman, F. L., *J. agric. Res.*, 1926, **32**, 133.
(*45*) Leach, L. D., *J. agric. Res.*, 1947, **75**, 161.
(*46*) Cook, W. C., *Tech. Bull. Minn. agric. Exp. Sta.* 12, 1923.

CLIMATIC FACTORS

Climate, the summation of weather, is of so complex a character that the study of its influence upon the prevalence of disease and pests has had to proceed upon analytical lines, each meteorological factor such as temperature, rainfall, intensity of sunlight, wind, being studied in relationship to the growth both of plant and parasite. As this work has already been reviewed by Uvarov (*1*) in the case of insect pests and by Foister (*2*) for plant diseases, it is possible to consider at once the collective influence of climate and weather on the incidence of parasitism.

Reference may, however, be made to Stoughton's analytical study of the influence of environmental conditions on the angular leafspot disease (Black-arm disease) of cotton caused by *Xanthomonas malvacearum* (*3*). He found that the importance of air temperature for leaf infection rested in its effect on the balance between the resistance of the host, conditioned by rate of growth and maturation, and the activity of the parasite; the main factor is therefore the average temperature throughout the incubation period. The importance of air humidity, on the other hand, was physical in character, affecting merely the time, after inoculation, during which the infection drop persists.

The importance of the collective factors which constitute climate is well shown by examples of their effects upon the distribution of pests and diseases. The distribution of the pest is governed primarily by the distribution of the host plant, but it may be that in a particular region the climatic conditions, though still favourable to the host plant, are unfavourable to the parasite.

The influence of temperature and relative humidity upon the distribution of fungus diseases is illustrated by the work of Peltier and Frederick (*4*) on the prevalence of citrus scab (*Sphaceloma fawcetti* Jenkins) and of citrus canker (*Xanthomonas citri* (Hasse) Dows.). They found that citrus scab is not serious wherever a mean monthly temperature of above 75° F prevailed, as in the Philippine Islands. Neither is the disease of economic importance in districts where the rainfall is below 50 inches, spread evenly over the year; nor where there is a dry season coincident with high temperature. They concluded that citrus canker occurs in regions where the temperature and rainfall curves are simultaneously ascending, as in the Gulf States of America, China and South Africa; if the rainfall curve is descending as the temperature curve ascends, as in California, it is absent.

The correlation of climate and the distribution of an insect pest may be illustrated from the work of Verguin (*5*), who pointed out that the regions in which the Mediterranean fruit fly, *Ceratitis capitata*, is a serious pest are within the January isotherm of 10° C on the world map. The insect appears to be incapable of establishment in areas where a suitable supply of host plants is not available throughout the year or where the mean monthly temperature falls to, or below, 50° F for three consecutive months of the year (Back and Pemberton, *6*).

The correct bio-climatic mapping of the distribution of a particular host-parasite complex forms a rational basis for the application of restrictive control measures. Between regions of dissimilar climatic features in so far as they affect a particular disease or pest, it is unnecessary to impose restrictions upon the interchange of host material. Between regions of similar ecological and climatic features there is, on the other hand, every reason to adopt restrictions such as the prohibition of imports or the inspection, quarantine or fumigation of imported plant products.

The phenomenon, that the climatic conditions may be unfavourable to the parasite while still favourable to the host plant, is observed more easily as a result of the climatic changes due to altitude. An example is given by Dowson (*7*) on the cultivation of coffee in Kenya. At altitudes of 6,000–7,000 ft., the climate is still favourable for coffee-growing but unfavourable to the organism (*Hemileia vastatrix* Berk. & Br.) causing coffee rust, and the disease has no serious consequences; at 5,000–6,000 ft. the disease is easily controlled by the application at the right time of any dilute fungicide; at 4,000–5,000 ft. regular spraying with stronger fungicides is necessary to secure control of the disease; below 4,000 ft. the conditions are so favourable to the rust organism that coffee-growing becomes unprofitable.

A prerequisite to the correlation of pest or disease distribution and climatic conditions is a suitable method of expressing graphically or statistically the variables concerned. The representation of climatic factors by a graph on which the mean temperature and mean relative humidity for each month are plotted on the two ordinates and the points are joined in the sequence of the months was adopted by Shelford (*8*) for entomological purposes. For phytopathological purposes, Tehon (*9*) employed a three-dimensional graph in which temperature and precipitation were plotted along the horizontal ordinates and the severity of attack was plotted vertically. By such a method, Tehon was able to trace interesting relationships as, for example, in curly top of sugar beet, a virus disease transmitted by the beet leaf-hopper, *Eutettix tenellus* Baker. The abundance of the insect is conditioned by relatively high temperatures and high rainfall, whereas the appearance of the disease is dependent on low rainfall. The range of the leaf-hopper is therefore more extensive than that of the disease it transmits.

There is a danger that the simplification effected by graphical and statistical methods may lead to erroneous generalizations because the climatic data used are but averages for a wide range of time and locality. An instructive example of the effect of local variations not expressed in the data is quoted by Uvarov. Graf (*10*) in attempting to correlate the distribution of the Mexican bean beetle, *Epilachna varivestis* Muls., used data supplied by the meteorological stations of Mexico and the neighbouring states of the United States. He concluded that climatic factors did not determine the distribution of the beetle. Sweetman (*11*) pointed out that, in arid regions, the beetle was confined to irrigated areas, the humidity of which differed from the average for the area.

Further, the numbers of short-lived organisms such as fungi and insects may undergo violent fluctuations dependent upon annual variations of climate. There are years in which the injury caused by a particular disease or pest is of extreme severity whereas, in other years, control measures are scarcely necessary. Much can be learnt of the reasons for these fluctuations by the study of past epidemics in relation to climatic conditions. Lambert (*12*), in such a study of the history of stem rust of wheat (*Puccinia graminis tritici* (Pers.) Erikss. & Henn.) in the Mississippi valley, was unable to obtain evidence of meteorological conditions specifically associated with severe outbreaks. Peltier (*13*) was more successful in a similar study in Nebraska where he found that the factors necessary for an epidemic were (1) suitable conditions for wind transmission of the uredo stage of the fungus from Texas and for the establishment of infection in Nebraska; (2) winter wheat in Nebraska at the "heading" stage of growth during or after the first week of June and (3) an extended fruiting period for the fungus, namely, optimum temperatures and an evenly distributed rainfall above the normal.

Knowledge of the relationships between climatic conditions and the degree of attack is applied to the protection of crops growing in the glass-house, where a control of climatic factors is possible. Outdoors, one practical application of epidemiology is in the forecasting of pest or disease outbreaks.

Meteorological Forecasting. Reliable forecasting of the likelihood of outbreaks of disease or of abnormally heavy insect infestation is of obvious value to the grower in enabling him to be better prepared for combating the outbreak. In the use of insecticides and fungicides for example, correct timing of the application is frequently a factor of efficiency. This timing is generally effected by employing some conspicuous event in the life-history of the crop plant, e.g., petal-fall, as a criterion. Occasionally an ecological short-cut is possible as in Darpoux' observation (*14*) that, in France, the first occurrence of anthracnose (*Gnomonia veneta* (Sacc. & Speg.) Kleb) generally precedes by a few days the appearance of vine mildew (*Plasmopara viticola* (Berk. & Curt.) Berk. & de Toni) and can be used as a spray warning.

Schemes for forecasting, from meteorological data, the optimum dates for the grower to begin control measures are in operation in many countries for a number of serious diseases. These schemes have been fully reviewed by Miller and O'Brien (*15*). The control of potato blight (*Phytophthora infestans*) is of such serious economic importance and as no phenological indicator has been found, spray warning services are now the rule in most potato-growing areas. The first successful service was in Holland where Van Everdingen (*16*) in 1926–27 evolved four criteria, the " Dutch rules ", for foretelling the appearance of blight : (1) a night temperature below dew-point for at least four hours, (2) a minimum temperature above 50° F, (3) a mean cloudiness on the next day of at least 0.8, (4) at least 0.1 mm. rain during the next 24 hours. The trustworthiness of these rules in S.W. England was tested by Beaumont and Staniland (*17*) who concluded that they may be reduced to two : (1) a minimum temperature of 50° F and (2) a relative humidity not falling below 75 per cent for at least two days. Attacks of blight may be expected to follow 15–22 days after these weather conditions have maintained for at least forty-eight hours. Clearly there must be prior to this period, conditions which permit the formation of spores by the overwintering mycelium of *P. infestans* and Grainger (*18*) assigned July 1st as the " zero time " for Ayrshire (Scotland) before which " Beaumont's rules " (*19*) are invalid. Grainger's graph for the years 1944–1952 inclusive reveals the usefulness of these criteria for the correct timing of the protective fungicides employed for the control of potato blight.

In localities where the primary source of infection of apple scab (*Venturia inaequalis*) is from perithecia in the overwintering leaves, the fact that the discharge of ascospores occurs only after the leaves are thoroughly wet enables the grower to time his first scab spray. For New York State and the adjacent apple-growing districts, Mills (*20*) constructed a graph enabling the operator to determine within what time he should spray after a wet period taking into account the temperature. Thus infection will follow, at 42° F, a continuous wet period of some thirty hours, a period reduced to fourteen hours at 50° F. Hence, in this region, the early scab programme can be timed for protection from the ascospores, a period spread over some six weeks. In England, the perithecia have not been shown to be the primary source of infection (*21, 22*) nor are the weather conditions during the spring months of so continuous a character that it is possible to time spray application on the basis of meteorological forecasts.

The Control of Temperature and Humidity Relationships. In the glasshouse, it is often possible to modify temperature and humidity in a manner such that the activities of the parasite are checked without severe interference with the well-being of the plant. Bewley (*23*) took advantage, in this manner, of the importance of the air and soil conditions in combating the Verticillium disease of tomatoes. The wilting of foliage, symptomatic of this disease, is due to toxins produced by the fungus and translocated in the water supply of the plant. As soon as it is proved that *V. albo-atrum* Reinke & Berth. is the pathogen, precautions are taken to reduce to a minimum water movement in the infected plants by shading, by increasing the humidity of the atmosphere by overhead damping yet maintaining the soil dry, by raising the temperature of the house above 25° C (77° F). These conditions held for a fortnight or so will permit the wilted plant to recover and produce a normal crop.

The damping off of tomatoes (*Phytophthora* spp.) is aggravated by high water content in air and soil. A regulated watering of the seed boxes to keep the soil uniformly moist, combined with efficient ventilation permitting the rapid drying of the surface soil, will help to keep this disease in check. The control of tomato leaf mould *Cladosporium fulvum* Cke, by cultural operations is discussed by Small (*24*). As 22° C is the optimum temperature for this fungus and higher temperatures are unfavourable to the tomato, the temperature should be kept below 22° C during March and April. Further, as the disease is favoured by high humidity, the relative humidity of the glasshouse should be kept below 75 per cent on warm nights and the ventilators should be opened during dull or rainy weather. The plants should be watered early in the day and preferably only on fine days. Finally, air circulation should be promoted by open ventilation and by pruning the lower foliage of the plants.

By cultural operations it is frequently possible to modify temperature and humidity relationships of crops growing in the open. The relative humidity may be reduced, for example, by more open planting, by clean cultivation or by pruning. In the apple orchard, the removal of low branches and of rank undergrowth provides " bottom " ventilation and tends to reduce the severity of scab attacks. The stripping of the lower leaves from the bines is a common practice in hop gardens, and designed to remove a habitat favourable to the development of aphis. Kirkpatrick, who quoted (*25*) the example of the pruning of coffee in order to promote air circulation and thereby to secure a lowering of the relative humidity within the crop as a defence against coffee mealy bug, *Pseudococcus citri* Risso, has investigated (*26*) the effect of various cultural practices upon the local climatic conditions within the coffee plantation.

It may be appropriate to close this section with a remarkable example, taken from outside crop protection, of the effect of environment on pest infestation. In work on the extermination of the tsetse fly (*Glossina morsitans* Westw.) from the southern shores of Lake Tanganyika, Glover and his colleagues (*27*) found that the clearing of trees from the narrow river valleys was unexpectedly successful in reducing the fly population from the 280 sq. miles of the experimental area, of which the total area of discriminative clearing came to only 3 per cent. " How it worked we still do not know " wrote the authors but their guess is that the trees provided the flies with a suitable habitat for mating.

REFERENCES

(*1*) Uvarov, B. P., *Trans. R. ent. Soc. Lond.*, 1931, **79**, 1.

(*2*) Foister, C. E., *Conf. Empire Meteorologists*, 1929, II, 168; *Bot. Rev.*, 1935, **1**, 497; 1946, **12**, 548.

(*3*) Stoughton, R. H., *Ann. appl. Biol.*, 1930, **17**, 90, 493; 1931, **18**, 524; 1932, **19**, 370; 1933, **20**, 590.

(*4*) Peltier, G. L. and Frederick, W. J., *J. agric. Res.*, 1926, **32**, 147.

(*5*) Verguin, J., *Rev. Zool. agric.*, 1928, **27**, 141.

(*6*) Back, E. A. and Pemberton, C. E., *Bull. U.S. Dept. Agric.*, 536, 1919.

(*7*) Dowson, W. J., *Ann. appl. Biol.*, 1921, **8**, 83.

(*8*) See Shelford, V. E., *Laboratory and Field Ecology*, London, 1929.

(*9*) Tehon, L. R., *Bull. Ill. nat. Hist. Surv.* 17, 1928, p. 321.

(*10*) Graf, J. E., *J. econ. Ent.*, 1925, **18**, 116.

(*11*) Sweetman, H. L., *Ecology*, 1929, **10**, 228.

(*12*) Lambert, E. B., *Phytopathology*, 1929, **19**, 1.

(*13*) Peltier, G. L., *J. agric. Res.*, 1933, **46**, 59.

(*14*) Darpoux, H., *Ann. Epiphyt.*, 1943, **9**, 177.

(*15*) Miller, P. R. and O'Brien, M., *Bot. Rev.*, 1952, **18**, 547; *Ann. Rev. Microbiol.*, 1957, **11**, 77.

(*16*) Van Everdingen, E., *Tijdschr. Plziekt.*, 1926, **32**, 129; 1935, **41**, 125.

(*17*) Beaumont, A. and Staniland, L. N., *Rep. Dept. Plant Pathology, Seale-Hayne Agr. College*, Newton Abbot, Devon, for year ending Sept. 30, 1936: 1937.

(*18*) Grainger, J., *West. Scot. Agric. Coll. Auchincruive Res. Bull.* 9, 1950; *Nature, Lond.*, 1953, **171**, 1012.

(*19*) Beaumont, A., *Trans. Brit. mycol. Soc.*, 1947, **31**, 45.

(*20*) Mills, W. D., *Cornell Ext. Bull.*, 630, 1944.

(*21*) Marsh, R. W. and Walker, M. M., *J. Pomol.*, 1932, **10**, 71.

(*22*) Dillon Weston, W. A. R. and Petherbridge, F. R., *J. Pomol.*, 1933, **11**, 185.

(*23*) Bewley, W. F., *Bull. Minist. Agric.*, 77, 1939.

(*24*) Small, T., *Ann. appl. Biol.*, 1930, **17**, 71.

(*25*) Kirkpatrick, T. W., *Proc. Tech. Conf. E. Africa Dependencies*, 1926, p. 184.

(*26*) Kirkpatrick, T. W., The Climate and Eco-climates of Coffee Plantations, *E. Africa agric. Res. Sta.*, 1935.

(*27*) Glover, P. E., Jackson, C. H. N., Robertson, A. G. and Thomson, W. E. F., *Bull. ent. Res.*, 1955, **46**, 57.

The Modification of Time of Sowing

Not infrequently the attack of the parasite occurs only during a period of short duration in the life of the crop plant. It may then be possible to ensure that, during this period, the plant is at a stage when infection or attack will cause least damage. The principle of such methods thus becomes the regulation of crop growth so that the period of greatest susceptibility is not coincident with the period of greatest liability.

The proper preparation of the seed bed to promote germination and rapid growth of the seedling, a suitable manuring, but more important, a regulation of the date of sowing are the main methods employed. The reduction of the amount of damage caused by the gout fly of barley (*Chlorops taeniopus*) by the use of phosphatic fertilizers which hasten maturity has already been mentioned. Frew (*28*) also suggested the early planting of spring barley for the same reason. A parallel example is provided by the work of Cunliffe and his colleagues (*29*) on the biology and cultural control of frit fly (*Oscinella frit* L.). This pest of oats, under English conditions, has three generations each year. Flies of the first generation oviposit on oat seedlings and attain maximum numbers towards the end of May. Ovipositing flies of the second generation are most abundant at the middle of July. The oat plant is most susceptible at the two- and three-leaf seedling stage and at the time of pollination, becoming resistant when it has reached the four-leaf stage and after the grain has grown to half its normal size. It follows that winter-sown or early spring-sown oats stand a better chance of resisting attack than oats sown late in March or in April. The early-sown oats reach the four-leaf stage prior to the period of maximum oviposition by the first-generation flies and escape the attack of the second swarm of flies until the grain has reached its resistant stage.

In a similar manner a delayed autumn sowing may enable the plant to escape infection. Thus McKinney (*30*) showed that early-sown winter wheat is more severely affected by the *Helminthosporium* disease than is late-sown winter wheat—an observation in agreement with trials in soil-temperature tanks, which have shown that high soil temperatures favour the disease. Howard (*31*) gave an analogous example in the destruction of wheat by termites at Bikar (India). If the wheat is sown a few days too early the seedlings are rapidly destroyed by termites, a fact attributed to the high soil temperature, for if planted twelve days later, the furrows being left open meanwhile to aid the cooling of the soil, damage becomes negligible. A much discussed example of this means of control is afforded by the relation between the date of sowing of winter wheat and the prevalence of bunt caused by two closely related species, *Tilletia caries* (DC.) Tul. and *T. foetida* (Wallr.) Liro. If sown either very early or very late, the crop is freer from the disease than the crop sown in the intermediate period (*32*).

The avoidance of insect attack by late autumn sowing is illustrated by American experience of the control of Hessian fly (*Phytophaga destructor*). The wheat is sown late so that the susceptible seedlings do not appear above ground until the autumn generation of the fly has died down. Yet the date of sowing must be early enough for the wheat to have made enough growth to enable it to survive winter. Hence, in certain States, the authorities suggest "fly-free" dates which, though the dates of maximum emergence of the flies, are those after which sowing will give the minimum infestation. In Illinois (*33*), the fly-free date in the northern part is 18 September and, in the south, 9 October. The fixing of the fly-free date to a definite day may fail in abnormal seasons (*34*).

In other cases, late spring sowing may be a means of avoiding attack. In England carrots sown in March are often more severely infested with carrot fly (*Psila rosae* F.) than those sown in May, attributed to the fact that the first brood of the fly has partially died off by the time the May-sown carrots are above ground—the stage of growth when the fly makes its attack.

The maturation of the crop before the attack of the pest may also be accomplished by the use of early-maturing varieties. Dowson (*35*) has shown that at Nairobi, where there are two rainy seasons in the year, it is possible to grow early-maturing varieties of wheat in both. By this means it is also possible to escape black rust (*Puccinia graminis*), which usually appears late in the season. The Australian wheat " Florence ", which matures in Nairobi in four months after sowing, when sown early enough escapes injury, whereas another Australian wheat " Bobs ", maturing in six months after sowing, is always attacked by the rust. If the Florence variety is planted two months later, so as to mature at the same time as the Bobs variety, it is equally infected.

The application of this principle to perennials is illustrated by the cultural control of blister blight of tea, due to *Exobasidium vexans* Mass. (*36*). The older leaves of the tea bush are resistant to infection; the basidiospores of the fungus are short-lived and require high humidities to establish infection. By timing the pruning operation, necessary to keep the bush at a height suitable for plucking, so that only mature growth is present during the monsoons and the younger growth develops during the dry season, infection is held in check.

REFERENCES

(*28*) Frew, J. G. H., *Ann. appl. Biol.*, 1924, **11**, 175.
(*29*) Cunliffe, N., *Ann. appl. Biol.*, 1929, **16**, 135.
(*30*) McKinney, H. H., *J. agric. Res.*, 1923, **26**, 195.
(*31*) Howard, A., *Ann. appl. Biol.*, 1921, **7**, 273.
(*32*) Reed, G. M., *Phytopathology*, 1924, **14**, 437.
(*33*) Flint, W. P. and Larrimer, W. H., *Bull. Ill. nat. Hist. Surv.* 17, 1928, p. 363.
(*34*) Drake, C. J. and Decker, G. C., *Ann. ent. Soc. Amer.*, 1932, **25**, 345.
(*35*) Dowson, W. J., *Ann. appl. Biol.*, 1921, **8**, 83.
(*36*) Portsmouth, G. B. and Loos, C. A., *Tea Quart.*, 1949, **20**, 77.

CHAPTER IV

BIOLOGICAL CONTROL

In the previous chapter, the use of the influence of external factors of a non-biological character on host-parasite relationships has been discussed as a principle of crop protection. The present subject is the use of biological factors to this end. Biological control may be interpreted broadly as the encouragement of beneficial organisms already existing in a locality and the introduction to that locality of new species of beneficial organisms.

The encouragmeent of beneficial organisms already established in a locality is, in general, limited to their protection from unfavourable conditions. The activities of insectivorous birds, for example, may be assisted by the provision of water or nesting-places and by protection from hawks, cats, egg-stealers—" hyperparasites " as it were. Compared with the astonishing results sometimes secured in the second group of biological-control methods—the introduction of new species—the success of such measures is small.

Those biological-control methods which involve the purposeful introduction of new species of organisms are subject to certain general features. The introduction of new plant species, an everyday occurrence in agriculture, horticulture and floriculture, offers no practical difficulties if the conditions of climate, nutrition, etc., are suitable for the healthy development of the plant. Indeed, in some cases, in which it has later been found that the pests of the exotic plant are absent from its new habitat, conditions have been so favourable that the introduced plant has exceeded the limits of its usefulness and has become a weed. In other cases, pests of the exotic plant have been introduced which, because of the absence of their usual hyperparasites, have thrived so well that they have threatened the cultivation, in their new sphere of action, not only of their original host but even of fresh hosts. By the introduction of the enemies of the pest from its old habitat a successful control in such cases has often been established.

It will be seen at once that such cases involve the absence of biological factors in a particular locality. With modern transport the chances that these factors will be unwittingly introduced into the new locality increase rapidly with time and it may well be assumed that in due course the missing factor will find its way into the region where its host has thrived. The scope for the use of this form of biological control, will therefore depend upon the time that has elapsed since the introduction of the host and/or its pests. Thus it is that the outstanding successes of the method have occurred in localities where the disturbance of the original ecological conditions has been comparatively recent—e.g., in the New World and in Australia.

Again the chance of absence of these factors will be dependent upon the relative isolation of the locality from those in which they are present. Such isolation is to be found, for example, in the Hawaiian Islands or New Zealand, where the surrounding sea provides the isolating barrier, or in California, or Western Australia, where the barrier on one side is the ocean, on the other, desert or mountain. " Ecological islands " of this type may also be on the smaller scale noted by Myers (*1*). In the West Indies and on the neighbouring mainland of South America, Myers found clearly defined ecological islands, bounded by forests, rivers or swamps, within which large areas of cultivated crops are of recent introduction. It is evident that, in such a circumscribed area, even an indigenous insect may become a pest of the cultivated crop under conditions where beneficial organisms, though present in the area, are helpless.

It is in ecological islands of these two types that the possibilities of biological control are greatest and where the widest scope has been found for the introduction and encouragement of beneficial organisms.

The principles of biological control have been discussed by Thompson (*2*) and by Sweetman (*3*) but, to avoid further abstractions, concrete examples of the methods will be given, special points being discussed as they appear.

Higher Animals. With higher animals, the scope of biological control is limited entirely to the encouragement of those which are beneficial because of their influence upon injurious organisms. For, with their migratory powers and their tolerance towards external conditions, the introduction of higher animals into new territory would follow, almost automatically, the penetration of civilization.

The value of certain birds as the natural enemies of voles and mice has been recognized from early times. These rodents, of which the short-tailed field " mouse " (really a vole), *Microtus agrestis*, is perhaps the most notorious, at times increase to such an extent that they cause great damage. For example, in 1892 in Scotland the plague was sufficiently serious to warrant a Government investigation. Under normal conditions, however, the kestrels and short-eared owls play a great part in the checking of these pests (*4*). Insectivorous birds contribute greatly to the natural control of insects. For instance, in the United States, thirty-six kinds of birds feed upon the codling moth. McAtee (*5*) calculated that, in some localities they destroy from 66–85 per cent of the hibernating larvae. In Holland (*6*) the value of birds in pest control is reflected in the establishment of a special section of Ornithology in the Phytopathological Service. Van Poeteren remarked that the " protection of insect-eating birds is the most important form of biological attack we apply in our country ".

That special encouragement of any particular species of bird should only follow an accurate knowledge of its life activities is shown by the case of the buff-backed egret (*Ardea ibis* L.) recorded by Kirkpatrick (*7*). This bird, once abundant in Egypt, was killed off so thoroughly by plume-hunters that, by 1912, only two small colonies of a few pairs remained. It was at that time thought that the bird contributed to the control of the cotton worm (*Prodenia litura* F.), for popular opinion claimed that, prior to the recognition of the value of its feathers, these birds were often seen picking up larvae and pupae from newly-turned soil (*8*). A law was then passed for the protection of insectivorous birds by the provision of sanctuaries and by the infliction of penalties upon bird-hunters. The egret, by 1924, had so increased that it threatened to become a pest. Stomach examination revealed that the claim of its value as a biological-control factor was, if not erroneous, most exaggerated.

The advocates of Bird Protection claim that graminivorous birds are beneficial in that they consume weed seeds. It would appear, however, that the substantiation of this claim is difficult, for mere stomach examination is an untrustworthy criterion. The hard seed-coats of the more pernicious of our weeds may prevent their digestion and the seed will pass through the intestine.

Mention must also be made of a system introduced by W. Rodier in New South Wales, for the suppression of rabbits which had become too numerous for control by trapping. Of the trapped animals, only the females were killed. The males were released in the hope that the remaining females would be pestered out of successful breeding by the superabundance of males. The method does not appear to have been successful.

REFERENCES

(*1*) Myers, J. G., *Bull. ent. Res.*, 1935, **26**, 181.
(*2*) Thompson, W. R., *Ann. appl. Biol.*, 1930, **16**, 306 ; *Annu. Rev. Ent.*, 1956, **1**, 379.
(*3*) Sweetman, H. L., *The Biological Control of Insects*, Ithaca, N.Y., 1936.
(*4*) See *Bull. Minist. Agric. Lond.*, 30, 5th Ed., 1939.
(*5*) McAtee, W. L., *Year-Book U.S. Dept. Agric.*, 1911, p. 237.
(*6*) Van Poeteren, N., *Rep. Int. Conf. Phytopath.*, 1923, p. 86.
(*7*) Kirkpatrick, T. W., *Bull. Minist. Agric. Egypt*, 56, 1925.
(*8*) Dudgeon, G. C., *Bull. imp. Inst. Lond.*, 1912, **10**, 584.

INSECTS

Insects versus Insects. The first suggestion that entomophagous insects could be used for crop protection appears to have been due to Noteriani (*1*) ; the first-recorded adaptation of the method, according to Trotter (*2*), was the use of carabid beetle *Calosoma sycophanta* L., by Boisgiraud in 1840, against gypsy moth (*Lymantria dispar* L.) on poplars growing around Poitiers. The early history of the subject has been reviewed by Howard and Fiske (*3*).

The classical example of the use of a beneficial insect is the control of the cottony cushion or fluted scale, *Icerya purchasi* Maskell, by the vedalia *Rodolia cardinalis* Muls. Marlatt (*4*) recorded how this scale insect was introduced into California from Australia, in 1868, on imported plants of *Acacia latifolia* Desf. The insect increased with such rapidity that by 1886 it threatened the entire citrus industry of the Pacific coast. The insecticidal sprays then available proved no check. In 1889 Koebele was sent to Australia to investigate the reason for the relative unimportance of the insect there. He returned with its natural enemy *R. cardinalis*. In short time the vedalia practically exterminated the fluted scale and, in 1900, Marlatt wrote, " The remedy for this scale is always and emphatically to secure at once its natural and efficient enemy the *R. cardinalis* ". The fluted scale has since appeared in most citrus-growing areas. Mendes (*5*) reported its occurrence on acacia in Portugal in 1910 ; *R. cardinalis* followed and within a year the fluted scale was destroyed by the predator. Berlese (*6*) in 1916 stated that at that time the fluted scale, which prior to the discovery of *R. cardinalis* was a great menace to citrus crops, was no longer of agricultural importance. The high efficiency of *R. cardinalis* is probably ascribable to its great activity both as a larva and adult, its freedom from hyperparasites and its high rate of reproduction in comparison with that of the scale insect.

A less direct but equally effective example of the use of beneficial insects was recorded by Taylor (*7*). In Fiji a hispid beetle, *Promecotheca reichei* Baly, rarely became a serious pest of the coconut palm as it was held in check by several native natural enemies. This check was effective because the generations of beetle then overlapped so that the various developmental stages occurred together and the several enemies were able to maintain a condition of " multi-stage " parasitism. In 1921 the mite *Pediculoides ventricosus* Newp. was recorded in certain parts of Fiji attacking the larvae, pupae and adults of the beetle so successfully that the number of the latter fell practically to zero. The mite population again rose, but this time with no overlapping stages, a one-stage condition in which the indigenous parasites were no longer able to maintain control. To restore control it was necessary to introduce a parasite effective in the presence of the mite and it was deduced that the successful parasite should have the following characteristics : it should tolerate the climate of Fiji ; it should be an internal parasite of the larval and pupal stages ; it should have a biotic potential (i.e., egg-capacity × sex-ratio) of at least twenty and complete a generation in not more than one month, thus ensuring a favourable ratio between numbers of parasite and pest ; finally when adult, it should live for at least three weeks. A survey of the parasites of an allied hispid beetle in Java led to the selection of *Pleurotropis parvulus* Ferrière, not the most effective in Java of the parasites found there, but of greater promise under Fijian conditions. Within a year of the introduction of *P. parvulus* into Fiji, *P. reichei* was no longer a pest.

The value of this method of pest control, which is most striking in the successful introduction of absent beneficial insects, requires no stressing. Once established, the activities of the beneficial insect will reduce or even eliminate the need for alternative control measures.

The difficulties which may attend the establishment of a beneficial insect are illustrated by the attempts to utilize, in Great Britain, *Aphelinus mali* Hald. for the control of woolly aphis. The introduction was begun in 1928 but, by 1946 (*8*) it was still uncertain whether the parasite would be effective in the English climate though it appeared to have become established in some areas.

The artificial carrying over of the beneficial insect has sometimes been found profitable as is recorded by Smith (*9*). The black scale (*Saissetia oleae* Bernard) is controlled in South Africa by parasites, amongst which *Aphycus lounsburyi* How. is prominent. In California, however, the adult black scale becomes scarce in July and the young insects surviving are too small to serve as hosts for this parasite until after the second moult, which does not occur till February. The parasite is therefore without food supply for six months and can be maintained only by annual liberation. Similarly, for the control of various of the citrus mealy bugs (*Pseudococcus* spp.) of California, the lady-bird beetle (*Cryptoloemus montrouzieri* Muls.) has been introduced. Armitage (*10*) showed that the permanent establishment of the beneficial insect is prevented because the winters are too cold for the mealy bug to breed yet not cold enough to force the lady-bird to hibernate. As in the former example, the parasite is without an available host during this period and it is necessary artificially to ensure its overwintering.

The principle that the greatest opportunities for the utilization of biological control are to be found under conditions of recent ecological disturbance is applicable to glasshouse crops. An example, in Great Britain, is afforded by the control of greenhouse whitefly *Trialeurodes vaporariorum* West. In 1915 (*11*), a chalcid, *Encarsia partenopea* Masi., was discovered by Fox Wilson and bred for the control of whitefly but, by mischance, was lost. In 1926 a related chalcid, *E. formosa* Gahan, was found by Speyer (*12*), and successfully reared for distribution to growers. It is necessary to maintain the chalcid through the winter, but so successful a control of whitefly is achieved that its use became routine.

The encouragement of a beneficial insect by means other than artificial rearing is illustrated by the example of the mealy bug *Pseudococcus lilacinus* Ckll., which had become a serious pest of coffee in Kenya (*13*). As its normal enemies were already present, it became a question why the mealy bug should be so serious a pest. The cause was ultimately found in the presence of a small myrmecine ant (*Pheidole punctulata* Mayr.). The association of ants with mealy bugs, scale insects and aphides has been the subject of much speculation and it is usually assumed that the ants attend and protect these insects for the sake of the " honey-dew " they secrete. In Kenya, it was observed that, in the absence of the ants, the mealy bugs only occurred spasmodically and it was proved that, if the ants were denied access to the infected coffee tree, the mealy bug was soon destroyed by its enemies. The success of biological control, in this case, becomes dependent upon the satisfactory control, by other means, of the ants. For this purpose, the banding of the coffee bushes with a special type of coal tar repellent (*14*) was recommended (*15*).

Mealy bugs have been shown by Posnette and Strickland (*16*) to be responsible for the transmission of the viruses causing the " Swollen Shoot " disease of cacao in W. Africa. The mealy bug-ant association is here complex (*25*) but the most prevalent of the bugs, *P. njalensis* Laing, rarely becomes successfully established unless attended by ants. In this case the predominant ants, which are of the genus *Crematogaster*, nest in the trees and would escape grease-bands, for they travel from tree to tree without descending the trunks. Attempts to break the ant-mealy bug link by the use of insecticides such as DDT or parathion were not successful (*17*) because it was thought that these sprays failed to kill those ants sheltered within their nests. In the similar case of the arboreal ant *Oecophylla longinoda* Latreille, however, treatment of the nests with benzene hexachloride was spectacular for the ants rapidly evacuated the nests even bringing out the queens. Hanna and his colleagues (*18*) also experimented with systemic insecticides and found dimefox sufficiently promising for field use for its use reduced by 99.9 per cent the number of mealy bugs found on the treated trees.

In these examples the insecticide has been used to promote biological control. Unfortunately, too often in the past the insecticide has hindered biological control, permitting a resurgence of the pest or another arthropod after the use of an insecticide so non-specific that it destroys both harmful and beneficial insects. Following the introduction of the tar oil winter washes into English orchards in 1921, infestations of the fruit tree red spider mite (*Metatetranychus ulmi* Koch) became serious. The tar oil has little effect on the eggs of this mite but killed its predators such as anthocorid bugs, ladybird beetles and predaceous mites (*19*). It is also possible that, by destroying lichen, the tar oil wash provides cleaner wood more favourable to the development of eggs of the mite (*20*). The introduction of DDT was rapidly followed by similar mishaps : in California, DeBach (*21*) reported that on citrus treated with DDT the cottony cushion scale, referred to above, reappeared as a pest. Following the restriction of the use of DDT in citrus groves in California in 1948 the vedalia regained control of the scale, according to Ripper (*22*) in his review of the effect of pesticides on the balance of arthropod populations.

A method of biological control, resembling in principle that proposed by Rodier for rabbit control, was suggested by Knipling (*23*) involving a novel insecticidal treatment. Flies of the screw worm *Callitroga hominivorax* Cqrl. survive winter in the United States only in sub-tropical regions such as Florida. It is from Florida that migration to the eastern and south-eastern states occurs and if the isolated and small winter populations of screw worm flies in Florida could be exterminated, the screw worm could be kept out of the eastern States. Under laboratory conditions the female fly mates but once and Knipling suggested that if males sterilized by irradiation with x-rays or gamma-rays were released in a limited area mating would produce offspring which, if surviving to adult age, would contribute lethal mutations to subsequent generations. The requisite conditions of irradiation were determined by Bushland and Hopkins and the first trial of the method in the island of Curaçao was highly successful (*24*).

The use of insecticidal methods to reinforce biological control was first emphasized in 1944 by Ripper (*25*) and Ullyett (*26*), who, basing his arguments on his S. African experiences on the control of the diamond-back moth (*Plutella maculipennis*), concluded that it would approach the ideal method of pest control. The first large-scale experiments were carried out in the apple orchards of Nova Scotia by Pickett and his colleagues (*27*) where the major pests are the oystershell scale *Lepidosaphes ulmi* L., the eye-spotted bud moth *Spilonota ocellana* D. & S., the codling moth *Carpocapsa pomonella* L. and the mite *Metatetranychus ulmi*. In a modified spray programme, they substituted, for sulphur, fungicides less harmful to predators of the scale, and instead of DDT and lead arsenate which are toxic to the predators of the mite, they used ryania which is specifically toxic to codling moth. This programme held the pest population of the orchards at a low level, but it appears that commercial success will depend upon the willingness of the grower to be satisfied with at least some loss of crop through insect damage.

Insects versus Weeds. The first attempt to control weeds by their insect pests was made by Koebele in 1898 (*28*). About fifty years previously the thorny species of lantana (*L. camara* Cham.) had been introduced into the Hawaiian Islands from Mexico as an ornamental shrub. The plant became a menace to the pastures of the low-lying regions. Koebele introduced many of the insects which in its original home contributed to its suppression with such success that now, in the drier regions, further control measures are unnecessary.

In Queensland and New South Wales, prickly pear (*Opuntia* spp.) introduced as a botanical curiosity, spread with such rapidity that it was estimated that sixty million acres of grazing country were rendered useless by the weed. In its original habitat in America, the prickly pear is held in check by many insect species and the Prickly Pear Board of Australia (*29*) imported selected pests from America. An early experiment, the introduction of the cochineal insect *Dactylopius ceylonicus* Green, from Ceylon, was successful in controlling *O. monacantha*, but was ineffective against *O. inermis* and *O. stricta*, the two most noxious species of prickly pear found in Australia. The related species of cochineal insect, *D. opuntiae* Ckll. proved more successful but is unfortunately itself attacked by the ladybird beetle *Cryptoloemus montrouzieri* which was mentioned above as a beneficial insect for the control of mealy bugs. Of the other beneficial insects imported during the campaign against prickly pear, the moth *Cactoblastis cactorum* Berg. and, to a lesser degree, the red spider *Tetranychus opuntiae* Banks, have proved the most effective and not only has the rate of spread of prickly pear been checked but about 25,000,000 acres of good agricultural land had been reclaimed by 1938.

It will be noted that the two examples quoted are concerned with the control of plants of somewhat unusual characters and that the beneficial insects used are specifically adapted to cacti-like hosts and are unlikely to attack plants of economic importance. In cases where the weed is related to crop plants, there is a danger that the phytophagous insect may infest the related crop plant in its new environment. The greatest caution is therefore necessary and importation should only follow a complete demonstration that the beneficial insect is unlikely to become a pest. The procedure of such trials has been discussed by Davies (*30*).

Miscellaneous Beneficial Animals. Although no successful example of the use of animal organisms lower than the insects can be recorded, there are indications that such organisms contribute to natural control. Filinger (*31*), for instance, observed that the centipedes *Lithobius forficatus* L. and *Poabius bilabiatus* Wood prey upon the symphylid *Scutigerella immaculata* Newport and, in one glasshouse, appeared to have controlled this reputed pest.

The records of nematodes parasitic upon insects have been summarized by Bodenheimer (*32*). Thorne (*33*) examined the status of the genus *Mononchus* in relation to the sugar-beet nematode, *Heterodera schachtii* Schmidt, but concluded that, because of their slow rate of reproduction and their food habits, *Monochus* spp. were of doubtful economic importance.

A remarkable example of parasitism, recorded by Goodey (*34*), is afforded by the nematode *Tylenchinema oscinellae* Goodey which infests the fritfly, *Oscinella frit*, causing sterilization in both sexes by preventing the development of the reproductive organs.

REFERENCES

(*1*) See *Redia, Giornale di Entomologia*, 1909, **6**, 193.
(*2*) Trotter, *ibid.*, 1908, **5**, 126.
(*3*) Howard, L. O. and Fiske, W. F., *Bull. U.S. Dept. Agric. Bur. Ent.* 91, 1912.
(*4*) Marlatt, C. L., *Year-Book U.S. Dept. Agric.*, 1900, p. 247.
(*5*) Mendes, C., abstr. in *Int. Inst. Agric., Mo. Bull. Agric. Intell.*, 1913, **4**, 1310.
(*6*) Berlese, A., *Int. Rev. Sci. Prac. Agric.*, 1916, **7**, 321.
(*7*) Taylor, T. H. C., *The Biological Control of an Insect in Fiji.* Imp. Inst. Ent., London, 1937.
(*8*) See *Adv. Leaflet, Minist. Agric.* 187, 1946.
(*9*) Smith, H. S., *Mo. Bull. California State Dept. Agric.*, 1921, **10**, No. 4, 127.
(*10*) Armitage, H. M., *ibid.*, 1922, **11**, No. 7, 45.
(*11*) Fox Wilson, G., *Gdnrs'. Chron.*, 1931, **89**, 15.
(*12*) Speyer, E. R., *Rep. exp. Res. Sta. Cheshunt*, 1928, p. 96.
(*13*) Kirkpatrick, T. W., *Bull. Dept. Agric. Kenya*, 18, 1927.
(*14*) Beckley, V. A., *Bull. Dept. Agric. Kenya* 7, 1930.
(*15*) James, H. C., *Repellent Banding to control the Ants attending the Common Coffee Mealybug*, Nairobi, 1930.
(*16*) Posnette, A. F. and Strickland, A. H., *Ann. appl. Biol.*, 1948, **35**, 53.
(*17*) *Ann. Rept. W. Afr. Cacao Res. Inst.*, 1947–48, p. 33.
(*18*) Hanna, A. D., Judenko, E. and Heatherington, W., *Bull. Ent. Res.*, 1955, **46**, 669 ; 1956, **47**, 219 : *Nature, Lond.*, 1952, **169**, 334.
(*19*) Massee, A. M. and Steer, W., *J. Minist. Agric.*, 1929, **36**, 253.
(*20*) Kearns, H. G. H. (*priv. comm.*).
(*21*) DeBach, P., *J. econ. Ent.*, 1946, **39**, 695.
(*22*) Ripper, W. E., *Annu. Rev. Ent.*, 1956, **1**, 403.
(*23*) Knipling, E. F., *J. econ. Ent.*, 1955, **48**, 459.
(*24*) Bushland, R. C. and Hopkins, D. E., *J. econ. Ent.*, 1951, **44**, 725 ; 1953, **46**, 648.
(*25*) Ripper, W. E., *Nature, Lond.*, 1944, **153**, 448.
(*26*) Ullyett, G. C., *J. econ. Ent.*, 1948, **41**, 337.
(*27*) Pickett, A. D., *Canad. Ent.*, 1949, **81**, 67 ; Pickett, A. D. and Patterson, N. A., *ibid.*, 1953, **85**, 472.
(*28*) See Perkins, R. C. L. and Swezey, O. H., *Ent. Bull. Hawaii Sugar Plant. exp. Sta.* 16, 1924.
(*29*) Dodd, A. P., *Bull. ent. Res.*, 1936, **27**, 503 ; The biological campaign against prickly pear, *Commonwealth Prickly Pear Board*, 1940.
(*30*) Davies, W. M., *Ann. appl. Biol.*, 1928, **15**, 263 ; *Bull. ent. Res.*, 1928, **19**, 267.
(*31*) Filinger, G. A., *J. econ. Ent.*, 1928, **21**, 357.
(*32*) Bodenheimer, F. S., *Zbl. Bakt.*, 1923, ii, **58**, 220.
(*33*) Thorne, G., *J. agric. Res.*, 1927, **34**, 265.
(*34*) Goodey, T., *Philos. Trans.*, 1930, B, **218**, 315.

FUNGI

Fungi versus Insects. The early observations of fungi parasitic upon insects are fascinatingly described by Steinhaus (*1*) in his historical account of the development of the principles of the microbial control of insects. The infectious nature of the muscardine disease of silkworm was recognized in 1821 though it was not until 1835 that Bassi (*2*) showed that the disease is caused by the fungus now known as *Beauveria bassiana* Balsamo. Among the first to examine the possibilities of using pathogenic fungi for insect control was Metchnikoff (*3*) who isolated, from diseased wheat cockchafers (*Anisoplia austriaca* Hobst.), the green muscardine (*Metarrhizium anisopliae* (Metch.) Sorokin) which he reared on sterilized beer mash.

The importance of entomogenous fungi as factors in the natural control of insects has been the subject of controversy though it is agreed that such fungi sometimes exert a great influence. For example, Vassiliev (*4*) estimated that, in 1902, in certain localities, 60–70 per cent of the larvae of *A. austriaca* were killed by the green muscardine. Burger and Swain (*5*) reported that, during a period of extreme heat in southern California, 88 per cent of the walnut aphis, *Chromaphis juglandicola* Kalt., were killed by a fungus *Entomophthora chromaphidis*, though the aphis again developed rapidly after the hot spell.

Attempts to apply such fungi for biological control by their introduction into localities where they are absent, have rarely met with success. There are, however, a few isolated examples : for instance, Dustan (*6*) was able to show that in Europe the chief factor contributing towards the natural control of the apple sucker (*Psyllia mali* Schmidb.) was a fungus disease caused by *Entomophthora spaerosperma* Fres. The sucker had caused much damage in Canada since its first appearance in 1919 and the liberation of the fungus resulted in a notable reduction of the pest, especially in the Maritime provinces where climatic conditions were highly favourable to the development of the pathogen.

The climate of Florida is particularly favourable to the growth of entomogenous fungi and attempts have been made artificially to increase fungal disease of citrus pests by spraying spore suspensions or by placing branches bearing whitefly (*Dialaurodes* spp.) infested with the fungus *Aegerita webberi* Fav. in fungus-free trees. The factors affecting the practical use of this fungus are detailed by Fawcett (*7*). In the same locality the value of fungi (*Nectoria diploae* B. & C., *Sphaerostilbe aurantiicola* (B. & Br.) Petch) parasitic on citrus scale insects, e.g., purple scale *Lepidosaphes beckii* Newm., is demonstrated by the increase of scale following application of fungicidal and other sprays leaving granular residues (*8*).

The practical importance of fungi in biological control would seem to be dependent, to an extent far greater than found in the use of beneficial insects, on the favourableness of external factors and those affecting spread. Petch (*9*) in 1921 summarized the position thus : " If entomogenous fungi exist in a given area, practically no artificial method of increasing their efficiency is possible. If they are not present, good may result from their introduction if conditions are favourable to their growth ; but, on the other hand, their absence would appear to indicate unfavourable conditions. It would seem that a fungus makes little progress until the insects are excessively numerous . . . when, for reasons not known, an epidemic of fungus disease breaks out . . . The problem which has yet to be solved . . . is how to create an epidemic at a time when such an epidemic would not occur naturally. The evidence indicates that it is not possible to accomplish that by the mere introduction of the fungus or by spraying spores from natural or artificial cultures ". These words hold true today.

Fungi versus Fungi. An early example of the practical utilization of the antagonistic action of one fungus upon another arose from the suggestion, in 1926, by Sanford (*10*) that the control of potato scab effected by ploughing in green rye was due to the inhibitory influence of microorganisms, encouraged by the green manuring, upon the development of *Streptomyces scabies*, the pathogen responsible for potato scab. The following year, Millard and Taylor (*11*) demonstrated that the development of scab on potatoes grown in sterilized soil inoculated with *S. scabies* was reduced by the simultaneous inoculation of the soil with saphrophytic actinomycetes with a more vigorous growth habit than *S. scabies*. They concluded that this result was not due to unfavourable soil reactions (see p. 278) but to a competitive action.

Competition between fungi proper was reported in 1928 by Machacek (*12*) in his study of the storage rots of fruit and vegetable. Although he found that saphrophytic fungi may invade host tissue already attacked by a pathogen, it was only rarely that two species of pathogenic fungi were found at the same time on the same host.

The ecological analysis of the soil fungi is most complex and it is impossible here to go beyond a reference to Garrett's masterly monographs (*13*) in which he reviewed the many factors involved. Of these factors, that which seems most significant from the aspect of biological control is microbiological antagonism. This aspect was first emphasized in 1931 by Sanford and Broadfoot (*14*). They found that the pathogenicity of *Ophiobolus graminis*, responsible for a foot rot of wheat, is suppressed by filtrates of cultures of many soil microorganisms and concluded that this suppression is due to toxins present in the filtrate. The reality of such antifungal toxins was rapidly confirmed by Weindling (*15*) who studied the methods by which the soil fungus *Trichoderma lignorum* (Tode) Harz (=*Trichoderma viride* (Pers.) Fr.) suppressed the activities of several soil fungi pathogenic to citrus seedlings. He found that *T. viride* produces in culture media a " lethal principle " which, with Emerson (*16*), he isolated and named gliotoxin for he then regarded the parent fungus as a species of *Gliocladium*.

The isolation and fungistatic properties of gliotoxin are described by Brian and Hemming (*17*) and its complex pentacyclic structure was elucidated by Johnson and his colleagues (*18*). The molecule contains a disulphide bridge which is responsible not only for the high fungicidal activity of gliotoxin but also for its extreme instability except in highly acid solution. Because of this instability, its activity in soils would be limited to the immediate vicinity of the *Trichoderma* hyphae.

The study of toxin production by fungi received an enormous impetus in 1940 when Fleming's (*19*) observation, eleven years before, that a strain of *Penicillium notatum* Westling produced in culture a powerful antibacterial substance, penicillin, was put to medicinal use by Florey and his associates. The number now known of antibiotics toxic to pathogenic fungi is too large for their enumeration. In 1951, Brian (*20*) listed ninety-six distinct antibiotics produced by fungi alone and of these, over half were well characterised. Some, such as griseofulvin and cycloheximide, have been commercially produced and used as fungicides ; these will be described later for their use is similar in principle to that of other fungicidal chemicals.

The acceptance that the antibiotics produced by soil microorganisms have significance in the biological control of soil pathogens has been slow. The controversy was reviewed by Brian (*21*) in 1949 ; there were two main arguments against acceptance. The first is that toxin production has been observed only in culture media rich in nutrients and with little resemblance to the natural substrate of the toxin-producing fungus. Secondly, the known toxins are either too unstable to persist in the soil or, if stable, will be rapidly inactivated by adsorption on soil colloids. Brian pointed out that the significant concentration of the antibiotic is that in the immediate vicinity of the cell producing it ; the habit of *Trichoderma viride* which coils around the hyphae of other fungi would ensure that this local concentration will have maximum biological effect. Garrett (*22*) also considered the argument misconceived and suggested that the substrate of a moribund plant might provide adequate nutrient for antibiotic production. Nevertheless, proof of the existence of antibiotics in normal soil proved elusive, until Wright (*23*) isolated gliotoxin from an acid soil inoculated with *T. viride* and from the seed coats of peas grown in potting compost.

The factors of instability and sorption doubtless have great significance in the practical utilization of antibiotics as pesticides but need not seriously intervene in the interactions of the soil microflora. The significance of these compounds under natural conditions is revealed by the work of Rayner and Neilson-Jones (*24*) on the failure of afforestation on certain acid heath soils. The conifers failed because of the absence of the mycorrhiza normally associated with these trees, an absence due to the development of a toxic agent in the soil. Brian, Hemming and McGowan (*25*) showed that this toxicity is due, in part at least, to the production of gliotoxin and other fungistatic substances by certain species of *Penicillium* characteristic of these soils.

In certain areas of Jamaica, bananas have a natural resistance to Panama disease due to *Fusarium oxysporum cubense* whereas nearby areas lack this resistance. Meredith (*26*) examined the microflora of these soils and isolated certain actinomycetes highly antagonistic to the pathogen. From cultures of these organisms, Arnstein, Cook and Lacey (*27*) isolated musarin, a powerful antifungal compound whose chemistry does not appear to have been further investigated.

The use of fungal antagonism for biological control rests mainly in the encouragement of the growth of the antibiotic-producing microorganism.

The encouragement has, to date, been accomplished almost unintentionally by the removal of competing organisms as in the practice of partial sterilization of the soil (see p. 275). It may be achieved also by a discriminate use of selective fungicides, an example being the use of fumigants for the control of the honey fungus *Armillaria mellea* (Vahl) Quel. This fungus, after becoming established on the roots of its host, develops a thick-walled pseudosclerotial layer resistant to the attack of *Trichoderma* in the presence of other soil-inhabiting organisms. Bliss (*28*) showed that the disturbance of the microbial soil population by treatment with, for example, carbon disulphide, permits a rapid development of *T. viride* which is then able to invade the pseudosclerotia. He concluded that the control of *Armillaria* in citrus soils by fumigation is due primarily not to the fumigant itself but to the antibiotic action of *Trichoderma* made possible by the effects of the fumigant on the microbial population of the soil. A second example is due to Richardson (*29*) who found that, after treatment of soil with the fungicide thiram, protection of pea seedlings from damping-off due to *Pythium ultimum* Trow was obtained long after the content of thiram had fallen below levels toxic to the pathogen. The thiram-treated soil was, however, rich in species of *Trichoderma* and *Penicillium* resistant to the fungicide and the knowledge that these fungi can produce antibiotics toxic to *P. ultimum* adds weight to the suggestion that the persistence of the fungicidal effects of thiram treatment is due to antibiotics produced by the thiram-resistant fungi.

A remarkable example of parasitism occurs among certain hyphomycetous fungi which capture and kill nematode larvae, a process first described by Zopf (*30*). Linford and Yap (*31*) from studies on the activity of such fungi in sterilized soil infested with the root knot nematode *Heterodera marioni* (Cornu) Goodey, concluded that nematode injury was reduced by the fungus *Dactylella ellipsospora* Grove. Duddington (*32*) has catalogued the nematode-destroying fungi of agricultural soils.

REFERENCES

(*1*) Steinhaus, E. A., *Hilgardia*, 1956, **26**, 107.
(*2*) Bassi, A., *Del mal del segno, calcinaccio o moscardino.* Orcesi, Lodi. 1835.
(*3*) Metchnikoff, E., *Diseases of the larva of the grain weevil*, Odessa, 1879.
(*4*) Vassiliev, I. V., abstr. in *Rev. appl. Ent.*, 1914, A, **2**, 260.
(*5*) Burger, O. F. and Swain, A. F., *J. econ. Ent.*, 1918, **11**, 278.
(*6*) Dustan, A. G., *J. econ. Ent.*, 1927, **20**, 68.
(*7*) Fawcett, H. S., *Bot. Rev.*, 1944, **10**, 327.
(*8*) Holloway, J. K. and Young, T. R., *J. econ. Ent.*, 1943, **36**, 453.
(*9*) Petch, T., *Trans. Brit. mycol. Soc.*, 1921, **7**, 18.
(*10*) Sanford, G. B., *Phytopathology*, 1926, **16**, 525.
(*11*) Millard, W. A. and Taylor, C. B., *Ann. appl. Biol.*, 1927, **14**, 202.
(*12*) Machacek, J. E., *Macdonald College* (*McGill Univ.*) *Tech. Bull.* 7, 1928.
(*13*) Garrett, S. D., *Root Disease Fungi*, Waltham, Mass., 1944; *Biology of Root-infecting Fungi*, Cambridge Univ. Press, 1956.
(*14*) Sanford, G. B. and Broadfoot, W. C., *Sci. Agric.*, 1931, **11**, 512.
(*15*) Weindling, R., *Phytopathology*, 1932, **22**, 837; 1934, **24**, 1153; 1937 **27**. 1175.
(*16*) Weindling, R. and Emerson, O. H., *Phytopathology*, 1936, **26**, 1068.
(*17*) Brian, P. W. and Hemming, H. G., *Ann. appl. Biol.*, 1945, **32**, 214.

(18) See Johnson, J. R. and Buchanan, J. B., *J. Am. Chem. Soc.*, 1953, **75**, 2103.
(19) Fleming, A., *Brit. J. exp. Path.*, 1929, **10**, 226.
(20) Brian, P. W., *Bot. Rev.*, 1951, **17**, 357.
(21) Brian, P. W., *Symp. Soc. exp. Biol.*, 1949, **3**, 357.
(22) Garrett, S. D., *Biology of root-infecting Fungi*, Cambridge Univ. Press, 1956, p. 18.
(23) Wright, J. M., *Nature, Lond.*, 1956, **177**, 896.
(24) Rayner, M. C. and Neilson-Jones, W., *Problems in Tree Nutrition*, Faber, London, 1944.
(25) Brian, P. W., Hemming, H. G. and McGowan, J. C., *Nature, Lond.*, 1945, **155**, 637.
(26) Meredith, C. H., *Phytopathology*, 1944, **34**, 426.
(27) Arnstein, H. R. V., Cook, A. H. and Lacey, M. S., *J. gen. Microbiol.*, 1948, **2**, 111.
(28) Bliss, D. E., *Phytopathology*, 1951, **41**, 665.
(29) Richardson, L. T., *Canad. J. Bot.*, 1954, **32**, 335.
(30) Zopf, W. F., *Nova Acta Leop. Carol.*, 1888, **52**, 314.
(31) Linford, M. B. and Yap, F., *Phytopathology*, 1939, **29**, 596.
(32) Duddington, C. L., *Nature, Lond.*, 1954, **173**, 500.

BACTERIA

Bacteria versus Insects. The existence of bacterial diseases of insects appears to have been first noted by Metchnikoff (1878), who in addition to finding pathogenic fungi (see p. 52), observed that diseased larvae of the beetle *Anisoplia austriaca* were infected with *Bacillus salutaris* Metchnikoff.

An early attempt to use such bacteria for pest control was by d'Hérelle (*1*) who, in 1911, isolated from diseased locusts (*Schistocerca pallens* Thumb.) bacteria which he named *Coccobacillus acridiorum.* He employed cultures of this organism with apparent success in Argentina and Tunisia but other workers listed by Steinhaus (*2*) have failed to confirm his results.

The value of *Bacillus thuringiensis* Berliner for the biological control of the alfalfa caterpillar *Colias philodice eurytheme* Biosduval was tested by Steinhaus (*3*) in California. This spore-forming bacterium was first isolated, in 1911, by Berliner from diseased larva of the Mediterranean flour moth *Ephestia kuehniella* Zeller. Steinhaus anticipated that a bacterium which formed endospores would be a better biological control agent than a non-spore former, that a speed of control faster than with a virus would be obtained and that because the virus-infected caterpillar remains on the plant, the alfalfa would be sweeter for the caterpillars killed by the bacterium drop to the ground. In his experiments the caterpillar population of the treated plots fell below economic level whereas alfalfa destruction persisted in the untreated plots. Spore-forming bacteria pathogenic to insects were so successful as insecticides in the trials reported by Métalnikov and Métalnikov (*4*) that a proprietary preparation consisting of 10 per cent by weight of such bacteria on bentonite was introduced in France under the name " Sporeine ". Jacobs (*5*) reported that treatment of flour with this product prevented the development of infestation by *E. kuehniella.*

In 1953, Hannay (*6*) observed that in *B. thuringiensis* the spore is invariably accompanied by a parasporal body, a highly refractile and well-formed diamond-shaped crystal. With Fitz-James (*7*) he isolated the crystal material which proved to be wholly protein dispersible in alkali. Following up Hannay's suggestion that crystal formation is in some way connected with the pathogenicity of the bacillus, Angus (*8*) showed that the crystal produced from *B. sotto* Ishiwata, a pathogen of silkworms, is toxic to silkworms producing septicaemia when injected into the larvae. Both *B. sotto* and *B. thuringiensis* are among variants of *B. cereus* Frankland & Frankland of which Steinhaus and Jerrel (*9*) examined a large number. All of those variants highly pathogenic to alfalfa caterpillar and to buckeye caterpillar (*Junonia coenia* Hübner) formed crystal-like inclusions on sporulation. It seems reasonable therefore to attribute the pathogenicity of the strain to the toxic action of the crystal protein and to expect that such bacteria will prove fatal only to those insects whose larvae have guts in which a region is alkaline enough to disperse the protein. An alternative mechanism was however suggested by Heimpel (*10*) from his examination of strains of *B. cereus* toxic to larvae of the larch sawfly *Pristiphora erichsonii* Htg. and to other insects whose midgut is not highly alkaline in reaction. He found a significant correlation between pathogenicity and the ability of the bacteria to produce the enzyme lecithinase.

Bacteria versus Higher Animals. Loeffler (*11*) in 1892 employed a bacterium which he isolated from diseased field mice and named *Bacillus typhi murium* (=*Salmonella typhimurium* Loeffler) to create an artificial epizootic among field and house mice. Other pathogenic bacteria were isolated from rodents the following year by Danysz (*12*), belonging mainly to the Gaertner group of *Salmonella enteritidis*, causing an inflammation of the intestines. Bacterial cultures of these organisms often called rat " virus " preparations have been marketed for use as poison baits (*13*, *14*). Savage and Read (*15*) pointed out that rats infected with the Gaertner group may infect meat, a possible cause of meat-poisoning outbreaks. This hazard (*14*) is so great that the use of bacterial rat poisons is now prohibited in most countries.

REFERENCES

(*1*) d'Hérelle, F. H., *C.R. Acad. Sci. Paris*, 1911, **152**, 1413 ; 1912, **154**, 623 ; 1915, **161**, 503.
(*2*) Steinhaus, E. A., *Insect Microbiology*, Comstock Publishing Co. Inc., Ithaca, N.Y., 1947, p. 181.
(*3*) Steinhaus, E. A., *Hilgardia*, 1951, **20**, 359.
(*4*) Métalnikov, S. and Métalnikov, S. S., *Ann. Inst. Pasteur*, 1935, **55**, 709.
(*5*) Jacobs, S. E., *Proc. Soc. appl. Bact.*, 1950, **13**, 83.
(*6*) Hannay, C. L., *Nature, Lond.*, 1953, **172**, 1004.
(*7*) Hannay, C. L. and Fitz-James, P., *Canad. J. Microbiol.*, 1955, **1**, 694.
(*8*) Angus, T. A., *Canad. J. Microbiol.*, 1956, **2**, 122.
(*9*) Steinhaus, E. A. and Jerrel, E. A., *Hilgardia*, 1954, **23**, 1.
(*10*) Heimpel, A. M., *Canad. J. Zool.*, 1955, **33**, 99, 311.
(*11*) Loeffler, L., *Zbl. Bakt.*, 1892, **11**, 129.
(*12*) Danysz, J., *C. R. Acad. Sci. Paris*, 1893, **117**, 869.
(*13*) Schander, R. and Götze, G., *Zbl. Bakt.*, 1930, **81**, ii, 260.
(*14*) Leslie, P. H., *J. Hyg.*, 1943, **42**, 552.
(*15*) Savage, W. G. and Read, W. J., *J. Hyg.*, 1913, **13**, 343.

VIRUSES

Virus Diseases of Insects. Insect larvae, particularly of the Lepidoptera, are subject to infectious disease characterized by the formation, in infected caterpillars, of polyhedral-shaped highly refractile granules. These polyhedra were described in 1898 in silkworms suffering from "jaundice" by Bolle (*1*) and the many early views of their nature are summarized by Bergold (*2*) who, in 1947 (*3*) showed that they consist largely of non-infective protein which may be dispersed by dilute alkali to reveal thin membranes containing rods which are the infectious virus particles. From his electron-micrographs he deduced (*4*) that the polyhedral viruses of *Bombyx mori* L., *Lymantria dispar* and *L. monacha* have life cycles and methods of multiplication.

A second type of polyhedrosis was observed, in 1950, by Smith and Wyckoff (*5*) in larvae of the moths *Arctia villaca* and *A. caja* L. in which the polyhedra do not dissolve completely in alkali but leave a shell-like residue full of spherical holes, the viruses being present apparently as small spheres and not as rods. The same observers (*6*) also reported that polyhedra obtained from the clothes moth *Tineola bisselliella* Hum. yielded no particles the size of virus.

Attempts to use the "Wipfelkranheit" of the nun moth (*L. monacha*) in Germany and the wilt disease of gypsy moth (*L. dispar*) in the United States have met with little success, though at times these diseases exert a tremendous toll. Apparently the disease is already present, and if conditions are right, will produce a natural epizootic whether or not the virus is purposely introduced. In Southern Ontario (*7*), however, the introduced European pine sawfly (*Neodiprion sertifer* Geoffr.) has been kept in check for three years on Scots pine plantations sprayed in 1951 with suspensions containing a virus obtained from Sweden. Success also followed the earlier introduction into Eastern Canada, in 1940, of a virus fatal to the introduced European spruce sawfly (*Diprion hercyniae* Htg.). This virus has frequently been recovered since that date from diseased sawfly and was associated with a severe epizootic in New Brunswick in 1945.

The epizootiology of insect virus diseases is complicated by an unusual and, as yet, unproved factor which indicates that the normal processes of infection may not be involved. Smith and Wyckoff (*6*) cited several instances of the sudden appearance of virus infection in stocks of caterpillars raised with every care through many generations. They suggested that virus in a latent condition had passed from generation to generation in such stocks. Bergold (*2*) also accepted the possibility of latency in insect viruses and suggested that it explains the observations of Yamafugi and his colleagues (*8*) who, in 1944, found that silkworms fed with hydroxylamine developed a typical polyhedral virus disease. This observation has since been extended to other chemicals including inorganic nitrites. If the latent virus is rendered pathogenic when the insect is exposed to unfavourable conditions, the problem of producing an epizootic is not solved merely by disseminating the virus.

Virus Disease of Higher Animals. The myxoma virus is fatal to the European rabbit *Oryctolagus cuniculus* causing a myxomatosis which, in 1898, wiped out stocks of laboratory rabbits in Uruguay. Aragão (*9*) suggested its use for rabbit control in Australia but the experiments met with little success until in 1950 the virus was liberated in several localities in the Murray Valley. The widespread epizootic which rapidly developed was described by Ratcliffe and his colleagues (*10*) who considered the phenomenal spread of the disease to the conditions which favoured the movement of the insects, particularly mosquitoes, by which the disease is transmitted.

The European history of the disease runs on similar lines. Earlier trials of the virus for rabbit control failed, as on the isolated island of Skokholm(*11*). But after a French doctor had released, in 1952, a few rabbits inoculated with the myxoma virus within the walls of his estate, the disease had by the end of 1953 spread throughout Europe west of the Rhine (*12*). Indeed, so devastating was the epizootic that measures for the protection of tame rabbits, such as the screening of hutches against insects and the use of insecticides, became necessary. One suggested method was the vaccination of rabbits by inoculation with the fibroma virus isolated from the American cottontail rabbit in 1932. Unlike myxoma, fibroma induces a benign disease in both *O. cuniculus* and in the American rabbit *Sylvilagus brasiliensis*, conferring protection against myxoma for six months or so. This phenomenon of the use of a mild strain to combat a more virulent strain of virus was described on page 21.

The success of myxomatosis for the biological control of rabbit is now threatened by the emergence of rabbits resistant to the disease. This development may be the result both of the selection of resistant strains or the acquisition of acquired immunity for in *S. brasiliensis*, which normally is immune, it has been possible by intradermal inoculation to produce a mild form of myxomatosis.

REFERENCES

(*1*) Bolle, J., In *Der Seidenbau in Japan*, Hartelbens Verlag, Budapest, 1898.

(*2*) Bergold, G. H., In *Advances in Virus Research*, Academic Press Inc., New York, 1953, p. 91.

(*3*) Bergold, G. H., *Z. Naturforsch*, 1947, **2b**, 122 ; **3b**, 35.

(*4*) Bergold, G. H., *Canad. J. Res.*, 1950, **E28**, 5.

(*5*) Smith, K. M. and Wyckoff, R. W. G., *Nature, Lond.*, 1950, **166**, 861.

(*6*) Smith, K. M. and Wyckoff, R. W. G., *Research*, 1951, **4**, 148.

(*7*) Bird, F. T., *Canad. Ent.*, 1955, **87**, 124.

(*8*) Yamafugi, K. and Yoshihara, I., *Enzymologia*, 1951, **15**, 10.

(*9*) Aragão, H. B., *Men. Inst. Osw. Cruz.*, 1927, **20**, 237.

(*10*) Ratcliffe, F. N., Myers, K., Fennessy, B. V. and Calaby, J. H., *Nature, Lond.*, 1952, **170**, 7.

(*11*) Lockley, R. M., *Nature, Lond.*, 1940, **145**, 767.

(*12*) Thompson, H. V., *Agriculture, Lond.*, 1954, **60**, 503.

CHAPTER V

FUNGICIDES AND INSECTICIDES

The idea of applying a poisonous substance to the leaf must have occurred to the first observers of foliage damage caused by leaf-eating insects. Probably, too, in some cases where the fungus is visible on the plant surface, it was found that it could be killed by the application of certain materials. The early history of insecticides and fungicides is therefore lost. Many materials must have been tested in an empirical manner, a method by which were slowly sifted out substances which found regular use as insecticides and fungicides.

Even if the insect were not killed by the treatment, various materials, such as decoctions of bitter or strong-smelling herbs, would be applied to the plant in the hope that the insect pest would be deterred or driven away by the objectionable taste or offensive odour. The realization that by the application of certain substances it is possible to protect the plant from fungus attack appears to date, however, from the accidental discovery by Millardet, in 1882, of the action of a lime-copper sulphate mixture (see p. 115). Since that time the list of protective fungicides has extended to many other substances effective when applied to the plant surface be it foliage or seed. With the early protective fungicides, their effect is localized and they served to protect only those parts of the plant in the vicinity of the fungicide; plant growth developing after application was not protected. In recent years however certain compounds have been discovered which are absorbed and translocated by the plant which thereby becomes systemically toxic to fungus or insect. In considering the principles underlying the use of chemicals for such protection, a differentiation of systemic and non-systemic chemicals is necessary.

According to their mode of action, the non-systemic fungicides may be classified into two groups: direct and protective. The direct or eradicant fungicide kills through contact with the fungus on the plant surface as in the case of lime sulphur used for the control of powdery mildews or of formaldehyde against bunt of wheat. The protective fungicide prevents the establishment of fungal infection as in the case of Bordeaux mixture against potato blight or of the organomercury derivatives against stripe disease of barley. The distinction between direct and protective fungicides is emphasized because the principles underlying the use of the two types of fungicide differ. Firstly, it is essential that the protective fungicide be applied to the plant before infection is established, whereas the direct fungicide is applied to the infected plant. Secondly, factors relating to the retention of the fungicide upon the plant surface determine protective fungicidal efficiency, whereas direct fungicidal efficiency is more dependent upon contact between fungicide and fungus.

In the case of insecticides, also, it is convenient to differentiate between direct and protective insecticides. The direct insecticides are used mainly against suctorial insects (see p. 24), killing by contact with and, probably, absorption through the cuticle. Hence they are often called " direct contact poisons ". The distinction between contact and stomach poisons seems first to have been made by Le Conte (*1*) who referred to " the poison by contact and the poison by food ". In the group of contact insecticides are also placed those materials which kill or inhibit the hatching of insect eggs and are therefore called " ovicides ". The protective insecticides include: (1) materials (" stomach poisons ") which, being eaten with the plant tissue by mandibulate insects, presumably kill by absorption through the alimentary system; (2) materials which, when deposited on the leaf surface, poison the insect by contact and presumably absorption through the cuticle; (3) materials the presence of which may deter the insect from feeding, or may interfere with normal activities, e.g., egg-laying, or may repel the insect by chemotropic action. As the efficiency of the stomach poison, the contact insecticide, the deterrent and the repellent, is related to the retention of the insecticide on the plant surface, these materials are essentially protectant in action.

This classification is quite arbitrary and often a given chemical will appear in more than one category. Nicotine, for example, is usually classified with the direct contact poisons though it may act through its vapour, i.e., as a fumigant. DDT is a typical protective contact insecticide though it can act as a stomach poison. Benzene hexachloride can kill insects by ingestion, by contact or by a fumigant action. The classification is nevertheless useful in that it enables generalizations to be made and affords a basis for the systematic discussion of the various groups of substances which find application as fungicides and insecticides.

Proceeding with generalizations concerning fungicides and insecticides applied to the growing plant, the toxic chemical should possess certain other properties. Firstly, it must be effective at concentrations which cause no injury of economic importance to the host plant. The sensitiveness of the plant to the insecticide or fungicide will, in general, vary with its stage of development; the dormant plant will often tolerate materials which if applied to the foliage would cause defoliation. A convenient distinction thus arises between materials applied during active growth and those applied during the dormant season.

Secondly, the material, if poisonous to man or stock, must be capable of use in a manner safe to the operator and to those in subsequent contact with the crop whether in harvesting or storage, or whether on food or fodder.

Thirdly, the use of the insecticide or fungicide must not have too unfavourable an effect on the biological environment of the treated crop (see p. 49). Such ecological effects increase vastly in importance as the area over which the material is used is increased and as the diversity of crops treated is reduced. Whereas the requirement might be overlooked if the chemical is to be applied to but one of a range of market-garden crops, it assumes high importance in the use of the chemical over large expanses of monoculture.

Fourthly, the chemical should be used in a manner which gives the greatest economic protection for, in common with all pest control methods, there is a risk that inadequate use may lead to an unwanted selection of resistant strains of pest or pathogen. Among examples of the development of a resistant population are those of hydrocyanic acid (see p. 248) and DDT (see p. 207), but the danger exists wherever the property of resistance is inherited.

Finally, the toxic material must be amenable to application in an effective manner. The problem, in the case of non-systemic materials, is to distribute the toxicant over the plant surface in such a way that it shall, if direct in action, come into contact with pest or fungus and, if protective in action, leave a uniform and persistent deposit upon the plant surface. It is rarely that the active insecticide or fungicide is itself in a form suitable for direct application for a small weight of material has to be distributed over a large area. Dilution is necessary and the various methods of application may be classified on the basis of the nature of the diluent, whether liquid (Spraying), solid (Dusting), or gaseous (Fumigating).

Spraying

The earliest diluent or "carrier" was water, the spray being applied either as an aqueous solution, or, if the active ingredient is water-insoluble, as a suspension or emulsion. In the case of suspensions and emulsions it is evident that such heterogenous systems should be stable enough to ensure the application, under practical conditions, of a dilution of uniform and known concentration. This stability may be conferred by supplementary materials ; thus the sedimentation of a suspension may be delayed by the addition of protective colloids and dispersing agents ; the coalescence of the dispersed droplets of an emulsion may be retarded by the addition of emulsifiers.

The dispersion of the spray, when water is used as the carrier, is usually achieved by its passage under pressure through special nozzles. The design of nozzle and specifications for pump, spray tank and ancillary equipment have been discussed in detail by Kearns (*2*) and, less technically, by McClintock and Fisher (*3*) and by Brown (*4*). By these means one operator can distribute quantities up to 9 gal. per minute with an accurate placement of the spray over a distance of fifteen feet in winds up to Beaufort scale 3 (10 m.p.h. breeze). But the older, so-called conventional, spray methods have severe drawbacks : the weight and cost of the machinery involved, the expense of supplying and transporting water, the high labour requirements and, in some cases, difficulties of terrain or of passage through the crop. For the treatment of large areas, alternative methods of application have been sought. The liquid may, for example, be dispersed by the action of a copious amount of air directed against a jet of the liquid. This

" scent-spray " principle has been used in various ways as in the dispersal of DDT-petroleum oil solutions from aeroplane where air speed alone is sufficient to shatter the jet of solution into droplets which can then be dispersed in the slip-stream. The mechanics of the formation of droplets in air streams was described by Lane (*5*) who found that the square of the difference of velocity between air and liquid necessary for droplet formation varied as the surface tension of the liquid divided by the diameter of the liquid jet. The down-draught of the rotor makes the helicopter a useful vehicle for such aerial spraying (*6*). For land use the required stream of air can be produced by fan, by propellor, or by turbine (axial flow) fan. These processes in which the energy required for dispersion is applied through the air stream and not, as in conventional spraying, through the liquid, are sometimes called by the misleading term " atomization ", for the droplets produced are much smaller than those from the older spray machines. For this reason smaller volumes of spray per acre are needed and the use of organic solvents such as kerosine or fuel oil in place of water becomes economical. If water is used as carrier the concentration of the active component can be increased, giving rise to the expression " concentrate " spraying. It has become usual to refer to the concentration used as ×5, ×8, etc., the factor being that used to convert the concentration normally used in the older spray machine to that used in the air-blast machine.

Dispersion may also be obtained by the use of heat. This principle was employed for the production of " smoke " screens from petroleum oils for military purposes ; the machines, e.g., the Hochberg-La Mer and the Todd generators, have been adapted for the production of insecticidal " fogs " (*7*). Similarly, the pyrotechnic methods for generating coloured smoke signals have been used for making insecticidal " smokes ". Unless used in conjunction with artificially-produced air streams, smokes and fogs find use mainly for the treatment of inaccessible areas such as forest canopy or of closed spaces such as the glasshouse.

Dusting

The use of a dust, in which the active component is diluted with a suitable finely-divided " carrier ", has potential advantages over spraying. The factor of water supply vanishes ; the dust may be purchased ready for use and is more easy to handle than the spray concentrate ; the dusting appliances, consisting essentially of fan or turbine blower, are lighter and more easily moved in difficult terrain than sprayers ; finally, dusting is less costly in time and labour than spraying. But in practice several less obvious difficulties emerge : the dust must flow freely ; it must not " cake " (presumably through the absorption of atmospheric moisture) ; it must not " ball " in the hopper or collect in the air-ducts of the duster (presumably through static electrification) ; it must be so mixed that the dust retains a uniform content of active component throughout its range of particle sizes and its distribution. Little is known either of the various physico-chemical factors which determine whether a particular dust will pass readily through the machine to give a good dispersion, or of the properties required to render dusting an effective means of applying insecticides and fungicides. Andreasen and his colleagues (*8*) evaluated the " dusting tendency " of various powders and the figures range from 100 for lycopodium to zero for powdered quartz. They found that contrary to expectation the figure for any one substance is increased if the extremely fine particles are removed. The effect of nature of surface on dust deposition was examined by Brittain and Carleton (*9*) who found more deposited on the lower than upper leaf surface presumably because of the more prominent veins of the lower surface. Greater deposition was obtained on leaves held parallel, rather than perpendicular, to the air-stream carrying the dust.

The general experience is that dusting is practicable only in the calmest weather and that a better protective action is obtained if the dust is applied when the plant is wet with dew or rain. Stanley (*10*) pointed out that quite large amounts of water are then involved ; cabbage plants will hold in their leaves 325 gal. (U.S.) dew per acre, and bean plants 1,591 gal. per acre. In so-called " liquidusters " a mixture of liquid droplets with the dust is applied. Potts (*11*) found the mere mixing of dust and oil droplets ineffective but by atomising the liquid in the dust-laden air stream, oil-coated particles were obtained which adhered to foliage better than the uncoated dust.

Fumigation

Haber (*12*) showed that the toxicity of a gas is proportional to its concentration (c) and to the time of exposure (t) of the organism to that concentration : $ct = \mathrm{K}$. The constant " K " is thus a measure of activity in the vapour phase ; the various estimates assembled by Peters (*13*) show that, at ordinary temperatures, K is rarely below 5 mg./l./hr. for effective insecticidal action. Waters (*14*) deduced that, for effective fumigant action in the open, K should be about 1 mg./l./hr. Hence unless volatile compounds of exceptional insecticidal potency are found, fumigation can only be successful in closed spaces or with special precautions to lengthen the time of exposure " *t* ". One device for producing the latter is the drag sheet by which the fumigant is confined below a light impervious sheet dragged at a rate dependent on its length behind the vaporizing appliance (*15*). This method is well-suited for the treatment of ground crops such as strawberries or peas, crops which present special difficulties in spraying because of their dense and low-growing habit. Another way of lengthening the time of exposure is by the use of fumigants of boiling point above ordinary temperatures. A deposit of such a fumigant on the plant surface may then also act directly as a contact or protective toxicant. The trick of painting sulphur on the hot water pipes of the glasshouse is an old example of this method ; modern versions include the thermostatically-controlled electric heater for the volatilization of the toxicant (*16*), the pyrotechnic smoke generator and the " Aerosol " method.

In the smoke generator, low combustion temperatures are required so that the destruction of the toxicant during volatilization is held as low as possible. Bateman and Heath (*17*) used sucrose and potassium chlorate to obtain a self-sustained reaction at a temperature of about 350° C ; Taylor

(*18*), in B.P. 592,788, used ammonium nitrate and potassium chromate which, it is claimed, give a self-sustained reaction around 330° C. For use without a suitably-designed container, Heath (*19*) recommended the addition of thiourea to the sugar-chlorate mixture giving a powder which will burn without flame in an open tin. A later recipe due to Johnstone and Marke (*20*) employed potassium peroxydisulphate ($K_2S_2O_8$), ammonium dichromate and guanidine nitrate as pyrotechnic components yielding non-phytotoxic combustion products.

In the Aerosol process the toxicant is dissolved in a liquid, gaseous at ordinary temperatures but liquifiable under pressure (*21*). On pressure release the solution is discharged through a nozzle, the solvent evaporates and the toxicant is dispersed in an extremely finely-divided state. Dichlorodifluoromethane (" Freon "), liquid at 90 lb. per sq. inch at ordinary temperature, and methyl chloride (80 lb. per sq. inch) are suitable solvents. Read (*22*) has used acetone (b.p. 56.5° C), though in this case a source of pressure other than the vapour pressure of the liquid is necessary to obtain dispersion. In the small " Sparklet " bulb, used by troops for the disinfestation of tents, compressed carbon dioxide expelled the acetone-DDT-pyrethrum solution from the bulb. The method then becomes one of atomization.

From this review of the methods of application of non-systemic toxicants, it is apparent that, for use in the open, two broad principles may be employed: firstly, conventional spraying in which relatively large droplets are discharged at high velocity from the nozzle; secondly, atomization and dusting in which finer droplets or solid particles are carried in an air stream. For plant protective purposes, the objective is a controlled and uniform application of a known concentration of the toxicant over an area limited to the acreage of crop plant treated; controlled because the treatment of a neighbouring crop is not always wanted; uniform because with most insecticides and fungicides the margin of safety between the concentration giving effective control and that liable to cause damage to the treated plant is small. These objects are attained in conventional spraying for, by applying dilutions in water, the amount of spray retained on the plant surface is automatically controlled by the interfacial tensions; the excess of spray runs off. Control over distribution is obtained by designing the nozzles so as to reduce to the lowest the spray emitted as mist and by spraying under conditions when spray drift is negligible. An important reason for not spraying in breezes stronger than 10 m.p.h. is that spray drift may reach previously sprayed and dried surfaces on which it will increase deposit beyond the normal run-off amount, increasing the risk of spray damage.

Droplet size in spraying is accordingly rather high, in the range of 0.1–2.0 mm. (100–2000 microns) diameter droplets. In atomization droplets of this size would quickly fall from the air stream and for this reason a particle size range below about 90 microns, dependent on the density of the carrier, is required for the effective distribution of dusts except to low-growing crops. In " fogs " a droplet diameter of 0.5–1 micron is the best for screening purposes but the appliances when adapted for the production of insecticidal fogs, give a range of 10–50 microns.

The use, in the open, of particles of such low dimensions creates difficulties. The rate of fall of water droplets in air indicates that a droplet diameter greater than 100 microns is necessary for an effective control over distribution; smaller droplets would be carried away by even a 5 m.p.h. wind. Moreover, unless a carrier of low volatility is used, evaporation will still further reduce droplet size. Waters (*14*) estimated that, given free evaporation, a water droplet initially of 100 microns diameter will have completely evaporated after falling eight feet through air of 75 per cent R.H. and at 90° F. Yet a further complication is that small droplets or particles are diverted from solid objects in their path and follow the lines of air flow. Potts (*11*) observed that deposition was low if the particles are of diameter smaller than 2 microns. Waters (*14*) calculated that only 2 per cent of particles of 8 micron diameter would be retained on wires of 1000 microns diameter. Townsend (*23*) similarly calculated that for mosquitoes in flight in a cloud of fine droplets, the impaction factor increases rapidly with droplet size, from 0.2 at 20 microns to 0.8 at 60 microns drop diameter. The non-impaction of the finer particles, evaporation and their low rate of fall result in uncontrollable drift which would severely prejudice the use of atomization, of smokes and of fogs for the protection of small areas of crop plants, unless some means be found of controlling distribution and of avoiding over-application to vegetation in close proximity to the machine.

Control over drift has been sought in the electrification of the particles. It has long been known that electrostatic charges are developed both during spraying and dusting. Moore (*24*) pointed out that the dust particle bearing a charge, either positive or negative, will induce an opposite charge on the leaf which, being an earthed conductor, leads to a mutual attraction of the leaf and particle. The sign and charge of the particles is dependent on several factors. Wilson and his co-workers (*25*) found that particle size is such a factor; particles of 40 micron diameter generally had high charges, particles of 2 micron diameter produced low charges. Atmospheric humidity also has a profound effect and though little change in voltage was found at 40 per cent R.H. or less, no charge was detectable when the relative humidity was increased to 50 per cent or over. Wilson and Jackson (*26*) concluded that the sign of the charge was controlled by the chemical nature of the particle; negative charges were acquired by materials exposing acidic groups on fracture, e.g., the SiO_2 tetrahedral lattice structure of quartz, talc, pyrophyllite; positive charges were acquired by alkaline materials or those silicates which exposed cations or hydroxyl groups on fracture. The nature of the tube through which the dust was discharged was found, in these experiments, to have little effect but MacLeod and Smith (*27*) found that its length and nature influenced the amount of dust retained on charged plates. Miller and his colleagues (*28*) showed the dust coating acquired by the inside of the tube to be of influence, especially when a second component is added to the dust. Practical experience has shown that the performance of a dust is often improved by the addition of a small proportion of another material. Blodgett (*29*), for example, found that the objectionable " balling " of ground sulphur is avoided if a little hydrated lime is mixed with it. The reasons for this effect are unknown ; MacLeod and Smith (*27*) were unable to predict the deposit which would result from a mixture of the two powders from a study of each of the components. For this reason and because of the absence of electrification at high humidities, an artificial charging of the particles may be required. Yadoff (*30*) claimed that, by passing the dust-laden air through tubes at ultrasonic speeds, high frictional charges are developed whereas Hampe (*31*) relied, for the same purpose, on electrostatic generators of the type of the Wimshurst machine.

Pending development of satisfactory methods of drift control, attention need be given only to spraying and dusting as practical methods of application of agricultural pest control materials. The effectiveness of field performance of the spray or dust will be dependent on certain physicochemical properties which, though already mentioned, require further examination. The importance of completeness of contact, conveniently described as " coverage" which has been stressed in the case of direct insecticides and fungicides, reappears in the protective materials, for the uneven deposit will be a less efficient protection than that uniformly distributed. It may be advantageous to improve coverage by the addition to the spray of supplementary components usually termed " spreaders ".

Field performance is also dependent on the actual amount of toxic material which is retained on the sprayed or dusted surface. This amount, the spray residue, is determined by that initially retained and by the ability of the spray residue to withstand agencies, e.g., subsequent growth of the substratum, rain, wind, etc., tending to its removal, a quality which it is convenient to call " tenacity ". It is again possible to add supplementary components, usually termed " stickers " to improve the retention of sprays and dusts. Finally, in the case of insecticides and fungicides not directly toxic themselves but from which, by various agencies, an active toxic product is formed after application, field performance will be governed by the inherent toxicity of the active derivative and the readiness with which lethal concentrations are liberated, i.e., the " availability ".

Summarizing, the main requirements of a satisfactory spray or dust are : (1) high field performance : determined by (a) the inherent toxicity, (b) availability of the active constituent, (c) coverage, (d) initial retention of spray or dust, and (e) the tenacity of the residue ; (2) low phytotoxicity ; (3) stability of the concentrate in storage ; and (4) a stability after dilution to spray strength, sufficient to ensure the application of a uniform known concentration of active constituents.

Items 1(a), 1(b) and 2 are characteristic of the particular insecticide or fungicide employed and must be discussed specifically. Similarly, reference to supplementary spray materials used as " activators " to improve availability or as " correctives " to reduce phytotoxic properties, must be reserved for particular treatment. But in the case of spreaders added to improve coverage, of stickers added to increase retention, of emulsifiers and protective colloids added to enhance stability, there are general points for discussion. It is therefore convenient to consider, somewhat out of order, the nature and action of these last four groups of supplementary agents.

These requirements are shared by the pesticide capable of systemic activity : the necessity for a uniform deposition over the plant surface is, however, reduced for the plant itself now co-operates in achieving distribution. The efficiency of translocation may be dependent on the crop plant and its stage of growth ; in the seedling, movement to the growing point will protect new growth ; in the mature plant, movement to the storage organs may permit protection of growth after dormancy.

REFERENCES

(*1*) Le Conte, J. L., *Proc. Amer. Ass. Adv. Sci.*, 1875, **2**, 202.

(*2*) Kearns, H. G. H., *Rep. agric. hort. Res. Sta. Bristol*, 1945, p. 110.

(*3*) McClintock, J. A. and Fisher, W. B., *Spray Chemicals and application equipment*, Hort. Press, LaGrange, Indiana, 1945.

(*4*) Brown, A. W. A., *Insect control by chemicals*, Chapman & Hall Ltd., London, 1951.

(*5*) Lane, W. R., *Industr. engng Chem.*, 1951, **43**, 1312.

(*6*) Ripper, W. E. and Tudor, P., *Bull. ent. Res.*, 1948, **39**, 1.

(*7*) Glasgow, R. D. and Collins, D. L., *J. econ. Ent.*, 1946, **39**, 227.

(*8*) Andreasen, A. H. M., Hofman-Bang, N. and Rasmussen, N. H., *Kolloidzschr.*, 1939, **86**, 70.

(*9*) Brittain, R. W. and Carleton, W. M., *Agric. Engng St. Joseph, Mich.*, 1957, **38** (1), 22.

(*10*) Stanley, W. W., *J. econ. Ent.*, 1948, **41**, 336.

(*11*) Potts, S. F., *J. econ. Ent.*, 1946, **39**, 716.

(*12*) Haber, F., *Fünf Vortrage aus den Jahren, 1920–23*, Berlin, J. Springer, 1924.

(*13*) Peters, G., *Chemie und Toxikologie der Schädlingsbekämpfung*, Stuttgart, F. Euhe, 1936.

(*14*) Waters, W. A., *Proc. XI Int. Congr. Chem.*, 1947, **3**, 437.

(*15*) Smith, R. H., *J. econ. Ent.*, 1938, **31**, 60.

(*16*) Stammers, F. M. G. and Whitfield, F. G. S., *Bull. ent. Res.*, 1947, **38**, 30 ; Fox Wilson, G., *J. R. hort. Soc.*, 1949, **74**, 444.

(*17*) Bateman, E. W. and Heath, G. D., *J. Soc. chem. Ind., Lond.*, 1947, **66**, 325.

(*18*) Taylor, J., *Research, Lond.*, 1949, **2**, 98 ; Marke, D. J. B. and Lilly, C. H., *J. Sci. Fd. Agric.*, 1951, **2**, 56.

(*19*) Heath, G. D., *J. Soc. chem. Ind. Lond.*, 1949, **68**, 40.

(*20*) Johnstone, C. and Marke, D. J. B., U.S.P. 2,695,258 (1954).

(*21*) Goodhue, L. D., *Industr. engng Chem.*, 1942, **34**, 1456.

(*22*) Read, W. H., *Rep. exp. Res. Sta. Cheshunt*, 1946, p. 59.

(*23*) Townsend, A. A., Appendix II to Kennedy, J. S., Ainsworth, M. and Toms, B. A., *Anti-Locust Bull.* 2, Anti-Locust Research Centre, London, 1948.

(*24*) Moore, W., *J. econ. Ent.*, 1925, **18**, 282.

(*25*) Wilson, H. F., Janes, R. J. and Campau, E. J., *J. econ. Ent.*, 1944, **37**, 651.

(*26*) Wilson, H. F. and Jackson, M. L., *Agric. Chemic.*, 1946, **1**, 32.

(*27*) MacLeod, G. F. and Smith, L. M., *J. agric. Res.*, 1943, **66**, 87.

(*28*) Miller, J. G., Heinemann, H. and McCarter, W. S. W., *Science*, 1948, **107**, 144.

(*29*) Blodgett, F. M., *Bull. Cornell agric. Exp. Sta.*, 328, 1913.

(*30*) Yadoff, O., *C. R. Acad. Sci., Paris*, 1946, **222**, 544 ; 1947, **224**, 1001.

(*31*) Hampe, P., *Rev. Vitic.*, Paris, 1947, **93**, 259 ; Ballu, T., *Rev. hort. Paris*, 1947, **30**, 333.

SPREADERS

The term " spreader " has been used to embrace all types of auxiliary spray materials (*1*) but it is advantageous to restrict the term to materials which directly facilitate contact between spray and sprayed surface. It is a common experience that water falling as rain or dew on a cabbage leaf collects to large droplets which run off leaving the leaf surface dry. In other words, no stable liquid/solid interface is formed when water is applied to cabbage foliage. The function of a spreader is to promote the formation of the liquid/solid interface by reducing the energy associated with its formation.

The Mode of Action of Spreaders

It is convenient to differentiate, in the present state of knowledge concerning the mode of action of spreaders, between three properties upon which, in a gas/liquid/solid system, the extent of liquid/solid interface is dependent. The first may be termed the wetting properties of the liquid for the solid, which determine the extent of the liquid/solid interface when excess of liquid is drained from the solid surface. Thus the droplet of dew runs off leaving the cabbage leaf dry because the leaf is not wetted by water. The second method of formation of the liquid/solid interface is by the extension of the liquid by capillary forces over the solid surface, a process which is best described as spreading. It is advantageous to retain the term "spreading" used by physicists for this phenomenon although, in spray practice, the term has the wider definition stated above. Thirdly, it is possible for the spray to possess the property of penetrating into the porous solid, a property of importance, for example, in the displacement of the air from the densely-packed conidiophores of a powdery mildew by a direct fungicide or, in the penetration by a contact insecticide, into the closely massed insects of a colony of woolly aphis.

The Comparison of Spreader Efficiencies

Upon the basis of these three properties it is possible to consider methods of evaluating the efficiency of various spreaders.

(a) **Wetting Properties.** The criterion of wetting being the formation of a stable non-retreating liquid-solid interface when excess of the liquid is drained from the solid, it is evident that in the ideal case the liquid will remain as a persistent film over the solid. If wetting is not perfect, the liquid will retreat to form a droplet showing a definite angle between the surfaces of liquid and solid at the air/liquid/solid circle of contact. This angle is called the receding contact angle, receding because the liquid is receding or retreating from the solid surface. Perfect wetting is attained when the receding contact angle is zero and this angle affords a measure of wetting properties.

The commonly used test for dipping a waxed card into the spray and observing, on withdrawal, that the liquid film does not retreat into droplets is, in fact, the observation that the receding contact angle of the spray on the waxed surface is zero. As such, the test is a simple method of assessing wetting properties for the waxed surface.

(b) **Spreading Properties.** If spreading properties are defined as the ability of a liquid to spread over a solid, they should be assessed by the comparison of the areas of spread of droplets of definite volume placed upon the test solid so that the area covered is determined only by capillary activity. Area of spread as a criterion of spreading properties has been examined by Woodman (*2*), by O'Kane, Westgate, Glover and Lowry (*3*) and by Evans and Martin (*4*).

The surface tension of the liquid at the air/liquid interface γ_1, and the interfacial tension at the liquid/solid interface $\gamma_{1,2}$ will oppose, whereas the solid/air tension γ_2 will favour, the extension of the droplet over the solid. If the latter force is greater than the sum of the other two, i.e., if

$$\gamma_2 > \gamma_1 + \gamma_{1,2} \qquad \text{.........(1)}$$

the droplet will spread. It is evident that the smaller the surface tension γ_1 the greater the likelihood of spread. Surface tension was used by Woodman (*1, 2*) for the evaluation of spreaders, but, as spreading properties involve the three tensions, the method has met with little success. No methods being known for the determination of the solid and interfacial tensions γ_2 and $\gamma_{1,2}$. Cooper and Nuttall (*5*) employed a thick oil in place of solid which enabled them to evaluate the " spreading coefficient " :

$$\text{S.C.} = \gamma_2 - (\gamma_1 + \gamma_{1,2}) \qquad \text{........(2)}$$

a procedure extensively used by Cupples (*6*).

It has been shown that, in a drop of liquid resting on a plane surface, the angle within the liquid between the liquid and solid surfaces at the air/liquid/solid circle of contact is a constant independent of the volume of the droplet. This angle, designated the advancing contact angle θ_a, for the liquid, in contradistinction to the case of wetting,* is advancing over the solid, is related to the three interfacial tensions involved by the following equation :

$$\gamma_2 = \gamma_1 \cos \theta_a + \gamma_{1,2} \qquad \text{........(3)}$$

As the contact angle can be determined experimentally (*7*) it has been used as an index of spreading properties (*3, 8, 9, 10*). If, in this method, leaf surfaces are to be used instead of artificial surfaces, the difficulties to be overcome include the uneven nature of the leaf surface and the diurnal variations in contact angle which Fogg (*11*) found were caused by changes in the degree of corrugation of the leaf surface produced by changes in the water content of the tissues.

Combining equations (2) and (3), an alternative expression of the " spreading coefficient " is obtained :

$$\text{S.C.} = \gamma_1 (\cos \theta_a - 1) \qquad \text{........(4)}$$

This equation, with the substitution of the mean of the cosines of the advancing and receding contact angles for the cosine of the advancing angle, was used by Wilcoxon and Hartzell (*12*) as a measure of spreading properties.

---

*The reasons for the observed difference between advancing and receding contact angles are still obscure and it is probable that the values obtained for each angle are dependent to some degree on the experimental conditions. Yarnold and Mason (*13*), for example, demonstrated that factors such as relative velocity of solid and liquid in advancing or receding, and the time they have been in contact affect the angle observed. Ray and Bartell (*14*), finding that both angles for water on a film of paraffin volatilized on a smooth glass surface are the same, 112°, suggested the magnitude of the difference is related to the roughness of the surface, a conclusion reached by theoretical argument by Good (*15*). Evans and Martin (*4*) associated the extreme difference in advancing and receding contact angles shown by saponin solutions to solidification of the surface layer.

Present evidence suggests that all the proposed methods, with the exception of those based only on the single surface tension γ_1, provide a useful indication of the spray performance of a particular spreader, for high degrees of correlation were obtained by Evans and Martin (*4*) between various pairs of methods. Mack (*7*) obtained satisfactory agreement between area of spread and contact angle but found that only small droplets assume the advancing contact angle, larger droplets exhibiting a fluctuating angle intermediate between the advancing and receding angles. Hoskins and his co-workers (*16*) suggested that the above-mentioned methods might not provide useful information of spray performance for the two following reasons. Firstly, the final equilibrium values for surface phenomena of solutions are not usually attained immediately the surface is formed. Only when equilibrium is established between the surface and bulk concentrations of the surface-active component, for example, does the surface tension reach a steady reproducible value. Laboratory assessments based upon such final static values may therefore give inadequate information of the dynamic value coming into play when the solution is sprayed. Secondly, the frictional electrostatic phenomena occurring when the liquid is broken into droplets by the spray appliance and when the droplets meet the surface may affect performance. Actually Wampler and Hoskins (*17*) were unable to trace any relationship between performance and electrostatic phenomena while Martin (*18*) established significant correlations between assessments of wetting and spreading properties based on spray performance and on contact angle estimates.

(c) **Penetrating Properties.** Theoretically the ability of a liquid to enter a tube of radius r is given by the expression :

$$P = \frac{2\ \gamma_1 \cos\ \theta_a}{r} \qquad \ldots\ldots\ldots(5)$$

in which θ_a is the advancing contact angle. Penetrating properties will then be proportional to the product $\gamma_1 \cos \theta_a$. In practice it is probable, in view of the rapid evaporation of the spray after application, that the rate of penetration is also of importance. The rate of penetration into a given porous solid will be inversely proportional to the absolute viscosity of the spray. Hoskins (*19*) demonstrated that the rate of penetration of a number of insecticidal oils was proportional to surface tension divided by viscosity. It is probable that the oils he examined exhibited a low advancing contact angle upon the solids used, but, as aqueous solutions have, in general, a higher advancing contact angle, this factor must receive consideration.

A simple laboratory assessment of penetration properties is obtained from time-of-sinking tests, in which the time is taken for the submergence of cotton-wool or of hanks of unbleached cotton placed on the surface of the spray. But as the theoretical basis of this test is still obscure it is better to rely on comparisons based on equation 5 or on biological tests of the types used by Greenslade (*20*) or by Kearns, Marsh and Martin (*21*). In such tests a standard concentration of a non-volatile active constituent is used, only the penetrant differing in the series of sprays. The test organism should be chosen from those in the killing of which penetration is known to be required, due allowance being made for the effect, on toxicity, of factors other than penetration.

As the precise relative importance of wetting, spreading and penetration properties in determining the efficiency of spray spreaders is governed to some extent by the nature and purpose of the active constituents with which they are used, a general conclusion regarding the merits of particular methods of their evaluation may be misleading. Further attention is given to this problem in the discussion of the action and efficiency of " stickers " (p. 77).

REFERENCES

(*1*) Woodman, R. M., *J. Soc. chem. Ind., Lond.*, 1930, **49**, 93T.
(*2*) Woodman, R. M., *J. Pomol.*, 1924, **4**, 38.
(*3*) O'Kane, W. C., Westgate, W. A., Glover, L. C. and Lowry, P. R., *Tech. Bull. N. H. agric. Exp. Sta.*, 39, 1930 ; 46, 1931.
(*4*) Evans, A. C. and Martin, H., *J. Pomol.*, 1936, **15**, 261.
(*5*) Cooper, W. F. and Nuttall, W. H., *J. agric. Sci.*, 1915, **7**, 219.
(*6*) Cupples, H. L., *Industr. engng Chem.*, 1935, **27**, 1219 ; 1936, **28**, 60, 434 ; *J. econ. Ent.*, 1938, **31**, 68 ; *U.S. Dep. Agric. Bur. Ent. Publ.* E.426, 1938.
(*7*) Mack, G. L., *J. phys. Chem.*, 1936, **40**, 159.
(*8*) Stellwaag, F., *Z. angew. Ent.*, 1924, **10**, 163.
(*9*) English, L. L., *Bull. Ill. nat. Hist. Sur.*, 1928, **17**, 235.
(*10*) Ebeling, W., *Hilgardia*, 1939, **12**, 665.
(*11*) Fogg, G. E., *Proc. roy. Soc.*, 1947, B, **134**, 503.
(*12*) Wilcoxon, F. and Hartzell, A., *Contr. Boyce Thompson Inst.*, 1931, **3**, 1.
(*13*) Yarnold, G. D. and Mason, B. J., *Proc. phys. Soc. Lond.*, 1949, B, **62**, 125.
(*14*) Ray, B. R. and Bartell, F. E., *J. colloid. Sci.*, 1953, **8**, 214.
(*15*) Good, R. J., *J. Amer. chem. Soc.*, 1952, **74**, 5041.
(*16*) Ben-Amotz, Y. and Hoskins, W. M., *J. econ. Ent.*, 1938, **31**, 879.
(*17*) Wampler, E. L. and Hoskins, W. M., *J. econ. Ent.*, 1939, **32**, 61.
(*18*) Martin, H., *J. Pomol.*, 1940, **18**, 34.
(*19*) Hoskins, W. M., *Hilgardia*, 1933, **8**, 49.
(*20*) Greenslade, R. M., *Rep. E. Malling Res. Sta.*, 1934, p. 185.
(*21*) Kearns, H. G. H., Marsh, R. W. and Martin, H., *Rep. agric. hort. Res. Sta. Bristol*, 1936, p. 99.

The Chemistry of Spray Spreaders

It is nowadays unlikely that the practical man will require to purchase spray spreaders : the older type of home-made spray has been, with the exception of Bordeaux mixture, replaced by formulated or compounded products in which the maker has already incorporated the necessary ancillary materials. Nevertheless, a knowledge of the properties of the many groups of synthetic surface active compounds now available is necessary for their correct use. For details, both theoretical and practical, useful references are the texts by Schwartz and Perry (*1*) and by Moilliet and Collie (*2*) and a recent review article by Baird (*3*).

If a compound has the property of yielding aqueous solutions of surface tension lower than that of water, Gibbs (*4*) has shown, by thermodynamic reasoning, that the solute is more concentrated in the surface layer of the solution, i.e., it is surface adsorbed. This phenomenon is shown by compounds with a particular type of molecular structure, namely, a bulky molecule not itself water-soluble but which is rendered miscible with water by the possession of certain water-soluble or " polar " groups. Such molecules (or ions) were conveniently described by Hartley (*5*) as " amphipathic ", one part of the molecule being water-attracting (having a sympathy for water), the other part being water-repelling (having an antipathy for water). Langmuir (*6*) and Harkins (*7*) demonstrated that surface adsorption is accompanied by an orientation of the molecules of the capillary-active compound at the solution/air interface. The non-polar part of the molecule is directed away from the water modifying the properties of the solution surface.

The soaps, for example, recommended as far back as 1821 by Robertson (*8*), for use with sulphur for the treatment of mildew on peaches, are the alkali salts of the fatty acids, the carboxylic acid derivatives of long-chain hydrocarbons. Surface activity appears to any marked extent only when the carbon chain exceeds a certain length, about C_{12}, which forms the non-polar water-insoluble group, the polar group being the carboxyl group (–COOH) in which the hydrogen is replaced by the appropriate alkali metal or a suitable organic base, such as triethanolamine or morpholine (tetrahydro-1 : 4-oxazine). The wetting properties of soaps are lost in hard water because of the precipitation of the non-alkaline metal soaps, for which reason they are nowadays replaced by synthetic detergents.

The long alkyl (R–) chain of the fatty acids is retained in the alkyl sulphates, $RO.SO_2.ONa$, or sulphated alcohols, but in order to retain solubility in cold water in the primary alcohol series, dodecyl alcohol is chosen, as in " Sulphonated Lorol " (*9*), the " Gardinols ", " Dreft " and others. The secondary alcohol sulphates derived from petroleum oil refinement and of general formula $R'R''CH.OSO_2.ONa$ are also readily soluble in cold water and are available under proprietary names such as " Teepol ", " Tergitols " (*10*).

Another group of fatty acid detergents are the sulphonates differing from the sulphates in that the sulphur is directly linked to carbon ; this group includes the " Mersolates " prepared by the sulphochlorination of a selected range of synthetic hydrocarbons produced by the Fischer–Tropsch process, and of general formula $R'R''CH.SO_2.ONa$. A common device has been the introduction of a short bridge between the hydrophobic hydrocarbon group and the polar sulphonic group as the " Igepon T " ($R.CO.N(CH_3).C_2H_4.SO_2.ONa$) ; " Breeze " ($R.CONH.C_2H_4.OCH_2.SO_2.ONa$) (*11*). In U.S. Patent 2,028,091 (1936) Jaeger used the ingenious method of reacting the esters of an unsaturated dibasic acid with sodium hydrogen sulphite to give the diesters of sulphosuccinic acid (*12*). The bis(2-ethylhexyl) ester, for instance, is " Aerosol OT " of the American Cyanamid Co. This reaction was later used in the manufacture of the organophosphate insecticide malathion (see p. 220).

In general the use of long chain hydrocarbons for the non-polar group yields compounds of high activity at laundering temperatures but of poor activity at spraying temperatures. This defect may be reduced by introducing unsaturated groups as in oleic acid, by branching the chain as in the

Aerosols or by incorporating aromatic groups which incidentally are readily sulphonated. The "Santomerse" group of the Monsanto Chemical Company, the "Nacconals" of National Aniline and Chemical Corporation, and "Dispersols" of I.C.I. Ltd., are such alkyl sulphonates of which a simple example is sodium dodecyl benzenesulphonate :

$$CH_3.(CH_2)_{10}.CH_2.C_6H_4.SO_2.ONa.$$

The compounds described above are all ionized in aqueous solution and as their surface activity is due to the amphipathic nature of the anion produced, they are called anionic detergents. Hartmann and Kagi (*13*) found that ethylene diamine will combine with oleic acid to form bases soluble in dilute acids yielding surface active salts. Certain of their compounds are marketed under the name "Sapamine"; "Sapamine CH", for instance, is diethylaminoethyloleylamide hydrochloride.

Quarternary ammonium salts such as cetyltrimethylammonium bromide ($CH_3(CH_2)_{14}.CH_2.N(CH_3)_3Br$) are not only highly surface-active but are strong disinfectants, and another of the group, 2-heptadecyl-2-imidazoline acetate, when tried as an emulsifier was found to be an excellent fungicide (see p. 134). In these compounds the surface-active ion is the cation, hence the group forms the cationic detergents which have been found of little use as spray spreaders for they are precipitated by soaps and other anionic detergents, a property useful for analytical purposes (*14*) and for the breaking of emulsions (*15*).

A new type of detergent was foreshadowed in 1935 by B.P. 439,435, protecting the fatty acid esters of polyglycerol. In these compounds the polar group is the hydroxyl group and the esters are not ionized on solution in water. Alternatively the hydroxyl groups may be those of the hexahydric alcohols, sorbitol and mannitol, which partly esterified with fatty acids yield the "Spans" of the Atlas Powder Company. But the rapid development of this group of non-ionized surface active compounds followed the use of the ethylene oxide to provide the requisite hydroxyl groups. These compounds have the general formula $R(OCH_2.CH_2)_nOR'$, the polar groups being derived from the polymeric ethylene oxide while R and R′ form the non-polar hydrocarbon. The "Igepals" developed during the 1939–45 war by the I.G. Farbenindustrie are the condensation products of polyethylene oxide with fatty acids or substituted phenols. The "Triton" series of Rohm and Haas include both non-ionic alkylated aryl polyether alcohols and the corresponding sulphonates ; in Triton X-100, for example, R is the *p*-(1 : 1 : 3 : 3-tetramethylbutyl) phenyl radical, R′ is hydrogen and $n = 10$. The "Tweens" of the Atlas Powder Company are the polyoxyethylene derivatives of the sorbitan fatty acid esters.

By a suitable choice of R, R′ groups, and of the value of n in these non-ionic types, compounds may be produced of the full range from highly water-soluble, giving exceptional wetting properties, to highly oil-soluble, giving emulsifiers suitable for the preparation of miscible oils (see p. 84).

A few of the older non-synthetic spray spreaders still survive :—

Sulphite Lye. In the sulphite process of wood pulp manufacture the wood is digested with sulphur dioxide in the presence of lime or caustic soda. The lignins are thereby sulphonated and pass into solution with the lye. The waste lye, which is too rich in sugars and other organic matter to be discarded as an effluent, is concentrated or is fermented to yield alcohol before concentration. The concentrated product is available either as a thick syrup or as a powder and is known as sulphite cellulose lye, lignin pitch, *Goulac*, *Bandarine*, *Bindex*, etc. The syrup (of specific gravity not greater than 1.3) and the powder (if of a non-deliquescent type) are suitable spray spreaders. The use of sulphite lye as a spray spreader was suggested by Martin (*16*) who, with Evans (*17*), showed that though sulphite lye is effective as a wetter, its spreading properties (in the limited sense) were poor. Sulphite lye has also proved useful as an emulsifier (*18*) and protective colloid, its great advantages being cheapness and the wide range of spray materials with which it can be used.

Casein. Vermorel and Dantony (*19*) proposed the use of casein, which is the protein separated from skim milk by the addition of rennet or acids, as a spreader for Bordeaux mixture. Casein, although insoluble in water, is readily dispersible in solutions of alkaline reaction. Thus the *Lime Caseins*, sometimes called calcium caseinates, are intimate mixtures of hydrated lime and casein. The " soluble caseins " are produced by dissolving casein in strong, hot solutions of alkali, borax (the sodium salt of tetraboric acid, $Na_2B_4O_7.10H_2O$) or disodium hydrogen phosphate ($Na_2HPO_4.12H_2O$) and are probably degradation products. The direct use of *skim milk* was advocated by Robinson (*20*). Casein, being a protein, is of high molecular weight, the polar groups present being the amino and carboxylic residues.

Gelatine. Gelatine was used as a spray spreader by David (*21*) and by Millardet and David (*22*), especially for Bordeaux mixture. Although not securing the popularity of lime casein, its use as a spray auxiliary has been urged, particularly with lead arsenate. Its advantages depend not so much on its properties as a wetter or spreader as on its high activity as a protective colloid and as a sticker of the adhesive type. As the purified grades of gelatine are expensive, the cheaper and impure grades, such as glue or size, are used. Woodman (*23*) indicated that fish glues, which are soluble in cold water, may be less effective than other forms of glue as " stickers ".

Saponin. The saponins form a group of glucosides, chemicals which by hydrolysis or by the action of specific enzymes, yield sugars and compounds of a non-carbohydrate character, the sapogenins. The saponins were early distinguished by their property of dissolving in water to form clear solutions which froth well on agitation. The use of saponins for spraying purposes appears to have been first made by Bedford and Pickering (*24*). Saponins are present in varying amounts in a wide variety of plants (see *25*), but the more important sources of commercial saponin are the inner layers of the bark of *Quillaja saponaria* Molina (soap bark), and the pericarp of the fruit of *Sapindus utilis*. The cost of extraction renders the price of the pure saponins prohibitive for commercial spraying practice, for which reason Bedford and Pickering recommended the use of ground Quillaja bark, whilst Gastine (*26*) advocated the use of the powdered fruits of *Sapindus utilis*, a tree common in Algeria.

Oils as Spreaders. The high spreading efficiency of oils on the waxy types of plant surface is associated with their chemical similarity to the non-polar hydrocarbon surface of the plant wax. The recognition that oil-containing sprays could function as wetters was belated for the water-soluble emulsifiers of the oil emulsion are themselves effective spreaders. The high insecticidal efficiency of cottonseed oil-pyrethrum-soap emulsions, for example, would have been associated with the excellent spreading properties of the soap. Austin, Jary and Martin (*27*) showed that cottonseed oil emulsified with Bordeaux mixture, an emulsifier of inferior spreading properties, was an effective carrier of nicotine, an efficiency associated with the excellent spreading properties of the oil which would become free to spread after the evaporation of the water present in the emulsion.

The penetrating properties of oils are perhaps their most valuable property. De Ong, Knight and Chamberlin (*28*) showed the penetration of the tracheae of red scale by petroleum oils and oil emulsions, and the penetration of oils into plant tissue has been repeatedly demonstrated in investigations of the phytotoxicity of petroleum oils (see p. 188). In this connection, de Ong (*29*) has suggested that improved fungicidal efficiency might be obtained by the use of oil-soluble fungicides and has given evidence of the entry into leaf tissue of copper resinate dissolved in pine oil. Pine oil is a fraction of the terpenes obtained by the distillation of pine wood and consists largely of the tertiary and secondary terpene alcohols (*30*). Though excellent spreaders, their use in sprays is limited by phytotoxicity (*31*).

REFERENCES

(*1*) Schwartz, A. M. and Perry, J. W., *Surface Active Agents*, Interscience Publ., New York, 1949.
(*2*) Moilliet, J. L. and Collie, B., *Surface Activity*, E. and F. N. Spon Ltd., London, 1951.
(*3*) Baird, W., *Synthetic Detergents* in *Progress in the Chemistry of Fats*, Pergamon Press, London and New York, 1955, **3**, 95.
(*4*) Gibbs, W., *Scientific Papers*, 1878, **1**, 230.
(*5*) Hartley, G. S., *Aqueous Solutions of Paraffin-Chain Salts*, Paris, 1936.
(*6*) Langmuir, I., *J. Amer. chem. Soc.*, 1917, **39**, 1848.
(*7*) Harkins, W. D., Davies, E. C. H. and Clark, G. L., *J. Amer. chem. Soc.*, 1917, **39**, 541.
(*8*) Robertson, J., *Trans. hort. Soc., London*, 1824, **5**, 175.
(*9*) Briscoe, M., *J. Soc. Dy. Col.*, 1933, **49**, 71.
(*10*) Wilkes, B. G. and Wickert, J. N., *Industr. engng Chem.*, 1937, **29**, 1234.
(*11*) *Anal. Methods Committee, Society of Public Analysts, Analyst*, 1951, **76**, 279.
(*12*) Caryl, C. R. and Cricks, W. P., *Industr. engng Chem.*, 1939, **31**, 44.
(*13*) Hartmann, M. and Kagi, H., *Z. angew. Chem.*, 1928, **41**, 127.
(*14*) Barr, T., Oliver, J. and Stubbings, W. V., *J. Soc. chem. Ind., Lond.*, 1948, **67**, 45.
(*15*) Batt, R. F., Martin, H. and Wain, R. L., *Ann. appl. Biol.*, 1944, **31**, 64.
(*16*) Martin, H., *Hort. Educ. Assoc. Yearb.*, 1932, **1**, 76.
(*17*) Evans, A. C. and Martin, H., *J. Pomol.*, 1935, **15**, 261.
(*18*) Kearns, H. G. H., Marsh, R. W. and Pearce, T. J. P., *Rep. agric. hort. Res. Sta. Bristol*, 1932, p. 66.
(*19*) Vermorel, V. and Dantony, E., *C. R. Acad. Sci. Paris*, 1912, **154**, 1300 ; 1913, **156**, 1475.
(*20*) Robinson, R. H., *J. econ. Ent.*, 1924, **17**, 396.
(*21*) David, E., *J. Agric. prat. Paris*, 1885, **49**, 659.

(22) Millardet, A. and David, E., *J. Agric. prat. Paris*, 1886, **50**, 764.
(23) Woodman, R. M., *Wye Provincial Conference* 29, Sept. 1931.
(24) Bedford, Duke of, and Pickering, S. U., *11th Rep. Woburn exp. Fruit Farm*, 1910, p. 159.
(25) Friend, H., *Gdnrs'. Chron.*, 1931, **90**, 412.
(26) Gastine, G., *C. R. Acad. Sci. Paris*, 1911, **152**, 532.
(27) Austin, M. D., Jary, S. G. and Martin, H., *Hort. Educ. Assoc. Yearb.*, 1932, **1**, 85.
(28) de Ong, E. R., Knight, H. and Chamberlin, J. C., *Hilgardia*, 1927, **2**, 351.
(29) de Ong, E. R., *Phytopathology*, 1935, **25**, 368.
(30) Palmer, R. C., *Industr. engng Chem.*, 1943, **35**, 1023.
(31) Martin, H. and Salmon, E. S., *J. agric. Sci.*, 1934, **24**, 469.

STICKERS

The amount of spray or dust residue is determined by the amount initially retained and by its tenacity (see p. 79). It is probable that the American expression " sticker " was originally given to materials added to improve tenacity only, but it is now convenient to apply the term more generally to substances improving the retention of spray or dust deposits.

Considering first initial retention, this amount will depend on the maximum quantity of spray or dust that the surface can hold, and in the case of the application of amounts insufficient to give this full load, on the proportion of the spray or dust applied which is retained by the surface. Maximum initial retention is more clearly defined in the case of sprays, but the problem is here complicated, for only rarely are simple homogeneous solutions used in spray practice. In most cases the sprays are the heterogeneous suspensions and emulsions of which one component may be retained to a greater extent than other components. De Ong, Knight and Chamberlin (*1*), for example, found that the percentage of oil in certain emulsions draining from sprayed foliage was lower than that of the emulsion before spraying, indicating a preferential retention of the oil phase. Smith (*2*) in studying this phenomenon applied the terms " initial deposit " to the oil which, apparently, is deposited on the surface the instant the spray strikes the surface and " secondary deposit " to the oil deposited as a result of evaporation of the water in the droplets or film of spray. Hoskins and his colleagues (*3, 4*) used the terms " primary " and " secondary deposits " for these two quantities, but it appears safer to adopt terms which do not imply the time sequence of deposition. Of such terms " preferential retention " used by Fajans and Martin (*5*) is preferred to " build-up " used by Marshall (*6*) to indicate the continuous increase of deposit with prolonged application of spray.

Maximum initial retention (M.I.R.) in the case of simple solutions, was shown by Martin (*7*) to be a function of wetting and spreading properties, decreasing with the improvement in these properties in accordance with the equation :

$$\text{M.I.R.} = k\sqrt{\gamma_1 (1 - \cos\theta)} \qquad \ldots\ldots(6)$$

It was not possible to judge from the correlation coefficients obtained whether the contact angle (θ) concerned is the receding contact angle (i.e., wetting properties) or the equilibrium angle intermediate between the advancing angle (spreading properties) and the receding contact angle.

As, however, the performance of the sprayed droplet is influenced by its kinetic energy, the receding angle probably determines the retention of the droplet which directly hits the surface.

Given a zero receding contact angle no part of the surface is left unwetted by the spray and "film coverage" (*8*) occurs. Theoretically the spray load would drain to a monomolecular film, but evaporation intervenes and the initial retention is determined by rate of evaporation and viscosity. The relationship between viscosity and the maximum initial retention of a spray completely wetting the surface was demonstrated by Woodman (*9*). Given imperfect wetting, droplets accumulate on the sprayed surface until coalescence and run-off occurs. Evans and Martin (*10*) showed that maximum initial retention occurs at the point of incipient run-off. Rich (*11*) found that the amount retained on foliage of Bordeaux mixture, sprayed until "run-off", increased as the logarithm of the spray concentration, but his results could be confirmed by Somers (*12*) only when a wetting agent had been added. Somers found that the copper deposit at run-off was severely reduced by an anionic wetting agent such as sodium dinonyl sulphosuccinate but that the addition of the cationic cetylpyridinium chloride gave but a slight reduction.

If a heterogeneous emulsion or suspension is sprayed, preferential retention may occur with consequent increase in the spray load of the preferred component beyond the quantity indicated by the maximum initial retention of the spray. The properties required for preferential retention to occur are obscure and Fajans and Martin (*13*) were unable to demonstrate its occurrence in the case of simple suspensions. In the case of emulsions, de Ong, Knight and Chamberlin (*1*) found that preferential retention of the oil phase increased with decrease in the stability of the emulsion as indicated by emulsifier content. Smith (*8*) suggested that the wetting properties of the aqueous phase of the emulsion also affect preferential retention, a view examined by Hoskins and his colleagues (*3, 4*). A slight improvement of wetting properties of the aqueous phase favours preferential retention but excessive wetting by this phase, generally accompanied by an increase in the stability of the emulsion, retards preferential retention presumably by inhibiting contact between the oil phase and the surface. This hypothesis does not appear to furnish a complete answer to the problem, for Fajans and Martin (*13*) were unable always to reduce preferential retention by the improvement of wetting properties without marked increase of emulsion stability.

If conditions are such that preferential retention of the oil phase occurs, this property can be used to enhance the retention of solids suspended in the emulsion (*13*). To obtain this effect the solid and emulsifier must be so chosen that an interaction, probably of chemical character, occurs resulting in a partial or complete adsorption of the oil phase by the solid. This phenomenon was also described by Marshall (*6*) who referred to the process as "inversion", for the suspended solid originally wetted by the aqueous phase becomes wetted by oil. This choice of term is unfortunate if it leads to confusion with the inversion of an emulsion (see p. 83). Oil-flocculation seems better and it is probable that conditions determining preferential retention are similar in the two systems. The extent to which preferential retention of oil or of oil-flocculated solid can be used in practice to augment spray loads is dependent on the spray machinery available. The unstable character of the spray systems showing preferential retention demands good agitation in the spray tank and alert attention by the spray gang, requirements which would render such sprays unpopular except among more enlightened growers. Fajans and Martin (*13*), indeed, questioned the practicability of recommending sprays showing marked oil-flocculation with the present type of spray machines but, with improved appliances, the phenomenon could be made of great practical use. Groves, Marshall and Fallscheer (*14*) devised an injector type of sprayer for the application of oil-flocculated suspensions.

Turning now to factors affecting the tenacity of spray residues, it is a general rule that the finer the particles the more tenacious the deposit. Coarse particles of sulphur, for example, are rapidly removed by rain or wind. Similarly if the spray is of poor wetting properties it leaves large droplets evaporating to thick blobs of deposit easily removed. For this reason it was generally recommended that the early protective sprays such as Bordeaux mixture, being of poor wetting qualities, should be applied in a fine mist-like spray in amounts just insufficient to cause drip from the foliage. The addition of spreaders, by improving coverage, tends to enhance tenacity and many of the older spray spreaders such as lime casein or gelatine can be classed as stickers. This effect is, however, offset firstly by the reduction of maximum initial retention following perfect wetting and, secondly, by the influence of the spreader upon the rain-resisting qualities of the spray deposit. Tenacity will, moreover, be dependent on such factors as the solubility and chemical stability of the active component of the spray or dust. Particle size will determine the total surface area of the particles and its effects on the persistence of a photosensitive protectant such as diclone have been theoretically examined by Burchfield and McNew (*15*).

In most climates rain and dew are the main agents reducing spray residues and, as a general rule, Fajans and Martin (*5*) found that resistance to leaching is determined by the relative ease of wetting of the spray deposit. Thus spray spreaders which remain in the spray deposit in a form readily soluble in cold water, such as sulphite lye and most detergents, reduce tenacity whereas those difficultly soluble, such as gelatine, or which decompose on drying to insoluble products (e.g., lime casein) enhance tenacity. It is perhaps a rain-proofing action which renders the oils of special merit as stickers, though the improvement of tenacity which they effect may be offset by the wetting properties of the emulsifier present. Fajans and Martin (*13*) showed, for example, that the adverse effect of sulphite lye, when employed at high concentrations as emulsifier, may outweigh the favourable effect on tenacity of oils. Of the oils they found cottonseed oil to be a better sticker than refined petroleum oil, a property perhaps associated with the semi-drying character of the former which acts not only as a water-proofer but as an adhesive.

The idea of gumming the spray residue to the foliage forms the basis of many older stickers. Flour paste, gums, dextrines (British gum) and soybean flour (*16*) have all been recommended. So efficient is flour paste as a sticker that Parker (*17*) found it effective against the red spider mite *Tetranychus bimaculatus* Harv., not because it is directly toxic to this organism, but because it sticks it to the leaf. The use of such adhesives never became popular with the exception of the polyethylene polysulphides (PEPS) first recommended by Stewart and Standen (*18*). Synthetic resins of the latex-type, such as polyvinyl acetate were found promising in the comprehensive trials of Somers (*19*) on stickers for use with copper fungicides. He suggested that the improved spray performance of cuprous oxide obtained by the addition of polyvinyl chloride was associated with an improved dispersion.

REFERENCES

(*1*) de Ong, E. R., Knight, H. and Chamberlin, J. C., *Hilgardia*, 1927, **2**, 351.
(*2*) Smith, R. H., *Bull. Calif. agric. Exp. Sta.*, 527, 1932.
(*3*) Hensill, G. S. and Hoskins, W. M., *J. econ. Ent.*, 1935, **28**, 942.
(*4*) Ben-Amotz, Y. and Hoskins, W. M., *J. econ. Ent.*, 1937, **30**, 879.
(*5*) Fajans, E. and Martin, H., *J. Pomol.*, 1937, **15**, 1.
(*6*) Marshall, J., *Bull. Wash. agric. Exp. Sta.*, 350, 1937.
(*7*) Martin, H., *J. Pomol.*, 1940, **18**, 34.
(*8*) Smith, R. H., *Hilgardia*, 1926, **1**, 403.
(*9*) Woodman, R. M., *J. Pomol.*, 1924, **4**, 38.
(*10*) Evans, A. C. and Martin, H., *J. Pomol.*, 1935, **13**, 261.
(*11*) Rich, S., *Phytopathology*, 1954, **44**, 203.
(*12*) Somers, E., *J. Sci. Food Agric.*, 1957, **8**, 520.
(*13*) Fajans, E. and Martin, H., *J. Pomol.*, 1938, **16**, 14.
(*14*) Groves, K., Marshall, J. and Fallscheer, H., *Bull. Wash. agric. Exp. Sta.*, 367, 1938.
(*15*) Burchfield, H. P. and McNew, G. L., *Contr. Boyce Thompson Inst.*, 1950, **16**, 131.
(*16*) Harman, J. W., *J. econ. Ent.*, 1937, **30**, 403.
(*17*) Parker, W. B., *Bull. U.S. Dep. Agric. Bur. Ent.*, 117, 1913.
(*18*) Stewart, W. D. and Standen, J. H., *Contr. Boyce Thompson Inst.*, 1946, **14**, 203.
(*19*) Somers, E., *J. Sci. Food Agric.*, 1956, **7**, 160.

PROTECTIVE COLLOIDS AND DISPERSING AGENTS

It is clearly essential that the spray applied shall be of uniform and known concentration but in the case of suspensions there is a risk that this result may be frustrated by a ready sedimentation of the particles. Although this danger can be avoided by effective agitation during spray application, it may also be reduced by adding auxiliary spray materials which function as protective colloids. An example may be taken from detergents. It was found that many surface-active substances failed as soap substitutes in laundering because, although they effectively removed the dirt from the textile fibre, the dirt was re-precipitated on the fabric. By the addition of protective colloids of the appropriate type, this re-precipitation could be prevented, for which purpose, water-dispersible derivatives of cellulose were first used in Germany (*1*).

These cellulose derivatives fall into two groups: the sodium carboxymethylcelluloses and the methyl celluloses. The former are prepared by the treatment of cellulose with caustic soda in the presence of alcohol followed by esterification with sodium monochloroacetate, a process by which carboxymethyl residues ($-OCH_2.COONa$) are added to the cellulose molecule. The methyl celluloses are prepared from cellulose by treatment with alkali and with methyl chloride whereby a proportion of the hydroxy groups of cellulose are converted to methoxy ($-OCH_3$) groups. For spray purposes these derivatives are usually incorporated in the compounded product.

The delayed sedimentation effected by protective colloids arises through two main reasons : firstly, the increased viscosity of the liquid medium to which the rate of fall is inversely proportional ; secondly, surface adsorption on the solid particle whereby the solid particle becomes surrounded by a liquid shell of similar density to the surrounding liquid.

Protective colloids also find an important application in the manufacture of concentrated pastes intended for dilution to spray strength by the grower. Familiar examples are the so-called " colloidal " sulphurs and copper pastes. Such products should contain as much of the active constituent as possible, should not be so viscous that they are difficult to mix and measure and should not be so fluid that, on storage, the solid particles present will sink to form a tough sediment difficult to remix to a uniform paste. By the incorporation of suitable protective colloids it is possible to prepare pastes of high concentration, yet easily stirred and poured. Early examples are the use of sulphite lye in certain copper pastes (B.P. 392,556) and of sulphonated naphthaleneformaldehyde condensation products (e.g.,dinaphthylmethane disulphonate (2)) with salicylanilide (B.P. 350,642). Use is sometimes made of the property possessed by some pastes of forming, on standing, a highly viscous and non-settling suspension but which when shaken becomes fluid and readily pourable. The cause of this property, known as thixotropy, is little understood and it arises almost by accident, though it is common in suspensions of bentonite and when the solid particles are flat or rod-like.

Paste formulations have nowadays been largely replaced by " wettable powders ", prepared by the incorporation of the active component with a finely divided mineral carrier to which has been added a surface-active component rendering the powder self-dispersing when added to water. Most of the synthetic detergents are effective surface-active components though, as mentioned on p. 79, their presence in the spray residues may have an adverse effect on tenacity.

REFERENCES

(1) Richardson, R. E., Kern, J. G., Murray, R. L. and Sudhoff, R. W., *Comb. Intell. Obj. Sub-comm., Item 22*, File XXVI-2, 1945.

(2) Stewart, A. W. and Bunbury, H. M., *Trans. Faraday Soc.*, 1935, **31**, 208.

EMULSIFYING AGENTS

Oils and other water-immiscible liquids on agitation with water break up to small droplets which, on standing, rapidly coalesce to form a separate layer. This coalescence may be prevented or retarded by the addition of auxiliary materials which, since they stabilize the emulsion, are called emulsifiers.

With two immiscible liquids such as oil and water, two types of emulsion are possible. Either the oil may be dispersed as fine droplets suspended in water, which is then the continuous phase, giving an oil-in-water (O/W) emulsion, or the water may be the disperse phase giving a water-in-oil (W/O) emulsion. The type of emulsion generally required in spraying practice is the O/W emulsion which, as water is the continuous phase, is readily diluted with water.

It is now generally accepted that the principal function of the emulsifier is to modify the properties of the interface between the disperse and continuous phases. In the case of soaps as emulsifiers for O/W emulsions, for example, adsorption of the soap occurs at the interface, the polar group being held by the water, whereas the non-polar group is held by the oil, thus producing a film which resists the tendency of the oil droplets to coalesce. In the same way, every surface-active substance is capable of functioning as an emulsifier, and many of the spreaders already mentioned have been suggested and used as emulsifiers. In addition, certain solids such as the freshly-prepared Bordeaux precipitate (*1*) are able to act as emulsifiers. The main requirement, according to Bancroft's explanation of the mode of action of solid emulsifiers, is that the solid should be more easily wetted by one liquid than by the other ; that liquid which yields the lower contact angle in contact with the solid becomes the continuous phase.

The Properties of Emulsions

The term stability of emulsion has been applied above in connection with the resolution of the emulsion to separate layers of the two liquids. This phenomenon is termed the " breaking " of the emulsion, but there are two other forms of instability of importance in spray practice, namely, the creaming and inversion of the emulsion.

The *creaming* of the emulsion is due to differences in specific gravity between the dispersed and continuous phases and is named from the analogous creaming of milk. Pickering (*1*) showed that dilute oil emulsions may be regarded as a mixture of the " cream " and the excess of the continuous phase, for the cream contained 65–82 per cent by volume of oil. If now undeformable spherical particles of the same size be tightly packed into a given volume they will occupy 74.05 per cent of that volume. The close agreement of this figure and the oil content of creams led Woodman (*2*) to consider the cream as the only truly stable emulsion.

If then less than 74 per cent by volume of oil be present the emulsion will tend to cream. The rate of creaming is governed by the difference between the densities of the two phases and by the size of droplets of the dispersed phase. Tar oil-water emulsions cream downwards, for the oil is of density greater than water, whilst the usual type of petroleum emulsion forms a surface cream layer. Creaming, if permitted in practice, would lead to the application of a spray of non-uniform concentration, but is easily prevented by efficient agitation in the spray tank.

The *breaking* of an emulsion is the usual method by which the toxic dispersed phase comes into play, breaking occurring immediately after application as in a quick-breaking emulsion or after the evaporation of the greater part of the water which forms the continuous phase. Examples will be quoted later (p. 189) of the relationship between toxicity and speed of breaking, which is, in general, dependent upon the stability of the emulsion. It is clearly imperative that neither the concentrated emulsion employed for the preparation of the spray nor the diluted emulsion in the spray tank should break before application. If, in spite of vigorous agitation, free oil is visible, the emulsion is unsuitable for use as a spray.

The *inversion* of an emulsion is the change in the type of emulsion. An O/W emulsion containing soaps as the emulsifying agent can be induced by the addition, for example, of calcium salts to invert to a W/O emulsion, a change which is shown by the appearance of a thick grease immiscible with water but which can be diluted with oil. An inverted emulsion is useless as a spray, and it implies that the particular emulsifier present must be replaced by one more suitable for the water used for the preparation of the spray. As a general rule, surface-active substances forming water-insoluble calcium or magnesium salts are unsuitable for use in emulsions to be diluted with hard water, by reason of their tendency to favour inversion.

The Preparation of Emulsions

At one time it was not unusual for the grower to prepare his own emulsions, either by the Tank-Mix Method using soap or Bordeaux mixture (*3*) as emulsifier, or by the 2-solution method (*4*) in which, for example (*5*), oleic acid is dissolved in the oil and the mixture added to water containing alkali. These methods are now largely replaced by the use of factory-made preparations which merely require dilution with water ; these preparations are of two types.

Stock Emulsions. The simpler type of product suitable for dilution by the grower to yield emulsions of spray strength are the stock emulsions which are concentrated emulsions obtained by the processing of oil-water-emulsifier mixtures in an emulsifying mill. The range of emulsifiers available is large, but practical requirements limit the number in actual use. These requirements are, firstly, that emulsions should be sufficiently stable after dilution with hard water, secondly, that the stock emulsion should be of the highest oil content yet should be of a consistency suitable for easy mixing and measuring. When admixture of other spray materials with the diluted emulsion is required, the emulsifier should be one which does not react chemically or interfere with the biological performance of the added materials. For this reason soaps and other derivatives yielding alkaline solutions are, on the whole, unsuitable for stock emulsion preparation for which purpose sulphite lye has been recommended (*6*). Finally, the emulsifier should not accelerate corrosion of the drums in which the stock emulsion is stored.

The main faults of stock emulsions, from the spraying point of view, are their tendency to cream on storage and the danger that, on exposure to frost, the emulsion may break. This danger may be reduced by the incorporation of anti-freezing materials such as glycerine in the aqueous phase, though some emulsifiers (e.g., sulphite lye) yield stock emulsions which after freezing, thaw without excessive oil liberation.

Miscible Oils. It would appear that the term " miscible " oil was first applied to certain proprietary petroleum oil preparations introduced in 1904. The term may be conveniently employed to cover all products which are clear (one-phase) solutions yielding emulsions when diluted with water to spray strength ; latterly, the expression " emulsifiable concentrate " has become usual in the United States. These preparations are essentially solutions of an emulsifier in the oil or water-immiscible liquid which ultimately forms the disperse phase of the emulsion.

Few of the earlier surface-active compounds were sufficiently oil-soluble for use but with the advent of the non-ionic detergents (see p. 74) with the right balance of polar and non-polar groups to confer oil solubility and retain emulsifying properties, miscible oils have become a frequent type of formulation.

Care must be taken to ensure that the range of temperatures over which perfect solutions are stable is adequate to prevent danger of layering or separation to two layers of different composition on exposure to frost. Ease of handling and measuring, their high oil content and, in many cases, the easy visibility of the diluted emulsion when sprayed, are practical advantages which render miscible oils popular with growers.

REFERENCES

(*1*) Pickering, S. U., *J. Chem. Soc.*, 1907, **91**, 2001.
(*2*) Woodman, R. M., *J. Pomol.*, 1925, **4**, 184.
(*3*) Martin, H., *Ann. appl. Biol.*, 1933, **20**, 342.
(*4*) Tutin, F., *Rep. agric. hort. Res. Sta. Bristol*, 1927, p. 81.
(*5*) Martin, H., *J. S-E. agric. Coll. Wye*, 1931, **28**, 181.
(*6*) Kearns, H. G. H. and Martin, H., *Rep. agric. hort. Res. Sta. Bristol*, **1936**, p. 118.

CHAPTER VI

THE MEASUREMENT AND MECHANICS OF TOXICITY

"When you can measure what you are speaking about, and express it in numbers, you know something about it, but when you cannot measure it, when you cannot express it in numbers, your knowledge is of a meagre and unsatisfactory kind."

Thomson, W., *Popular Lectures and Addresses*. London, 1889. Vol. 1, p. 73.

MEASUREMENT is a process of comparison and only by the use of methods of comparing the values of the various chemicals used as insecticides, as fungicides, as seed disinfectants or as soil partial sterilizing agents, can the full benefit of the application of chemistry to Crop Protection be secured.

The acid test of any crop-protective chemical is its performance under practical conditions, and field trial is the ultimate criterion. But comparisons by the method of field trial are expensive, laborious and lengthy; expensive for the provision of adequate biological material is necessary, laborious because of the work involved in obtaining quantitative results, lengthy because repetition in order to obtain a sufficient variation in environmental factors is required to justify a generalization applicable to average conditions. It is, in fact, advantageous to obtain, in the series of field trials, the maximum variation of the many factors, such as degree of infestation, the variety and condition of the host plant, climatic conditions and methods of application, which affect the field performance. The greater the extent of these variations, the more generally applicable will be the result of the comparison.

Progress by the method of field trial is so slow that many attempts have been made to simplify, by analysis, the method. The influence of one or more of the variable factors is eliminated and, by the combination of the results of a series of trials in which different variables have been held constant, an attempt is made to synthesize a comparison which will hold good under field conditions. This analytical procedure is the principle of the method of laboratory trial.

The main factors determining field efficiency have already been summarized on p. 67 and some can be examined by conventional chemical or physical methods. Tenacity, for example, could be assessed by the estimation of the residue on the foliage or artificial surfaces exposed to simulated rain, wind and sunlight. Indeed it has been possible, in many groups of pest control products to erect specifications (*1*) in which the criteria determining field efficiency are listed and methods are given for their physico-chemical examination. But the purpose of the specification is to define and it can seldom be based on a full knowledge of the mode of action of the product specified. Generally those factors affecting biological activity such as inherent toxicity or availability have still to be examined by methods involving the use of the living organism. It is to the principles of such bioassays that attention must now be given.

The primary object of the bioassay is to determine the response of the individual organism exposed to the toxicant under conditions which reduce to the lowest the influence of all but one of the factors affecting that response. The variable factors are usually the amount of the toxicant to which the organism is exposed and the time of exposure. Variation in other factors may arise through differences in the resistance of the individual organisms, in their exposure to the toxicant and in the environmental conditions.

Variation in the resistance of the individual organisms is reduced not only by standardization of growth conditions but, where possible, by the elimination of hereditary and sex differences. Thus in the comparison of the abilities of protective fungicides to prevent spore germination, the American Phytopathological Society (*2*) suggest suitable fungi, the conditions of culture and the age of the spores used. For the comparison of direct fungicidal properties, a difficulty may arise through inability to grow such highly specialized fungi as the Erysiphaceae on artificial media. Salmon (*3*) overcame this difficulty by using only plants propagated vegetatively from one parent plant, and selecting for treatment, leaves and mildew patches in the same stage of growth. A similar precaution was described by Staudermann (*4*) using *Plasmopara viticola* as test organism.

In their studies upon contact insecticides, Tattersfield and his colleagues (*5*) used only aliencolae (see p. 283) of one generation reared from a single fundatrix of *Aphis rumicis*. But, for continuous tests, insect species which can be bred successfully all the year round are required. Steer (*6*) and Crauford-Benson (*7*) have discussed the choice of test insect which nowadays usually falls on a stored-products pest such as the saw-toothed grain beetle *Oryzaephilus surimanensis* L. (*8*), the flour beetle *Tribolium castaneum* Hbst. (*9*) or the granary weevil *Calandra granaria* L. (*10*).

To reduce differences in exposure to the toxicant, the test organism must be subjected to a known concentration of the toxic material applied in a standardized manner. For some purposes such as the examination of direct insecticides or fungicides which, in practice are applied in a medium which wets completely the surfaces involved, uniformity of application may easily be obtained by an immersion or dipping method. This method is well suited for the testing of ovicides using insect eggs laid on bark or other easily wetted surface (*11*). If it is used for insects in active stages, Tattersfield (*12*) considered that there is a risk that the insect may swallow the liquid thus adding stomach poison effects to contact action, but Crauford-Benson (*13*) accepted this risk because of the greater convenience of the method over spraying. Simple dipping would be dangerous to use if the liquid were a quick-settling suspension or especially a rapidly-creaming emulsion ; McIntosh (*14*) devised an end-over-end shaker to overcome these dangers. Of spraying methods the simplest is that used by Salmon (*3*) who, using sprays of good wetting properties found that application continued until the mildew was thoroughly wetted gave concordant results. This method would fail to give equal spray deposits if used with sprays of different and inferior wetting properties (*15*). If the purpose is to apply a spray or dust to insects exposed on a surface, the insects may be placed in a falling mist of spray droplets or dust particles produced under standardized conditions. Tattersfield and Morris (*16*) devised a suitable apparatus, later calibrated by Tattersfield (*17*), in which a small known amount of the insecticide solution is atomized into a cylinder at the base of which the insects are exposed. The same principle is used in Campbell's " turn-table " method, the various modifications and limitations of which are discussed by Tattersfield (*18*). Potter (*19*) added an earthed metal tower above the settling chamber so that the cone of spray would be better distributed by turbulence and the area receiving a uniform spray deposit thereby increased. Recognition of the ability of the insects to pick up the insecticide from the deposit on which they walked made it necessary to distinguish between the toxic effect of the direct spray and that of the residual film of insecticide. This " film " effect may be assessed by spraying the substrate before the insects are placed on it ; both Parkin and Green (*20*) and Tattersfield and Potter (*21*) found hardened filter paper a suitable substrate. Alternatively the filter paper may be treated directly with a solution of the insecticide in a volatile or non-volatile solvent just sufficient to saturate the paper (*22*). If the solvent is non-volatile, another variable, the concentration of toxicant, must be considered in addition to the amount of deposit ; if the toxicant is volatile, the possibility of fumigant action must be allowed for.

For the comparison of protective fungicidal efficiencies a simpler method of obtaining a known deposit is to expose a surface of standard wetting properties at right angles to the axis of a spray cone for a fixed time, less than that needed to give a run-off of the spray from the surface (*15*). Droplets of spore suspension, of equal volume, spread and spore concentration, are then placed on the dried surface and, after incubation in a moist chamber, the percentage of germinated spores is determined (*2*). If the spray deposits differ in surface activity, the spore droplets may be placed in small areas delimited by circles cut in the surface, as suggested by Montgomery and Moore (*23*).

The direct application of the toxicant to the organism by means of a micro-pipette was used by O'Kane (*24*) and by Nelson (*25*) and their colleagues. This method is better suited to the examination of the site of insecticidal action than to the routine testing of insecticides. If, however, the site of action is known, methods such as that devised by Lowenstein (*26*) become possible. He assayed pyrethrum extracts by direct application to the exposed insect nerve and, by amplification of the potential differences set up by stimulation, determined the time taken for the response of the nerve to disappear after treatment with the insecticide. This technique would seem especially useful for the determination of the extent to which penetration factors affect the response of the insect to toxicants applied externally as in the " film " method.

Variations in the response of the test organism due to environmental factors are reduced by holding such factors constant, as a general rule, at the best conditions for growth or for the establishment of infection. The laboratory methods for the comparison of fungicidal efficiencies against apple scab (*Venturia inaequalis*), devised by Keitt and Jones (*17*) and used by Hamilton (*28*), and that of McCallan and Wellman using various fungal pathogens of tomato (*29*) afford examples of the latter case. In the spore germination method of evaluating protective fungicides (*2*), the recommendation to add stimulants such as orange juice (*30*), potato dextrose (*31*) or Coenzyme R (= biotin) (*32*) to improve germination may be misleading for the precise nature of the toxicant under test may become indeterminate. Thus the spore germination curves ascribed by McCallan (*33*) to copper sulphate were in fact responses to a copper sulphate-orange juice mixture, the composition of which was complicated because, though the copper sulphate content of his spore suspensions increased logarithmically, the orange juice content remained constant.

The influence of external factors on the response of the test organism may be further illustrated by Muskett's (*34*) work on the effect of soil moisture on the relative fungicidal efficiencies of seed disinfectants in the control of *Helminthosporium avenae*. Morrison (*35*) traced variation in the response to nicotine sprays of the fruit fly, *Drosophila melanogaster*, to different numbers of flies per container, different sized containers and to differences in the larval and adult nutrition of the flies. The influence of pre-history on the response of the beetle *Ahasversus advena* Waltl. to rotenone-containing sprays is illustrated by Crauford-Benson (*7*) and Gough (*36*) traced variations in the effect of hydrocyanic acid on the flour beetle *Tribolium confusum* Duv. to the handling of the insects before fumigation. Swingle (*37*) observed that the toxicity of lead arsenate to larvae of *Prodenia eridania* Cram. was affected by the food plant on which the larvae were reared. Potter and Gillham (*9*), studying the effects of atmospheric environment on the toxicity of contact insecticides to *Tribolium castaneum*, found greater differences due to conditions after treatment than to pre-treatment. In physiological studies of toxic action, the effect of environmental factors on relative toxicity may be determined by bioassay in which one of these factors, e.g., temperature, becomes the variable factor (see *38*).

Finally, there should be uniformity in the method of recording results and deducing a figure representing toxicity. In some cases, suitable methods are obvious ; in tests of ovicides, for example, the proportion of eggs hatched to total eggs may be taken if the proportion of viable eggs per test is constant and known. Salmon accepted the minimum concentration of toxic constituent necessary to prevent the regrowth of conidiophores within ten days after treatment, a period long enough to indicate the death of the mycelium, as the index of direct fungicidal properties. In other cases, complications arise, as in spore germination tests when there may be germinated spores of which the germ tubes show abnormalities suggestive of a toxic action ; a figure based on the proportion of spores germinated will not reflect this type of toxic action. Further, the effect of the material under test may not be to inhibit germination but to extend the latent period before germination, a phenomenon shown in Tomkins' experiments (*39*). Adequate measures must be taken to ensure that a retardation of germination is not interpreted as non-germination, although, in practice, a fungistatic action may achieve the same purpose as a fungicidal action. In insecticide tests, complications may arise through the difficulty of determining precisely the death or time of death of the insect. Campbell (*40*) examined the effect of the substitution of " knock-out " point for death point, defining " knock-out " as the inability of the silkworm to regain its feet after being pushed over, and death as the absence of response to touch. He showed that the relative toxicities of lead hydrogen arsenate and sodium silicofluoride depended on the choice of end-point, but he regarded the " knock-out " point as the more practical because it is more easily determined. Richardson (*41*) also employed paralysis as an index of toxicity, comparing the results with the proportion of dead and live flies after a suitable interval. He selected paralysis by reason of its greater sensitiveness. Tattersfield (*42*) pointed to the need for observations of the ultimate effect of the insecticide, for, with the pyrethrins for example, narcosis rapidly follows treatment and recovery may occur with sub-lethal doses. He and his colleagues (*5*) have successfully used the percentage of moribund and dead insects as the criterion of insecticidal properties of contact insecticides. A further point is that the criteria chosen should be such that different observers reach similar conclusions. Burchfield, Hilchey and Storrs (*43*), to avoid differences due to the personal judgment of the operator, used the negative phototopism of mosquito larvae as the basis for an objective method of insecticidal bioassay.

REFERENCES

(*1*) *Tech. Bull., Minist. Agric. Lond.*, 1, 1951.
(*2*) *Phytopathology*, 1943, **33**, 627.
(*3*) Salmon, E. S. *et al.*, *J. agric. Sci.*, 1919, **9**, 283 ; 1922, **12**, 269 ; 1926, **16**, 302 ; 1929, **19**, 405 ; 1930, **20**, 18, 489.
(*4*) Staudermann, W., *Mitt. Biol. Abt.* (*Anst. Reichsanstalt*) *Berl.*, 1937, **55**, 43.
(*5*) Tattersfield, F. *et al.*, *Ann. appl. Biol.*, 1925, **12**, 61, 218 ; 1926, **13**, 424 ; 1927, **14**, 217.
(*6*) Steer, W., *J. Pomol.*, 1938, **15**, 338.
(*7*) Crauford-Benson, H. J., *Bull. ent. Res.*, 1938, **29**, 119.
(*8*) Potter, C., *Ann. appl. Biol.*, 1941, **28**, 142.
(*9*) Potter, C. and Gillham, E. M., *Ann. appl. Biol.*, 1946, **33**, 142.
(*10*) Parkin, E. A., *Ann. appl. Biol.*, 1946, **33**, 97.
(*11*) Kearns, H. G. H. and Martin, H., *Rep. agric. hort. Res. Sta., Bristol*, 1935, p. 49.
(*12*) Tattersfield, F., *J. Soc. chem. Ind. Lond.*, 1937, **56**, 79T.
(*13*) Crauford-Benson, H. J., *Bull. ent. Res.*, 1938, **29**, 41.
(*14*) McIntosh, A. H., *Ann. appl. Biol.*, 1947, **34**, 233.
(*15*) Evans, A. C. and Martin, H., *J. Pomol.*, 1935, **13**, 261.
(*16*) Tattersfield, F. and Morris, H. M., *Bull. ent. Res.*, 1924, **14**, 223.
(*17*) Tattersfield, F., *Ann. appl. Biol.*, 1934, **21**, 691.
(*18*) Tattersfield, F., *Ann. appl. Biol.*, 1939, **26**, 365.
(*19*) Potter, C., *Ann. appl. Biol.*, 1941, **28**, 142.
(*20*) Parkin, E. A. and Green, A. A., *Ann. appl. Biol.*, 1943, **30**, 279.
(*21*) Tattersfield, F. and Potter, C., *Ann. appl. Biol.*, 1943, **30**, 259.
(*22*) Stringer, A., *Ann. appl. Biol.*, 1949, **36**, 213.
(*23*) Montgomery, H. B. S. and Moore, M. H., *J. Pomol.*, 1938, **15**, 253.

(*24*) O'Kane, W. C., Walker, G. L., Guy, H. G. and Smith, O. J., *Tech. Bull. N.H. agric. Exp. Sta.*, 54, 1933.
(*25*) Nelson, F. C., Buc, H. E., Sankowsky, N. A. and Jernakoff, M., *Soap, N.Y.*, 1934, **10**, (10), 85.
(*26*) Lowenstein, O., *Nature, Lond.*, 1942, **150**, 760.
(*27*) Keitt, G. W. and Jones, L. K., *Res. Bull. Wis. agric. Exp. Sta.*, 73, 1926.
(*28*) Hamilton, J. M., *Phytopathology*, 1931, **21**, 445.
(*29*) McCallan, S. E. A. and Wellman, R. H., *Contr. Boyce Thompson Inst.*, 1943, **13**, 93; McCallan, S. E. A., *Contr. Boyce Thompson Inst.*, 1948, **15**, 71.
(*30*) McCallan, S. E. A. and Wilcoxon, F., *Contr. Boyce Thompson Inst.*, 1939, **11**, 5.
(*31*) Peterson, P. D., *Phytopathology*, 1941, **31**, 1108.
(*32*) Goldsworthy, M. C. and Green, E. L., *J. agric. Res.*, 1938, **56**, 489.
(*33*) McCallan, S. E. A., *Contr. Boyce Thompson Inst.*, 1948, **15**, 77.
(*34*) Muskett, A. E., *Ann. Bot. Lond.*, 1938, **2**, 699.
(*35*) Morrison, F. O., *Canad. J. Res.*, 1943, **21**, D, 35.
(*36*) Gough, H. C., *Ann. appl. Biol.*, 1939, **26**, 533.
(*37*) Swingle, M. C., *J. econ. Ent.*, 1939, **32**, 884.
(*38*) Busvine, J. R., *Ann. appl. Biol.*, 1938, **25**, 605.
(*39*) Tomkins, R. G., *Rep. Fd. Invest. Bd., Lond.*, 1930, p. 48.
(*40*) Campbell, F. L., *J. econ. Ent.*, 1930, **23**, 357.
(*41*) Richardson, H. H., *J. econ. Ent.*, 1931, **24**, 97.
(*42*) Tattersfield, F., *Ann. appl. Biol.*, 1932, **19**, 281.
(*43*) Burchfield, H. P., Hilchey, J. D. and Storrs, E. E., *Contr. Boyce Thompson Inst.*, 1952, **17**, 57.

PROBIT ANALYSIS

In toxicological work the response of an organism to a given treatment may be assessed in a variety of ways. If rate of growth is the criterion, the statistics involved relate to a continuous variable; but if the number of deaths among the treated organisms is observed, the variable is discontinuous for clearly the number counted cannot include a fraction of an organism. Emmens (*44*) has discussed the principles underlying both types of bioassay but, in the examination of insecticides and fungicides, the discontinuous variable is the more usual. The organism either responds or not, it lives or dies; such a head-or-tail response is called a quantal response. Suppose that the bioassay is a series of exposures of batches of organisms from the same population to different amounts of toxicant. When plotted on squared paper, the results yield a mortality-concentration curve which is generally of a sigmoid character. The sigmoid nature of the curve, as Henderson Smith (*45*) pointed out, is a reflection of the variations in the susceptibility of the individual organisms to the toxicant. As, at a given concentration, the mortality figure will include those individuals which succumb to lower concentrations, it is an integration; hence, the curve may be regarded as a cumulative frequency curve. If the concentrations are plotted on a linear scale, the frequency curve is usually unsymmetrical but, by a suitable transformation of the concentration scale, it is often possible to obtain the normal curve. It is convenient to call the concentration actually used the dose, and the transformed concentration the dose metameter.

Assuming that a normal frequency curve can be fitted to the data, the curve can be converted to a straight line by transforming the percentage mortality figures to suitable units. Gaddum (*46*) chose as this unit the "normal equivalent deviation" (N.E.D.), derived from the normal frequency curve symmetrical about the zero ordinate and with scale of abscissa such that the variance is one unit. The N.E.D. corresponding to a given mortality percentage is then the abscissa of the line which divides the area below the curve into two areas in the ratio of percentage mortality to percentage survival. Thus 90 per cent mortality would be 1.645 in units of the variance. As kills below 50 per cent would yield negative numbers, Bliss (*47*) suggested that 5 be added to the units to give probability units which he named "probits". The probit corresponding to 95 per cent mortality is therefore 6.645.

In most cases it is found that the transformation of dose to its logarithm yields a linear regression on probit mortality. But exceptions are known and Parker-Rhodes (*48*) suggested that, in general, some power (α) of the dose has to be taken to render the mortality curve a normal cumulative frequency curve. He called α the "index of variation" and showed that the logarithmic transformation is the case in which $\alpha = 0$. The selection of the most appropriate transformation of concentration, based on the method of Maximum Likelihood, is by no means easy, for Finney (*49*) showed that the variance of the estimate is often too great for any reliable estimate to be made; but Parker-Rhodes (*50*) considered that, on *a priori* grounds, the index of variation is more likely to be a simple than a complicated fraction. He interpreted his own bioassay results by the use, in addition to the common cases of 0 and 1, of values of 1/2, 1/4, −1/3 and −2/3.

By means of the probit and the dose metameter it should now be possible to calculate a linear equation which will represent the relationship between mortality and the concentration of toxicant to which the organism is exposed. This equation:

$$y = a + bx^{\alpha}$$

where y = probit mortality, x = dose and α = the index of variation, reduces where $\alpha = 0$, to

$$y = a' + b' \log_e x$$

and methods of computation are given by Finney (*49*).

The advantages of the statistical treatment are that the toxicity data are reduced to a simple equation of three, or usually two, terms and that the accuracy with which the estimates of these terms represent the data can be determined. The first of these terms is usually stated as the concentration corresponding to 50 per cent mortality (i.e., y = 5) called by Trevan (*51*) the median lethal dose (LD 50) or less restrictedly by Finney (*49*) the median effective dose (ED 50).

The second of these terms is "b", the regression coefficient of probit mortality on dose metameter, more usually called, when the latter is a logarithm of the dose, the slope of the regression line. This value provides a measure of the spread of the differences between the responses of individual organisms of the test population for it is by definition the reciprocal of the standard deviation of the logarithms of the individual lethal doses. Clearly the more uniform the effect of the toxicant on the individuals of the test sample the narrower the difference in the concentrations of toxicant required to increase the mortality by one probit unit. If, however, the dose metameter is not the logarithm of the dose, or, in Parker-Rhodes's terms, if the index of variation is not zero, a more complicated expression is necessary to measure uniformity of response or " variability " $W\alpha(x)$:

$$W_{\alpha}(x) : 1/b^2m^2\alpha^2$$

where m is the ED 50. As stated above if $\alpha = 0$, this expression reduced to :

$$W'_0(x) = 1/b^2$$

where b is the regression coefficient expressed in Naperian logarithms.

The main use of the ED 50 is for the comparison of relative potencies which is simple only when the test organism displays similar variability to the toxicants being compared. This condition is shown, when $\alpha = 0$, by the parallelism of the regression lines from which it follows that there is a constant difference between the dose metameters producing the same mortality. This qualification is not as restricted as it might seem for parallelism appears frequently in comparative bioassay, especially of insecticides. As an example of the method, Martin's (*52*) comparison of the relative toxicities of rotenone and of four different derris roots may be cited for this example was worked out in detail by Finney (*49*, p. 76).

Estimates of variability, or in cases (where $\alpha = 0$) of the slope of the regression line, are often most informative. If, for example, in a series of tests with organisms from the same population, different variabilities are shown to two toxicants, it may be inferred that the two toxicants differ in the mechanism by which they produce their physiological effects. Moreover, variability may be expected to be independent of external factors such as temperature at which the tests are carried out (*53*). It is independent of such factors as particle size, which usually have a great influence on the ED 50. An example is provided by Burchfield and McNew's (*54*) examination of particle size effects on the fungicidal activity of dichlone. These tests were conducted with samples of different range of particle sizes sprayed on tomato plants which were subsequently exposed to infection by *Alternaria solani* (Ell. & Mart.) Jones & Grout, a technique admittedly subject to high experimental error, for which reason the comparison was made not of the ED 50 but the ED 95 values. The slopes of the regression lines obtained with the different sized particles range from 2.5–3.5, with one atypical value of 1.7, whereas the range of ED 95 is from 46–1050 ppm. Now if G = the ED 95 in g. per sq. cm. of leaf surface, r = the particle radius in cm., d = density of dichlone, and N = number of particles per sq. cm. leaf area required to attain the ED 95 :

$$N = \frac{3G}{4\pi\ r^3d}$$

Burchfield and McNew found that if the reciprocal of G is plotted against the logarithm of the number of particles N, a straight line results, i.e.,

$$1/G = m \log_e N + q$$

where m = slope of line and q its position.

The implications of variability have been theoretically examined by Parker-Rhodes (*48*) for the case of fungus spores immersed in solutions of different compounds of the same element. If the compound cannot penetrate the spore wall it will be non-fungicidal, but suppose that a fungal diffusate converts it by a series of reactions to a permeative form which is fungicidal. Parker-Rhodes deduced that the variability shown to any non-permeative compound which can be so converted to a permeative form will be greater than that shown to a permeative compound. Further, each additional reaction required to convert the compound to a permeative form will increase the variability. Thus, if a metal can only be absorbed by the spore in the form of a coordination complex with some organic compound diffusing from the spore, then the variability to the simple ion will be greater than that to the complex added ready-made to the spore suspension. If a different but non-permeative complex is added instead, the variability to this complex will be greater still for the spore diffusate has first to break down this complex before forming the permeative complex.

Suppose now that the compound is one which contains several atoms of an element but which breaks down in the test to simpler molecules, one of which is permeative and fungicidal. Parker-Rhodes deduced that in this case the variability of the spores to the compound will be proportional to the square of the number of atoms of the effective element in the molecule of that compound and that the index of variation (if not zero) will be inversely proportional to that number. He applied his theory to the fungicidal action of sulphur on *Stemphylium sarcinaeforme* (Cav.) Wilts. using those compounds of sulphur which might arise by hydrolysis. His task was simplified by the observation that the spores of this fungus germinate freely in acidified solutions of sodium thiosulphate. Bassett and Durrant (*55*) had previously shown that such solutions contained not only the thiosulphate ion (S_2O_3'') but also the hydrothiosulphate ion and the ions of the oxy-sulphur acids (except tetrathionic acid) containing three or more sulphur atoms. Elementary sulphur is also present but is presumably not fungicidal because it is in an acid solution (see p. 108). Sulphate may be eliminated for sodium sulphate is non-fungicidal.

Bioassays of solutions of the remaining possible compounds gave results for which Finney (*49*, p. 178), by an improvement of Parker-Rhodes' mathematics, calculated the following figures given in Table I.

TABLE I

Variability of spores of S. sarcinaeforme *to sulphur compounds*

	Index of Variation	abm	Variability
Hydrogen sulphide H_2S	1	3.114 ± 0.234	0.103
Sodium dithionite $Na_2S_2O_4$ ...	0.5	1.895 ± 0.110	0.278
Sodium tetrathionate $Na_2S_4O_6$ ...	0.25	0.712 ± 0.059	1.97

Parker-Rhodes's predictions that, provided the permeative compounds all have the same number of sulphur atoms, the index of variation will vary inversely as the number of atoms present in the molecule tested and that the variabilities shown will be proportional to the square of that number, are confirmed. Because of the latter relationship it may be assumed that of the tested compounds those which can permeate the cell wall have the same number of sulphur atoms which, by elimination, has been shown to be two. The only omission from the list above of possible sulphur compounds is sulphoxylate, a compound too unstable for direct test. But as sulphoxylate (HSO_2') is a first product of the hydrolysis of sulphur it would almost certainly be present in the sulphur suspension. Variability of the spore to the sulphur suspension was estimated by Parker-Rhodes to be 0.78, a much higher value than would be expected if sulphoxylate is the agent responsible for the fungicidal activity of sulphur. He therefore concluded that only certain sulphur ions containing two sulphur atoms to the ion can penetrate the cell wall of *S. sarcinaeforme*; that these ions include hydrodisulphide, dithionite and perhaps pyrosulphite; that if only one such ion is permeative it is thiosulphite; that any sulphur-containing ion or compound capable of being transformed into one of these ions will be toxic to the spores and that any ion or compound not so capable will be non-toxic.

Other examples of the information concerning the route of toxic action to be gained by the probit analysis of the results of bioassay will be given later under the individual toxicant. But it is appropriate to refer here to the case when, as often happens, the relationship between probit mortality and dose metameter yields, not one regression line, but two or more lines of different slope (see, e.g., Bliss, *47*). When the population of test organisms can be assumed to be uniform, the general explanation of such "breaks" is that the toxicant is non-uniform, each slope representing one constituent of a mixture of poisons.

Bliss (*56*) examined statistically the problem of poisons applied jointly. He postulated three types of joint action: (1) similar joint action, when the constituents act independently and similarly; (2) independent joint action, when the constituents act independently and diversely; (3) synergistic action, when the toxicity of the mixture is greater than that predicted from studies of the individual constituents. If the toxicity of the mixture is less than the predicted value, antagonistic action is indicated. The application of these postulates to actual examples requires therefore a prediction of the joint effect from information on the effects of the individual toxicants.

If the two poisons act similarly, for example by causing a failure of the same physiological system, it is probable that the organism will exhibit the same variability to each and that their respective regression lines will be parallel. One component of the mixture can then be substituted in a constant proportion by the other and the toxicity of the mixture can be predicted directly if the proportion of the two poisons is known. Finney (*49*, p. 126) provided an example based on tests of rotenone, a deguelin concentrate and a mixture of the two in 1 : 4 proportion.

If the two poisons act diversely, for example, by causing a failure of different physiological systems, their respective regression lines will probably have different slopes. The theoretical regression lines for the mixtures will then form curves resembling hyperbolas of which the arms rapidly become asymptotic to lines, the slopes of which are determined by the separate constituents (see Finney, *49*, Figure 15). Plackett and Hewlett (*57*), considering that lines of different slope may be given by poisons of similar action if they differ in respect of rate of movement from site of application to site of action, deduced a set of dosage-response curves conforming to the above description.

Plackett & Hewlett (*58*) pointed out that Bliss' treatment of independent action was incomplete in that he failed to cover completely the degree of correlation between the resistance of the organisms to the two poisons. Suppose that those organisms of the test population highly resistant to A are those highly resistant to B (i.e., the correlation coefficient is $+1$) and that doses A and B applied separately kill proportions p_1 and p_2, p_1 being greater than p_2. If now doses A and B are applied together the proportion surviving will be $1 - p_1$, for the organisms remaining are all resistant to poison B. If, on the other hand, the resistance to A and B are uncorrelated (i.e., the correlation coefficient is 0) the proportion killed by the mixture A and B will equal $p_1 + p_2 - p_1p_2$. But if those organisms resistant to poison A are those susceptible to poison B to a degree such that the correlation is complete and negative (i.e., the correlation coefficient $= -1$) then the proportion killed by the mixture will $= p_1 + p_2$.

Applying these considerations to the prediction of the joint effects of the two poisons, Plackett and Hewlett demonstrated that if the correlation is 0 the slope of the probit-dose metameter line will increase, at low doses, from that of the poison giving the less steep regression line to a maximum equal to the square root of the sum of the squares of the separate regressions. But if the correlation coefficient is -1, the slope of the line will increase rapidly attaining infinity at a finite dose. The complete population will succumb to the concentration at which $p_1 + p_2 = 1$. Unfortunately no practical example of this valuable property of mixed toxicants is yet available.

It will now be evident that the prediction of the expected toxicity of a mixture on the basis of the probit equations of the components is so complicated that a clear-cut demonstration of synergism in the sense used by Bliss is extremely difficult. It is not surprising therefore that the term has frequently been applied to mixtures without statistical proof, for many cases are known in which the addition of a second compound results in an apparent increase in toxicity. The added compound need not by itself display marked toxicity for there are many ways in which it could potentiate or augment the physiological activity of the poison to which it is added. It might, for example, increase the proportion of the poison which reaches the site of action or intervene in the reactions by which the poison interferes with vital processes.

REFERENCES

(*44*) Emmens, C. W., *Principles of Biological Assay*, Chapman and Hall, London, 1948.
(*45*) Henderson Smith, J., *Ann. appl. Biol.*, 1921, **8**, 27.
(*46*) Gaddum, J. H., *Spec. Rep. Ser. Med. Res. Coun., Lond.*, 183, 1933.
(*47*) Bliss, C. I., *Science*, 1934, **79**, 38, 409; *Ann. appl. Biol.*, 1935, **22**, 134, 307.
(*48*) Parker-Rhodes, A. F., *Ann. appl. Biol.*, 1942, **29**, 126.
(*49*) Finney, D. J., *Probit Analysis*, Cambridge University Press, 2nd, 1952.
(*50*) Parker-Rhodes, A. F., *Ann. appl. Biol.*, 1943, **30**, 170.
(*51*) Trevan, J. W., *Proc. roy. Soc.*, 1927, B, **101**, 483.
(*52*) Martin, J. T., *Ann. appl. Biol.*, 1940, **27**, 274.
(*53*) Parker-Rhodes, A. F., *Ann. appl. Biol.*, 1941, **28**, 389.
(*54*) Burchfield, H. P. and McNew, G. L., *Contr. Boyce Thompson Inst.*, 1950, **16**, 131.
(*55*) Bassett, H. and Durrant, R. G., *J. chem. Soc.*, 1927, p. 1401.
(*56*) Bliss, C. I., *Ann. appl. Biol.*, 1939, **26**, 585.
(*57*) Plackett, R. L. and Hewlett, P. S., *J. R. Statist. Soc.*, 1952, **14**, 141.
(*58*) Plackett, R. L. and Hewlett, P. S., *Ann. appl. Biol.*, 1948, **35**, 347.

HYPOTHESES ON THE MECHANICS OF TOXIC ACTION

It is remarkable that so early in the history of toxicology it has been possible to suggest and test a range of hypotheses on the nature of the reactions by which a chemical intervenes in biological processes. These hypotheses being general in character, may be conveniently discussed before dealing with the reasons for the activity of the individual toxicants employed for crop protection.

Death by poisoning is the result of an intervention in some vital process, an intervention described noncommitally as a biochemical lesion. Clearly, the poison must have properties enabling it to penetrate and perhaps to concentrate at its site of action within the organism it kills. In early quantitative studies it was often tacitly assumed that this concentration could be inferred from the concentration of the toxicant to which the test organism was exposed. Comparisons of toxicity had, indeed, to be made on the basis of the concentrations applied to produce similar mortalities. In this work correlations were frequently observed between toxicity and certain physical properties. Holt (*1*) noted that the times of exposure required to kill cockroaches exposed to simple volatile organic compounds were inversely related to the boiling points of the compounds, though compounds boiling above a certain critical temperature had little toxicity. Moore (*2*), working with houseflies, and Tattersfield and Roberts (*3*), using the wireworm *Agriotes*, obtained a similar result though the latter pointed out that the relationship failed with some highly toxic fumigants such as hydrocyanic acid.

Boiling point is but one of many physical properties of which the correlation with chemical structure has been much investigated. These relationships emerge more distinctly in groups of related compounds such as homologous series. An early example is due to Traube (*4*) who found that the reciprocals of the equivalent concentration necessary to lower the surface tension of their aqueous solutions by a fixed amount (the specific surface activity) of the simpler fatty alcohols rose in the ratio $1 : 3 : 3^2$ as the homologous series was ascended. A like ratio was observed also in the alkyl acetates and in the fatty acids by Traube, by Szyszkowski (*5*) and by Forch (*6*).

The same geometric progression has been observed in the toxicity of some homologous series. Fühner (*7*) showed that each member of the series of normal paraffins, pentane C_5H_{12} to octane C_8H_{18}, is three times as effective as its lower homologue in producing narcosis in rats and mice. Uppal (*8*) found that the relative toxicities of ethyl alcohol, normal propyl alcohol and normal butyl alcohol to spores of *Phytophthora colocasiae* Rac. were in the ratio $1 : 3 : 3^2$. Dagley and Hinshelwood (*9*) assembled other instances of this threefold increase with successive members of the homologous series and showed, on thermodynamic grounds, that the numerical value of the factor is governed by the energy required for the toxic molecule to be removed from the aqueous medium.

Traube, in extending his Rule to other physical properties, showed that the partition coefficients in the systems benzene/water, carbon disulphide/water and carbon tetrachloride/water of methyl, ethyl and propyl alcohols were again in the approximate ratio of $1 : 3 : 3^2$. That toxicity is related to the oil/water partition coefficient is an hypothesis originally advanced by Overton (*10*) and by Meyer (*11*) who suggested that narcotic action will be governed by the solubility of the narcotic in the lipoids of the organism.

These various relationships between narcosis and physical properties shown in the homologous series were considered by Meyer and Hemmi (*12*) to indicate that narcosis follows when the substance reaches a certain concentration in the immediate surroundings of the cell. They noted that all the physical properties mentioned involve a distribution between different phases : solubility is the expression of distribution of the substance between solid and liquid ; boiling point is determined by vapour pressure which indicates the distribution between solid or liquid and its vapour ; surface activity is an expression of the distribution of the solute between the air-solution interface and the bulk of the solution. Ferguson (*13*) therefore suggested the effects of this equilibrium could be eliminated in certain cases if, instead of the external concentration, the thermodynamic activity at that concentration is used. When a substance is in equilibrium between the different phases of a polyphase system, it has the same thermodynamic activity in the different phases. Consequently if the activity is known in the external phase it is also known at the internal biophase of the organism which is the seat of toxic action.

A close approximation to the thermodynamic activity of a fumigant is given by the relative saturation of the vapour. Thus if p_t is the pressure of the vapour in the fumigation chamber and p_s is the saturation vapour pressure at the temperature of the experiment, the thermodynamic activity is given by the ratio p_t/p_s. Similarly, if the toxicant is of limited solubility and is applied in solution, its activity is approximately equal to s_t/s where s_t is the molar concentration of the solution tested and s its solubility in mol./litre at the temperature of the experiment.

Ferguson and Pirie (*14*) have applied this hypothesis to the toxicity of the vapours of a range of volatile compounds to the grain weevil *Calandra granaria.* A few of their results are given in Table II, in which the median lethal dose derived by probit analysis is expressed both as concentration and as the thermodynamic activity (p_t/p_s) at the temperature of the experiment (25° C). It will be seen that, in accordance with Traube's Rule, the vapour pressures of the three homologous series descend in geometric progression. Similarly, the median lethal doses, when expressed as mg./l. and with the exception of methyl chloride, follow the same rule. Yet when the median lethal concentrations are expressed as thermodynamic activities, toxicity decreases as the series are ascended. The reason for the correlation between physical properties and toxicity emerges as an expression of the relative amounts of toxicant which reach the seat of action within the organism. Moreover an explanation is forthcoming for the frequent observation that, as the homologous series is ascended, toxicity, when expressed as external concentration, increases until a certain number is reached when it falls rapidly (e.g., the alkyl thiocyanates, p. 201, or the critical boiling point observed by Holt, p. 96). This "cut-off" is the point when the thermodynamic activity is unity, i.e., at saturation and is illustrated in Table II by amyl alcohol.

TABLE II

Toxicity of vapours to Calandra granaria *from Ferguson and Pirie* (14)

Compound		V.P. 25° C (mm.)	MLD (mg./l.)	p_t/p_s
Aliphatic hydrocarbons :				
Pentane	C_5H_{12}	511	897	0.45
Hexane	C_6H_{14}	151	353	0.50
Heptane	C_7H_{16}	45.6	137	0.56
Decane	$C_{10}H_{22}$	1.6	12	1.00
Monochloro paraffins :				
Methyl chloride	CH_3Cl	4200	166	0.014
Ethyl chloride	C_2H_5Cl	1170	1124	0.28
Propyl chloride	C_3H_7Cl	339	428	0.30
Butyl chloride	C_4H_9Cl	107	200	0.38
Amyl chloride	$C_5H_{11}Cl$	32	73	0.40
Alcohols :				
Methyl alcohol	CH_3OH	124.0	100	0.47
Ethyl alcohol	C_2H_5OH	58.6	85	0.59
Propyl alcohol	C_3H_5OH	20.1	50	0.77
Butyl alcohol	C_4H_9OH	6.78	30	1.00
Amyl alcohol	$C_5H_{11}OH$	2.50	<50% kill at saturation	

It will be noted that in Table II the median lethal thermodynamic activities appear to increase regularly within each homologous series. Ferguson and Pirie (*14*) showed that this relationship could be expressed by the equation $p_t = kp_s$ and that it is held for the paraffins, the alkyl benzenes, the normal alkyl chlorides (except methyl chloride), the normal alcohols, the chloroethylenes, certain polychloroethanes, the ketones and (possibly) certain esters. No such relation was found in the aliphatic amines, alkyl formates, aldehydes or alkyl bromides and iodides.

This use of thermodynamic activity is only justified if equilibrium is established between internal and external phases. Ferguson and Pirie suggested that substances with which the equation $p_t = kp_s$ holds reach such an equilibrium; they do not therefore undergo chemical change within the organism and their toxic action must be of a physical character. On the other hand with substances which do not conform to this rule and which, like methyl chloride (Table II), are toxic at low thermodynamic activity, equilibrium is not attained; toxic action follows some chemical reaction preventing the establishment of equilibrium; they are "chemically toxic" substances. With such substances the concentration of toxicant applied externally has no relationship to the internal concentration attained and the use of the median lethal concentration as a criterion of toxicity becomes dangerous except where the comparison is made between toxicants of similar action and permeativity. The investigations of the Boyce Thompson Institute (*15*) of the take up of radio-labelled fungicides by fungus spores have borne out this conclusion, revealing wide discrepancies between toxicity (ED 50) expressed as concentration in external phase and ED 50 expressed as p.p.m. of spore weight. McCallan (*16*) went further and gave evidence that comparison of slopes of the dosage response curves may be misleading when derived from external concentrations.

Turning now to the general hypotheses concerning the nature of "chemical toxicity", it is necessary to devote a sentence or two to the events which occur in a chemical reaction. Broadly speaking, a chemical reaction involves, first, the spacial separation of two or more atoms held by the interatomic forces which stabilize the reacting compounds and, second, the formation of the assemblage of atoms which are the products of reaction. The interatomic forces are usually termed "bonds" and the chemical reaction therefore involves the rupture of existing bonds and the formation of new bonds. In a crystal, for example, the distance separating the atoms or ions is that at which the weak electrostatic attraction of two atoms balance the more powerful repulsive forces which come into interplay when the electron distribution of one atom overlaps that of a second. These repulsive forces have to be overcome to result in a chemical bond. When the two interacting atoms differ widely in their affinity for electrons, the atom with the greater affinity may rob the other of an electron and such a discrete transfer of electron gives rise to an ionic bond. In the interaction of sodium and chlorine, for instance, the lone outer electron of the sodium atom is transferred to the outer shell of the seven electrons of the chlorine atom giving rise to a positively-charged sodium ion and a negatively-charged chlorine ion. In gaseous sodium chloride the ions are held together as Na^+Cl^-, but in solution the ions become hydrated and exist as separate hydrated ions. If the two reacting atoms do not differ greatly in their affinity for electrons, the electrons will be shared in such a way that the greatest number of the extremely stable 8 electron shells are formed. If the shared electrons are supplied by both atoms a covalent bond is produced; if the shared electrons are supplied entirely by one atom a semipolar bond results.

As the coulombic energy of attraction of ions and the energy required to break covalent and semipolar bonds are high, these bonds were early recognized but, later, biochemical evidence revealed other and weaker bonds which have become of great significance in toxicological reactions. The hydrogen bond became necessary to explain the curious physical properties of water and it arises from the attraction of two electro-negative ions for the proton, i.e., the hydrogen atom without its electron. The energy involved in hydrogen bond formation is so low that the bond becomes apparent only with the most electro-negative atoms such as fluorine, oxygen, nitrogen; one of its important biological functions is the holding together of long-chain molecules. Another series of attractive forces comes into play through the electrical asymmetry due to the separation of electrostatic charges in the molecule, an asymmetry measured by the dipole moment. Those atoms which contribute electrons to the semipolar bond, for example, become electrically positive and exert an attraction on a negatively-charged ion or atom, or may create such an attraction by inducing a dipole in a susceptible neighbouring molecule. Even in neutral molecules the displacement of electrons by the approach of a second molecule will create a momentary dipole capable of inducing a dipole on a neighbouring atom and thus giving rise to an attractive force. Such forces, first recognized by London (*17*) and known by his name, include the weak electrostatic attraction which, as mentioned above imparts stability to the crystal. The induced dipole forces and the London forces become powerful only when the atoms involved are in close proximity, for these forces vary inversely as the sixth power of the distance between the atoms. They will only become evident, therefore, if the reacting atoms or groups in the two molecules involved can approach to within this close distance.

It is now possible to provide an explanation of the remarkable specificity which is a feature of many biochemical reactions such as the action of enzyme on substrate, or the precipitation of antigen and antibody. The protein comprising the enzyme or the antigen may be pictured as a large molecule possessing a surface on which are disposed reactive atoms or groups of atoms. If now the antibody or the substrate has reactive groups complementary to those of the enzyme or antigen, a reaction will follow only if these reactive groups are so disposed spacially that they can fit closely to the complementary reactive groups of the active antigen or enzyme surfaces. It is evident that a high degree of specificity can be obtained if the surface area over which this complementariness in structure is exercised is great enough to include several reacting pairs or groups able to approach so closely that forces such as the hydrogen bond and the London forces can come into interplay to stabilize the antigen-antibody compound or enzyme-substrate complex. The concept of structural complementariness

(*18*) translates, into terms more amenable to test, the older " lock and key " simile which Fischer (*19*) suggested to illustrate the specificity of enzyme reactions. It is also evident that the reactive groups brought into close proximity by the complementary nature of the surface of the active molecules represents a more concise version of the receptor and toxaphore hypothesis employed by Ehrlich in his pioneer work in chemotherapy.

An essential part of any hypothesis concerning toxic action is that the poison reacts with groupings on a surface or an interface within the organism and thereby interferes with the vital processes carried out at that surface or interface. The surface may be a membrane to which the toxicant is held by one or more of the interatomic forces which have been sketched in the previous paragraph, with the consequence that the normal functions of that membrane are lost. Or the surface may be that of an enzyme to which the toxicant is held by groups which normally would have functioned in the formation of the enzyme-substrate complex. Or it may be that the attraction of the toxicant to the enzyme so changes the electron distribution that the groupings which normally unite with the substrate are no longer able to do so. The number of such biochemical lesions which could be postulated is legion but experimental tests may indicate those more likely to be nearer reality.

Because of the circumstance that it is possible to examine the properties of enzymes *in vitro* and, in some cases, to isolate and purify specific enzymes by crystallization, the hypotheses concerning enzyme inhibition as a mechanism of toxic action have received greater experimental attention. As an example of the application of this theoretical argument, the inhibition of the enzymic hydrolysis of acetylcholine may be taken, for this inhibition is probably the reason why certain organophosphorus compounds are powerful insecticides. It is now generally agreed that in most animals the transmission of nerve impulses across synapses and from nerve endings to muscle depends upon the momentary release of acetylcholine and that its rapid removal is necessary for normal transmission. This removal is effected by hydrolysis to acetic acid and choline, an hydrolysis catalysed by cholinesterase. If the activity of this enzyme is inhibited, acetylcholine accumulates, nerve conduction is stopped and death follows. The interaction of the enzyme cholinesterase with its substrate, acetylcholine, can be studied in the Warburg respirometer and it has been shown that the rate of hydrolysis of the substrate declines when the concentration of acetylcholine is increased beyond a certain point ; in other words, excess of substrate produces inhibition. The accepted explanation of Murray (*20*) and Haldane (*21*) of such a humped curve is that the substrate is attached at more than one point to the enzyme ; there are at least two active groups on both enzyme surface and on substrate. When one molecule of substrate is so attached, the enzyme-substrate complex (ES) will break down to enzyme and hydrolytic products. If excess substrate is present the chances are increased that each of the two active centres on the enzyme surface will be occupied by a separate molecule of substrate forming the complex ES_2, which does not readily break down to enzyme and hydrolytic products.

What then are these two active groupings of the substrate acetylcholine ?

$$(CH_3)_3N(OH).CH_2.CH_2.O.CO.CH_3$$

The chemist will at once suspect that the quarternary nitrogen with its positive charge is one such group and the ester group with its electrophilic carbon is probably the second group. Hence it would be postulated that there is on the enzyme surface an ionic site and a site which Wilson and Bergmann (*22*) termed the esteratic site. These two active centres are separated by a distance which can be represented, in acetylcholine, by $N–CH_2–CH_2–O–CO$.

Suppose that another compound has in its structure on electrophilic carbon and cationic group so arranged that these groups can approach and match the corresponding groups of the enzyme surface, combination will follow, perhaps with the result that the ability of the enzyme to combine with and to hydrolyse acetylcholine is lost. Many compounds have this suggested structure and shape ; among them are most of the drugs whose physiological symptoms are those of acetylcholine poisoning. As this subject has been well reviewed by Nachmansohn and Wilson (*23*), and by Whittaker (*24*) it is not necessary to cite the many examples.

In vitro experimentation enables a mathematical verification of the above-mentioned ideas of the mechanics of enzyme inhibition. The mathematical treatment of the kinetics generally followed is that of Lineweaver and Burk (*25*), permitting a distinction between competitive and non-competitive inhibition. In competitive inhibition the inhibitor competes with the substrate for the same specific groups of the enzyme, hence the degree of inhibition is dependent on the relative concentration of both substrate and inhibitor. In non-competitive inhibition the extent of inactivation of the enzyme depends only on the concentration of the inhibitor indicating that the latter becomes attached to the enzyme at groups not involved in the enzyme-substrate complex and that the complex enzyme-inhibitor-substrate does not break down.

So far consideration has been given mainly to those toxicants which compete with substrate for position on the enzyme surface. It was early recognized that many enzymes consist of protein to which is attached a non-protein entity readily separated by processes such as dialysis. The latter entity is known as the coenzyme, many of which have been shown to be structurally similar to vitamins or growth factors. Fildes (*26*) defined a growth factor as an essential metabolite which the organism itself cannot synthesize. If now the organism is presented with a compound which is able to replace the growth factor giving a stable but ineffective pseudo-enzyme complex, that compound may prove toxic. This hypothesis was brilliantly applied to the discovery of the sulphonamide drugs which compete with *p*-aminobenzoic acid, a growth factor essential to many organisms for the biosynthesis of the vitamins, folic and folinic acids and pteridines. These vitamins function as the coenzymes of many important enzymes including some thought to be required for the synthesis of nucleic acids.

Certain metal ions including those of magnesium, copper, zinc, and manganese are required by a wide range of enzymes, the ion functioning as a coenzyme. If the metal ion can be removed from participating in enzyme action, for example, by combination with a powerful chelating agent, that chelator will be toxic.

To conclude this discussion of general hypotheses on enzyme inhibition as a mechanism of toxic action, it is appropriate to extend the theory of two-point attachment which was developed above from the example of cholinesterase inhibition. Suppose that a three point attachment of toxicant to either enzyme surface or interface is necessary to achieve enzyme inhibition or disruption of the normal functions of the membrane. Let these three points be represented by a triangle :

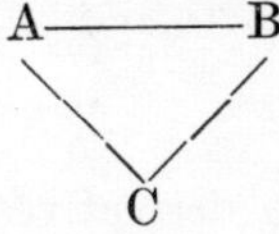

The toxicant is supposed to have three complementary reactive groups so arranged on the molecule that they may be represented by the thumb, and the first two fingers of each hand. Now the thumb (A), forefinger (B) and middle finger (C) of the right hand may be easily placed on the points A, B and C above, but it is impossible to place the thumb and two fingers of the left hand on these points, at least from above the surface of the page. Examples of a pair of compounds having the required spacial arrangement of groups A, B and C are the stereoisomers. It is a frequent occurrence that, of a pair of stereoisomers, only one is toxic. Its optical isomer can become attached by two groups only and, if the attraction is strong enough, it can compete with its toxic isomer for position on the surface and thereby function as an antagonist.

There are many ways in which these various hypotheses of toxic action may be put to test and at least two methods will be met later, though general principles may be outlined here. Often the probable active groupings of the toxicant are apparent on inspection, but it may be possible to determine whether a particular group or atom is involved by a purposeful substitution by another group of similar or dissimilar molecular dimensions. For example, the substitution of the *p*-chlorine atoms of DDT by the larger bromine or iodine atoms results in a marked reduction of insecticidal properties, yet their substitution by methyl or methoxy groups, of similar volumes to that of the chlorine atom, does not seriously reduce insecticidal activity. Clearly, the replacement grouping should not be such that physical properties such as solubility are drastically changed. Incorporation of sulphonic acid groups into DDT, for instance, enhances water solubility but abolishes insecticidal activity. Similarly, the replacement of the chlorine atoms by hydroxyl groups eliminates insecticidal activity—which is not surprising, for the hydroxy group is so strongly polar that the forces which bind the DDT to its biological interface would be drastically changed. This evidence indicates that the size of the grouping in the position occupied by the *p*-chlorine atoms of DDT is an important factor in determining its union with the interface.

The hydrogen atom, particularly if mobile as in keto-enol tautomerism, i.e., the change $R.CO.CH_2- \rightleftharpoons R.C(OH){:}CH-$, may be replaced by a non-mobile group often with a striking change in biological activity as in the dithiocarbamate group of fungicides (see p. 131). Its replacement by the fluorine atom, of closely similar dimensions to hydrogen, will sometimes reveal that the occupation of the position by hydrogen is required for toxicity. This device was brilliantly exploited by the Millers (*27*) in their study of the carcinogenetic activities of the dyestuff butter-yellow, 4-dimethylaminoazobenzene :

$$C_6H_5{-}N = N{-}C_6H_4{-}N(CH_3)_2$$

Of the mono- and polyfluoro derivatives which they prepared, all were active, many more active than the parent compound, in producing liver tumours in the rats to which they were fed. Exceptions were those in which fluorine was attached in both 2- and 6-positions of the benzene ring asterisked above. One of these two positions must be occupied by hydrogen for the compound to be active from which it is inferred that this hydrogen is involved in the attachment of the dyestuff to the protein, a reaction which is among the first stages of the carcinogenic activity of these compounds.

These two examples will suffice to illustrate the devices open to the organic chemist by which he may help the toxicologist not only to elucidate the reaction by which the compound exerts its toxicity but in the search for new compounds of possible use in crop protection.

REFERENCES

(*1*) Holt, J. J. H., *Lancet*, 1916, **5**, 1136.
(*2*) Moore, W., *J. agric. Res.*, 1917, **10**, 365.
(*3*) Tattersfield, F. and Roberts, A. W. R., *J. agric. Sci.*, 1920, **10**, 199.
(*4*) Traube, I., *Liebigs Ann.*, 1891, **265**, 27.
(*5*) Szyszkowski, B. von, *Z. phys. Chem.*, 1908, **64**, 385.
(*6*) Forch, C., *Ann. Phys. Lpz.*, 1905, IV, **17**, 744.
(*7*) Fühner, H., *Biochem. Z.*, 1921, **115**, 235.
(*8*) Uppal, B. N., *J. agric. Res.*, 1926, **32**, 1069.
(*9*) Dagley, S. and Hinshelwood, C. N., *J. chem. Soc.*, 1938, p. 1942.
(*10*) Overton, E., "*Studien über Narkose*", Jena, 1901.
(*11*) Meyer, H. H., *Arch. exp. Path. Pharmak.*, 1899, **42**, 109.
(*12*) Meyer, K. H. and Hemmi, H., *Biochem. Z.*, 1935, **277**, 39.
(*13*) Ferguson, J., *Proc. Roy. Soc. B.*, 1939, **127**, 387.
(*14*) Ferguson, J. and Pirie, H., *Ann. appl. Biol.*, 1948, **35**, 532.
(*15*) Miller, L. P., McCallan, S. E. A. and Weed, R. M., *Contr. Boyce Thompson Inst.*, 1953, **17**, 151, 173, 283.
(*16*) McCallan, S. E. A., Plant Protection Conference, 1956, *Butterworth's Scientific Publ., London*, 1957, p. 77.
(*17*) London, F., *Z. Physik.*, 1930, **63**, 245.
(*18*) Pauling, L., *IX Int. Congr. Chem.*, 1947, Congress Lecture I.
(*19*) Fischer, E., *Ber. dtsch. Chem. Ges.*, 1894, **27**, 2985.
(*20*) Murray, D. R. P., *Biochem. J.*, 1930, **24**, 1890.
(*21*) Haldane, J. B. S., *Enzymes*, Longmans, London, 1930.

(*22*) Wilson, I. B. and Bergmann, F., *J. biol. Chem.*, 1950, **186**, 683.
(*23*) Nachmansohn, D. and Wilson, I. B., *Advanc. Enzymol.*, 1951, **12**, 259.
(*24*) Whittaker, V. P., *Physiol. Rev.*, 1951, **31**, 312.
(*25*) Lineweaver, H. and Burk, D., *J. Amer. chem. Soc.*, 1934, **56**, 658.
(*26*) Fildes, P., *Lancet*, 1940, (i), 955.
(*27*) Miller, E. C. and Miller, J. A., In *Origins of Resistance to Toxic Agents*, Academic Press Inc., New York, 1955, p. 287.

CHAPTER VII

FUNGICIDES

THE SULPHUR GROUP

Sulphur

THE early history of the use of sulphur as a fungicide is lost for even in 1803 Forsyth (*1*) gave the recipe for a tobacco-sulphur-quick lime-elder bud spray for application to fruit. The recommendation by Robertson (*2*) in 1821 of a sulphur-soap spray for use against peach mildew has already been mentioned. Its use as a fungicide became established through its success in combating vine powdery mildew (*Uncinula necator* (Schw.) Burr.) which caused heavy loss in European vineyards until Duchartre, in 1848, recommended sulphur as the remedy.

For dust application " flowers of sulphur ", produced by sublimation, was at one time deemed especially suitable because of its fine state of division. Ground sulphurs of particle size controlled by fractionation by means of a current of air are now available and to these products the name of flowers of sulphur has also been applied, although the designations " Flour Sulphur " or " Wind-blown Sulphur " are happier. Finely-ground sulphurs have an objectionable tendency to " ball ", forming small aggregates difficult to disintegrate. This tendency may be overcome by the addition of a small percentage of inert material, such as kaolin, lead arsenate (*3*), lamp black (U.S.P. 2,037,090), iodine (U.S.P. 2,061,185), dicalcium phosphate (U.S.P. 2,069,568) or zinc oxide (U.S.P. 2,069,710). In a proprietary dust " Kolodust ", ground sulphur is mixed with bentonite-sulphur (" Kolofog " U.S.P. 1,550,650), prepared by grinding the products of the fusion of sulphur and bentonite (*4*).

In addition to sublimed and ground sulphurs, there are sulphur-containing by-products which have been used as fungicides. " Green Sulphur "* (Soufre noir) is the spent oxide from a process by which impure coal-gas is passed through hydrated ferric oxide which is afterwards " revivified " in air, when the oxide is re-formed and free sulphur is deposited. A development of this process (*5*), using suspensions of ferric oxide, yields the " flotation " sulphur known as " Ferrox " sulphur. This product, after washing to remove thiosulphate and thiocyanate, may contain 2–6 per cent ferric oxide. The Thylox liquid purification process, in which the ferric oxide suspension is replaced by solutions of certain thioarsenites, yields " Thylox " flotation sulphur. " Gray flotation " sulphur is also produced in

*The term " green sulphur " has also been applied to sulphurs to which a green pigment (e.g., malachite green) has been added to render the deposit on foliage less conspicuous.

the liquid purification of water gas. These three types of flotation sulphur have been widely investigated by American workers (*6, 7, 8*).

For use as sprays, the finely-divided sulphurs require the addition of a wetter, such as soap used by Robertson. " Wettable " sulphurs have been widely introduced in the United States, an early example being " Dry Mix Sulphur-Lime ", containing sulphur, casein and hydrated lime, used by Farley (*9*) as a substitute for lime sulphur (see p. 112).

A third method of applying sulphur is by volatilization. The practice of painting the hot-water pipes of vine-houses with sulphur, recommended by Bergmann (*10*) in 1852, and proved of practical value by Barker, Gimingham and Wiltshire (*11*), is dependent upon this process. Among the various types of apparatus which have been proposed for the more efficient volatilization of sulphur, are Campbell's Patent Sulphur Vaporizer, the Rota method due to Rupprecht (*12*) in which steam is passed over molten sulphur, and Barker and Wallace's (*13*) method of blowing air over molten sulphur at a temperature below its ignition point (170–230° C).

Action on the Fungus. The mode of action of sulphur upon fungi has been the subject of much investigation and speculation. The earlier views concern physical properties, for Mangani (*14*) considered that electricity generated by the contact of plant and sulphur particle was responsible. Mach (*15*) suggested that the fine sulphur particles focus the sun's ray on the fungus and kill it by heat. Hence, why, in sunlight sulphured vines appear burnt at those places covered by the fungus !

Attention was later directed to the mechanism of the action of sulphur " at a distance ". That fungicidal activity is shown by sulphur applied to a heated surface led to the suggestion that the toxic agent is a volatile derivative formed in the absence of the host plant. The results of Muth (*16*), Vogt (*17*), Goodwin and Martin (*18*) and Sempio (*19*) indicated that this volatile agent is non-acidic and discredited earlier suggestions (*20, 21*) that sulphur dioxide or other oxidation products are responsible. An alternative hypothesis, suggested by Sestini and Mori (*22*), is that the active agent is sulphur vapour. Barker, Gimingham and Wiltshire (see *23*), however, found that the active agent could be removed by filtration through a cellulose pad. As Barker could detect no translocation of the sulphur as a gas or vapour, he considered that the sulphur emanated in the form of minute solid particles, " particulate sulphur ", from the parent mass. Goodwin and Martin (*18*) found that, if the filtering medium was at the same temperature as the parent sulphur, the active agent passed the filter. It is, therefore, unnecessary to postulate any mechanism other than the condensation of sulphur vapour to account for the formation of particulate sulphur. That sulphur exerts a definite vapour pressure at ordinary temperatures was also shown by Tucker (*24*) who suggested that admixture with lamp black or other dark-coloured, heat-absorbent substances, would enhance the fungicidal properties. Taillade (*25*) estimated that the vapour pressure of sulphur at 30.4° C was 3.96×10^{-6} mm. Hg.

It was observed by Pollacci (*26*) and by Selmi (*27*) that sulphur in contact with living plant tissue gave rise to hydrogen sulphide, an observation extended by Barker (*28*) who gave lists of fungi and flowering plants able and unable to reduce sulphur in this manner. McCallan and Wilcoxon (*29*) found that all the species of plants tested evolved hydrogen sulphide when in contact with sulphur.* The fungicidal activity of hydrogen sulphide has been established by many observers, including Pollacci. Marsh (*30*) and McCallan and Wilcoxon found that moist hydrogen sulphide inhibited the germination of the spores of the various fungi examined by a closed-cell technique.

Against actively-growing fungi, hydrogen sulphide has not been proved so effective a fungicide. The observations of Eyre and Salmon (*31*) that hydrogen sulphide solutions are without effect upon *Sphaerotheca humuli* were extended and confirmed by Martin and Salmon (*32*) who, to secure greater sulphide sulphur concentrations, used sodium sulphide solutions. The fungicidal activity of the alkali sulphide was not greater than that of equivalent alkali solutions. Of the simpler sulphur derivatives examined by these investigators, only sulphur in polysulphide form was found to possess toxicity great enough to account for the fungicidal properties of elementary sulphur. In the presence of excess sulphur, or under certain conditions, of oxygen, hydrogen sulphide reacts to form polysulphide sulphur.

The formation of hydrogen sulphide from sulphur in contact with living tissue was considered by de Rey-Pailhade (*33*) to be due to a specific enzyme which he named "Philothion". This product has since been identified with glutathione, a tripeptide of glycine, glutamic acid and cysteine, first isolated by Hopkins (*35*). Barker (*23*) has suggested that hydrogen sulphide is formed by the action of a reducing agent, probably glutathione, the reaction taking place outside the organism. McCallan and Wilcoxon (*29*), on the other hand, adduced evidence that the hydrogen sulphide is produced on or within the fungus spore and not on the sulphur. Martin (*36*) pointed out that hydrogen sulphide is an initial product of the hydrolysis of sulphur, i.e., its interaction with water. The hydrolysis of sulphur in the presence of alkali or lime is the usual method of preparing the polysulphide solutions (see p. 112) and the reaction is readily detected under alkaline conditions or under conditions in which one of the products of hydrolysis is removed as, for example, in the presence of copper foil. The formation of hydrogen sulphide from sulphur in the presence of living tissue may therefore be explained by the intervention of the organism to remove a product of hydrolysis.

In the re-examination of the hypothesis that hydrogen sulphur is the toxic agent responsible for the fungicidal action of sulphur, Miller, McCallan and Weed (*37*) made a direct comparison between the toxicity of hydrogen sulphide and of colloidal sulphur. They found that of eight fungus species tested, five were more sensitive to colloidal sulphur than to hydrogen

*The method of hydrogen sulphide detection employed in the investigation was the blackening of lead acetate paper. The possibility that the agent responsible was a volatile organic sulphide, on the parallel of the production of alkyl arsenides by the action of fungi on arsenical pigments, may be judged improbable for such compounds are not encountered in the comprehensive studies of Challenger and his colleagues (see 34) on biological methylation.

sulphide ; the latter compound could therefore not be responsible for the fungicidal action of sulphur.

If the acidity of the products of hydrolysis of sulphur is incompletely neutralized, oxy-acids of sulphur are formed. The suggestion that the fungicidal activity of sulphur is due to sulphur dioxide and sulphuric acid (see *20, 38, 39*) has already been mentioned. That the toxic derivative is an oxidation product was indicated by the work of Doran (*40*), Young (*21*) and Sempio (*19*), who showed that oxygen is necessary for the prevention of the germination of various fungus spores by sulphur. Sempio suggested that oxygen made the cell membrane more permeable to sulphur vapour and increased the sensitivity of the spore. If, however, the toxic agent is a product of hydrolysis, oxygen may play the part of an accelerator of the reaction. Young (*21*), observing that sulphur suspensions rich in pentathionic acid were more fungicidal than those containing little pentathionic acid, considered that this oxidation product was the active fungicide. His suggestion was critically examined by Roach and Glynne (*41*) and by Wilcoxon and McCallan (*42*) ; in neither study were indications obtained that the toxicity of pentathionic acid solutions is greater than that due to their acidity.

So the theorist has completed the circle and now rejoins either Sempio (*19*) by concluding that sulphur is itself the fungicidal agent, or Salmon and his colleagues (*31, 32*) by concluding that the direct toxicity of sulphur to the powdery mildews is due to sulphur transported in polysulphide form. From the application of his theory of variability (see p. 93) Parker-Rhodes (*43*) concluded only certain sulphur ions, containing two sulphur atoms, can penetrate the spore wall of *Stemphylium sarcinaeforme* ; the hydrodisulphide ($-S_2H'$) is such an ion. On the other hand, Horsfall (*44*) found it reasonable to assume that, because of the close proximity of sulphur to oxygen in the Periodic Table, sulphur vapour would permeate the fungus cell as easily as oxygen. He suggested that, once within the cell, sulphur competes with oxygen as a hydrogen acceptor, disrupting the normal hydrogenation and dehydrogenation reactions of the cell, a view put forward by Miller and his co-workers (*37*) and supported by the study of Sciarini and Nord (*45*) on the effect of elementary sulphur in the dehydrogenation processes of Fusaria. But no explanation of the toxic action of sulphur can be satisfying until it provides for the remarkable selectivity of action shown by sulphur, for this chemical is fatal only to fungi and to a limited range of higher plants and of the Arachnida.

Sulphur provides an excellent example of the significance of particle size in relation to toxicity. Wilcoxon and McCallan (*46*) found, as would be expected on *a priori* reasoning, that the more firmly divided the sulphur the greater its fungicidal effect ; they also established that a comparison of different sulphur samples on the basis of an equal number of particles per unit area revealed no differences in toxicity. Later, Feichtmeir (*47*) extended the tests to include different allotropic forms of sulphur. His mortality-concentration graphs reveal a remarkable series of parallel lines proving that, though the LD50 of the samples is determined by degree of fineness and is lower with those allotropic forms insoluble in carbon disulphide, his spore population had the same variability to all his sulphur preparations. Degree of fineness will also affect the field efficiency of the dust by its influence on adherence to the fungus or plant surface. The greater adherence of finer particles was established by the work of Thatcher and Streeter (*48*), Wilcoxon and McCallan (*46*), Hamilton (*49*) and White (*50*).

Action on the Plant. It is convenient in discussing phytotoxicity to distinguish between acute injury, characterized by the localized killing of plant tissue (necrosis) and popularly termed " scorch " or " burn ", and chronic injury which involves deep-seated physiological changes causing for example, the stunting and premature drop of fruit or leaves. This distinction, although not always clear-cut, is permissible because in many cases the two types can be ascribed to unrelated properties of the chemical responsible for injury.

Acute injury by sulphur is rare in temperate climates but, in hotter climates as in Texas (*51*), severe burning is sometimes caused by the sulphuring of cucurbits for the control of powdery mildew (*Erysiphe cichoracearum* DC.) a danger avoided, in part, by the development of sulphur-resistant varieties (see *52*). Sulphured apples grown in semi-arid districts may develop lesions on the sun-exposed cheek of the fruit, an injury attributed to " sulphur sun scald ". A similar injury to lemons was attributed by Turrell (*53*) to a lowering of the critical temperature at which the fruits are injured by the absorption of sunlight.

On the other hand, elementary sulphur is responsible, even in temperate climates, for the premature defoliation and drop of fruit of " sulphur-shy " varieties. Among apply varieties, Stirling Castle and Lane's Prince Albert are particularly susceptible ; among gooseberries the use of sulphur for the control of American gooseberry mildew (*Sphaerotheca mors-uvae* (Schw.) Berk.) on the varieties Leveller, Early Sulphur and others, results in rapid defoliation. This phenomenon has been studied mainly in connection with the polysulphide sprays (see p. 114).

Phytotoxic action also arises when sulphur-containing fungicides are applied during the blossoming period. MacDaniels and Furr (*54*) have shown that sulphur placed on the stigma of apple blossoms inhibits pollen germination and hence reduces the setting of the fruit. Hamilton (*55*) concluded, however, that, as a good fruit set follows a limited period, at the most two days, of favourable pollinating conditions, the practice of spraying during bloom for apple-scab control is commercially feasible. Instances have been recorded of the beneficial effect of sulphur on the host plant, apart from the prevention of injury by the fungus. It has been claimed that, since the introduction of sulphuring into French vineyards, the average date of picking has been advanced by about a fortnight. De Castella (*56*) urged a return to sulphur in Victoria, not only for the control of Oidium but also on account of its " tonic " effect on the vine. Fryer (*57*) stated that the sulphuring of hops just as the " burr " forms leads to an increased growth of bine, whilst Bobilioff (*58*) observed a stimulus to flowering in Hevea rubber plantations sulphured for mildew control.

REFERENCES

(*1*) Forsyth, W., *A Treatise on the culture and management of fruit trees.* London, 2nd Ed., 1803, p. 358.
(*2*) Robertson, J., *Trans. hort. Soc., London*, 1824, **5**. 175.
(*3*) Streeter, L. R., *Tech. Bull. N.Y. St. agric. Exp. Sta.*, 125, 1927.
(*4*) McDaniel, A. S., *Industr. engng Chem.*, 1934, **26**, 340.
(*5*) Powell, A. R., *Industr. engng Chem.*, 1939, **31**, 789.
(*6*) Sauchelli, V., *Industr. engng Chem.*, 1933, **25**, 363.
(*7*) Smith, M. A., *Phytopathology*, 1930, **20**, 535.
(*8*) Hamilton, J. M., *Phytopathology*, 1931, **21**, 445.
(*9*) Farley, A. J., *Bull. N. J. agric. Exp. Sta.*, 379, 1923.
(*10*) See *Gdnrs'. Chron.*, 1852, p. 419.
(*11*) Barker, B. T. P., Gimingham, C. T. and Wiltshire, S. P., *Rep. agric. hort. Res. Sta. Bristol*, 1919, p. 57.
(*12*) Rupprecht, G., *Angew. Bot.*, 1921, **3**, 253.
(*13*) Barker, B. T. P. and Wallace, T., *Rep. agric. hort. Res. Sta. Bristol*, 1921, p. 122.
(*14*) Mangini, —., *Weinlaube*, 1871, **3**, 18.
(*15*) Mach, E., *Weinlaube*, 1879, **11**, 113.
(*16*) Muth, F., *Wein. u. Rebe*, 1920, **2**, 411.
(*17*) Vogt, E., *Angew. Bot.*, 1924, **6**, 276.
(*18*) Goodwin, W. and Martin, H., *Ann. appl. Biol.*, 1928, **15**, 623 ; 1929, **16**, 93.
(*19*) Sempio, C., *Mem. R. Accad. Ital.*, 1932, **3**, 1.
(*20*) Mach, E. and Portele, K., *Weinlaube*, 1884, **16**, 433.
(*21*) Young, H. C., *Ann. Mo. bot. Gdn.*, 1922, **9**, 403.
(*22*) Sestini, F. and Mori, A., *Staz. sper. agr. Ital.*, 1890, **19**, 257.
(*23*) Barker, B. T. P., *Ann. appl. Biol.*, 1926, **13**, 308 ; *Rep. agric. hort. Res. Sta. Bristol*, 1927, p. 71.
(*24*) Tucker, R. P., *Industr. engng Chem.*, 1929, **21**, 44.
(*25*) Taillade, M., *C. R. Acad. Sci., Paris*, 1944, **218**, 836.
(*26*) Pollacci, E., *Gazz. chim. ital.*, 1875, **5**, 451.
(*27*) Selmi, F., abstr. in *Just's Jber.*, 1876, **4**, 116.
(*28*) Barker, B. T. P., *Rep. agric. hort. Res. Sta., Bristol*, 1929, p. 130.
(*29*) McCallan, S. E. A. and Wilcoxon, F., *Contr. Boyce Thompson Inst.*, 1931, **3**, 13.
(*30*) Marsh, R. W., *J. Pomol.*, 1929, **7**, 237.
(*31*) Eyre, J. V. and Salmon, E. S., *J. agric. Sci.*, 1916, **7**, 473.
(*32*) Martin, H. and Salmon, E. S., *J. agric. Sci.*, 1932, **22**, 595.
(*33*) Rey-Pailhade, J. de, *C. R. Acad. Sci., Paris*, 1888, **107**, 43.
(*34*) Challenger, F., *Adv. in Enzymol.*, 1951, **12**, 429.
(*35*) Hopkins, F. G., *Biochem. J.*, 1921, **15**, 286.
(*36*) Martin, H., *J. agric. Sci.*, 1930, **20**, 32.
(*37*) Miller, L. P., McCallan, S. E. A. and Weed, R. M., *Contr. Boyce Thompson Inst.*, 1953, **17**, 151.
(*38*) Lee, H. A. and Martin, J. P., *Industr. engng Chem.*, 1928, **20**, 23.
(*39*) Reckendorfer, P., *Z. PflKrankh.*, 1935, **45**, 537.
(*40*) Doran, W. L., *Tech. Bull. N. H. agric. Exp. Sta.*, 19, 1922.
(*41*) Roach, W. A. and Glynne, M. D., *Ann. appl. Biol.*, 1928, **15**, 168.
(*42*) Wilcoxon, F. and McCallan, S. E. A., *Phytopathology*, 1930, **20**, 391.
(*43*) Parker-Rhodes, A. F., *Ann. appl. Biol.*, 1942, **29**, 136.
(*44*) Horsfall, J. G., *Principles of Fungicidal Action*, Chronica Botanica Co., Waltham, Mass., 1956, p. 166.
(*45*) Sciarini, L. J. and Nord, F. F., *Arch. Biochem.*, 1943, **3**, 261.
(*46*) Wilcoxon, F. and McCallan, S. E. A., *Contr. Boyce Thompson Inst.*, 1931, **3**, 509.
(*47*) Feichtmeir, E. F., *Phytopathology*, 1949, **39**, 605.
(*48*) Thatcher, R. W. and Streeter, L. R., *Tech. Bull. N. Y. St. agric. Exp. Sta.*, 116, 1925.
(*49*) Hamilton, J. M., *Tech. Bull. N. Y. St. agric. Exp. Sta.*, 227, 1935.
(*50*) White, R. P., *Bull. N. J. agric. Exp. Sta.*, 611, 1936.
(*51*) 39th *Ann. Rep. Tex. agric. Exp. Sta.*, 1926, p. 45.

(52) Yarwood, C. E., *Bot. Rev.*, 1957, **23**, 235.
(53) Turrell, F. M., *Plant Physiol.*, 1950, **25**, 13.
(54) MacDaniels, L. H. and Furr, J. R., *Bull. Cornell agric. Exp. Sta.*, 499, 1930.
(55) Hamilton, J. M., *Tech. Bull. N. Y. St. agric. Exp. Sta.*, 227, 1935.
(56) de Castella, F., *J. Dep. Agric. Vic.*, 1927, **25**, 732.
(57) Fryer, P. J., *Insect Pests and Fungus Diseases of Fruit and Hops*, Cambridge, 1920, p. 627.
(58) Bobilioff, W., abstr. in *Rev. appl. Mycol.*, 1930, **9**, 485.

Lime Sulphur and the Polysulphide Group

The sulphides of the alkali and alkaline earth metals combine with further sulphur to form polysulphides. Of this group the surviving practical fungicide is lime sulphur. The concoction of Forsyth referred to on p. 106, prepared by adding hot water to a mixture which included quick lime and sulphur, was a forerunner of " Eau Grison ". Grison (*1*) in 1852, boiled a suspension of slaked lime and sulphur in water and diluted the supernatant liquid for use against vine powdery mildew. A similar preparation but with salt added was used as a sheep dip and this mixture was found effective by Dusey (*2*) in 1886 against the San José scale, an insect pest. To confer fungicidal properties to this mixture, Lowe and Parrott (*3*) added copper sulphate but, in one case, recorded some control of apple scab (*Venturia inaequalis*) by the lime-sulphur-salt mixture. The addition of salt was later found unnecessary and Parrott, Beach and Sirrine (*4*) found this modified wash effective in the control of apple scab.

Home-prepared lime sulphurs are now replaced by factory-made products, permitting the use of a standardised product, generally of specific gravity not less than 1.28 (60° F). The main compounds present are calcium polysulphides and calcium thiosulphate. The structure of the calcium polysulphide is best represented by $CaS.S_x$, a formula which distinguishes the sulphur combined as true sulphide (the " monosulphide " sulphurs) from the added polysulphide sulphur S_x. On acidification, by the action of carbon dioxide or by the action of a water-soluble salt of a metal yielding an insoluble sulphide, the polysulphide sulphur is precipitated as elementary sulphur. Hence the diluted lime sulphur on application to the plant surface will leave a residue of elementary sulphur, the amount of which is determined by the content of polysulphide sulphur of the spray.

A practical disadvantage of lime sulphur is its large proportion of water adding cost to transport and to containers. An early attempt to remedy this defect was to substitute lime sulphur by liver of sulphur, prepared by the fusion of alkali carbonate and sulphur. The evaporation of lime sulphur leads to extensive decomposition prevented, in U.S.P. 1,254,908, by the addition of a stabilizer such as sugar. The product, Dry Lime Sulphur, not to be confused with Dry Mix Sulphur-Lime (see p. 107) guaranteed to contain 70 per cent calcium polysulphide and 5 per cent sulphur, was considered by Hamilton (*5*) to be the best of the lime sulphur substitutes that he tested. Barium tetrasulphide forms a moderately stable crystalline solid, $BaS_4.H_2O$, which was recommended by Scott (*6*) in 1915 primarily for use against scale insects. A surviving product containing barium polysulphides is " Solbar " (*7*) marketed in Germany for fungicidal and acaricidal use.

A second defect of lime sulphur is a ready reactivity causing it to be "incompatible" with many other pest control chemicals, disadvantageous because the simultaneous application of a fungicide and insecticide is often a practical requirement. This defect is particularly serious with lead arsenate, where the main source of trouble is a reaction leading to the formation of highly phytotoxic thioarsenates (*8*). As this reaction is due to hydrogen sulphide formed by the decomposition of the polysulphides, the prior precipitation of the monosulphide as an insoluble metal sulphide should prevent it. Kearns, Marsh and Martin (*9*) used ferrous sulphate for this purpose and showed that the production of water-soluble arsenical compounds is inhibited. The addition of ferrous sulphate to lime sulphur had been previously recommended, first by Waite (*10*), to reduce lime-sulphur injury; the black tenacious precipitate will rapidly oxidise to yield ferric oxide rendering the spray deposit conspicuous and Dutton (*11*) remarked in 1932 that its use as a "marker" was practised in certain apple-growing districts of the United States.

The application of oil sprays either in combination with lime-sulphur or to foliage already bearing deposits of sulphur is not recommended because of a tendency to leaf damage. The reasons for this phytotoxicity, first recorded by Talbert (*12*) have not been determined; the interaction of sulphur and hydrocarbons under the influence of heat or sunlight is well known though Hoskins (*13*) reported that the addition of sulphur or certain sulphur compounds affected the oxidation of the hydrocarbons neither in sunlight nor in dark, yet leaf injury resulted from the addition of sulphur or the sulphur compounds used.

Action on the Fungus. The direct fungicidal activity of the polysulphide sprays was shown by Salmon and his co-workers (*14*) to depend on their polysulphide sulphur content. Against the hop mildew, the amount of polysulphide sulphur necessary to prove fungicidal was equal in every class of polysulphide compound and was independent of the base present.

The protective fungicidal action of lime sulphur is due to the elementary sulphur formed by the decomposition of the polysulphides after spraying. Because of the rapid action of carbon dioxide on the polysulphides the amount of sulphur deposited will be determined mainly by this reaction and will therefore be equal to the amount of polysulphide sulphur in the spray applied. The fungicidal efficiency, both direct and protective, is thus determined by the content of polysulphide sulphur, a criterion accepted (*15*) for the evaluation of lime sulphurs as spray materials.

Action on the Plant. Lime sulphur damage may be of both acute and chronic types. The former, described by Young and Walton (*16*), takes the form of a scorch, browning the tips and edges of the young leaves and producing necrotic patches adjoining the larger veins of older leaves. When occurring at pre-blossom stages, the check to leaf growth may seriously aggravate frost damage. Wallace (*17*) in his pioneer work on lime sulphur injury, concluded that it occurred before the spray had dried on the leaf and he hinted that the soluble forms of sulphur were more caustic than the dried spray residue. His contemporaries, Waite (*10*) and Volck (*18*), also suggested that the soluble sulphides were the prime cause of injury and recommended the precipitation of the lime sulphur with ferrous sulphate prior to application. Wallace employed carbon dioxide for the same purpose and Safro (*19*) experimented with a number of metallic salts yielding insoluble sulphides. A similar idea seems to underlie the " catalytic sulphur " devised by Peterson (U.S.P. 2,098,257). By such means the monosulphide sulphur is removed as sulphide and the polysulphide sulphur is precipitated as free sulphur ; these mixtures may be regarded as wettable sulphurs rather than polysulphide sprays. Comparison of lime sulphur damage with hydrogen sulphide injury reveals similarities (*20*) and the observations of Berry (*21*) support the view that hydrogen sulphide is a factor responsible for lime sulphur damage.

The phytotoxic action of lime sulphur is aggravated by high temperatures (*22*) and Howlett and May (*23*) stated that 26.5° C is generally regarded as the danger-point. It is probable also that weather conditions prior and subsequent to application affect the response of the plant by their influence on the physiological condition of the foliage. Thus McCallan, Hartzell and Wilcoxon (*20*) found, in some cases, that wilted plants withstood hydrogen sulphide better than turgid plants. Similarly the nutritional status of the plant influences response which, in the case of lime sulphur injury, is generally more apparent upon apple trees deficient in nitrogen. Wallace (*24*) cited evidence of the greater susceptibility to sulphur sprays of trees suffering magnesium deficiency.

Chronic lime sulphur damage has a peculiar characteristic in that the spray causes a premature abscission of leaves and fruit without necessarily causing marked leaf scorch. This phenomenon, already mentioned under sulphur (see p. 110) was originally considered specific to polysulphides, but Martin (*25*) found defoliation of sulphur-shy gooseberries was not prevented by the precipitation of lime sulphur by aluminium sulphate prior to application, in which case, elementary sulphur is the responsible agent. The drop of young fruit when lime sulphur is used against apple scab may prove serious. According to Sanders (*26*) the substitution of Bordeaux mixture for lime sulphur in the Annapolis valley apple-growing district resulted in a doubled output.

The reasons for abscission and sulphur sensitivity are obscure, but the process was shown by Volck (*18*) to be analogous to that causing normal leaf-fall. An entry of the responsible constituent into the leaf seems necessary and Sanders (*27*) concluded that this is effected at the stomata, for when he sprayed lime sulphur only upon the fruit and upper apple leaf surfaces, which bear no stomata, fruit-drop did not occur. He suggested, on finding a degradation of the chlorophyll of the palisade cells, that the fruit and leaves are starved off by the reduction in photosynthetic activity, a reduction of more serious consequences in regions or seasons of low sunlight. Berry (*21*) questioned the application of this hypothesis in cases where the shed leaves show no visible injury and suggested that penetration of hydrogen sulphide is the first part of the process, the hydrogen sulphide accelerating enzyme activity and causing premature defoliation. One merit of this hypothesis is that it brings into line lime sulphur and sulphur damage, for hydrogen sulphide is a product of the decomposition of the former material and is a product of biological activity in the case of sulphur.

REFERENCES

(*1*) See *Gdnrs'. Chron.*, 1852, p. 419.
(*2*) Dusey, F., see Quayle, H. J., *Bull. Calif. agric. Exp. Sta.*, 166, 1905.
(*3*) Lowe, V. H. and Parrott, P. J., *Bull. N. Y. St. agric. Exp. Sta.*, 228, p. 389, 1902.
(*4*) Parrott, P. J., Beach, S. A. and Sirrine, F. A., *Bull. N. Y. St. agric. Exp. Sta.*, 262, p. 37, 1905.
(*5*) Hamilton, J. M., *Tech. Bull. N. Y. agric. Exp. Sta.*, 277, 1935.
(*6*) Scott, W. M., *J. econ. Ent.*, 1915, **8**, 206.
(*7*) "*Bayer*" *Pflanzenschutz Compendium.* Farbenfabriken Bayer A.G., Leverkusen, Undated, p. 47.
(*8*) Goodwin, W. and Martin, H., *J. agric. Sci.*, 1925, **15**, 307, 476.
(*9*) Kearns, H. G. H., Marsh, R. W. and Martin, H., *Rep. agric. hort. Res. Sta. Bristol*, 1934, p. 109.
(*10*) Waite, M. B., *U.S. Dep. Agric., Bur. Plant Ind. Circ.* 58, 1910.
(*11*) Dutton, W. C., *Spec. Bull. Mich. agric. Exp. Sta.*, 218, 1932.
(*12*) Talbert, T. J., *Bull. Mo. agric. Exp. Sta.*, 236, 1926.
(*13*) Hoskins, W. M., *J. econ. Ent.*, 1941, **34**, 791.
(*14*) Eyre, J. V. and Salmon, E. S., *J. agric. Sci.*, 1916, **7**, 473 ; Eyre, J. V., Salmon, E. S. and Wormald, L. K., *ibid.*, 1919, **9**, 283 ; Horton, E. and Salmon, E. S., *ibid.*, 1922, **12**, 269 ; Goodwin, W., Martin, H. and Salmon, E. S., *ibid.*, 1926, **16**, 302 ; 1930, **20**, 489 ; *Ann. appl. Biol.*, 1930, **17**, 127.
(*15*) *Tech. Bull., Minist. Agric. Lond.*, 1, 1951.
(*16*) Young, H. C. and Walton, R. C., *Phytopathology*, 1925, **15**, 405.
(*17*) Wallace, E., *Bull. Cornell agric. Exp. Sta.*, 288, 1910.
(*18*) Volck, W. H., *Bett. Fruit*, 1911, **5**, No. 9, 60.
(*19*) Safro, V. I., *Res. Bull. Ore. agric. Exp. Sta.*, 2, 1913.
(*20*) McCallan, S. E. A., Hartzell, A. and Wilcoxon, F., *Contr. Boyce Thompson Inst.*, 1936, **8**, 189.
(*21*) Berry, W. E., *Rep. agric. hort. Res. Sta. Bristol*, 1938, p. 124.
(*22*) Dutton, W. C., *Spec. Bull. Mich. agric. Exp. Sta.*, 218, 1932.
(*23*) Howlett, F. S. and May, C., *Phytopathology*, 1929, **19**, 1001.
(*24*) Wallace, T., *J. Pomol.*, 1939, **17**, 150.
(*25*) Martin, H., *J. S.-E. agric. Coll. Wye*, 1930, **27**, 182.
(*26*) Sanders, G. E., *Res. Bull. Dosch. Chemical Co., Louisville*, 8, 1922.
(*27*) Sanders, G. E., *Ann. Rep. Nova Scotia Fruit Growers' Assoc.*, 1915, **54**, 72.

THE COPPER FUNGICIDES

Although the fungicidal activity of copper sulphate was first recognized in 1807 by Prévost (*1*) it was not until 1885 that this property was utilized for the protection of foliage from fungal pathogens. The success of the lime-copper sulphate spray then used led quickly to alternative copper-containing fungicides few of which have survived. Indeed, no factory-made substitute has yet succeeded in displacing the first member of the series :—

Bordeaux Mixture. In 1882, Millardet, investigating the downy mildew of the vine (*Plasmopara viticola*), a disease then recently introduced from America, observed that along the roadside at Medoc in the Gironde, certain vines bore leaves though elsewhere they had been defoliated by the disease. He attributed this persistence of the foliage to the practice of daubing the leaves of these particular vines with verdigris or with a mixture of lime and copper sulphate. For this reason, passers-by, thinking the fruit below would be similarly treated, would not touch it for fear of poison. Millardet followed up this slender clue, and in 1885, was able to announce the successful use of a lime-copper sulphate mixture as a fungicide against *Plasmopara* (*2*).

The value of the new fungicide, called Bordeaux mixture from the locality in which it originated, was rapidly established and improvements in the original formula followed at once. Millardet had proposed the admixture of 100 l. of water in which 8 kg. of bluestone was dissolved with 30 l. of a lime suspension prepared from 15 kg. quicklime, giving a thick slush which had to be applied with a brush made of twigs. With the help of a brass-worker named Vermorel, he improved the spraying apparatus then available —the Riley or Barnard nozzle—by introducing the degorger, but spraying with his mixture was still difficult. Experience showed that field efficiency was affected not only by the proportions of copper sulphate and lime used but by the method of mixing. Pickering (*3*) considered that the best method was to add the copper sulphate, in as concentrated a solution as possible, to the lime suspended in the bulk of the water. Butler (*4*) obtained an unsatisfactory spray only when concentrated solutions were mixed direct. The usual stock solution of bluestone, $CuSO_4.5H_2O$, is 1 lb. per gallon, as first suggested by Ricaud (*5*) but nowadays granulated (" Snow ") bluestone is added direct to water in the spray tank and, after agitation, the required amount of lime is washed into the tank through a sieve, giving a product known as " Instant Bordeaux ". The troublesome slaking of quicklime became unnecessary when good grades of hydrated lime became available, due allowance being made for its lower content of calcium oxide. Bordeaux mixture should be applied as soon as possible after preparation for, on standing, crystallization and chemical changes produce a less satisfactory spray.

The correct proportion of copper sulphate and calcium oxide are determined by the chemical reactions which proceed on mixing. An aqueous solution of copper sulphate has a slightly acid reaction, due to small amounts of sulphuric acid produced by hydrolysis. When lime is added the acid is removed as calcium sulphate and hydrolysis proceeds until all the copper sulphate is reacted. Millardet and Gayon (*6*) thought that this reaction proceeded according to the equation :—

$$CuSO_4 + Ca(OH)_2 = Cu(OH)_2 + CaSO_4$$

As late as 1907 Bell and Taber (*7*) claimed that the precipitate of Bordeaux Mixture consists of lime, gypsum (calcium sulphate) and cupric hydroxide. Millardet and Gayon observed, however, that neutralization of the acidity is completed before the full amount of lime required by this equation has been added. Pickering (*3*), Sicard (*8*) and Wöber (*9*) and Martin (*10*) showed neutrality and complete precipitation of the copper are reached at 0.75 equivalent. The initial product of the precipitation of copper sulphate by calcium hydroxide at ordinary temperatures is therefore the trioxy-sulphate $4CuO.SO_3.3H_2O$. The precise degree of hydration of this compound is unknown, but the formula cited is that derived from the phase rule studies of Posnjak and Tunell (*11*) at 50° C.

The ratio of bluestone to quicklime required to precipitate the trioxysulphate is approximately 1 : 0.3; the older so-called neutral Bordeaux mixtures were prepared from this ratio. The ratio more usually employed was, however, 1 : 1 giving a distinctly alkaline Bordeaux mixture. A further interaction is then involved for when lime is added to a suspension of the trioxysulphate its alkalinity slowly disappears. Pickering (*3*) ascribed this change to the production of sulphates more basic than the $4CuO.SO_3$ compound, but his conclusions were disputed by Sicard (*8*) and by Wöber who advanced divergent suggestions concerning the nature of these more basic sulphates. Martin (*11*) returned to the older view that the precipitate of freshly-prepared Bordeaux mixture is cupric hydroxide. His conclusion is based on the observation that permanent alkalinity is not reached until an equimolecular quantity of lime has been added and upon the analogous cupric chloride-lime interaction. Cupric hydroxide is a blue gelatinous precipitate which, when prepared from the chloride, undergoes dehydration to form black cupric oxide. Martin suggested that the blue precipitate of Bordeaux mixture is stabilized by adsorbed calcium sulphate for, if the moist or dried precipitate is washed with carbon dioxide-free water, it undergoes dehydration. When applied to foliage, however, the Bordeaux deposit does not blacken, and Martin indicated that the cupric hydroxide-calcium sulphate complex undergoes an *in situ* change, associated with atmospheric carbon dioxide, to a compound not converted to cupric oxide on leaching with water.

The weathering of Bordeaux mixture was studied by Wilcoxon and McCallan (*12*), who support the view that the precipitate is an adsorption complex (or solid solution).* They found a continual increase in the copper content of the spray deposit, a change in composition accompanied by an increase in the amount of copper dissolved by the leaching water, a change not displayed in the washing of Bordeaux mixture in bulk.

Evidence concerning the chemical nature of the Bordeaux precipitate was also derived by Martin, Wain and Wilkinson (*13*) from the study of the inhibition of germination of spores of *Stemphylium sarcinaeforme*. Most of the simple uncoordinated cupric salts gave probit inhibition—log concentration lines of parallel slope indicating a common variability (see p. 92) of the spores to these compounds which included cupric sulphate and Bordeaux mixture. Towards basic salts, however, the variability shown by the spores was higher and in the rates anticipated by the Parker-Rhodes hypothesis, namely, $4^2 : 4^2 : 2^2 : 1$ for the basic sulphate $[Cu\{(OH)_2Cu\}_3]SO_4$, the basic chloride $[Cu\{(OH)_2Cu\}_3]Cl_2$, the basic carbonate $[Cu\{(OH)_2Cu\}]CO_2$, and the sulphate $CuSO_4$, respectively. It will be noted that in this series the copper atoms of the respective molecules are in the ratio 4 : 4 : 2 : 1. Because the variability shown to Bordeaux mixture is that shown to cupric sulphate, the inference is that the freshly-prepared product of Bordeaux mixture is not a basic salt.

The Bordeaux precipitate, if allowed to stand with excess lime, ultimately

*If, by solid solution, Wilcoxon and McCallan infer a distribution of the calcium and sulphate ions in the space lattice of the cupric hydroxide, the molecular dimensions of the invading ions seem to require too large a strain of the lattice to warrant this inference.

forms a violet-coloured precipitate which is probably a calcium cuprite of the type CuO.3CaO postulated by Pickering (*3*) and Sicard (*8*). The changes which proceed are described by Burchfield and Schechtman (*14*).

Burgundy Mixture. In 1887 Masson (*15*), finding difficulty in procuring a suitable supply of quicklime for the preparation of Bordeaux mixture, proposed its substitution by washing soda (sodium carbonate). This substitute was named Burgundy mixture, from the district in which it was first employed. A similar mixture was recommended for use in Ireland for potato spraying (*16*) under the name Soda Bordeaux.

The main chemical reaction taking place when copper sulphate and sodium carbonate solutions are mixed is one of neutralization, the sodium carbonate reacting with the free acid formed from the hydrolysis of the copper sulphate. Carbon dioxide is evolved and produces secondary reactions. Pickering (*17*) studying the relative amount of sodium carbonate required to secure (almost) complete precipitation of the copper and also permanent alkalinity, found that these points occur simultaneously when 1.6 moles of sodium carbonate are added to 1 mole of copper sulphate. Because of the coincidence of these two points, a feature not shown by the lime-copper sulphate mixture, Pickering concluded that the reaction results in the formation of the $10CuO.4CO_2$ basic carbonate and sodium sulphate, while the carbon dioxide evolved is combined with the sodium carbonate to form the bicarbonate. Mond and Heberlein (*18*) concluded that neutrality is reached at the molar ratio of 1 : 1. Such a mixture contains copper in solution, due, as these workers showed, to the presence of carbon dioxide. When the latter was removed no soluble copper could be detected. The ratio proposed by Pickering as representing the point of neutralization was shown by these workers to be the point of minimum soluble copper, due to the removal of carbon dioxide by excess sodium carbonate, the copper in solution thereby being precipitated.

According to Pickering, the precipitate of the 1 : 1.84 mixture is the basic carbonate $5CuO.2CO_2$, a bulky blue precipitate which gradually changes on standing in the mixed liquid to malachite, $2CuO.CO_2$, a dense green precipitate unsuitable as a spray material. Mond and Heberlein considered that the precipitate obtained at 15° C with the 4 : 4.25 mixture contained various basic copper sulphates and a basic copper carbonate. They thought that no well-defined compound of copper appears but that the composition of the precipitate varies according to the conditions of precipitation. The precipitate undergoes the change from colloidal to crystalline, the change being retarded by the presence of free sodium carbonate. Hsu (*19*) concluded that Pickering's ratio 1 : 1.6 is reduced to 1 : 1.1 if the carbon dioxide is removed from the reaction mixture, that the gelatinous precipitate is a mixture of basic carbonate and sulphate, and that the ultimate product corresponds in composition to malachite. The mechanism of the formation of malachite from the basic precipitate was studied by Hepburn (*20*).

The Cuprammonium Group. When ammonium hydroxide is added to a solution of copper sulphate, a basic copper compound is thrown down which, with excess of ammonia, is re-dissolved to form the deep blue solution of cuprammonium sulphate. This salt is subject to electrolytic dissociation to cuprammonium and sulphate ions; the cuprammonium ion may be represented by the formula $Cu(NH_3)_4^{++}$ (see, however, *21*). The cuprammonium salts are not hydrolysed to form sulphuric acid. The two factors to which the spray injury characteristic of solutions of copper salts are ascribed, namely, free soluble copper and free sulphuric acid, are therefore absent in the cuprammonium sulphate. As, however, cuprammonium sulphate is decomposed after spraying to form basic copper compounds, it will possess fungicidal properties.

The deep blue solution obtained by the addition of excess of ammonium hydroxide to solutions of copper sulphate was first proposed as a spray by Audoynaud (*22*) in 1885. This solution was named "Eau Celeste"; it has since been called "**Azurin**". The solution is unstable and requires the presence of a large excess of ammonia to prevent precipitation. Audoynaud suggested this material for spraying late in the season, for it does not leave the objectionable deposit upon the fruit which results with Bordeaux mixture. Unfortunately the material often caused severe foliage injury, attributed to the ammonium sulphate in solution, a known cause of leaf "scorch".

To obviate this scorching Patrigeon (*23*) proposed the use of a solution of the precipitate of Burgundy mixture in excess of ammonium hydroxide. This wash was at one time widely used in the United States under the name of "**Modified Eau Celeste**". To simplify its manufacture, Gastine (*24*) proposed the direct solution of basic copper carbonate in ammonium hydroxide. A closely-related product, obtained by the solution of copper carbonate in ammonium carbonate solution, was suggested by Chester (*25*). The basic copper carbonate of commerce, $Cu(OH)_2.CuCO_3$, dissolves in ammonium hydroxide in the presence of ammonium carbonate to give a cuprammonium carbonate. According to Ephraim (*26*), the cuprammonium carbonate stable at ordinary temperatures yields the ion $Cu(NH_3)_2^{++}$. Solutions of basic copper carbonate in ammonium hydroxide or carbonate were at one time popular in the United States under the name "**Cupram**".

Ammonium carbonate may also be used in place of the hydroxide with copper sulphate. A mixture of copper sulphate and ammonium carbonate, ground ready for solution, was suggested by Johnson (*27*) in 1890. Johnson's mixture is the forerunner of **Cheshunt Compound** (*28*), a mixture of 2 parts by weight bluestone and 11 parts ammonium carbonate, recommended for the control of "damping-off" diseases. The chemistry of these mixtures is complicated for the hard vitreous ammonium carbonate of commerce is a mixture of the bicarbonate, NH_4HCO_3, and the carbamate $(NH_4)CO_2(NH_2)$. On exposure to air, the decomposition of the latter compound to form the bicarbonate enables the material to be easily powdered. The cuprammonium solutions, after spraying, all decompose to form residues of basic copper compounds of properties akin to those of the Bordeaux and Burgundy precipitates.

Miscellaneous Basic Copper Sprays. Of uncompounded basic copper derivatives, **verdigris** has been long used in France where it is made by the action of the acid " marc " of grapes on sheet copper. Verdigris consists mainly of basic cupric acetates, the composition of which is determined by conditions during manufacture. It should not be confused with the patina of basic sulphate, carbonate or chloride on copper exposed to atmospheric corrosion. Although the verdigris preparations (" Vert de Montpelier " and " Verdet gris extra sec ") appear to be good fungicides they are too expensive for extensive use. Butler and Smith (*29*), however, recommended the basic acetates, especially where a colourless spray deposit is required.

Of compounded products containing basic copper derivatives, the earliest were pastes. Bordorite, based on Pickering's work (*30*), has been largely supplanted by " colloidal " pastes such as **" Bouisol "** (B.P. 392,556) made from the basic chloride $CuCl_2.3Cu(OH)_2$. This compound, frequently called the oxychloride, is readily formed by the action of air on cupric chloride solutions or scrap copper (see *31*). The disadvantage that the pastes, being corrosive, require packing in expensive glass or wooden containers led to the introduction of the dispersible powders of which many are now on the market.

Normal Copper Derivatives. By accident or on the basis of hypotheses of the mode of action of the copper fungicides, it has been found that a wide range of fungicidal copper compounds exists outside the range of basic derivatives used by earlier workers. It was, for example, observed that the use, with Bordeaux mixture, of supplements such as sulphite lye or molasses (*32*), which contain reducing sugars, led to a slow reduction of the precipitate to cuprous oxide. Contrary to Holland, Dunbar and Gilligan's suggestion (*33*), Martin (*34*) found that this change was not accompanied by a reduction of fungicidal efficiency and tests (*35*) showed that cuprous oxide is itself fungicidal. **Cuprous oxide** was, at the same time, found by Horsfall (*36*) to be an effective fungicide for the control of " damping-off " of vegetable seedlings. Dispersible cuprous oxide powders are now in wide use as foliage protectants, a familiar British example being **" Perenox "**.

Of cupric derivatives the phosphate $Cu_3(PO_4)_2$, silicate and oxide were used by Goldsworthy and Green (*37*) who found the incorporation of bentonite and lime advantageous to improve tenacity (U.S.P. 1,954,171, 1,958,102, 2,004,788). Sessions (*38*) selected the silicate as the most suitable buffer to counter acidity, adjusting its buffering power by the introduction of the ammonium ion to form complex copper ammonium silicates. The product, introduced commercially under the name **" Coposil "** (U.S.P. 2,051,910) also contained a small proportion of zinc ammonium silicates. A somewhat similar product resulted from Nikitin's work (*39*) upon the copper zeolites. The zeolites are complex silicates possessing base exchange properties utilized, for example, in the softening of water. The sodium zeolite of the water softener removes the calcium and magnesium ions from the compounds responsible for water hardness which are converted to the corresponding sodium salts. In the same way foreign bases will liberate cupric ions from the copper zeolite.

It is evident that the range of copper compounds toxic to fungi is inexhaustible but their practical use is limited by phytotoxicity or by lack of means of securing a spray deposit of adequate tenacity. Martin, Wain and Wilkinson (*40*) concluded that the compounds tested fell into four broad groups :—

(a) Compounds active through the cupric ion and including cupric chloride, dinitrocresylate, phosphate, sulphanilate, sulphate and sulphide, cuprous oxide, Bordeaux and Burgundy mixtures.

(b) Compounds in which copper is held by coordination and from which the cupric ion (or other fungicidal radical) is liberated by the intervention of the spore and of which fungicidal value is governed by the stability of the coordination complex. The basic compounds fell into this group.

(c) Compounds which appear to inhibit spore germination without the intervention of the cupric ion, e.g., copper sebacate and phthalate.

(d) Compounds so insoluble or so stable that they are virtually non-fungicidal; e.g., cuprous iodide and thiocyanate, cupric ferrocyanide, quinaldinate, the copper salt of salicylaldoxime, ethylenediaminocupric dinitrocresylate. The basic arsenate used by Waters *et al.* (*41*) as an insecticide for which fungicidal properties were claimed, came into this group.

The present trend is towards the discovery of copper fungicides in which other fungicidal radicles are incorporated, or of special properties such as oil solubility. In the first group, for example, may be placed the attempted combination of the fungicidal but phytotoxic chromate with copper in the complex **copper-zinc-chromates**, described by Harry *et al.* (*42*), or the utilization of the known fungicide 8-hydroxyquinoline (oxine) in the co-ordinated **copper oxinate**, first used on foliage by Meyer (*43*) and later recommended by Powell (*44*).

Of oil-solubles copper compounds the naphthenates are powerful fungicides but their phytotoxicity limits their use to growing plants. de Ong (*45*) used copper resinate dissolved in pine oil, a mixture found phytotoxic by Hickman, Marsh and Wilkinson (*46*) who selected the coordinated complex, copper 3 : 5-di-*iso*propyl salicylate, for field trials. The chemistry of this compound, developed for use in lubrication under the name DIPS, though fascinating, is almost unknown, an ignorance which will delay its development as a fungicide. In general, however, the use of oil-soluble compounds is handicapped by the phytotoxicity of both oil carrier and copper fungicide.

Copper Dusts. Copper preparations for application as dusts may be grouped into two categories. Firstly, there are the basic copper derivatives in finely-divided form which were introduced early in the history of the copper fungicides, e.g., David's powder. In general, dusts of this type have proved of inferior adherent properties and liable to cause damage, disadvantages which are countered, to some extent, by the addition of lime.

The second group of copper dusts are usually called copper-lime dusts and were originated by Sanders and Kelsall (*47*), whose first dusts contained arsenicals which were later omitted. Sanders (*in litt.* 9/1/31) was interested in the production of a form of Bordeaux which could be made by direct addition to water in the spray tank and experimented with a mixture of hydrated lime and mono-hydrated copper sulphate. On addition to water, however, this product curdled and he then tried the mixture as a dust. In the presence of moisture upon the plant surface, the components of the dust interact with the *in situ* formation of a Bordeaux precipitate. Streeter, Mader and Kokoski (*48*) investigated the optimum conditions of humidity for the formation of an adherent film and demonstrated the importance of applying the copper-lime dust to moist foliage.

The relative efficiencies of copper dusts and sprays in the control of potato diseases was investigated by Muskett (*49*), Murphy and McKay (*50*) and Boyd (*51*). Boyd found that, under field conditions, about 44 per cent more copper per unit amount used per acre was retained by the foliage sprayed with Bordeaux mixture than by the foliage dusted with a copper-lime dust. Experience indicates that, under general conditions, the copper dusts are unable to replace sprays in the routine fungicide programme but that they are excellent supplementary fungicides for use should weather or other conditions render spraying impracticable (*52*). Copper dusts may also be of particular value for late application to tree fruit, for, at this stage, the leaves and fruit appear to be less susceptible to damage by copper than earlier in the season.

REFERENCES

(*1*) Prevost, B., *Mémoire sur la cause immédiate de la Carie ou Charbon des Blés,* Montauban, 1807.
(*2*) Millardet, A., *J. Agric. prat.*, 1885, **49**, 513, 801.
(*3*) Pickering, S. U., *J. chem. Soc.*, 1907, **91**, 1981, 1988.
(*4*) Butler, O., *Phytopathology*, 1914, **4**, 125.
(*5*) Ricaud, J., *J. Agric. prat.*, 1887, **51**, 517.
(*6*) Millardet, A. and Gayon, U., *J. Agric. prat.*, 1885, **49**, 707 ; 1887, **51**, 698, 728, 765.
(*7*) Bell, J. M. and Taber, W. C., *J. phys. Chem.*, 1907, **11**, 632.
(*8*) Sicard, L., *Ann. Éc. Agric. Montpellier*, 1915, **14**, 212.
(*9*) Wober, A., *Z. PflKrankh.*, 1919, **29**, 94.
(*10*) Martin, H., *Ann. appl. Biol.*, 1932, **19**, 98.
(*11*) Posnjak, E. and Tunell, G., *Amer. J. Sci.*, 1929, **18**, 1.
(*12*) Wilcoxon, F. and McCallan, S. E. A., *Contr. Boyce Thompson Inst.*, 1938, **9**, 149.
(*13*) Martin, H., Wain, R. L. and Wilkinson, E. H., *Ann. appl. Biol.*, 1942, **29**, 412.
(*14*) Burchfield, H. P. and Schechtman, J., *Contr. Boyce Thompson Inst.*, 1955, **18**, 215.
(*15*) Masson, E., *J. Agric. prat.*, 1887, **51**, (1) 814.
(*16*) *Leaflet Bd. Agric. Tech. Inst. Ireland*, 14.
(*17*) Pickering, S. U., *J. chem. Soc.*, 1909, **95**, 1409.
(*18*) Mond, R. L. and Heberlein, C., *J. chem. Soc.*, 1919, **115**, 908.
(*19*) Hsu, C. T., *J. appl. Chem.*, 1956, **6**, 84.
(*20*) Hepburn, J. R. I., *J. chem. Soc.*, 1925, **127**, 1007.
(*21*) Dawson, H. M., *J. chem. Soc.*, 1906, **89**, 1666.
(*22*) Audoynaud, A., *Progr. agric. vitic.*, 1885, **6**.
(*23*) Patrigeon, G., *J. Agric. prat.*, 1887, **51**, (1) 882.
(*24*) Gastine, G., *Progr. agric. vitic.*, 1887, **8**, 114.
(*25*) Chester, F. D., *Rep. Del. agric. Exp. Sta.*, 1891, **4**, 71.
(*26*) Ephraim, F., *Ber. dtsch. chem. Ges.*, 1919, **52**, B, 940.
(*27*) Johnson, S. W., *Rep. Conn. agric. Exp. Sta.*, 1890, p. 113.
(*28*) Bewley, W. F., *J. Minist. Agric.*, 1921, **28**, 653.
(*29*) Butler, O. and Smith, T. O., *Phytopathology*, 1922, **12**, 279.
(*30*) Bedford, Duke of, and Pickering, S. U., *11th Rep. Woburn exp. Fruit Farm*, 1910, p. 59.
(*31*) *Brit. Intell. Obj. Sub-comm., Final Report*, 1480, 1947.
(*32*) Raleigh, W. P., *Phytopathology*, 1933, **23**, 29.

(33) Holland, E. B., Dunbar, C. O. and Gilligan, G. M., *Bull. Mass. agric. Exp. Sta.*, 252, 1929.
(34) Martin, H., *Ann. appl. Biol.*, 1933, **20**, 342.
(35) Horsfall, J. G., *Tech. Bull. agric. Exp. Sta.*, 198, 1932 ; *Bull. agric. Exp. Sta.*, 615, 1932.
(36) Horsfall, J. G. and Hamilton, J. M., *Phytopathology*, 1935, **25**, 21.
(37) Goldsworthy, M. C. and Green, E. L., *Phytopathology*, 1933, **23**, 561.
(38) Sessions, A. C., *Industr. engng Chem.*, 1936, **28**, 287.
(39) Nikitin, A. A., Thesis Columbia Univ., 1937 ; U.S.P. 2,157,861 (with Adams, J. F.).
(40) Martin, H., Wain, R. L. and Wilkinson, E. H., *Ann. appl. Biol.*, 1942, **29**, 412.
(41) Waters, H. A., Witman, E. D., and DeLong, D. M., *J. econ. Ent.*, 1939, **32**, 1426.
(42) Harry, J. B., Wellman, R. H., Whaley, F. R., Thurston, H. W. and Chandler, W. A., *Contr. Boyce Thompson Inst.*,- 948, **15**, 195.
(43) Meyer, A., *Rev. vitic., Paris*, 1932, **77**, 117.
(44) Powell, D., *Phytopathology*, 1946, **36**, 572.
(45) de Ong, E. R., *Phytopathology*, 1935, **25**, 368.
(46) Hickman, C. J., Marsh, R. W. and Wilkinson, E. H., *Ann. appl. Biol.*, 1943, **30**, 179.
(47) Sanders, G. E. and Kelsall, A., *Proc. Nova Scotia Ent. Soc.*, 1918, p. 32.
(48) Streeter, L. R., Mader, E. O. and Kokoski, F. J., *Phytopathology*, 1932, **22**, 645.
(49) Muskett, A. E., *J. Minist. Agric. N. Ireland*, 1929, **2**, 54 ; 1931, **3**, 117.
(50) Murphy, P. A. and McKay, R., *J. Dep. Agric. Ireland*, 1933, **32**, 30.
(51) Boyd, O. C., *Bull. Cornell agric. Exp. Sta.*, 451, 1926.
(52) Petherbridge, F. R., *J. Minist. Agric.*, 1933, **40**, 209.

Action of the Copper Fungicides on the Fungus

As the copper fungicides are mainly used as protective fungicides, there is no need to consider separately their direct activity. The problem is reduced to the prevention of infection by spores settling on the spray-coated plant surface.

The process of the establishment of fungal infection, discussed in Chapter II, is so complicated that many ways could be suggested in which the fungicides might play an indirect part in preventing infection. Rumm (*1*) suggested that the more vigorous state of the sprayed vine might enable it to withstand fungal attack. Swingle (*2*), among many hypotheses of the mode of action of Bordeaux mixture, included the possibility that copper absorbed by the plant tissue might prevent infection either by killing the germ tube or by preventing penetration by some chemotropic influence or by interference with enzymatic processes operating in penetration.

Hypotheses based on an indirect action through the host plant fail, however, to explain inhibition of germination in the absence of the host plant. This action was known to Prévost (*3*) who showed that copper sulphate, at a dilution of 1 : 400,000, prevented the germination of " smut " spores. This high toxicity of copper in solution has encouraged many hypotheses which suggest the formation of soluble copper from the relatively insoluble copper fungicide. The solubility of the Bordeaux precipitate might, for example, be sufficient of itself to prevent germination, a hypothesis examined by Branas and Dulac (*4*) who concluded that, although freshly-prepared Bordeaux mixture contained enough soluble copper to be toxic to spores of *Plasmopara viticola*, the dried precipitate is too insoluble in distilled water to give toxic copper concentrations. Delage (*5*) also favoured this hypothesis and re-introduced the idea of cumulative action put forward by earlier workers (e.g., *6*), that minute traces of copper appearing in solution would be absorbed by the spore until a toxic accumulation is eventually reached. But, even if this possibility of accumulation be accepted the small solubility product of copper compounds known to be fungicidal does not encourage the acceptance of hypotheses based solely on solubility in water. Most investigators have considered it necessary to look for other agencies by which copper is brought into an active fungicidal form.

Millardet and Gayon (*7*) observed the formation from Bordeaux deposits, after carbonation of the excess lime, of soluble copper by the action of the carbon dioxide and slight traces of ammonium salts present in rain water and dew. Many investigators, in particular Pickering (*8*), accepted the view that atmospheric agencies provided a satisfactory explanation for the formation of the active fungicide. But the role of carbon dioxide in the fungicidal action of cuprous oxide, for example, is obscure and, even in the case of basic copper compounds, there are reasons for regarding the hypothesis as inadequate. Gimingham (*9*), for example, in repeating Pickering's experiments, found that the whole of the copper dissolved from the Bordeaux precipitate by the action of carbon dioxide is reprecipitated on removal of the carbon dioxide. Gimingham therefore concluded that, although the action of excess carbon dioxide brings copper into solution, it appears impossible to assign the fungicidal activity of Bordeaux mixture to copper sulphate formed by atmospheric carbon dioxide. More recently Reckendorfer (*10*) has supported the carbon dioxide hypothesis suggesting that copper appears in solution as the sulphate and bicarbonate but, as Wilcoxon and McCallan (*11*) pointed out, since the normal cupric carbonate, $CuCO_3$, is not obtained by ordinary procedures, it is unlikely that the bicarbonate, $Cu(HCO_3)_2$, could be formed under the low carbon dioxide pressure existing in the atmosphere. The latter investigators confirmed the formation of soluble copper from weathered Bordeaux precipitate and suggested that this arises through changes in composition of the adsorption complex on leaching. The removal of calcium and sulphate from the complex leaves a precipitate relatively richer in copper, the solubility of which might well be expected to increase to a point when it might become a factor in the biological activity of Bordeaux mixture.

The other agency concerned in the *in vitro* action of copper fungicides is the spore itself and the suggestion that spore excretions are involved in the fungicidal action of Bordeaux mixture was first made by Barth (*12*) and by Swingle (*2*) receiving support later from Clark (*13*), Ruhland (*14*) and Schander (*15*). This hypothesis was examined by Barker and Gimingham (*16*) who concluded that such a solvent action cannot be denied while McCallan (*17*) concluded that agents other than spore excretions are non-essential for the functioning of Bordeaux mixture. In conjunction with Wilcoxon (*18*) he demonstrated the presence of malic acid and of amino acids in washings from spores of *Neurospora sitophila* (Mont.) Shear and Dodge. As the spore excretions are almost neutral, contrary to the suggestion of Dubacquié (*19*) that the solution of copper is due to acid phosphates, McCallan and Wilcoxon considered that the salts of hydroxy-acids such as malate and of amino-acids present in spore excretions, act on the

Bordeaux deposit to form soluble toxic copper hydroxy- and copper amino-salts. Wain and Wilkinson (*20*) confirmed the solution of copper from dried Bordeaux deposit by the lysate of spores of *N. sitophila* and examined the solvent action of a range of compounds likely to be present in the lysate. They found solution to be independent of acidity and to occur only when the formation of coordinated complexes was possible. Active solvents included amino-, hydroxy-, and certain dicarboxylic acids and their salts, of which, in view of the earlier work of McCallan and Wilcoxon (*18*) and of Martin, Wain and Wilkinson (*21*), the malates are of special interest. Wain and Wilkinson suggested that the soluble cupric-complex enables transportation of the copper to the cell-wall and that dissociation of the complex then makes it possible for the active toxicant to be removed by the spore and enables the reversible reactions involved to continue until a toxic concentration is reached within the spore.

In practice, however, the host plant may well play a part in the mechanism of fungicidal action, either directly, or indirectly through modification of the processes concerned in *in vitro* action. Barth (*12*) suggested that host plant secretions might render soluble or available the copper of the spray deposit while Clark (*13*) and Bain (*22*) regarded the solution of copper by plant secretions the cause of spray damage by Bordeaux mixture. Barker and Gimingham (*16*) showed that substances exuded from mechanical injuries to the leaf are capable of dissolving copper from Bordeaux precipitates, but being unable to secure leaves certainly free from previous injury, they could not decide whether soluble copper would be formed on healthy leaves. DeLong, Reid and Darley (*23*) by the examination of the action of expressed plant juices and of sugar solutions on the Bordeaux precipitate, obtained evidence in support of the host plant excretion hypothesis, an idea on which Sessions (*24*) seems to have based his work leading to the development of the complex silicates as fungicides. He traced some interesting relationships between the acidity of the cell sap of the host plant and its response to the fungicide.

If it be accepted that fungicidal action follows the formation of soluble or available copper by any of three main agencies : (1) atmospheric carbon dioxide and ammonium salts dissolved in rain water or dew ; (2) secretions from the fungus ; (3) secretions from the healthy or wounded surface of the host plant ; in what forms does the fungicidally active copper appear ? It seems unnecessary to look further than the cupric ion but for indirect evidence which suggests more fungicidal copper derivatives. Horsfall, Marsh and Martin (*25*) pointed out that the cupric ion concentration of the solutions of the copper glycine complex examined by McCallan and Wilcoxon (*18*) is probably below that of equivalent cupric sulphate solutions. Yet at low concentrations the solutions of the copper-glycine complex exhibited greater fungicidal powers than the copper sulphate solutions, an indication that there exists, in the former, a material of greater inherent toxicity than the cupric ion. Martin, Wain and Wilkinson (*21*) not only found that the sebacate and phthalate are fungicidal apparently without the intervention of the cupric ion, but obtained some evidence that the cuprimalates are of greater direct availability to the spore than is the cupric ion.

In the case of the organomercury compounds it was suggested as long ago as 1928 by Walker (*26*) that phenylmercuric chloride is more toxic to the protozoon *Colpidium colpeda* than mercuric chloride because it is more lipoid soluble. It is perhaps significant that copper compounds such as the oxinate and the sebacate which exhibit a greater inherent fungitoxicity than the cupric ion are also fat soluble. Durkee (*27*) therefore examined a range of coordinated copper compounds and found that those which are fungicidal are lipoid soluble. Hence it may be suggested that the function of the spore exudate is to provide for the formation of a lipoid-soluble copper coordination complex which can penetrate the cell wall ; dissociation of the complex within the cell will give rise to cupric ions which would most certainly intervene in metabolism in innumerable ways capable of inhibiting the germination of the spore.

REFERENCES

(*1*) Rumm, C. *Ber. dtsch. bot. Ges.*, 1895, **13**, 189.
(*2*) Swingle, W. T., *Bull. U.S. Dep. Agric. Div. Veg. Phys. Path.*, 9, 1896.
(*3*) Prévost, B., *Mémoire sur la cause immédiate de la Carie ou Charbon des Blés*, Montauban, 1807.
(*4*) Branas, J. and Dulac, J., *C. R. Acad. Sci. Paris*, 1933, **197**, 938, 1245.
(*5*) Delage, B., *Chim. et Industr.*, 1932, **27**, 853.
(*6*) Pickering, S. U., *J. agric. Sci.*, 1912, **4**, 273.
(*7*) Millardet, A. and Gayon, U., *J. Agric. prat.*, 1887, **51**, 56, 698.
(*8*) Bedford, Duke of, and Pickering, S. U., *11th Rep. Woburn exp. Fruit Farm*, 1910, p.22.
(*9*) Gimingham, C. T., *J. agric. Sci*, 1911, **4**, 69.
(*10*) Reckendorfer, P., *Z. PflKrankh.*, 1936, **46**, 418.
(*11*) Wilcoxon, F. and McCallan, S. E. A., *Contr. Boyce Thompson Inst.*, 1938, **9**, 149.
(*12*) Barth, M., *Die Blattfallkrankheit der Reben und ihre Bekämpfung*, 4th Ed. Gebweiler, 1896.
(*13*) Clark, J. F., *Bot. Gaz.*, 1902, **33**, 26.
(*14*) Ruhland, W., *Arb. biol. Abt.* (*Anst. Reichsanst.*) *Berlin*, 1904, **4**, 157.
(*15*) Schander, R., *Landw. Jb.*, 1904, **33**, 517.
(*16*) Barker, B. T. P. and Gimingham, C. T., *J. agric. Sci.*, 1911, **4**, 76.
(*17*) McCallan, S. E. A., *Mem. Cornell agric. Exp. Sta.* 128, p. 25, 1930.
(*18*) McCallan, S. E. A. and Wilcoxon, F., *Contr. Boyce Thompson Inst.*, 1936, **8**, 151.
(*19*) Dubacquié, J., *C. R. Acad. Agric. Fr.*, 1934, **20**, 944, 1063.
(*20*) Wain, R. L. and Wilkinson, E. H., *Ann. appl. Biol.*, 1943, **30**, 379.
(*21*) Martin, H., Wain, R. L. and Wilkinson, E. H., *Ann. appl. Biol.*, 1942, **29**, 412.
(*22*) Bain, S. M., *Bull. Tenn. agric. Exp. Sta.* 15, p. 21, 1902.
(*23*) DeLong, D. M., Reid, W. J. and Darley, M. M., *J. econ. Ent.*, 1930, **23**, 383.
(*24*) Sessions, A. C., *Industr. engng Chem.*, 1936, **28**, 287.
(*25*) Horsfall, J. G., Marsh, R. W. and Martin, H., *Ann. appl. Biol.*, 1937, **24**, 867.
(*26*) Walker, E., *Biochem. J.*, 1928, **22**, 292.
(*27*) Durkee, A. B., *J. agric. Food Chem.*, 1958, **6**, 194.

The Effect of Copper Fungicides on the Plant

The phytotoxicity of water-soluble copper compounds is utilized in weed killing. But with the less soluble derivatives applied as fungicides to foliage, symptoms of acute damage may arise, starting as small purple flecks on leaves and fruit. On leaves these spots have little ill-effect unless numerous when the intervening tissue may yellow and the leaf drop. On some plants, such as peach and hop, the collapse of the injured tissue produces a " shot-hole " effect. On fruit the killing of the epidermis at these localized areas may induce cork formation producing a " russet ". In severe cases, cracking and malformation may follow, a stage well illustrated by Moore, Montgomery and Shaw (*1*). The belief that soluble copper is the agent responsible for acute copper injury naturally led to attempts to reduce damage by the addition, to the spray, of components tending to inhibit the formation of soluble copper. The addition of more lime to Bordeaux mixtures has frequently been recommended. The fungicidal efficiency of excess lime Bordeaux was demonstrated by Grubb (*2*) but no reduction of phytotoxicity was observed either by Grubb or by Adams (*3*). Indeed, Horsfall and his colleagues (*4*) showed that lime itself was capable of causing severe injury to certain plants such as tomato.

Spraying conditions may influence the extent of acute damage by copper sprays. Hedrick (*5*) stated that wet weather favoured the development of injury, a statement confirmed by Howlett and May (6). There is also evidence (*7*) that a heavy and blotchy spray deposit, such as results from the over application of a poorly-wetting spray, will augment injury, a reason for the early recommendation that Bordeaux mixture should be applied in limited amounts insufficient to cause drip from foliage.

Even when no acute injury results from the application of copper fungicides, profound physiological effects may be exerted on the plant. In part the study of these effects belongs to the general problem of the physiological response of the plant to the spray residue left by most protectant insecticides and fungicides. But as this general problem has been attacked mainly by the more specific study of the influence of Bordeaux mixture, it may be discussed alongside that of the effects peculiar to copper sprays.

The importance of the general physiological effects of spray residues was first stressed by Schander (*8*) who attributed the differences he found between the rates of transpiration and assimilation of sprayed and unsprayed leaves to the shade effects of the spray deposit. Early work on the effects of the spray deposit on rate of transpiration gave conflicting results (see, e.g., *9–12*) but in general favoured an increase. Butler (*13*) suggested that, owing to the opaqueness of the spray coating to radiation, sprayed plants cooled less rapidly than unsprayed plants and therefore transpired more fully under conditions favourable to radiation. This view was supported by the later work of Tilford and May (*14*) and of Kroemer and Schanderl (*15*). A practical application of this effect is in the reduction of " Tip-burn " of potato by spraying with Bordeaux mixture. The browning of the tips and margins of the leaf, due in part to high temperature and, in part, to insect injuries (*16*) was found by Lutman (*17*) to be delayed by the application of Bordeaux mixture. He suggested that, though the spray may result in a general increase in transpiration, the opaqueness of the spray coating will smooth the transpiration curve. Though more water may be lost by the greater average transpiration, the sudden demand for water which brings about the death of the potato tissue around the water pores is lessened.

The general effect of the spray coating is to decrease carbon assimilation for, as Amos (*18*) suggested, the stomata are partially blocked by the spray particles. As diffusion into the intracellular spaces of the leaf is thereby decreased, less carbon dioxide is available for photosynthetic activity.

The presence of water-soluble material in the spray may increase water loss from the sprayed leaves through osmosis, a possibility investigated by Berry (*19*). He concluded, from the results of the application of sprays containing sucrose or calcium chloride, that desiccation by osmosis was not an important factor of spray damage. The idea that sensitivity to spray damage is correlated with a cell content of low osmotic value, adopted by Menzel (*20*) and by Tilemans (see *19*), has not been widely confirmed by experiment, neither has the simple relationship between acidity of cell sap and sensitivity to copper or sulphur fungicide put forward by Sessions (*21*).

Beneficial physiological responses to copper-containing sprays follow if the plant is growing under conditions of copper deficiency, a reason Muskett (*22*) suggested, of the success of Bordeaux mixture in the control of apple scab in Ulster. The causes of the so-called " tonic " effect of copper sprays employed on coffee in Kenya are obscure.

REFERENCES

(*1*) Moore, M. H., Montgomery, H. B. S. and Shaw, H., *Rep. E. Malling Res. Sta.*, 1936, p. 259
(*2*) Grubb, N. H., *J. Pomol.*, 1921, **2**, 93 ; 1924, **3**, 157.
(*3*) Adams, J. F., abstr. in *Exp. Sta. Res.*, 1924, **50**, 449.
(*4*) Horsfall, J. G., Magie, R. O. and Suit, R. F., *Tech. Bull. N. Y. St. agric. Exp. Sta.*, 251, 1938.
(*5*) Hedrick, U. P., *Bull. N. Y. St. agric. Exp. Sta.* 287, 1907, p. 103.
(*6*) Howlett, F. S. and May, C., *Phytopathology*, 1929, **19**, 1001.
(*7*) McAlpine, D., *Bull. Dep. Agric. Victoria*, 17, 1904.
(*8*) Schander, R., *Landw. Jb.*, 1904, **33**, 517.
(*9*) Duggar, B. M. and Cooley, J. S., *Ann. Mo. bot. Gard.*, 1914, **1**, 351.
(*10*) Martin, W. H., *J. agric. Res.*, 1916, **7**, 529.
(*11*) Duggar, B. M. and Bonns, W. L., *Ann. Mo. bot. Gard.*, 1918, **5**, 153.
(*12*) Lutman, B. F., *Bull. Vt. agric. Exp. Sta.* 196, 1916.
(*13*) Butler, O., *Tech. Bull. N. H. agric. Exp. Sta.* 21, 1922.
(*14*) Tilford, P. E. and May, C., *Phytopathology*, 1929, **19**, 943.
(*15*) Kroemer, K. and Schanderl, H., *Die Gartenbauw.*, 1934, **8**, 672.
(*16*) Mader, E. O., Rawlins, W. A. and Udey, E. C., *Amer. Potato J.*, 1938, **15**, 337.
(*17*) Lutman, B. F., *Phytopathology*, 1922, **12**, 305.
(*18*) Amos, A., *J. agric. Sci.*, 1907, **2**, 257.
(*19*) Berry, W. E., *Rep. agric. hort. Sta. Bristol*, 1938, p. 124.
(*20*) Menzel, K. C., *Angew. Bot.*, 1935, **17**, 225.
(*21*) Sessions, A. C., *Industr. engng Chem.*, 1936, **28**, 287.
(*22*) Muskett, A. E., *Nature, Lond.*, 1950, **165**, 900.

THE DITHIOCARBAMATE AND THIURAM DERIVATIVES

The suggestion that the fungicidal action of sulphur may involve a reaction akin to the combination of sulphur and the rubber molecule, which is the basis of the vulcanization process, was followed by a study of the fungicidal properties of rubber accelerators (*1*). Of the latter materials Marsh (*2*) found tetramethylthiuram disulphide of some promise. It will be noted that this compound resembles a polysulphide in possessing a sulphur atom in excess of the monosulphide. It is a derivative of dithiocarbamic acid of which a number were, from 1931, under examination by the du Pont Company for fungicidal and insecticidal properties (*3*).

These compounds may be regarded as derivatives of carbonic acid (I) in which an hydroxyl group is replaced by an amino group to give carbamic acid (II), the replacement of the two oxygen atoms by sulphur giving the

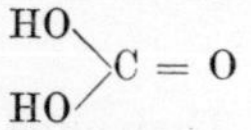

$$\begin{matrix} HO \\ HO \end{matrix}\!\!>\!C=O \qquad \begin{matrix} HO \\ H_2N \end{matrix}\!\!>\!C=O \qquad \begin{matrix} HS \\ H_2N \end{matrix}\!\!>\!C=S$$

(I) 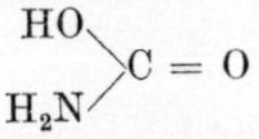 (II) 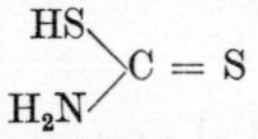 (III)

hypothetical dithiocarbamic acid (III). The substituted dithiocarbamates $R_2N.CS.SH$ arise by the interaction of an amine with carbon disulphide and the thiuram disulphides are formed by gentle oxidation of the dithiocarbamates thus :

$$2(R_2N.CS.SH) \rightarrow R_2N.CS.S.S.CS.NR_2$$

The alkali dialkyldithiocarbamates are readily water-soluble and of little use as protective fungicides but the ferric and zinc salts, christened **ferbam** and **ziram**, respectively, are now established in practical use.

Of the thiuram derivatives, tetramethylthiuram disulphide (**thiram**) being readily available as a rubber accelerator, was in early use and today survives for seed treatment (see p. 266) and, in Europe and Australia, for foliage protection.

The rule that water-solubility is incompatible with protective fungicidal efficiency breaks down with the alkali dithiocarbamate of ethylene diamine, disodium ethylene bisdithiocarbamate (**nabam**), $NaS.CS.NH.CH_2CH_2.NH.CS.SNa$, of which Dimond, Heuberger and Horsfall (*4*) found that water solutions left a rain-resistant and fungicidal deposit. When introduced into practice, nabam sprays proved unreliable until Heuberger and Manns (*5*) reported the better field performance, against *Phytophthora infestans*, obtained by adding zinc sulphate and lime to the spray. The original intention of this addition was to improve tenacity, but Barratt and Horsfall (*6*) showed that there is an increased stability of the dithiocarbamate due to the formation, in the presence of carbon dioxide, of zinc ethylene bisdithiocarbamate (**zineb**), which has displaced nabam. The manganous salt (**maneb**) was later successfully introduced.

Systematic studies of the fungicidal properties of the metallic dialkyldithiocarbamates and related derivatives were made by Goldsworthy, Green and Smith (*7*), by Parker-Rhodes (*8*) and by Barratt and Horsfall (*6*). Alkyl groups higher than methyl were found not to improve fungicidal properties and, contrary to the earlier suggestion that the higher sulphides may simulate polysulphides, Barratt and Horsfall showed the tetrasulphides to be of no greater toxicity than the disulphides to spores of *Stemphylium sarcinaeforme*.

It became evident that the range of fungi to which thiram and the metallic dimethyldithiocarbamates were markedly toxic differed from that to which the nabam and zineb were effective ; thus Heuberger (*9*) referred to the specificity of the metallic dithiocarbamates in the control of certain vegetable diseases. The distinction was reinforced by the laboratory tests of Klöpping (*10*) who suggested that the antifungal activity of the dialkyldithiocarbamates depended on their ability to form the canonical structure :

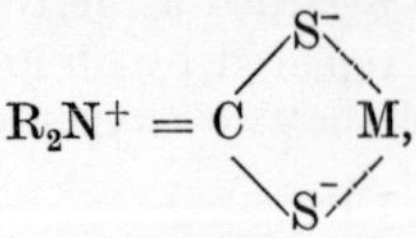

$$R_2N^+ = C \begin{matrix} \diagup S^- \diagdown \\ \\ \diagdown S^- \diagup \end{matrix} M,$$

an idea extended by van der Kerk and Klöpping (*11*). They cited a number of related compounds, of lower fungicidal activity because, they suggested, of their inability to produce this " thioureide " ion, of whose existence evidence was obtained by Chatt, Duncanson and Venanzi (*12*). The latter workers attributed a strong infrared absorption at about 1500 cm.$^{-1}$ and a high dipole moment to this ion which, however, was shared by a wide range of derivatives of the dialkyldithiocarbamic acids, including methyl dimethyldithiocarbamate and the higher analogues. Indeed, this grouping appeared to be an important canonical form in the structure of all dithiocarbamates, so that other factors such as lipoid and water solubility must be sought as an explanation of the decreased fungitoxicity of the ester and higher members. The analogous grouping

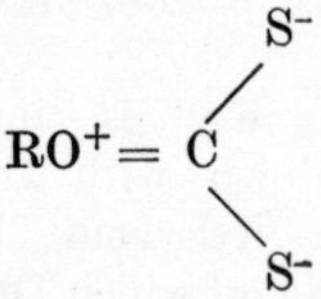

$$RO^+ = C \begin{matrix} \diagup S^- \\ \\ \diagdown S^- \end{matrix}$$

is by no means prominent in the xanthates, the fungicidal action of which on the whole is insignificant. The anomaly of " tetramethylthiuram oxide " found by Klöpping and van der Kerk (*13*) to be highly fungicidal, although apparently unable to produce the thioureide ion, was removed by White (*14*) who showed that this compound is more correctly named as thiocarbamylcarbamyl sulphide, $(CH_3)_2NCS.S.CO.N(CH_3)_2$, which contains the required grouping.

In contradistinction to this hypothesis concerning the dialkyldithiocarbamates, Klöpping (*10*) considered that the antifungal activity of the bisdithiocarbamates was dependent upon the formation of ethylene di*iso*thiocyanate, a compound toxic to a similar range of fungi as is nabam from which, theoretically, it should be formed by the loss of 2 moles of hydrogen sulphide. This hypothesis will be examined further below.

A peculiar feature of thiram and the dimethyldithiocarbamates, not shared by nabam and its derivatives, is the so-called " bimodal " bioassay curve obtained when probit-inhibition is plotted against the logarithm of the dosage and first observed by Dimond and his colleagues (*15*). As the dose is increased toxicity rises, then declines but eventually gives a linear regression line. Having found that the fungitoxicity of thiram is reduced on synthetic media to which an acid hydrolysate of casein (casamino acids) is added, Kaars Sijpesteijn and van der Kerk (*16*) traced the effect to

L-histidine present in the acid hydrolysate. Histidine eliminates that portion of the dosage-response curve below the normal linear regression line and has no effect on the fungicidal activity of nabam or of tetramethylene di*iso*thiocyanate.

In the course of investigations on the effect of dithiocarbamyl compounds on the enzymic processes of yeasts, Goksøyr (*17*) found that the oxygen uptake by yeast supplied with sodium acetate as a carbon source was inhibited by sodium dimethyldithiocarbamate, an inhibition strongly increased when a complete mineral salt solution is added. An examination revealed that zinc was predominantly responsible for this increased inhibition which he regarded as synergism, and that copper salts produced an effect comparable to the bimodal bioassay curve obtained by Dimond and his colleagues. Goksøyr found that at the inversion point at which, in the presence of traces of copper sulphate, the inhibition caused by dithiocarbamates is at a maximum, the molar ratio of copper to dithiocarbamate is 1 : 1. He found spectrographic evidence of the existence of a stable 1 : 1 complex. At dithiocarbamate concentrations higher than this ratio, the extent of inhibition is decreased because of the formation of cupric dimethyldithiocarbamate (in which the ratio is 1 : 2), a compound which Goksøyr found without effect on acetate oxidation, a non-toxicity which he ascribed to its insolubility. Cupric dimethyldithiocarbamate is, however, readily soluble in organic solvents and a more probable explanation of this surprising result is the extreme stability of the compound which is most certainly strongly coordinated.

Kaars Sijpesteijn and her colleagues (*18*) reproduced Goksøyr's (*17*) results with the fungus *Aspergillus niger* van Tiegh. growing in liquid culture; they accepted the conclusion that the inversion of growth inhibition occurs at the 1 : 1 ratio of copper to dithiocarbamate and is due to the appearance of a non-toxic 1 : 2 complex. With the complete precipitation of the copper as the 1 : 2 complex, the second zone of inhibition appears and is due to the toxicity of the dithiocarbamate ion. Hence it may be concluded that the bimodal bioassay curve of Dimond and his colleagues is ascribable to the presence of the traces of copper salts which are necessary and inescapable in culture media required to support fungus growth. The effect of histidine in removing the first zone of inhibition is then due to the removal of the copper by precipitation. The fungitoxicity of the dithiocarbamates arising through the formation of the "thioureide" ion is attributed to enzyme inhibition probably by a competition with thiol-containing coenzymes; the enzyme triosephosphate dehydrogenase was thought by Sisler and Cox (*19*) to be that most affected.

The difference between the mechanics of the toxic action of the dithiocarbamates and the bisdithiocarbamates have already been mentioned and it now seems logical to associate this difference with the fact that whereas the former are derived from secondary amines, the latter are produced from primary diamines. As a consequence the latter compounds possess a reactive hydrogen atom linked to the nitrogen rendering them somewhat unstable. Sodium methyl dithiocarbamate, for example, is except in concentrated solutions, too unstable for fungicidal use, but the range of fungi against which it is highly toxic coincides with the range of the bis-dithiocarbamates rather than that of thiram and the alkyl dithiocarbamates. It has already been mentioned that nabam in solution readily decomposes, as do maneb and zineb (in alkaline solution) (*20*). The aeration products of dilute nabam solutions were shown by Ludwig and Thorn (*21*) to consist of about 15 per cent elementary sulphur, 10–20 per cent **ethylene thiuram monosulphide** (hexahydro-1 : 3 : 6-thiadiazepine-2 : 7-dithione) and a large amount of its polymer. The fungicidal properties of ethylene thiuram monosulphide would account for the action of sodium, manganese and zinc ethylene bisdithiocarbamates and Kaars Sijpesteijn and van der Kerk (*22*) confirming the results of Ludwig and his colleagues, considered that this activity was due to the rearrangement of the sulphide yielding an *iso*thio-cyanate group thus :—

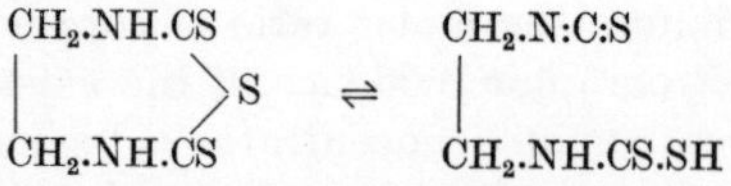

$$\begin{array}{l} CH_2.NH.CS \\ | \qquad\qquad\quad \rangle S \\ CH_2.NH.CS \end{array} \rightleftharpoons \begin{array}{l} CH_2.N{:}C{:}S \\ | \\ CH_2.NH.CS.SH \end{array}$$

The hypothesis is supported by the identity of the range of fungi against which the bisdithiocarbamates and the di*iso*thiocyanates are active but experimental proof of the existence of the *iso*thiocyanate grouping in aqueous media has proved elusive. Ludwig and his co-workers (*20*), however, found absorption bands characteristic of the *iso*thiocyanate grouping in a non-aqueous solution of ethylene thiuram monosulphide to which anhydrous ferric or zinc sulphate had been added. The fumigant action in soil of sodium methyldithiocarbamate is due to a volatile toxicant which most probably is methyl *iso*thiocyanate.

The discovery by Kaars Sijpesteijn and van der Kerk (*23*) that thiol compounds such as cysteine and thioglycollic acid reduce the fungicidal activity of bisdithiocarbamate is a strong indication that the biochemical lesion produced by the latter fungicides is a reaction with the sulphydryl groups essential for the functioning of the sulphydryl enzymes.

The hypothesis that the fungicidal action of nabam arises through the formation of the *iso*thiocyanate group has been used as a signpost to other fungicides. The ring closure by acid treatment of the condensation product of a sodium dithiocarbamate and the sodium salt of monochloroacetic acid, for example, was used by van der Kerk and his colleagues for the preparation of a new series of compounds of which **3-(*p*-chlorophenyl)-5-methyl rhodanine** (IV) was highly fungicidal. This compound was introduced in 1952 as an experimental fungicide and nematicide :—

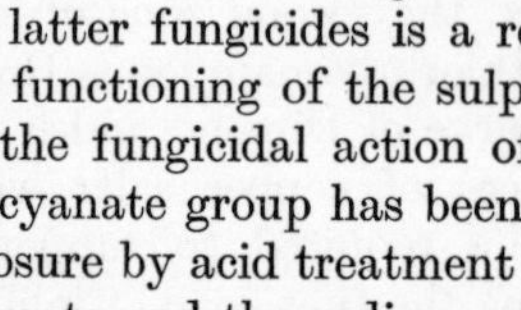

$$Cl\text{-}C_6H_4\text{-}\begin{array}{l} N - C = O \\ | \qquad\ \ | \\ S{:}C \quad CH.CH_3 \\ \quad \diagdown \ \diagup \\ \quad\ \ S \end{array}$$

(IV)

Van der Kerk (*24*) observed that those compounds of low fungicidal activity were stable whereas the more active compounds underwent degradation in solution, with spectrographic evidence of the formation of the *iso*thiocyanate group. A series of 5-substituted rhodanines with mildew-proofing properties was described by Brown and her associates (*25*). The reaction of formaldehyde and methylammonium methyldithiocarbamate yields 2-thio-3 : 5-dimethyl tetrahydro-1 : 3 : 5-thiadiazine, (V) " **Mylone** ", a

```
            S
          /   \
       CH2     C = S
        |      |
  CH3.N        N.CH3
        \     /
          CH2
          (V)
```

compound introduced in 1952 from two different commercial sources as an experimental fungicide and nematicide. Although the nitrogen atoms of the compound bear no hydrogen atoms, van der Kerk (*24*) pointed out that the methylene group between the nitrogen groups is readily removed by hydrolysis to form the parent ester which is readily amenable to rearrangement to give the *iso*thiocyanate group.

REFERENCES

(*1*) Martin, H., *J. S. E. Agric. Coll., Wye*, 1934, **33**, 38.

(*2*) Marsh, R. W., *Ann. appl. Biol.*, 1938, **25**, 583.

(*3*) Tisdale, W. H. and Flenner, A. L., *Industr. engng Chem.*, 1942, **34**, 501.

(*4*) Dimond, A. E., Heuberger, J. W. and Horsfall, J. G., *Phytopathology*, 1943, **33**, 1095.

(*5*) Heuberger, J. W. and Manns, T. F., *Phytopathology*, 1943, **33**, 113.

(*6*) Barratt, R. W. and Horsfall, J. G., *Bull. Conn. agric. Exp. Sta.*, 508, 1947.

(*7*) Goldsworthy, M. C., Green, E. L. and Smith, M. A., *J. agric. Res.*, 1943, **66**, 277.

(*8*) Parker-Rhodes, A. F., *Ann. appl. Biol.*, 1943, **30**, 170.

(*9*) Heuberger, J. W., *Phytopathology*, 1947, **37**, 439.

(*10*) Klöpping, H. L., Chemical constitution and antifungal action of sulphur compounds. Dissert. Univ. of Utrecht, Utrecht, 1951.

(*11*) van der Kerk, G. J. M. and Klöpping, H. L., *Rec. Trav. chim. Pays-Bas*, 1952, **71**, 1179.

(*12*) Chatt, J., Duncanson, L. A. and Venanzi, L. M., *Nature*, 1956, **177**, 1042 ; *Acta chem. fenn.*, 1956, B **29**, 75.

(*13*) Klöpping, H. L. and van der Kerk, G. J. M., *Rec. Trav. chim. Pays-Bas*, 1951, **70**, 917.

(*14*) White, R. W., *Can. J. Chem.*, 1954, **32**, 867.

(*15*) Dimond, A. E., Horsfall, J. G., Heuberger, J. W. and Stoddard, E. M., *Bull. Conn. agric. Exp. Sta.*, 451, p. 635, 1941.

(*16*) Kaars Sijpesteijn, A. and van der Kerk, G. J. M., *Leeuwenhoek ned. Tijdschr.*, 1952, **18**, 83.

(*17*) Goksøyr, J., *Physiol. Plant.*, 1955, **8**, 719.

(*18*) Kaars Sijpesteijn, A., Janssen, M. J. and van der Kerk, G. J. M., *Biochim. biophys. Acta*, 1957, **23**, 550.

(*19*) Sisler, H. D. and Cox, C. E., *Amer. J. Bot.*, 1954, **41**, 338 ; 1955, **42**, 351.

(*20*) Ludwig, R. A., Thorn, G. D. and Unwin, C. H., *Canad. J. Bot.*, 1955, **33**, 42.

(*21*) Ludwig, R. A. and Thorn, G. D., *Plant Dis. Reptr.*, 1953, **37**, 127.

(*22*) Kaars Sijpesteijn, A. and van der Kerk, G. J. M., *Biochim. biophys. Acta.*, 1954, **13**, 545.

(*23*) Kaars Sijpesteijn, A. and van der Kerk, G. J. M., *Biochim. biophys. Acta.*, 1954, **15**, 69.

(*24*) van der Kerk, G. J. M., *Meded. LandbHoogesch. Gent.*, 1956, **21**, 305.

(*25*) Brown, F. C. and Bradsher, C. K., *Nature, Lond.*, 1951, **168**, 171 ; Bradsher, C. K., Brown, F. C. and Grantham, R. J., *J. Amer. chem. Soc.*, 1954, **76**, 114.

GLYODIN

In the course of a study of various organic compounds as fungicides, Wellman and McCallan (*1*) discovered the effectiveness of certain glyoxalidines, or, in more correct chemical nomenclature, 2-imidazolines, of which 2-heptadecyl-2-imidazoline (VI) was chosen. This compound protected by B.P. 598,927, is slowly hydrolysed by water to give the less fungicidal N-(2-aminoethyl)stearamide :

$$\begin{array}{ccc} & C_{17}H_{35} & \\ & | & \\ & C & \\ & /\!\!/ \quad \backslash & \\ N & & NH \\ | & & | \\ CH_2 & - & CH_2 \end{array} \quad \rightleftharpoons \quad C_{17}H_{35}.CO.NH.CH_2.CH_2.NH_2$$

(VI)

for which reason the acetate (glyodin) is employed, lime being added to the spray dilution to liberate the free base.

The choice of the heptadecyl derivative arose from the examination of the homologous series by Wellman and McCallan ; maximum fungistatic activity appeared when the side chain was 13–17 carbons long whereas phytotoxicity was greatest with side chain 11–13 carbons long, especially if unsaturated.

It is not surprising that Miller, McCallan and Weed (*2*) found that the amount of heptadecylimidazoline taken up by spores should be up to 10,000 times that of the concentration of the solution in which the spores were placed, for the compound is highly surface-active. Its fungistatic activity is reduced in the presence of the purines guanine and xanthine but not by other purines, by histidine or allantoin. Consequently West and Wolf (*3*) suggested that its biological activity rested in an interference in the biosynthesis of these purines, which are components of the nucleic acids.

On this hypothesis it would be tempting to examine other alkyl-substituted pyrimidines and purines. A series of 2-alkyl 4 : 4 : 6-trimethyltetrahydropyrimidines was examined by Rader and his colleagues (*4*) ; and again a greater fungitoxicity was obtained with the alkyl chain of 17 carbon atoms, at which level phytotoxicity is decreased. Horsfall and Rich (*5*) reported **2-hendecyl-2-oxazoline** a powerful spore inhibitor and suggested that the hendecyl group enables the compound to permeate into the cell enabling the 2-oxazoline to exert its toxic action. 2-Oxazoline is structurally like 2-imidazoline for the imido group (–NH) of the latter compound is replaced by oxygen.

REFERENCES

(*1*) Wellman, R. H. and McCallan, S. E. A., *Contr. Boyce Thompson Inst.*, 1946, **14**, 151.
(*2*) Miller, L. P., McCallan, S. E. A. and Weed, R. M., *Contr. Boyce Thompson Inst.*, 1953, **17**, 173.
(*3*) West, B. and Wolf, F. T., *J. gen. Microbiol.*, 1955, **12**, 396.
(*4*) Rader, W. E., Monroe, C. M. and Whetstone, R. R., *Science*, 1952, **115**, 124.
(*5*) Horsfall, J. G. and Rich, S., *Contr. Boyce Thompson Inst.*, 1946, **14**, 151.

CAPTAN

Captan is the common name adopted for *N*-trichloromethylmercapto-4-*cyclo*hexene-1 : 2-dicarboximide (VII), one of a series of condensation products of perchloromethyl mercaptan with compounds possessing an imido

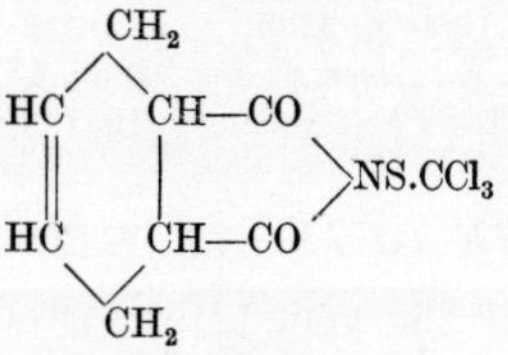

$$\begin{array}{c} CH_2 \\ HC \quad CH{-}CO \\ \| \quad | \qquad \rangle NS.CCl_3 \\ HC \quad CH{-}CO \\ CH_2 \end{array}$$

(VII)

group sufficiently acid to form a stable sodium salt. As all sixteen of the compounds which he prepared by this reaction proved highly fungicidal, Kittleson (*1*) considered that the activity is attributed to the $>NSCCl_3$ group. This suggestion has been quickly followed up and Waeffler *et al.* (*2*) reported successful fungicidal treats with certain condensation products of perchloromethyl mercaptan with sulphonamides, of which *N*-methyl-sulphonyl-*N*-trichloromethylthio-4-chloroaniline gave the most promising results. Sosnovsky (*4*) examined a series of comparatively simple tri-chloromethylthio derivatives to ascertain to what extent fungicidal activity was affected by the replacement of the $>NSCCl_3$ by the $-OSCCl_3$ radical. He found marked activity, at least against *Pythium ultimum* and *Botrytis allii* in most of his compounds, even in trichloro-*S*-methoxymethanethiol, $CH_3OS.CCl_3$. Uhlenbroek and his colleagues (*3*) prepared and examined a series of condensation products of trichloromethanesulphenyl chloride and the alkali salts of various sulphinic acids. The aliphatic trichloro-methylthiolsulphonates they found strongly phytotoxic but this property was diminished by introducing a substituted phenoxy group into the β-position of the aliphatic chain, particularly *p*-carboxyphenyltrichloro-methylthiolsulphonate (VIII), which was selected for field tests.

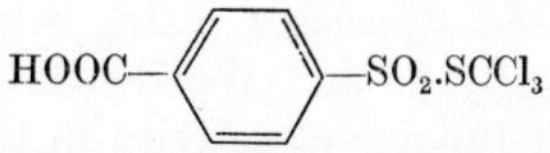

$$HOOC{-}C_6H_4{-}SO_2.SCCl_3$$

(VIII)

In spite of the extremely successful use of captan as a fungicide little is yet known of the biochemical reasons for its activity. Hochstein and Cox (*5*) found that captan competes with cocarboxylase (thiamine pyro-phosphate) for sites on the co-enzyme-free carboxylase in the decarboxylation of pyruvate, indicating that the biochemical lesion produced by captan is an interference with those decarboxylation reactions which require this co-enzyme. Lukens and Sisler (*6*) found that only compounds such as cysteine and glutathione containing sulphydryl groups were effective in reducing the toxicity of captan to the yeast *Saccharomyces pastorianus* Hansen and suggested that this toxicity may be due to a reaction with sulphydryl groups or that the reaction may release toxic products from captan within the fungal cell.

REFERENCES

(*1*) Kittleson, A. R., *Science*, 1952, **115**, 84 ; *J. agric. Food Chem.*, 1953, **1**, 677.
(*2*) Waeffler, R., Gasser, R., Margot, A. and Gysin, H., *Experientia*, 1955, **11**, 265.
(*3*) Uhlenbroek, J. H., Koopmans, M. J. and Huisman, H. O., *Rec. Trav. chim. Pays-Bas*, 1957, **76**, 129.

(4) Sosnovsky, G., *J. chem. Soc.*, 1956, p. 3139.
(5) Hochstein, P. E. and Cox, C. E., *Amer. J. Bot.*, 1956, **43**, 437.
(6) Lukens, R. J. and Sisler, H. D., *Phytopathology*, 1957, **47**, 22.

THE QUINONE GROUP

In 1940 Cunningham and Sharvelle (*1*) reported the successful use of two organic compounds for the protection of lima bean seedlings from damping-off. It emerged that one of these compounds was tetrachloro-*p*-benzoquinone (IX), a compound long in use as an oxidising agent in the dyestuffs industry under the name **chloranil**. It was rapidly adopted for the treatment of leguminous seed but proved unsatisfactory as a foliage protectant, doubtless because it is decomposed by light. The naphthoquinones are generally less photosensitive and in 1943, ter Horst and Felix (*2*) reported that 2 : 3-dichloro-1 : 4-naphthoquinone (X) was some four to eight times as effective as chloranil in the protection of legume and cotton seed. This compound, christened **dichlone**, has been recommended as a foliage spray against apple scab, though it is phytotoxic to many plants and is apt to cause dermatitis.

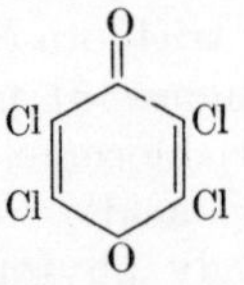

(IX) 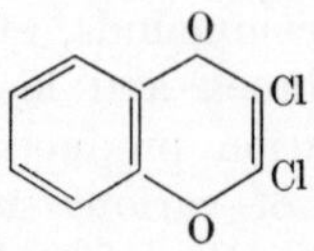 (X)

The effects of substitution on the benzo- and naphthoquinones are discussed by McNew and Burchfield (*3*) and, in general, halogenation enhances fungitoxicity and decreases water solubility and phytotoxicity. Emphasis was placed by Byrde and Woodcock (*4*) on the necessity of two vicinal chlorine atoms for fungicidal activity in the naphthoquinone derivatives for both 2 : 3-dimethyl- and 2-chloro-3-hydroxy-1 : 4-naphthoquinone, in which the chlorines are replaced by methyl or hydroxy groups of dimensions similar to the chlorine atom, are non-fungicidal. They observed that acetylation of the hydroquinone produced by a reduction of dichlone produced a compound of the same order of fungitoxicity yet of reduced phytotoxicity. Having established a close relationship between fungitoxicity and ease of hydrolysis of the esters of 2 : 3-dichloro-1 : 4-naphthahydroquinone, they suggested that the lack of phytotoxicity of the esters is due to their stability on the leaf surface whereas their fungitoxicity is due to the formation of an active fungicide through hydrolysis of the ester by a fungal esterase.

A large variety of quinones occur as natural components of higher plants and as metabolic products of microorganisms ; many are of high biological activity. Little and his colleagues (*5*) recorded the antifungal action of 2-methoxy-1 : 4-naphthoquinone isolated from the garden balsam (*Impatiens balsamina* L.) and the greater potency of its sulphur analogue, 2-methylmercapto-1 : 4-naphthoquinone.

The biological activity of the quinones is not surprising in view of their known role in oxidation-reduction reactions, attributed to the stability of the semiquinone free radical. Woolley (*6*), noting that vitamin K is a naphthoquinone, showed that the toxicity of dichlone to yeasts could be reduced by the addition, over a moderate range, of vitamin K in the form of 2-methylnaphthoquinone. Foote *et al.* (*7*) traced a parallelism between the fungistatic activities of hydroxy-free naphthoquinones and their ability to inhibit the enzyme carboxylase.

REFERENCES

(*1*) Cunningham, H. S. and Sharvelle, E. G., *Phytopathology*, 1940, **30**, 4.

(*2*) ter Horst, W. P. and Felix, E. L., *Industr. engng Chem.*, 1943, **35**, 1255.

(*3*) McNew, G. L. and Burchfield, H. P., *Contr. Boyce Thompson Inst.*, 1951, **16**, 357.

(*4*) Byrde, R. J. W. and Woodcock, D., *Nature, Lond.*, 1952, **169**, 503 ; *Ann. appl. Biol.*, 1953, **40**, 675.

(*5*) Little, J. E., Sproston, T. J. and Foote, M. W., *J. biol. Chem.*, 1948, **174**, 335 ; *J. Amer. chem. Soc.*, 1949, **71**, 1124.

(*6*) Woolley, D. W., *Proc. Soc. exp. Biol. Med.*, 1945, **60**, 225.

(*7*) Foote, M. W., Little, J. E. and Sproston, T. J., *J. biol. Chem.*, 1949, **181**, 481.

FUNGICIDAL ANTIBIOTICS

The study of the biological activity of the metabolic products of micro-organisms received tremendous impetus from the successful development of penicillin for medicinal use. No less than 340 antibiotics are listed in the Handbook of Biological Data (*1*) ; some are highly fungicidal ; most, like gliotoxin (see p. 53) are too unstable for practical use in crop protection. Indeed, doubts were long current that these toxic metabolic products played an important role in biological control (see p. 53), not only because of their instability and ready sorption but also because their production was only detected under the high nutrient status of a culture medium. But in the fermentation vat their presence is undeniable and commercial production requires only knowledge of the strains and conditions for workable yields. Clearly, however, in the exploration of the potential use of antibiotics in crop protection, attention was first directed to those either already in production for medicinal purposes or which are present in the mother liquor from which a commercial antibiotic has been recovered.

Streptomycin. The early history of the discovery, in 1942, of this important drug in the culture filtrate of certain strains of *Streptomyces griseus* (Krainsky) Waksman & Schatz, has been given by Waksman (*2*) in his Nobel Prize address. Its chemical structure (XI) elucidated by cooperative effort (see *3*) is that of a glycoside in which the aglycone streptidine is linked to *N*-methyl glucosamine through an unusual sugar streptose. The two guanido and the amino-groups render the molecule strongly basic and the compound is usually marketed as sulphate or hydrochloride. In the United States a frequent formulation is " Agri-mycin " containing, in addition to streptomycin, a proportion (about one-tenth) of " Terramycin ", the trade name of Chas. Pfizer & Co. Inc. for oxytetracycline.

Streptomycin, though of little toxicity to true fungi, is effective against

```
                    NH
                    ||
                HN.C.NH2
                    |
                   CH
                  /  \
            HOCH   CH — O — CH
              |     |       / \
  H2N.C.NH.CH  CHOH   O    CH — O — CH
     ||          \  /     |     |      / \
     NH          CHOH     CH—COH     O   CH.NH.CH3
                          |   |      |   |
                         CH3 HC:O    CH  CH.OH
      Streptidine                     \  /
                      Streptose   H2C.OH CH.OH
                                  N-methyl glucosamine
                             (XI)
```

both Gram-positive and Gram-negative bacteria. The Gram-positive bacterial phytopathogens are few and usually seed-borne, and seed treatment with streptomycin at 100 ppm. or greater has given control of bacterial wilt of maize (*Xanthomonas stewarti* (E. F. Smith) Dows.). Against the Gram-negative bacteria its greatest success in crop protection has been against fire blight (*Bacterium amylovorum*) (see *4*). The ability of seedling plants to take up streptomycin, first noted by Anderson and Nienow (*5*), indicates that systemic activity is possible though phytotoxicity limits the concentration which can be so used.

Cycloheximide. In addition to streptomycin, the culture filtrates of *S. griseus* were found by Whiffen, Bohonos and Emerson (*6*) to contain a fungicidal antibiotic subsequently called cycloheximide, though it is frequently known by the trade name of the Upjohn Company " Acti-dione ". From the work of Kornfeld (*7*) and of Whiffen (*8*) and their respective colleagues, it was considered that cycloheximide is β-[2-(3 : 5-dimethyl-2-oxy*cyclo*hexyl)-2-hydroxyethyl]glutarimide (XII).

```
        CH3
        |
        CH—CO                        CH2—CO
       /      \                     /      \
    CH2        CH.CH(OH).CH2.CH             NH
       \      /                     \      /
        CH—CH2                       CH2—CO
        |
        CH3

                         (XII)
```

Wallen (*9*) reported a promising control of stem rust of wheat but even at concentrations of 50 ppm. evidence of plant damage appeared. Phytotoxicity and a considerable mammalian toxicity would limit the usefulness of cycloheximide in crop protection.

Griseofulvin was first isolated and described, as far back as 1939, by Raistrick and his school (*10*) as a metabolic product of *Penicillium griseofulvum* Dierckx. It was subsequently identified (*11*) as the substance produced by *P. janczewskii* Zal. responsible for the abnormal curling of the germ tubes of *Botrytis allii*. Its structure was deduced by Grove and his associates (*12*) to be 7-chloro-4 : 6-dimethoxycoumaran-3-one-2-*spiro*-1′-(2′-methoxy-6′-methyl*cyclo*hex-2′-en-4′-one) (XIII).

```
           OCH3
            |
            C           OCH3
          //  \          |
        CH     C — CO    C = CH
        |      ||      \ /      \
        |      ||       C        CO
        |      ||      / \      /
CH3O.C \\  C / C — O    CH—CH2
           |             |
           Cl            CH3
```

(XIII)

Griseofulvin is not fungicidal in the usual sense of the word but the peculiar feature of its biological activity is its profound effect on the morphogenesis of many fungi. At concentrations of 10 to 1,000 ppm. it produces a stunting of the hyphae which acquire a spiral twist. Brian (*13*) noted that only fungi with chitinous cell walls were sensitive to griseofulvin ; fungi with non-chitinous cell walls, actinomycetes and bacteria are unaffected. Grove and his colleagues (*12*) suggested that this curling is associated with the spiro structure of the molecule.

The initial tests of griseofulvin against fungal pathogens were most promising (*14*). In sand culture experiments Stubbs (*15*) secured tomatoes resistant to *Alternaria solani* (Ell. & Mart.) Jones and Grout by watering with solutions containing 0.05 per cent griseofulvin. Brian (*16*) obtained a similar control of *Botrytis* on lettuce and of powdery mildew on barley. Griseofulvin was detected in gutation drops from the treated plants (*17*), evidence of its translocation by the plant and its capacity to function systemically. Crowdy and Pramer (*18*) incidentally observed that part of the griseofulvin of the plant sap is recoverable only by extraction with an organic solvent indicating that it is in some way " bound ". Wright (*19*) found it to be of comparatively low phytotoxicity.

REFERENCES

(*1*) *Handbook of Toxicology, Vol. II, Antibiotics*, W. B. Saunders Co., Philadelphia, 1957.

(*2*) Waksman, S. A., *Science*, 1953, **118**, 259.

(*3*) Kuehl, F. A., Peck, R. L., Hoffhine, C. E. and Folkers, K., *J. Amer. chem. Soc.*, 1948, **70**, 2325.

(*4*) Dunegan, J. C., Kienholz, J. R., Wilson, R. A. and Morris, W. T., *Plant Dis. Reptr.*, 1954, **38**, 666.

(*5*) Anderson, H. W. and Nienow, I., *Phytopathology*, 1947, **37**, 1.

(*6*) Whiffen, A. J., Bohonos, N., and Emerson, R. L., *J. Bact.*, 1946, **52**, 610.

(*7*) Kornfeld, E. C., Jones, R. G. and Parke, T. V., *J. Amer. chem. Soc.*, 1949, **71**, 150.

(*8*) Leach, B. E., Ford, J. H. and Whiffen, A. J., *J. Amer. chem. Soc.*, 1947, **69**, 474.

(*9*) Wallen, V. R., *Plant Dis. Reptr.*, 1955, **39**, 124.

(*10*) Oxford, A. E., Raistrick, H. and Simonart, P., *Biochem. J.*, 1939, **33**, 240.

(*11*) Brian, P. W., Curtis, P. J. and Hemming, H. G., *Trans. Brit. mycol. Soc.*, 1946, **29**, 173.

(*12*) Grove, J. F., MacMillan, J., Mulholland, T. C. P. and Thorold Rogers, M. A., *J. chem. Soc.*, 1952, 3977.

(*13*) Brian, P. W., *Ann. Bot. Lond.*, 1949, **13**, 59.

(*14*) *Report on Griseofulvin*, Glaxo Laboratories Ltd., Stoke Poges, Bucks., England, 1955.

(*15*) Stubbs, J., *Ann. appl. Biol.*, 1952, **39**, 439.

(*16*) Brian, P. W., *Ann. appl. Biol.*, 1952, **39**, 434.
(*17*) Brian, P. W., Wright, J. M., Stubbs, J. and Way, A. M., *Nature, Lond.*, 1951, **167**, 347.
(*18*) Crowdy, S. H. and Pramer, D., *Ann. Bot. Lond.*, 1955, **19**, 79.
(*19*) Wright, J. M., *Ann. Bot. Lond.*, 1951, **15**, 493.

MISCELLANEOUS FUNGICIDES

To avoid degeneration into a mere catalogue, only those materials which have found practical employment or which offer special points of interest as fungicides, will be mentioned.

Metallic derivatives. The apparent uniqueness of copper, among the metals provoked research for it is to be expected that at least the related metals should yield compounds of fungicidal activity. Wüthrich (*1*) found that mercuric chloride was more effective than copper salts in inhibiting the germination of fungus spores but was, he concluded, too phytotoxic for foliage use. Though doubtless the organomercury compounds, developed for seed treatment (see p. 260) were tried for the purpose, it is later that suitable compounds were developed. The simplest is **phenylmercuric chloride** chosen because of its water insolubility and its non-volatility and found by Montgomery, Moore and Shaw (*2*) to be highly toxic to spores of *Venturia inaequalis*, but not entirely safe on apples, when applied in sulphite lye suspension. By incorporation on lead arsenate or on china clay, phytotoxicity seems to be lowered and these products find use in certain localities, e.g., on Bramley in Ulster (*3*) though damage has been reported on Cox. In the United States successful trials were made by Howard, Locke and Keil (*4*) of "**Puratized N5D**", of which the active component is phenylmercuric triethanolammonium lactate, but Hamilton and Mack (*5*) reported this material to be injurious in some areas and to be incompatible with dirty water, the usual diluent available to growers. Nevertheless, such organomercury compounds are now in commercial use not so much as protectant fungicides but, because of the rapidity with which the fungus is destroyed, as eradicant fungicides preventing the extension of an already-established infection.

The fungicidal properties of metallic salts were studied by Wöber (*6*) on the basis of the Periodic Law, and he correlated the position of copper, silver and mercury upon the atomic weight-specific gravity curve with fungicidal properties. The possibilities of silver fungicides were explored by Nielsen (*7*). McCallan and Wilcoxon (*8*) extended Wöber's work, finding silver and osmium compounds the most fungicidal and suggesting that compounds of cerium, cadmium, lead, thallium and arsenic should be of promise. Mention of cerium recalls the use of "**Perozid**" as a substitute for copper sulphate in Germany during the 1914–18 war. Perozid, a residue from gas-mantle manufacture, is a mixture of the sulphates of the cerium group of rare earth metals. It was first used by Appel (*9*) but did not survive when copper sulphate again became available for making Bordeaux mixture. A zinc sulphate-lime mixture was used successfully on peach by Roberts and Pierce (*10*) against *Bacterium pruni* E. F. Sm.

The fungicidal properties of certain chromium salts are used in wood preservation but their high solubility renders them toxic to foliage. Harry *et al.* (*11*) concluded that for safety to foliage the solubility of the chromate should be below 0.03 per cent and above 0.0001 per cent to ensure fungicidal action. They accordingly prepared complex **copper-zinc-chromates,** basic compounds not of stoichiometric proportions but of the nature of solid solutions in which copper replaces zinc in the crystal lattice of zinc chromate. These compounds are reversibly decomposed by water and their performance was correlated with the ratios of zinc to copper to chromium. Such chromates at about one-third the copper content of Bordeaux mixture were shown to be more effective as potato fungicides.

The fungicidal properties of arsenical compounds are of interest in relation to the action of the combined lime sulphur-lead arsenate spray at one time widely used in apple growing. That **lead arsenate** possesses fungicidal properties was first observed by Waite (*12*) and Morse (*13*) claimed that arsenate of lead paste controlled apple scab as well as did Bordeaux mixture or lime sulphur. Whetzel, McCallan and Loh (*14*) found **calcium arsenate** of promise for the control of *Alternaria* spp. In trials upon the hop powdery mildew, Horton and Salmon (*15*) found that the soluble arsenates are highly toxic. Continuing this work, Goodwin, Martin and Salmon (*16*) showed lead arsenate to be less fungicidal than disodium arsenate or dicalcium arsenate, but that solutions of the lime sulphur and lead arsenate below fungicidal strength, when mixed together, were fungicidal. This result was ascribed to the formation of thioarsenates, which were found to have marked direct fungicidal properties but, owing to their higher solubility, more liable to damage foliage than the arsenates. Dilute solutions of arsenite or arsenate of soda were employed with success by Garbowski and Leszczenko (*17*) for the control of American gooseberry mildew and by Szembel (*18*) against cucumber powdery mildew.

The use of **potassium permanganate** as a fungicide has often been suggested, notably by Guozdenović (*19*), who advocated its addition to Bordeaux mixture. He found this spray to be effective against both the powdery and downy mildews of vine, and it has been recommended in New Zealand (*20*). Difficulty in the use of potassium permanganate arises through its ready decomposition in the presence of organic matter. It is presumably to its oxidizing properties that its fungicidal action is related and, in contact with leaf and fungus, its rapid decomposition will not permit it to function as a protective fungicide. Further, the entire list of organic spreaders react with permanganate. Guozdenović employed either lime or Bordeaux mixture to improve adherence.

The **alkali carbonates** were found of value against American gooseberry mildew by Dorogin (*21*); washing soda ($Na_2CO_3.10\ H_2O$) has been used with success both in Holland (*22*) and in England (*23*), particularly upon sulphur-shy varieties of gooseberry. It is probable that fungicidal activity is associated with the alkalinity of the solution. Martin and Salmon (*24*) showed that solutions of sodium carbonate, hydroxide and sulphide had similar fungicidal properties to the hop powdery mildew but that all were highly phytotoxic to hop foliage, the least injurious being sodium sulphide.

Organic Compounds. Of organic derivatives, **oils** have attracted attention for fungicidal purposes, mainly on account of their excellent spreading properties. Kerosine emulsions were used by Halstead and Kelsey (*25*) and, for the control of American gooseberry mildew, by Barker and Lees (*26*). Although the 2 per cent emulsion was found to cause but little foliage damage, it did not protect the sprayed plant from re-infection. The introduction of highly-refined lubricating oils (see p. 186) which may be applied to foliage, resulted in tests of their action upon fungi. McWhorter (*27*) found one such product successful against rose powdery mildew (*Sphaerotheca pannosa* (Wal.) Lév.). The relative non-volatility of the oil would suggest that this fungicidal action is solely protective, though McWhorter considered the spray possessed a direct action. This conclusion was supported by the work of Martin and Salmon (*28*), who found that emulsions of highly-refined petroleum oils were effective against hop powdery mildew. Under certain conditions, however, the oil emulsions caused foliage damage, and Martin and Salmon showed that **vegetable oils** were less phytotoxic and more fungicidal than the petroleum oils. Later Martin and Salmon (*29*) proved that direct fungicidal activity to *S. humuli* was a property common to the glyceride structure and hence shared by all animal and vegetable oils. Castor oil, a glyceride of the hydroxylated fatty acid, ricinoleic acid, proved less effective. The high degree of correlation between instability of the emulsion and its fungicidal efficiency is of interest in connection with the influence of type of emulsion upon insecticidal efficiency (see p. 188).

Soaps, which are chemically related to the glyceride oils, have found use against *Botrytis* spp. (*30, 31*). The marked lytic action of surface-active substances upon fungal zoospores was observed by Goodwin, Salmon and Ware (*32*), who suggested that soap dusts may be of practical value as fungicides in cases where rapid but not persistent protective action is required.

An early survey of the fungicidal properties of synthetic organic compounds was made by Morris (*33*) and by Fargher, Galloway and Probert (*34*) at the Shirley Institute. Their purpose was the discovery of antiseptics suitable for the protection of textile fabrics from mould fungi. Fungicidal efficiency was determined by the concentration necessary to prevent the growth of saphrophytic fungi upon flour paste. Their results showed that, in general, the toxicity of phenolic derivatives was increased by the introduction of alkyl, nitro or halogen groups and was decreased by the introduction of acrylyl, sulphonic acid or additional hydroxyl groups. Acetanilide was found to be about half as toxic as phenol and a survey of the anilides revealed the high fungicidal activity of the salicyl derivative. **Salicylanilide,** under the trade name " Shirlan ", was used by Bewley and Orchard (*35*) for the control of tomato leaf mould (*Cladosporium fulvum*), and of certain powdery mildews. Martin and Salmon (*36*) found it effective against hop powdery mildew when applied with an alkaline spreader such as soap or when the water-soluble sodium salt of salicylanilide was used.

The application of **dyestuffs** as fungicides has been investigated, mainly in France by Truffaut and his colleagues (*37*), who have found certain auramines and phosphines efficacious against powdery mildews. Among points of interest in this work are the inefficiency of phosphine ACR when tested by *in vitro* methods as contrasted with its high activity against vine mildew when in contact with the leaf or fungus, and the generalizations, deduced by Pastac (*38*), concerning fungicidal activity and molecular structure. He showed, firstly, that activity increased with molecular weight up to a maximum and then fell rapidly when the molecule is presumably too large to pass readily through the cell wall, secondly, the marked reduction of toxicity following sulphonation and, thirdly, the increased activity which follows the alkylation of certain amino compounds. Thus dimethylaniline was found to be more toxic than aniline and Violet 5BO was more fungicidal than Fuchsine. In England, Bennett (*39*) successfully controlled the Fusarium patch disease of turf by watering with dilute solutions of **malachite green,** a diaminotriphenylmethane dye found by Takahashi (*40*) to inhibit the multiplication of tobacco mosaic virus.

A survey of the highly substituted benzenes, carried out by I.G. Farbenindustrie, revealed that 1 : 3 : 5-trichloro-2 : 4 : 6-trinitrobenzene (I) and 1 : 2 : 4-trichloro-3 : 5-dinitrobenzene (II) are toxic to a limited number of fungi. The former, though dangerous to make (*41*), was effective against *Cladosporium fulvum* on tomato ; the latter was used for the control of *Plasmodiophora brassicae* against which pentachloronitrobenzene (III) was also used by Smieton (*42*) who, with Brown (*43*) found this compound effective against *Botrytis* disease of lettuce. For this purpose it was marketed in " Folosan " but the active component of this preparation was later changed to 2 : 3 : 5 : 6-tetrachloronitrobenzene (= 1 : 2 : 4 : 5-tetrachloro-3-nitrobenzene) (IV), since given the common name **tecnazene.** Compound (III) and the more volatile tecnazene were found particularly useful for the prevention of dry rot due to *Fusarium caeruleum* (Lib.) Sacc. in seed potatoes. In the course of this work, Brown (*43*) discovered the

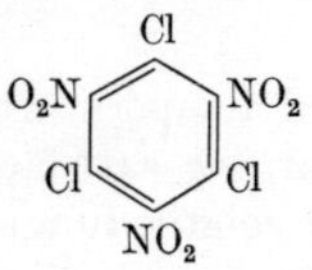

(I) 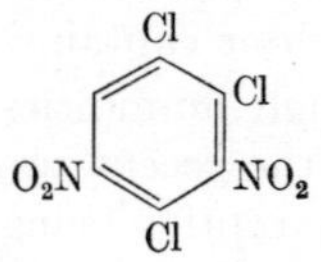 (II) 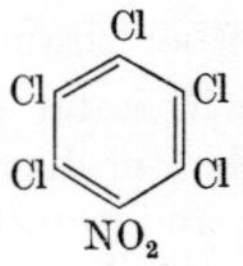 (III) 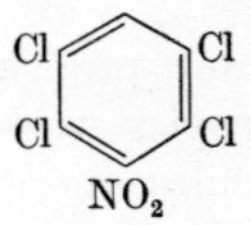 (IV)

marked effect of these two compounds in delaying the sprouting of potatoes when stored in clamps. 2 : 3 : 4 : 5-Tetrachloronitrobenzene was reported by Brook (*44*) to be more fungicidal than its isomer tecnazene but also to be more phytotoxic though less effective in the inhibition of sprouting of potato tubers. It was also toxic to tecnazene-resistant mutants of *F. caeruleum* produced by McKee (*45*). Interest in pentachloronitrobenzene was revived in 1956 in the United States when it was recommended for the control of certain soil-borne diseases, a feature being its long persistence in soil.

The fungicidal use of the high biological activity of the nitrated phenols, already employed as insecticides and herbicides, is frustrated by their generally intense phytotoxicity. The latter properties may be sufficiently reduced by the incorporation of suitable groupings as in **2 : 4-dinitro-1-thiocyanobenzene** (V), a compound which Staudermann (*46*) found effective for the control of downy mildew of vine. This compound is the active component of the commercial product " Nirit ". **2 : 4-Dinitro-6-(2-octyl)-phenyl crotonate** (VI), known more popularly as dinitrocaprylphenyl crotonate or by the trade names, " Karathane " and " Iscothan ", when tried in the field in 1945 as an acaricide was found to be an effective fungicide. It has survived mainly because of its toxicity to powdery mildews, first reported by Sprague (*47*). Rich and Horsfall (*48*) found that on glass

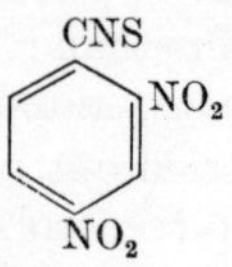

CNS
NO_2
NO_2
(V)

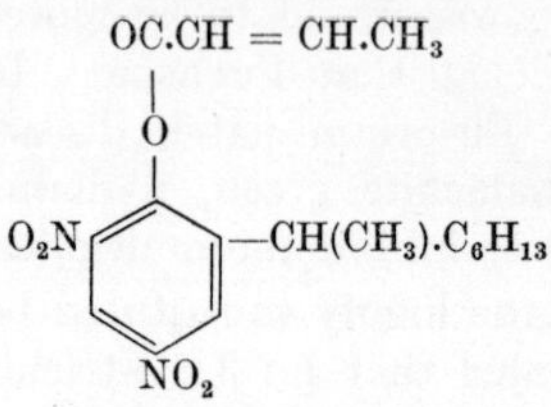

OC.CH = CH.CH_3
O
O_2N
—$CH(CH_3).C_6H_{13}$
NO_2
(VI)

slides the deposit loses its fungitoxicity on drying, which they suggested was due to hydrolysis and loss of crotonic acid. For this reason they considered that the latter radical was responsible for the fungicidal activity of the molecule, and that on the leaf hydrolysis is prevented whereby the formation of a highly phytotoxic alkylated dinitrophenol is avoided. Crotonaldehyde was found, however, by McGowan and his colleagues (*49*) to be but feebly toxic to the fungus spores of their tests ; they concluded that in ethylenic compounds fungistatic activity was associated with the tendency of the substituents to withdraw electrons from the double bond. A later recruit to this series is **1-chloro-2 : 4-dinitronaphthalene**, reported by Soenen and Werotte (*50*) to be an effective foliage protectant and more potent in laboratory tests than either ziram, zineb or captan.

Mildew-proofing to withstand the high humidities and temperatures of the tropics was an urgent problem of the recent war but few satisfactory fungicides emerged (*51*), an interesting exception being the volatile fungicide, *m*-cresol acetate (" **Cresatin** "). This need promoted an examination of the bisphenols (*52*). The parent compound, diphenyl, C_6H_5–C_6H_5, was used by Farkas (*53*) to impregnate paper wraps to prevent fungal spoilage of citrus fruits. Marsh, Butler and Clark (*54*) in a systematic study of this group of compounds found that those most active tended to a general

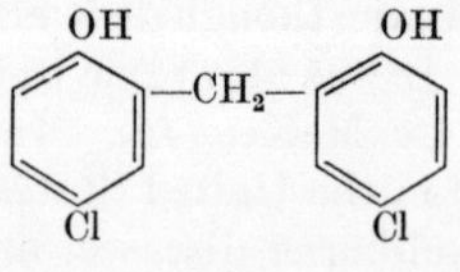

OH OH
—CH_2—
Cl Cl
(VII)

structure, represented by 2 : 2′-methylene bis(*p*-chlorophenol), Compound G4 (VII), though the substitution of the –CH_2 bridge by the –S–, –CO–, –$CH(CH_3)$– or –CH:CH– groups gave effective fungicides. That with the sulphur bridge, bis(2-hydroxy-5-chlorophenyl) sulphide was found by Horsfall and Rich to give good control of apple scab but failed in field tests because it produced a russet of the fruit (*55*).

REFERENCES

(*1*) Wüthrich, E., *Z. PflKrankh.*, 1892, **2**, 16, 81.
(*2*) Montgomery, H. B. S., Moore, M. H. and Shaw, H., *Rep. E. Malling Res. Sta.*, 1942, p. 26.
(*3*) Martin, H., *Sci. Hort.*, 1949, **9**, 143.
(*4*) Howard, F. L., Locke, S. B. and Keil, H. L., *Proc. Amer. Soc. hort. Sci.*, 1947, **45**, 131.
(*5*) Hamilton, J. M. and Mack, G. L., *Proc. N.Y. St. hort. Soc.*, 1947, p. 9.
(*6*) Wöber, A., *Z. PflKrankh.*, 1920, **30**, 51.
(*7*) Nielsen, L. W., *Mem. Cornell. Univ. agric. Exp. Sta.*, 248, 1942.
(*8*) McCallan, S. E. A. and Wilcoxon, F., *Contr. Boyce Thompson Inst.*, 1934, **6**, 479.
(*9*) Appel, O., *Flugbl. biol. Reichsanst. Berl.*, 63, 1917.
(*10*) Roberts, J. W. and Pierce, L., *Phytopathology*, 1932, **22**, 415.
(*11*) Harry, J. B., Wellman, R. H., Whaley, F. R., Thurston, H. W. and Chandler, W. A., *Contr. Boyce Thompson Inst.*, 1948, **15**, 195.
(*12*) Waite, M. B., *Circ. U.S. Dep. Agric. Bur. Plant Ind.*, 58, 1910.
(*13*) Morse, W. J., *Bull. Maine agric. Exp. Sta.*, 223, 1914.
(*14*) Whetzel, H. H., McCallan, S. E. A. and Loh, T. C., *Phytopathology*, 1929, **19**, 83.
(*15*) Horton, E. and Salmon, E. S., *J. agric. Sci.*, 1922, **12**, 269.
(*16*) Goodwin, W., Martin, H. and Salmon, E. S., *J. agric. Sci.*, 1926, **16**, 302.
(*17*) Garbowski, L. and Leszczenko, P., abstr. in *Rev. appl. Mycol.*, 1926, **5**, 505.
(*18*) Szembel, S. J., abstr. in *Rev. appl. Mycol.*, 1931, **10**, 500.
(*19*) Guozdenović, F., abstr. in *Z. PflKrankh.*, 1902, **12**, 242.
(*20*) Woodfin, J. C., *N.Z. J. Agric.*, 1927, **35**, 298.
(*21*) Dorogin, G., *Z. PflKrankh.*, 1913, **23**, 334.
(*22*) Patkaniane, A., *Rep. int. Cong. Phytopath., Holland*, 1923, p. 275.
(*23*) Nattrass, R. M., *J. Minist. Agric.*, 1926–27, **33**, 265, 1017.
(*24*) Martin, H. and Salmon, E. S., *J. agric. Sci.*, 1932, **22**, 595.
(*25*) Halstead, B. D. and Kelsey, J. A., *Bull. New Jersey agric. exp. Sta.*, 167, 1903.
(*26*) Barker, B. T. P. and Lees, A. H., *A. R. agric. hort. Res. Sta. Bristol*, 1914, p. 73.
(*27*) McWhorter, F. P., *Phytopathology*, 1927, **17**, 201.
(*28*) Martin, H. and Salmon, E. S., *J. agric. Sci.*, 1931, **21**, 638.
(*29*) Martin, H. and Salmon, E. S., *J. agric. Sci.*, 1933, **23**, 228.
(*30*) Schmidt, E. W., *Ber. dtsch. bot. Ges.*, 1924, **42**, 131.
(*31*) Kramer, O., abstr. in *Rev. appl. Mycol.*, 1928, **7**, 221.
(*32*) Goodwin, W., Salmon, E. S. and Ware, W. M., *J. agric. Sci.*, 1929, **19**, 185.
(*33*) Morris, L. E., *Mem. Shirley Inst.*, 1926, **5**, 321.
(*34*) Fargher, R. G., Galloway, L. D. and Probert, M. E., *Mem. Shirley Inst.*, 1930, **9**, 37.
(*35*) Bewley, W. F. and Orchard, O. B., *Ann. appl. Biol.*, 1932, **19**, 185.
(*36*) Martin, H. and Salmon, E. S., *J. agric. Sci.*, 1934, **24**, 469.
(*37*) Truffaut, G. and Pastac, I., *C. R. Acad. Agric., France*, 1929, **15**, 1058 ; Truffaut, G., *Hort. Abst. Imp. Bur. Fruit Prod.*, 1935, **5**, Ab. 387.
(*38*) Pastac, I., *Chim. et Industr.*, 1932, **27**, 851.
(*39*) Bennett, F. T., *J. Board greenkeeping Res.*, 1933, **3**, 79.
(*40*) Takahashi, W. N., *Science*, 1948, **107**, 226.
(*41*) Tanner, C. C., Greaves, W. S., Orrell, W. R., Smith, N. K. and Wood, R. E. G., *Brit. Intell. Obj. Sub-comm. Final Report*, 1480, 1947.
(*42*) Smieton, M. J., *J. Pomol.*, 1939, **17**, 195.
(*43*) Brown, W., *Ann. appl. Biol.*, 1947, **34**, 422.
(*44*) Brook, M., *Nature, Lond.*, 1952, **170**, 1022.
(*45*) McKee, R. K., *Nature, Lond.*, 1951, **167**, 611.
(*46*) Staudermann, W. cited by Martin, H. and Shaw, H., *B.I.O.S. Final Report*, 1095, H.M. Stationery Office, London, 1946.

(*47*) Sprague, R., *Proc. Wash. St. hort. Ass.*, 1949, **45**, 47.
(*48*) Rich, S. and Horsfall, J. G., *Phytopathology*, 1949, **39**, 19.
(*49*) McGowan, J. C., Brian, P. W. and Hemming, H. G., *Ann. appl. Biol.*, 1948, **35**, 25.
(*50*) Soenen, A. and Werotte, L., *Agricultura, Louvain*, 1956, **4**, 241.
(*51*) " Report on Tropic Proofing," *Ministry of Supply*, London, 1945.
(*52*) Marsh, P. B. and Butler, M. L., *Industr. engng Chem.*, 1946, **38**, 701.
(*53*) Farkas, A., *Hadar*, 1939, **12**, 227.
(*54*) Marsh, P. B., Butler, M. L. and Clark, B. S., *Industr. engng Chem.*, 1949, **41**, 2176.
(*55*) Horsfall, J. G., *Principles of Fungicidal Action.* Chronica Botanica Co., Waltham, Mass., 1956, p. 185.

CHAPTER VIII

INSECTICIDES (I)

INORGANIC COMPOUNDS

Broadly speaking, inorganic compounds are not soluble in fats and can be insecticidal only if digested, for which reason this group of insecticides are stomach poisons. With the discovery of synthetic compounds such as DDT highly toxic to insects both by contact and by ingestion, the growers' reliance on these older stomach poisons has diminished but they include several insecticides still in wide use.

THE ARSENIC GROUP

Although arsenic was long known to be poisonous to insects, its use as an insecticide only became general about the middle of the last century. In France, the use of arsenicals for the destruction of insects was prohibited by the French Ordinance of 1846, Article 10, but so successful were they in the United States that their reintroduction into Europe was inevitable.

Paris Green. Following the appearance of Colorado beetle in the eastern parts of the United States during the early part of the twentieth century, desperate remedies were tried, and in 1867 Markham (*1*) found in Paris Green, an arsenical pigment, a successful means of control. In 1872 Le Baron (*2*) used this material against the canker worm, and it was in repeating these experiments in 1878 that E. P. Haynes observed that Paris Green simultaneously controlled the codling moth (*Carpocapsa pomonella*).

The main constituent of Paris Green, known in Germany as Schweinfurtergrün, and also called Emerald Green, French Green and Mitis Green, is a compound of copper acetate and copper arsenite of formula $(CH_3COO)_2Cu.3Cu(AsO_2)_2$. The material, made originally from the variable raw material verdigris, was far from constant in composition (*3*). Avery (*4*), in an examination of Paris Greens prepared by different methods, found that the ratio of copper meta-arsenite to copper acetate was usually nearly 3 : 1 but sometimes approached 2 : 1. The methods of manufacture of Paris Green have since become more standardized. To distinguish the improved products, the older name of Schweinfurtergrün, which had become associated with an inferior insecticide, was abandoned in Germany in favour of such names as Uraniagrün, Silesiagrün, Titaniagrün, Frukusgrün, St. Urbansgrün, and Elafrosin. The specification proposed by Hilgendorff (*5*), to which these products conform, requires a definite degree of fineness and limits the water-soluble arsenic to below 3.5 per cent expressed as arsenic trioxide. The product must also contain at least 55 per cent total arsenic oxide, at least 30 per cent cupric oxide and at least 10 per cent acetic acid.

The specification adopted in Gr. Britain (*6*) requires similar chemical properties but the content of water-soluble arsenic must not exceed 1.5 per cent arsenic trioxide.

Copper aceto-arsenite, the main constituent of Paris Green, is readily decomposed. Avery and Beans (*7*) showed that even the purest sample available breaks down on treatment with water to form water-soluble arsenic. Further, Holland and Reed (*8*) showed that arsenic goes into solution on treatment with either carbon dioxide or ammonia. This ready hydrolysis is, as will be shown later, the most objectionable feature of an excellent insecticide.

London Purple. Because of the success of Paris Green against Colorado beetle, an arsenical residue from the manufacture of magenta was sent in 1878 by Hemingway & Co., a London firm, to C. E. Bessey for trial as a substitute. The material proved successful and was given the name London Purple. The active component is calcium arsenite together with some arsenate. The material often contained a higher percentage of arsenic than Paris Green, but proved so variable that its use as an insecticide was abandoned.

Between the years 1868 and 1892, the date of the first use of lead arsenate, a number of other arsenicals were tried as substitutes for Paris Green. Products similar in nature to London Purple, such as Paris Purple and the arsenite of copper known as Scheele's Green, met with no marked success. It was, however, realized that the successful arsenical would be of more constant composition and would give rise less easily to soluble arsenic, in the form of arsenite or arsenate, to which the foliage injury was attributed. Work in this direction finally led to the introduction of lead arsenate.

Lead Arsenate. This arsenical was first employed by Moulton in 1892, who, experimenting on the control of the gipsy moth (*Lymantria dispar*) proposed the precipitation of sodium arsenate by lead acetate in order to obviate the foliage injury caused by the soluble arsenical compounds. Tests with this material, to which the name " Gypsine " was given, were reported by Fernald (*9*) and it soon became established.

The lead arsenates most commonly employed for insecticidal purposes and described in America as " acid " lead arsenates, approximate in composition to diplumbic hydrogen arsenate $PbHAsO_4$. This compound is comparatively stable, almost insoluble in water and though containing the heavy lead atom remains well in suspension.

In the United States, lead arsenates of a more basic nature have been employed under the name of neutral or " basic " lead arsenates, or triplumbic arsenate (T.P. arsenate). It was at one time considered that triplumbic ortho-arsenate, $Pb_3(AsO_4)_2$, was the main constituent of such products, but the work of Robinson and Tartar (*10*) and Streeter and Thatcher (*11*) has shown that the compounds present are members of a series of ill-defined basic arsenates.

Calcium Arsenate. The need to replace the lead of lead arsenate by a less poisonous metal led to the commercial production of calcium arsenates free from water-soluble arsenic. Apart from London Purple, arsenites of lime had previously been suggested as insecticides in 1889 by Gillette (*12*), and independently recommended by Kilgore (*13*). The history of the use of arsenate of lime as an insecticide is not known, but Pickering (*14*) stated that it had been in use in the United States prior to 1907.

The chemistry of the calcium arsenates is in many ways similar to that of the lead arsenates. It was at one time thought that besides the dicalcium hydrogen arsenate $CaHAsO_4$, there existed a definite tricalcium salt $Ca(AsO_4)_2$. In repeating the methods given by Robinson (*15*) for the preparation of these two compounds, Goodwin and Martin (*16*) found that, whereas the dicalcium hydrogen arsenate may be prepared as the crystalline monohydrate $CaHAsO_4,H_2O$, the preparation of the tricalcium salt presented difficulties. They showed that, as with the lead arsenates, there is probably a continuous series of basic calcium arsenates formed and that no definite break corresponding to one definite basic arsenate is shown. Clifford and Cameron (*17*) also concluded, from phase rule studies, that there is no evidence that tricalcium arsenate can persist in contact with aqueous solutions. On the other hand, Pearce and his colleagues (*18*) found phase-rule evidence of the existence, at 90° C of four compounds, dicalcium hydrogen arsenate, pentacalcium dihydrogen arsenate $Ca_5H_2(AsO_4)_4$, tricalcium arsenate and a basic arsenate $[Ca_3(AsO_4)_2]_3.Ca(OH)_2$, the last named compound being absent in the system at 35° C.

With lead arsenate, the material in general use as an insecticide is diplumbic hydrogen arsenate. With the calcium arsenates, however, dicalcium hydrogen arsenate is too soluble for safe application to foliage. Commercial calcium arsenates are therefore prepared so as to contain an excess of lime. Those examined by Cook and McIndoo (*19*) were all more basic than the tricalcium arsenate. Such preparations may be considered to consist of mixtures of indefinite basic calcium arsenates and calcium hydroxide. Goodwin and Martin (*16*) showed that the basic calcium arsenates are hydrolysed in aqueous suspension, yielding calcium hydroxide which will be re-precipitated as calcium carbonate. The ultimate result is a slow transformation of the basic calcium arsenates to dicalcium hydrogen arsenate and calcium carbonate. Thus the solubility of the dicalcium salt is temporarily reduced by the addition of lime, to which is due in part the increased freedom from phytotoxicity of calcium arsenates containing excess lime.

Miscellaneous Arsenicals. Attempts to use **arsenious oxide** (white arsenic) direct as a spray insecticide have met with little success. Although the oxide is not readily soluble in cold water, solution is, in the presence of impurities likely to be present on the leaf surface, sufficiently rapid to produce foliage damage. Reports are to be found of its successful use without injury to the plant (*20*).

Of the commoner arsenicals tried by Volck and Luther (*21*), at the California Experiment Station, **zinc arsenite** was recommended for general use; but, in other localities, caused severe spray damage (*22*). Its ready decomposition by carbon dioxide to liberate soluble forms of arsenic was demonstrated by Schoene (*23*). In Sweden, a **zinc fluoroarsenate** has been introduced as a foliage protectant by the Bolinden Mining Co. (*24*). This product appears to be a basic zinc arsenate precipitated in the presence of fluoride ions when the hydroxy groups of the basic arsenate are partly replaced by fluoride ions, yielding a product less readily decomposing to water-soluble compounds of arsenic. Kirby (*25*) concluded that it was no better than lead arsenate for codling moth control.

In 1919, Patten and O'Meara (*26*) reported unsuccessful trials of **magnesium arsenate**, made by a process then recently patented (U.S.P. 1,344,018). Later patents (U.S.P. 1,420,978, 1,466,983 ; B.P. 251,330) were the results of attempts to prepare magnesium arsenates safe for foliage application. The physical and chemical properties of these products were examined by Dearborn (*27*) but it would appear (see *28*) that there is risk of spray damage even with these forms of magnesium arsenate.

For use on tobacco, on which the white deposit left by lead and calcium arsenates is objectionable because it resembles mould and reduces market value, a product "**Manganar**" was introduced. This material is manufactured under patent from pyrolusite and white arsenic in the presence of oxidizing agents. The mixed manganese arsenates so formed are mixed with lime and burnt umber may be added as a diluent, the final product containing approximately 40 per cent. arsenic pentoxide, 32 to 40 per cent manganese oxide and about 16 per cent lime. The general conclusion of trials reviewed by Dearborn (*29*) is that Manganar is not as effective an insecticide as lead arsenate.

Action of Arsenicals on the Insect

In the arsenical compounds which so far have been used as insecticides, the arsenic is present either as arsenite or arsenate, i.e., in trivalent or pentavalent form. Although Cook and McIndoo (*19*) were unable to find differences in the toxicity of arsenite and arsenate of the same base and arsenic content, Campbell (*30*) showed that trivalent arsenic is more toxic to the tent caterpillar *Malacosoma americana*. Fink (*31*) concluded, from studies of oxygen consumption, that arsenious acid is 57 per cent. more toxic than arsenic acid but Munson and Yeager (*32*), from statistical studies on the concentration-survival time relationships for cockroaches (*Periplaneta americana*), found no differences between the ions formed from sodium arsenate and sodium metarsenite, differences in toxicity between these compounds being largely explained by differences in degree of dissociation. Fink's use of the depression of oxygen consumption of the poisoned insect arose from the hypothesis, developed mainly by Voegtlin and his associates (*33*), that the toxic action of arsenical compounds is associated with an inactivation of the oxidising enzymes, in particular, by combination with the thiol groups (–SH) of reduced glutathione. The more general hypothesis that the toxic action of arsenic is due to combination with the thiol groups of enzyme proteins led Stocken and Thompson (*34*) to seek a thiol compound which would form a more stable complex with the arsenical compound than did the enzyme proteins. One such compound was 2 : 3-dimercaptopropanol subsequently introduced as BAL (British anti-Lewisite) as an antidote for lewisite, an arsenical chemical warfare agent.

In comparative tests of arsenical insecticides on Colorado beetle, Wilson

(*35*) found that Paris Green was the most toxic; arsenate of lead, zinc arsenite and calcium arsenate were a trifle slower in action. Cook and McIndoo (*19*), from experiments on a range of insects, arrived at the following conclusions: Paris Green is more toxic than zinc arsenite; acid lead arsenate is more toxic than basic lead arsenate, but on equivalent arsenic contents Paris Green is no more toxic than acid lead arsenate.

As a working hypothesis it may be taken that the relative toxicity of an arsenical is closely related to its chemical instability. This hypothesis has been examined by Fulmek (*36*), who found the order of toxicity of metallic arsenites to various species of caterpillars to be: magnesium, lead = calcium = copper, iron and zinc; the metallic arsenates being of the order: lead, copper, calcium, magnesium, zinc and iron. The arsenites proved more toxic than the arsenates and the order of toxicity of the latter ran roughly parallel to their solubilities in buffer solutions of *p*H 9.0, which is approximately that of the digestive juices of the insects. Work upon the hydrogen ion concentration of the insect intestine has been surveyed by Uvarov (*37*) and variations are shown according to insect species and part of gut examined. Trappmann and Nitsche (*38*), for example, found the mid-gut of representative species of insects to be of *p*H 7.58–9.07. Swingle (*39*) obtained evidence of a wider range, from *p*H 5.9 to 9.6, and of the importance of phosphates in controlling the solution of arsenic by the more alkaline digestive secretions, especially from lead hydrogen arsenate. He attributed the decomposition to the precipitation of lead as lead phosphate. The relative toxicities of lead hydrogen arsenate, calcium arsenate and magnesium arsenate towards nine species of leaf-eating insects were correlated with the relative amounts of soluble arsenic formed from these arsenates in phosphate buffer solutions of the hydrogen ion concentration found in the mid-gut of these insects. The solubilities of the three arsenates were approximately equal in phosphate buffer solutions of *p*H 6.0–6.5; in more alkaline solutions, lead hydrogen arsenate was more soluble than the magnesium arsenate; in more acid solutions, the magnesium arsenate was more soluble than the lead arsenate. In a parallel manner, the lead arsenate was more toxic than magnesium arsenate to larvae with alkaline mid-guts, e.g., *Protoparce sexta* Johan., magnesium arsenate was more toxic than the lead arsenate to larvae of acid reaction in their mid-gut, e.g., *Epilachna varivestis*.

The possibility of devising a solubility test, similar in principle to the determination of "available" phosphate in phosphatic fertilizers, to determine the toxicity of a commercial arsenical has been reviewed by Borchers and May (*40*) and by Trappmann and Nitsche (*38*). The latter were sceptical of the value of such laboratory tests, for the practical worth of an arsenical will be determined, not only by its inherent toxicity, but by physical properties which affect retention on foliage. Further, impurities present in commercial products may affect biological performance by deterrent or emetic properties. Thus O'Kane (*41*) mentioned the case of the Japanese beetle (*Popillia japonica* Newm.), a pest difficult to poison with arsenicals which appear to act more as deterrents. By mixing with the spray an intestinal sedative such as bismuth subcarbonate, more of the poisoned leaves were eaten and a greater kill secured. Then again, the insect may vomit the irritant poison, as shown by Cook and McIndoo (*19*), who recorded that for honey bees, which voided none of the arsenical eaten, the fatal dose of arsenic was about six times less than that of the silkworm, which voided 90 per cent of the amount eaten.

The effect of sub-lethal doses of an arsenical upon the insect is of interest, for Friederichs and Steiner (*42*) and Voelkel (*43*) observed that the larvae of certain moths, when not killed, produced females of impaired fertility.

Action of Arsenicals on the Plant

Injury to foliage by arsenical sprays or dusts first appears as a wilting followed by a browning and shrivelling of the tissue often with interveinal necrosis. This damage was attributed to water-soluble arsenic compounds and Swingle, Morris and Burke (*44*) subdivided the arsenicals they tested into the " soluble " and therefore phytotoxic class and the " insoluble " class, e.g., lead arsenate, zinc arsenite. Owing to variance between different samples of the " insoluble " class, they gave no generalization on the relative " safety " of these arsenicals though it was realized that the presence of water-soluble arsenical compounds in these products was a cause of spray damage. Early in the history of lead arsenate, restrictions were imposed on the content of water-soluble arsenic and the same test has been applied, with equal or less success to other arsenical compounds of the " insoluble " class. With the calcium arsenates, the direct estimate of water-soluble arsenic proved an unreliable indication of phytotoxic properties and, following up earlier work by Kelsall and Herman (*45*), Pearce, Norton and Chapman (*46*) devised a chemical test for the " safeness to foliage " of commercial calcium arsenates involving a preliminary carbonation of the free lime present.

Another corollary of the observation that phytotoxic properties are associated with water-soluble arsenic is the attempt to reduce the latter by precipitation with " correctives " such as lime. Swingle, Morris and Burke concluded that the addition of lime reduces injury of calcium arsenite and Paris Green. Lime added to the calcium arsenates precipitates the soluble arsenic as basic calcium arsenates which slowly yield soluble arsenic under the influence of carbon dioxide (*16*).

The interactions of free lime and the more insoluble arsenicals are less simple. Campbell (*47*), investigating the cause of severe arsenical injury by acid lead arsenate to which lime had been added concluded that an interaction occurred resulting in the formation of basic lead arsenates and basic calcium arsenates. The latter compounds, on exposure, gave rise to the dicalcium arsenate, yielding more soluble arsenic than the original lead arsenate. On the other hand, trials by Van der Meulen and Van Leeuwen (*48*) showed that the addition of slaked lime to lead arsenate prevented injury to peach foliage. They showed, by laboratory experiments, that the reaction between calcium hydroxide and diplumbic hydrogen arsenate results in a decomposition of part of the latter to form lead hydroxide and basic calcium arsenates. The reaction is however too slow to produce an amount of soluble arsenic sufficient to cause injury.

Correctives added to lessen arsenical injury have been discussed by Ginsburg (*49*). Goodwin and Martin (*50*) observed that Bordeaux mixture appeared to be a more effective corrective than lime. This observation was confirmed by Moznette (*51*) in the case of Paris Green and calcium arsenate and by Parfentjev and Wilcoxon (*52*) for calcium arsenite.

Arsenical damage may arise through the presence, in the water used for spraying, of compounds reacting to form soluble arsenic. Ginsburg (*53*) found that hard waters rich in bicarbonates reacted with lead hydrogen arsenate to form water-soluble arsenates. With such waters, an anomalous situation appears, for Ginsburg found that the addition of soap, a spreader not generally recommended for lead arsenate because of interaction to produce soluble arsenates, reduced the tendency to spray damage through the preliminary precipitation of the soap by the hard water, the insoluble soap preventing the interaction of lead arsenate with the alkali carbonate simultaneously formed.

Patten and O'Meara (*54*) suggested that the carbon dioxide given off by the leaves may give rise to soluble arsenic, especially from calcium arsenates. The dew collected from cotton foliage was shown by Smith (*55*) to contain relatively large quantities of salts favouring the formation of soluble arsenic from calcium arsenate. Swingle, Morris and Burke (*44*) noted that the injury to the foliage occurs *via* the lower epidermis regardless of the number of stomata on the two surfaces. This would indicate a direct penetration of the thinner cuticle as distinct from the stomatal penetration of the sulphur sprays.

Supplements for Arsenical Insecticides

The arsenical insecticides depend, for efficiency, on the initial amount and tenacity of the spray residue and on degree of coverage. The addition of supplements to improve these properties was not at first practised, but soaps have been (*19*) and are still sometimes recommended, in spite of their interaction with the arsenical, to produce water-soluble arsenic compounds.

Lime casein was, at one time, frequently recommended and the free lime present will function as a corrective for injury from the more soluble types of arsenical. It was, however, the more uniform coating of the spray deposit which prompted the recommendation of lime casein by Lovett (*56*). He found that petroleum oil emulsions enhanced the tenacity of lead arsenate sprays. Spuler (*57*), who examined the effect of adding various types of petroleum oils to lead arsenate sprays against codling moth, found that, in one case, the combination of high-boiling petroleum oil and lead arsenate was as effective an insecticide as a spray containing four times the amount of lead arsenate alone. The increased difficulty of arsenical residue removal when applied in combination with oils has frequently been reported.

The use of glyceride oils as stickers for lead arsenate was suggested by Hood (*58*), who selected a refined fish oil for recommendation, emulsifying the oil by agitation with lead arsenate paste. The stability of the emulsion so prepared when petroleum oils are substituted for the glyceride oil is probably insufficient to ensure uniform application. Further, petroleum oils are found to cause a flocculation of the lead arsenate suspension (*59*), a disadvantage overcome by the addition of lime casein or other spreaders which act as deflocculators probably through their ability to function as emulsifiers. The relative merits of fish oil and petroleum oil were examined by Webster and Marshall (*60*) who also noted that tenacity was improved by the addition of oleic acid to the oils. From this work was evolved the so-called " inverted " spray mixture in which advantage is taken of preferential retention (see p. 78) to build up a spray load of oil-flocculated residue of high tenacity.

REFERENCES

(*1*) See *Amer. Ent.*, 1869, **1**, 219.
(*2*) Le Baron, W., *2nd Ann. Rep. Noxious Insects of Illinois*, 1872, p. 116.
(*3*) Van Slyke, L. L. and Andrews, W. H., *Bull. N. Y. St. agric. Exp. Sta.*, 222, 1902.
(*4*) Avery, S., *J. Amer. chem. Soc.*, 1906, **28**, 1155.
(*5*) Hilgendorff, G., *NachrBl. dtsch. PflSchDienst. Berl.*, 1930, **10**, 28.
(*6*) *Tech. Bull. Minist. Agric.*, 1, *2nd Ed.*, 1951.
(*7*) Avery, S. and Beans, H. T., *J. Amer. chem. Soc.*, 1901, **23**, 111.
(*8*) Holland, E. B. and Reed, J. C., *Rep. Mass. agric. Exp. Sta.*, 1911, p. 177.
(*9*) Fernald, C. H., *Bull. Mass. (Hatch) agric. Exp. Sta.*, 24, 1894.
(*10*) Robinson, R. H. and Tartar, H. V., *Bull. Ore. agric. Exp. Sta.* 128, 1915.
(*11*) Streeter, L. R. and Thatcher, R. W., *Indust. Engng Chem.*, 1924, **16**, 941.
(*12*) Gillette, C. P., *Bull. Iowa agric. Exp. Sta.* 10, 1890, p. 401.
(*13*) Kilgore, B. W., *Bull. N. C. agric. Exp. Sta.* 77B, 1891.
(*14*) Pickering, S. U., *J. chem. Soc.*, 1907, **91**, 307.
(*15*) Robinson, R. H., *J. agric. Res.*, 1918, **13**, 281.
(*16*) Goodwin, W. and Martin, H., *J. agric. Sci.*, 1926, **16**, 596.
(*17*) Clifford, A. T. and Cameron, F. K., *Industr. engng Chem.*, 1929, **21**, 69.
(*18*) Pearce, G. W. and Norton, L. B., *J. Amer. chem. Soc.*, 1936, **58**, 1104 ; Pearce, G. W. and Avens, A. W., *J. Amer. chem. Soc.*, 1937, **59**, 1258.
(*19*) Cook, F. C. and McIndoo, N. E., *Bull. U.S. Dept. Agric.*, 1147, 1923.
(*20*) e.g., Cooley, R. A., *Bett. Fruit*, 1920, **15**, No. 5, p. 9.
(*21*) Luther, E. E., *Bett. Fruit*, 1911, **5**, No. 8, p. 65.
(*22*) See *Rep. Ore. agric. Exp. Sta.*, 1913–14, p. 137.
(*23*) Schoene, W. J., *Tech. Bull. N.Y. St. agric. Exp. Sta.* 28, 1913.
(*24*) Lundbäck, S. V., *Proc. VIII Int. Congr. Ent.* 1950, p. 924.
(*25*) Kirby, A. H. M., *Rep. E. Malling Res. Sta.*, 1951, 160.
(*26*) Patten, A. J. and O'Meara, P., *Mich. agric. Exp. Sta. Quart. Bull.*, 1919, **2**, 83.
(*27*) Dearborn, F. E., *J. econ. Ent.*, 1930, **23**, 758.
(*28*) *J. econ. Ent.*, 1928, **21**, 36.
(*29*) Dearborn, F. E., *J. econ. Ent.*, 1930, **23**, 630.
(*30*) Campbell, F. L., *J. agric. Res.*, 1926, **32**, 359.
(*31*) Fink, D. E., *J. agric. Res.*, 1920, **33**, 993 ; *J. econ. Ent.*, 1927, **20**, 794.
(*32*) Munson, S. C. and Yeager, J. F., *J. econ. Ent.*, 1945, **38**, 634.
(*33*) Voegtlin, C., Dyer, H. A. and Leonard, C. S., *J. Pharmacol.*, 1925, **25**, 297.
(*34*) Stocken, L. A. and Thompson, R. H. S., *Biochem. J.*, 1946, **40**, 529, 535.
(*35*) Wilson, H. F., *Bull. Wis. agric. Exp. Sta.*, 303, 1919.
(*36*) Fulmek, L., *Fortsch. Landw.*, 1929, **4**, 209.
(*37*) Uvarov, B. P., *Trans. ent. Soc. Lond.*, 1929, **76**, 255.
(*38*) Trappmann, W. and Nitsche, G., *Mitt. biol. Abt. (Anst. Reichsanst.) Berl.*, 1933, **46**, 61.
(*39*) Swingle, H. S., *J. econ. Ent.*, 1938, **31**, 430.
(*40*) Borchers, F. and May, E., *Z. PflKrankh.*, 1931, **41**, 417.
(*41*) O'Kane, W. C., *Industr. Engng Chem.*, 1923, **15**, 911.
(*42*) Friederichs, K. and Steiner, P., *Z. angew. Ent.*, 1930, **16**, 189.
(*43*) Voelkel, H., *NachrBl. dtsch. PflSchDienst.*, 1930, **10**, 44.
(*44*) Swingle, D. B., Morris, H. E. and Burke, E., *J. agric. Res.*, 1923, **24**, 501.

(*45*) Kelsall, A. and Herman, F. A., *Sci. Agr.*, 1927, **7**, 207.
(*46*) Pearce, G. W., Norton, L. B. and Chapman, P. J., *Tech. Bull. N.Y. St. agric. Exp. Sta.*, 234, 1935.
(*47*) Campbell, F. L., *J. agric. Res.*, 1926, **32**, 77.
(*48*) Van der Meulen, P. A. and van Leeuwen, E. R., *J. agric. Res.*, 1927, **35**, 313.
(*49*) Ginsburg, J. M., *Bull. N. J. agric. Exp. Sta.*, 468, 1929.
(*50*) Goodwin, W. and Martin, H., *J. agric. Sci.*, 1928, **18**, 460.
(*51*) Moznette, G. F., *J. econ. Ent.*, 1930, **23**, 691.
(*52*) Parfentjev, I. A. and Wilcoxon, F., *Anz. Schadlingsk.*, 1929, **5**, 107, 123.
(*53*) Ginsburg, J. M., *J. econ. Ent.*, 1937, **30**, 583.
(*54*) Patten, A. J. and O'Meara, P., *Quart. Bull. Mich. agric. Exp. Sta.*, 1919, **2**, 83.
(*55*) Smith, C. M., *J. agric. Res.*, 1923, **26**, 191.
(*56*) Lovett, A. L., *Bull. Ore. agric. Exp. Sta.*, 169, 1920.
(*57*) Spuler, A., *Bull. Wash. St. agric. Exp. Sta.*, 232, 1929.
(*58*) Hood, C. E., *J. econ. Ent.*, 1925, **18**, 280 ; *Bull. U.S. Dep. Agric.*, 1439, 1926.
(*59*) Marshall, J., *Bull. Wash. St. agric. Exp. Sta.*, 350, 1937.
(*60*) Webster, R. L. and Marshall, J., *Bull. Wash. St. agric. Exp. Sta.*, 293, 1934.

THE FLUORINE GROUP

Sodium fluoride is an old-established cockroach poison which, according to Fulton (*1*) is not less toxic to the earwig *Forficula auricularia* L. than arsenious oxide. Ripley (*2*) recommended its use in poison baits. As it is too soluble in water to act as a protective insecticide, Marcovitch (*3*) recommended instead the less soluble sodium silicofluoride (sodium fluosilicate, Na_3SiF_6). Marcovitch also used calcium silicofluoride and cryolite (sodium aluminofluoride, Na_3AlF_6). His claims were substantiated by Gimingham and Tattersfield (*4*) and the use of natural cryolite from Greenland was recommended by Bovien (*5*).

A difficulty met early in the use of the silicofluorides as sprays or dusts was due to their high apparent density. Thus Walker and Mills (*6*) found that sodium silicofluoride was more toxic to the boll weevil *Anthonomus grandis* Boh. than calcium arsenate when compared on a volume basis, but that two to four times the weight of silicofluoride was required to cover effectively the area treated with calcium arsenate. Similarly the weight required per acre of barium silicofluoride or of cryolite was greater than that needed of calcium arsenate. Less dense forms of cryolite can be produced synthetically and trials of the synthetic product are reviewed by DeLong (*7*). Barium silicofluoride proved unduly corrosive to spray pumps and valves, a disadvantage which it is claimed (U.S.P. 1,931,367) is corrected in " Dutox " by adding a small proportion of a more soluble fluoride.

The fluorides were thought to act mainly as stomach poisons. Shafer (*8*) suggested that the cockroach, to rid itself of the irritating sodium fluoride, drew its legs through its mouth parts thereby ingesting the poison. Sweetman (*9*), however, showed that cockroaches dusted with sodium fluoride, were killed within twenty-four hours, even when their mouth parts were sealed. Griffiths and Tauber (*10*) considered that although sodium fluoride acts as a stomach poison, its contact effect is primarily and mainly responsible for insecticidal action. The effect of fluoride on the insect intestine was examined by Pilat (*11*). The mechanism of the toxic action of fluorides after penetration into the insect body is obscure but an interference with the enzymes associated with cellular oxidation is suspected. The literature of this subject has been reviewed by Borei (*12*).

The mammalian toxicity of the inorganic fluorides has received special study because of the connection, observed in 1892 by Sir James Crichton-Browne (*13*), between fluorine and dental caries. Abnormally high contents of inorganic fluorine in drinking water produced a mottling of the enamel of teeth; indeed Smith and Smith (*14*) became concerned that a danger existed at levels as low as 2–3 ppm. (expressed as fluorine). Nevertheless a content of 1 ppm. (see *15*) in drinking water has been shown to reduce the incidence of dental caries and fluoridation of water supplies is practised in many localities.

REFERENCES

(*1*) Fulton, B. B., *J. econ. Ent.*, 1923, **16**, 369.
(*2*) Ripley, L. B., *Bull. ent. Res.*, 1924, **15**, 29.
(*3*) Marcovitch, S., *Industr. engng Chem.*, 1924, **16**, 1249; *Bull. Tenn. agric. Exp. Sta.*, 131, 1914; Marcovitch, S. and Stanley, W. W., *Bull. Tenn. agric. Exp. Sta.*, 140, 1929.
(*4*) Gimingham, C. T. and Tattersfield, F., *Industr. engng Chem.*, 1925, **17**, 323; *Ann. appl. Biol.*, 1928, **15**, 649.
(*5*) Bovien, P., *Verh. VII Int. Kongr. Ent.*, 1939, **4**, 2961.
(*6*) Walker, H. W. and Mills, J. E., *Industr. engng Chem.*, 1927, **19**, 703.
(*7*) DeLong, D. M., *Ohio J. Sci.*, 1934, **34**, 175.
(*8*) Shafer, G. D., *Tech. Bull. Mich. agric. Exp. Sta.*, 21, 1915.
(*9*) Sweetman, H. L., *Canad. Ent.*, 1941, **73**, 31.
(*10*) Griffiths, J. T. and Tauber, O. E., *J. econ. Ent.*, 1943, **36**, 536.
(*11*) Pilat, M., *Bull. ent. Res.*, 1935, **26**, 165.
(*12*) Borei, H., *Ark. Kemi Min. Geol.*, 1945, **20**, A, 1.
(*13*) Crichton-Browne, J., *Lancet*, 1892, **2**, 6.
(*14*) Smith, H. V. and Smith, M. C., *Tech. Bull. Ariz. agric. Exp. Sta.*, 43, 1932.
(*15*) See *The fluoridation of domestic water supplies in North America as a means of controlling Dental Caries.* Her Majesty's Stationery Office, London, 1953.

MISCELLANEOUS STOMACH INSECTICIDES

It is a general rule that protective contact insecticides will function as stomach poisons if ingested but these materials find wider practical use as contact poisons. A direct insecticide such as nicotine which has little protective ability because of its volatility can be made a stomach poison by conversion to a suitable non-volatile derivative (see p. 161). The present section is therefore confined mainly to those substances, some not inorganic compounds, which have been suggested in place of lead arsenate or other stomach poisons.

The ground rhizome of **White Hellebore*** (*Veratrum album* L.) was used as far back as 1842 (*1*) against gooseberry sawfly, *Pteronidia ribesii* Scop. The powder, on exposure to air, rapidly loses its poisonous properties and was therefore of special value for the treatment of crop plants at times too near harvest for the use of arsenicals. The causes of the instability of hellebore powder are obscure and the chemistry of the active constituents

*The common name hellebore is misleadingly applied to plants of the genus *Veratrum*; the genus *Helleborus* does not yield veratrum alkaloids.

is little known though similar alkaloids are present in **Green American hellebore** (*Veratrum viride* Ait.) and **Sabadilla** (*Schoenocaulon officinale* Gray) both of which have been used as insecticides. Interest in sabadilla was revived by the observation of Allen, Dicke and Harris (*2*) that heat or alkali treatment greatly increased the toxicity of oil extracts of the freshly-powdered seeds to the housefly *Musca domestica* L. (U.S.P. 2,348,949 and 2,390,911). The mixed alkaloids, collectively called veratrine, which are presumably present in the seeds as salts and liberated by the alkali treatment, are known to contain cevadine ($C_{32}H_{49}N$), veratridine, sabadilline, sabadine and cevine of which cevadine proved the most toxic to adult milkweed bugs, *Oncopeltus fasciatus*, Dallas (*3*). Against houseflies, however, veratridine was the more toxic (*4*). From *V. viride*, Seiferle and his co-workers (*5*) isolated several alkaloids of the cevadine group but were unable to trace the cause of the greater insecticidal action of the crude alkaloid fraction. The complex chemistry of these alkaloids was reviewed by Prelog and Jeger (*6*).

In 1945 Pepper and Carruth (*7*) reported that the wood of the shrub *Ryania speciosa* Vahl. was toxic to the European corn borer *Pyrausta nubilalis* and it has since been found a promising stomach poison for codling moth control (*8*), where its special virtue is a less drastic action on beneficial insects than have less specific insecticides such as DDT. The nature of its slow action on insects is unknown but considered to be due to the alkaloid ryanodine, the chemistry of which was discussed by Kelly *et al.* (*9*).

Of inorganic substitutes for the arsenicals, **lead chromate** was suggested by Lefroy (*10*). He claimed that this compound was comparable to lead arsenate in insecticidal properties ; that, being yellow, it is more easily seen on the sprayed foliage ; that, being extremely insoluble, it does not decompose readily nor is easily washed off by rain. Tests by Johnson (*11*) against Colorado beetle gave unsatisfactory results. Moore and Campbell (*12*), who examined a number of inorganic and organic derivatives, found **cuprous cyanide** as toxic as lead arsenate to Japanese beetle and **cuprous thiocyanate** effective against the tent caterpillar *Malacosoma americana*, the latter product failing against Japanese beetle. The efficiency of cuprous cyanide as a stomach insecticide was later confirmed by Bulger (*13*), Fleming and Baker (*14*) and Speyer (*15*), but it is phytotoxic (*16*). Cuprous cyanide was included in Moore and Campbell's tests because it was thought that, by slow decomposition, hydrogen cyanide would be liberated as the active insecticide. When it was used as a poison bait, admixed with bran or dried blood, Speyer found that decomposition of cuprous cyanide does occur with the result that the efficiency of the bait is lost.

An ingenious idea (U.S.P. 2,062,911) was that the insecticidal properties of chromium derivatives and of the thiocyanate group might be coupled in the co-ordination complex obtained by the fusion of ammonium dichromate and ammonium thiocyanate. This product, known as Reinecke's salt, is ammonium diammonochromium tetrathiocyanate $NH_4[Cr(NH_3)_2Cr(SCN)_4]$ or, for convenience, **ammonium reineckate**. The insecticidal properties of this compound and of the reineckates of organic bases of known toxicity were examined by Guy (*17*) who found that the most potent of those tested was **piperidinium reineckate**, as toxic as lead arsenate to Mexican bean beetle (*Epilachna varivestis*) and with greater speed of kill. In field trials in combination with protective fungicides, spray damage and a lower insecticidal efficiency were obtained, the result apparently of the decomposition of the reineckate in the presence of lime to form ammonium thiocyanate, a compound of intense phytotoxicity.

Among the fifty compounds selected by McAllister and Van Leeuwen (*18*) as worthy of further trial as lead arsenate substitutes against codling moth was **triphenyl phosphine**. Guy (*17*) tested this compound and some of its phosphonium derivatives against Mexican bean beetle and Colorado beetle and found both the methyl triphenyl phosphonium chloride and iodide approached lead arsenate in toxicity.

Of organic compounds suggested as substitutes for the arsenicals, one of the oldest is **dinitro-o-cresol** (2-methyl-4 : 6-dinitrophenol) which was the active constituent of " Antinnonin ", a product introduced in 1892 (*19*) against the Nun moth (*Lymantria monacha*). The subsequent development of the dinitrocresols as acaricides and contact insecticides is discussed on p. 198.

The success of certain organic sulphur compounds as fungicides led to a testing of a series of these products against mosquito larvae (*20*). The most successful compound to be tested was **thiodiphenylamine** (dibenzo-1 : 4-thiazine or phenothiazine), a powerful helminthicide. The insecticidal properties of this compound were established by contemporary investigators (see *17*, *21*) but in practical trials, surveyed by Zukel (*22*), the commercial product gave erratic results. Minor difficulties are its hydrophobic character which make compounding to sprayable products difficult, oxidation to dark-coloured compounds of objectionable appearance on sprayed fruit and a tendency to cause skin irritation to the spray operators through photosensitization (*23*). The addition of suitable antioxidants (e.g., B.P. 488,428–9) retards oxidation but attempts to find more stable derivatives have been unsuccessful (*24*). Indeed, the introduction of an *N*-methyl group to replace the labile hydrogen halved the toxicity to mosquito larvae. Zukel (*22*), in examining the cause of the variable field results, concluded that, of thiodiphenylamine (I) and its oxidation products phenothiazone (II) and thionol (III), I and II are effective contact insecticides to the American

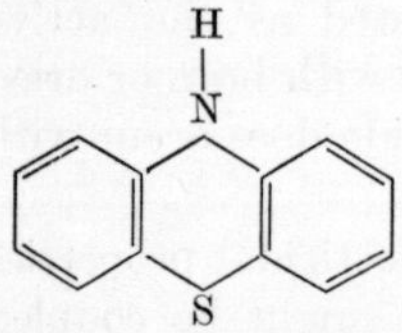

(I) 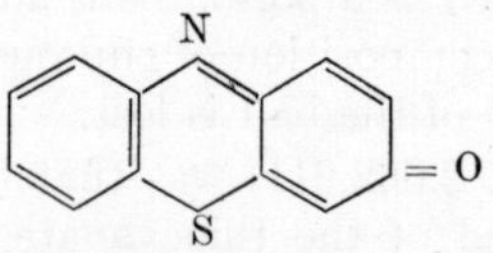 (II) 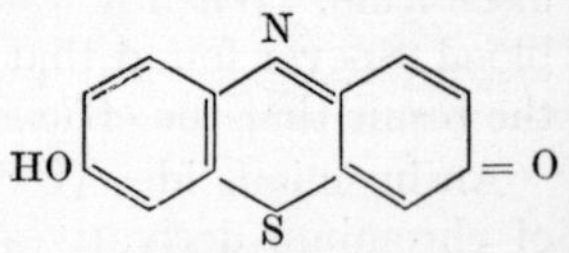 (III)

cockroach *Periplaneta americana*, but that none acted as a stomach poison, due, he suggested, to the impermeability of the intestinal wall to these compounds. He attributed the toxicity of I and II to the inhibition of the respiratory enzyme, cytochrome oxidase.

Because of its similarity in structure to thiodiphenylamine, carbazole was selected for systematic study by the I.G. Farbenindustrie (*25*). Two derivatives, 1 : 3 : 6 : 8-tetranitrocarbazole and 1 : 8-dichloro-3 : 6-dinitrocarbazole, subsequently appeared on the market as the active components of "**Nirosan**" and "**Nirosit**" respectively. In field trials disappointing results were again obtained but both products found use as selective insecticides against the vine moths *Clysia ambiguella* Hl. and *Polychrosis botrana* Schiff.

An unusual compound, S_4N_4, known by the misleading name of **sulphur nitride**, was reported by Fulton (*26*) to be a stomach poison (U.S.P. 2,101,645) The chemistry and probable constitution of the compound are discussed by Arnold, Hugill and Hutson (*27*) and its structure by Clark (*27*). A probable disadvantage is its explosive character, not conspicuous in the pure product, but which may appear in the compounded products.

REFERENCES

(*1*) *Gdnrs'. Chron.*, 1842, June 18th, p. 397.
(*2*) Allen, T. C., Dicke, R. J. and Harris, H. H., *J. econ. Ent.*, 1944, **37**, 400.
(*3*) Allen, T. C., Link, K. P., Ikawa, M. and Brunn, L. K., *J. econ. Ent.*, 1945, **38**, 293.
(*4*) Ikawa, M., Dicke, R. J., Allen, T. C. and Link, K. P., *J. biol. Chem.*, 1945, **159**, 517.
(*5*) Seiferle, E. J., Johns, I. B. and Richardson, C. H., *J. econ. Ent.*, 1942, **35**, 35.
(*6*) Prelog, V. and Jeger, O., In *The Alkaloids*, ed. by Manske, R. H. F. and Holmes, H. L., Academic Press Inc., 1953, Vol. III, pp. 270–309.
(*7*) Pepper, B. B. and Carruth, L. A., *J. econ. Ent.*, 1945, **38**, 59.
(*8*) Pickett, A. D. and Patterson, N. A., *Canad. Ent.*, 1953, **85**, 472.
(*9*) Kelly, R. B., Whittingham, D. J. and Wiesner, K., *Canad. J. Chem.*, 1951, **29**, 905; *Chem. & Ind.*, 1952, 857.
(*10*) Lefroy, H. M., *Agric. J. India*, 1910, **5**, 138.
(*11*) Johnson, F. A., *Bull. U.S. yep. Agric. Bur. Ent.*, 109, p. 53, 1912.
(*12*) Moore, W. and Campbell, F. L., *J. agric. Res.*, 1924, **28**, 395.
(*13*) Bulger, J. W., *J. econ. Ent.*, 1932, **25**, 261.
(*14*) Fleming, W. E. and Baker, F. E., *J. agric. Res.*, 1934, **49**, 39.
(*15*) Speyer, E. R., *Rep. exp. Res. Sta., Cheshunt*, 1934, p. 70.
(*16*) Marsh, R. W., Martin, H. and Munson, R. G., *Ann. appl. Biol.*, 1937, **24**, 853.
(*17*) Guy, H. G., *Bull. Del. agric. Exp. Sta.*, 206, 1937.
(*18*) McAllister, L. C. and Van Leeuwen, E. R., *J. econ. Ent.*, 1930, **23**, 907.
(*19*) Cooper, W. F. and Nuttall, W. H., *Ann. appl. Biol.*, 1915, **1**, 273.
(*20*) Campbell, F. L., Sullivan, W. N., Smith, L. E. and Haller, H. L., *J. econ. Ent.*, 1934, **27**, 1176.
(*21*) Smith, L. E., Munger, F. and Siegler, E. H., *J. econ. Ent.*, 1935, **28**, 727.
(*22*) Zukel, J. W., *J. econ. Ent.*, 1944, **37**, 796.
(*23*) DeEds, F. and Thomas, J. O., *J. Parasitol.*, 1941, **27**, 143.
(*24*) Schaffer, P. S., Haller, H. L. and Fink, D. E., *J. econ. Ent.*, 1937, **30**, 361.
(*25*) Martin, H. and Shaw, H., *Brit. Intell. Obj. Sub-comm., Rep.* 1095, 1946.
(*26*) Fulton, R. A., *J. econ. Ent.*, 1938, **31**, 545.
(*27*) Arnold, M. H. M., Hugill, J. A. C. and Hutson, J. M., *J. chem. Soc.*, **1936**, **1645**.
(*28*) Clark, D., *J. chem. Soc.*, 1952, 1615.

CHAPTER IX

INSECTICIDES (II)

NATURALLY-OCCURRING CONTACT INSECTICIDES

It is still customary to distinguish between those contact insecticides which occur in nature and those of synthetic origin. But the dividing line has disappeared for the organic chemist has produced compounds so akin to the natural product that their separate discussion is impossible. The distinction serves only for the separation into groups of, (1) naturally-occurring insecticides and their synthetic relatives, the former being either of vegetable origin, e.g., nicotine or pyrethrum, or of mineral origin, e.g., the petroleum oils and, (2) synthetic insecticides which bear no obvious relationship to naturally-occurring compounds and which will be the subject of Chapter X.

NICOTINE

The value of tobacco, a plant introduced into Europe about 1560,* as an insecticide was known to Peter Collinson, who in 1746 wrote from England advising Bartram, an American botanist, to use water in which tobacco leaves had been soaked against the plum curculio (*1*). The poisonous property of tobacco was traced to the presence of nicotine, an alkaloid discovered as far back as 1828 by Posselt and Reimann.

Nicotine is a mobile, colourless liquid boiling at 247° C, which on ageing takes on a brownish colour, deepening until almost black and becoming more viscous. It is soluble in water and of disagreeable smell. Appreciably volatile, with a vapour pressure of 0.0425 mm. Hg/25° C (*2*), it may be distilled in steam, a property to which no doubt its early isolation was due. It is a well-defined base and in the tobacco plant it is combined with malic and citric acids. It appears on the market both as the free alkaloid and as the sulphate, sold under the name " Black Leaf 40 ", containing 40 per cent of the base.

Both commercial nicotine and the sulphate keep well, apart from the " ageing " changes already mentioned. These changes do not appear, however, to be accompanied by any decline of the toxic properties of the materials. Upon exposure nicotine, being hygroscopic, absorbs moisture. According to McDonnell and Young (*3*) equilibrium is reached at a content of approximately 89 per cent of nicotine, whereas with the sulphate equilibrium is established, by evaporation, at a content of about 45 per cent nicotine.

*It is reported that in 1560 John Nicot sent seeds to the French king, describing them as germs of a medicinal plant of great value.

Action on the Insect. McIndoo (*4*) concluded that nicotine acts, whatever be the manner of application, as a fumigant and that nicotine vapour passes into the tracheae of the insect paralysing the nervous system. De Ong (*5*) established a close parallelism between the volatility of nicotine and its toxic action to aphids ; the salts of nicotine are much less volatile and are correspondingly less toxic.

The possibility that the greater volatility of nicotine is not the only factor responsible for its greater toxicity over its salts was suggested by Richardson and Shepard (*6*), who showed that, in solution, the free base is five to seven times as toxic as nicotine sulphate or hydrochloride to mosquito larvae. The acidity of the spray should therefore not be below *p*H 5 to ensure that the nicotine is present as the undissociated molecule and not as nicotinium ion.

Evidence has since accumulated that nicotine can penetrate directly through the insect integument. Glover and Richardson (*7*) recovered nicotine from body parts of cockroaches exposed to nicotine vapour in such a way that entry through the spiracles was excluded and Richardson (*8*) later suggested that the degree of susceptibility of insect species to nicotine is largely due to differences in rate of penetration, a detoxication process enabling those species in which penetration is slow to survive. Wigglesworth (*9*) noted the greater toxicity of nicotine to the bug *Rhodnius prolixus* Stål., of which the outer cuticle layers had been damaged by abrasion or by the action of surface-active substances.

The need for a substitute for lead arsenate in codling moth control to reduce residue hazards prompted much research on the use of nicotine as a stomach poison. In this case volatility is a disadvantage and attention was given to methods of " fixing " nicotine to reduce the rapidity of loss by volatilization. Of the fixed nicotines, the tannate was found of promise by Headlee, Ginsburg and Filmer (*10*). Combinations of nicotine and bentonite have also become popular since their introduction by Driggers and Pepper (*11*), and a variety of other nicotine compounds and complexes have appeared in the patent literature, e.g., nicotine-peat (U.S.P. 2,107,058) (*12*), Quebracho-fixed nicotine (U.S.P. 2,152,236). Hansberry (*13*) tested many " fixed nicotines " against codling moth larvae and found water-insoluble derivatives such as the cupricyanide toxic as stomach poisons although ineffective as aphicides (*14*).

The ability of nicotine to function as an ovicide against codling moth was reported by De Sellem in 1916 (*15*) and Feytaud (*16*) observed that eggs of *Polychrosis botrana* failed to hatch after nicotine treatment in spite of the complete development of the embryo. A similar " aborting " action of nicotine has been reported on apply sawfly, *Hoplocampa testudinea* Klug. (*17*), and was examined in detail by Shaw and Steer (*18*) and Misaka (*19*). Possible explanations are the fixation of the nicotine by the egg-shell and surrounding plant tissue, the insect being killed by a stomach poison action at emergence, and the penetration of nicotine to the embryo which is killed only when the development of the nervous system is sufficiently advanced. The phenomenon has practical value for it reduces the difficulty of correct timing of the application of nicotine sprays against pests such as apple sawfly.

Action on the Plant. No record can be found of any damage to the plant caused by nicotine itself in the use of nicotine washes and dusts. In this respect nicotine, like many of the plant products used as insecticides, approaches the ideal.

Supplements for use in Nicotine Sprays. An intimate contact between spray and insect is an obvious requisite for a contact insecticide and if, as McIndoo (*20*) reported, nicotine acts in the vapour phase by penetration into the insect tracheae, the addition of supplements to aid penetration should enhance its efficiency. Moore and Graham (*21*) examined the ability of liquids to enter the tracheae of insects and found that, of the aqueous solutions tested, only soap solutions were capable of this penetration. O'Kane and his co-workers (*22*) and Wilcoxon and Hartzell (*23*) also demonstrated the penetration of soap solutions, the latter showing that penetration occurred with live but not with dead larvae of *Protoparce quinquemaculata* Haw. and suggesting that factors other than capillary forces are involved in penetration.

Of the penetrants, soap has been recommended in a most emphatic manner. Moore (*24*) stated that the efficiency of free nicotine is sometimes increased 50 per cent by the addition of soap. Its special virtue appears to lie in the fact that soap is a most efficient spreader and that, being alkaline in reaction, it does not affect the volatility of the nicotine. Further, soap itself has an insecticidal action, a point discussed on p. 194.

A number of soap preparations containing nicotine ready for use have been marketed, but it was found that such products deteriorated with an apparent loss of nicotine. This deterioration was investigated by McDonnell and Nealon (*25*), who showed that it is accompanied by the formation of an insoluble resinous product with which the nicotine is carried down. McDonnell and Graham (*26*) found that this insoluble product was formed by oxidation and occurred only in preparations made from drying oils (i.e., unsaturated glyceride oils). When packed so that air was excluded, the preparations suffered no loss in nicotine content during two years of storage.

As soaps are compounds of fatty acids and alkalis, the suggestion has been made of employing nicotine itself as the base, giving nicotine " soaps ". Moore (*27*), for example, recommended nicotinium oleate. A number of nicotine " soaps " were examined by Hoyt (*28*), who found marked differences according to the type of fatty acid employed but such compounds have not found practical utilization. An idea of a similar character is the attachment of long side chains in a manner such that the nicotine molecule is made capillary active (B.P. 401,707). Austin, Jary and Martin (*29*) recorded preliminary trials of " Tinocine D ", a long-chain nicotinium bromide, which indicated that insecticidal properties were not markedly reduced by this drastic change of the molecular structure. Dodecyl nicotinium bromide was found, by Hansberry and Norton (*14*) to be somewhat less toxic than nicotine-soap to *Aphis rumicis*.

Nicotine Dusts. Dusts, prepared by the adsorption of nicotine or nicotine sulphate upon a finely divided " carrier " were first used experimentally by Smith (*30*) in 1917. As the efficiency of the dust would appear to depend, within limits, upon the rate of volatilization of the nicotine, their success may be connected with the relatively large surface area of the particles, upon which the nicotine is exposed to evaporation. Headlee and Rudolfs (*31*) have shown that this volatilization is controlled by the adsorptive character of the carrier and that the more finely divided " colloidal " carriers are inferior to the crystalline carriers. This conclusion was confirmed by Thatcher and Streeter (*32*), who classed kaolin, kieselguhr and talc as " absorbent " carriers, the crystalline gypsum, sulphur, etc., as " inert " carriers.

With the nicotine sulphate dusts, the chemical factor of the liberation of the nicotine comes into play. Headlee and Rudolfs recommended for use as the carrier a mixture of dolomite (magnesium and calcium carbonates) and hydrated lime, both of which will react with the nicotine sulphate. Thatcher and Streeter classed such carriers, called by Headlee and Rudolfs " chemical accelerators ", as " active ", recommending the use of hydrated lime or precipitated chalk.

REFERENCES

(*1*) Waite, M. B. *et al.*, *Yearb. U.S. Dep. Agric.*, 1925, p. 453.
(*2*) Norton, L. B., Bigelow, C. R. and Vincent, W. B., *J. Amer. chem. Soc.*, 1940, **62**, 261.
(*3*) McDonnell, C. C. and Young, H. D., *Bull. U.S. Dep. Agric.*, 1312, 1925.
(*4*) McIndoo, N. E., *J. agric. Res.*, 1916, **7**, 89.
(*5*) de Ong, E. R., *J. econ. Ent.*, 1923, **16**, 486.
(*6*) Richardson, C. H. and Shepard, H. H., *J. agric. Res.*, 1930, **41**, 337.
(*7*) Glover, L. H. and Richardson, C. H., *Iowa St. Coll. J. Sci.*, 1936, **10**, 249.
(*8*) Richardson, C. H., *J. econ. Ent.*, 1945, **38**, 710.
(*9*) Wigglesworth, V. B., *J. exp. Biol.*, 1945, **21**, 97.
(*10*) Headlee, T. J., Ginsburg, J. M. and Filmer, R. S., *J. econ. Ent.*, 1930, **23**, 45.
(*11*) Driggers, B. F. and Pepper, B. B., *J. econ. Ent.*, 1934, **27**, 432.
(*12*) Markwood, L. N., *Industr. engng Chem.*, 1936, **28**, 561, 648.
(*13*) Hansberry, R., *J. econ. Ent.*, 1942, **35**, 915.
(*14*) Hansberry, R. and Norton, L. B., *J. econ. Ent.*, 1941, **34**, 80.
(*15*) See Headlee, J. T., *J. econ. Ent.*, 1935, **28**, 172.
(*16*) Feytaud, J., *Ann. Epiphyt.*, 1915, **2**, 109.
(*17*) Kearns, H. G. H., Marsh, R. W. and Martin, H., *Rep. agric. hort. Res. Sta., Bristol*, 1935, p. 37.
(*18*) Shaw, H. and Steer, W., *J. Pomol.*, 1939, **16**, 364.
(*19*) Misaka, K., *Bull. Imp. agric. Exp. Sta.*, 1932, **3**, 225 ; *J. agric. Exp. Sta., Tokyo*, 1938, **3**, 239.
(*20*) McIndoo, N. E., *J. agric. Res.*, 1916, **7**, 89.
(*21*) Moore, W. and Graham, S. A., *J. agric. Res.*, 1918, **13**, 523.
(*22*) O'Kane, W. C., Westgate, W. A., Glover, L. C. and Lowry, P. R., *Tech. Bull. N. H. agric. Exp. Sta.*, 39, 1930.
(*23*) Wilcoxon, F. and Hartzell, A., *Contr. Boyce Thompson Inst.*, 1931, **3**, 1.
(*24*) Moore, W., *J. econ. Ent.*, 1918, **11**, 443.
(*25*) McDonnell, C. C. and Nealon, E. J., *Industr. engng Chem.*, 1924, **16**, 819.
(*26*) McDonnell, C. C., and Graham, J. J. T., *Industr. engng Chem.*, 1929, **21**, 70.
(*27*) Moore, W., *J. econ. Ent.*, 1918, **11**, 341.
(*28*) Hoyt, L. F., *Industr. engng Chem.*, 1924, **16**, 1171.
(*29*) Austin, M. D., Jary, S. G. and Martin, H., *Hort. Educ. Assoc. Yearb.*, 1932, **1**, 85.

(*30*) Smith, R. E., *Bull. Calif. agric. Exp. Sta.*, 336, 1921.
(*31*) Headlee, T. J. and Rudolfs, W., *Bull. N.J. agric. Exp. Sta.*, 381, 1923.
(*32*) Thatcher, R. W. and Streeter, L. R., *Bull. N.Y. St. agric. Exp. Sta.*, 501, 1923.

Compounds of Structure akin to Nicotine

In addition to nicotine, several alkaloids have been isolated from tobacco and related plants but of these, two only, nornicotine and anabasine, have been found to be markedly insecticidal. A number of related compounds have also arisen in attempts to discover a substitute more suited to commercial synthesis than nicotine.

The molecular structure of nicotine was established by its successful synthesis in 1904 by Pictet and Rotschy (*1*) as 3-(1-methyl-2-pyrrolidyl)-pyridine (I). The molecule thus contains two nuclei, pyridine and pyrollidine, the hydrogenated derivative of pyrrol. Pyridine and pyrrol are but

CH_2—CH_2
—CH CH_2
N $N.CH_3$

(I)

feebly toxic to insects (*2*). Richardson and Smith (*3*), however, found that a crude dipyridyl oil was highly insecticidal. The dipyridyls are compounds of two pyridine molecules linked by the elimination of one atom of hydrogen from each molecule. Six isomers are possible, all present in the crude oil, but the crude oil was found to be more insecticidal than any of the isomers. This high toxicity was traced to the presence of 3-(2-piperidyl)-pyridine (II), subsequently named neonicotine.

CH_2
CH_2 CH_2
—CH CH_2
N NH

(II)

The same compound was later isolated from a Turkestan weed, *Anabasis aphylla* L. (Chenopodiaceae) by Orékhov and Menschikov (*4*) who named it anabasine. The identity of anabasine and neonicotine, except that the former is the laevorotatory (—) form (see p. 17) whereas the latter is the optically inactive (±) form, was confirmed by Smith (*5*).

Anabasine is a water-soluble liquid of b.p. 280.9°/760 mm. and of vapour pressure 2.5 mm./79° (*6*). It was extracted commercially from *A. aphylla* and marketed as a solution of the sulphate containing about 40 per cent total alkaloids of which about 70 per cent was anabasine. Smith (*7*) found it also present in the leaves and roots of the tree tobacco, *Nicotiana glauca*.

Nornicotine differs in structure from nicotine only in the absence of the methyl group on the pyrollidine nucleus. It is a colourless liquid, b.p. 270–271°, and is less volatile than nicotine in steam. Chemically it is 3-(2-pyrollidyl)-pyridine (III). Markwood (*8*) found, among the strains of *N. tabacum* L. which he examined, one of which (—)-nornicotine comprised

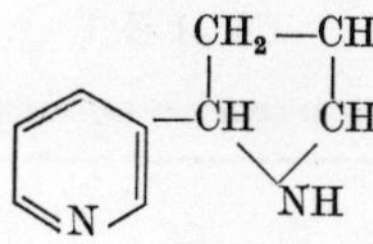

(III)

95 per cent of the total (0.73 per cent) alkaloids. The (—)-form had previously been isolated by Smith (*9*) from *N. sylvestris* Spegaz and Comes and was found by Späth, Hicks and Zajic (*10*) in the Australian plant *Duboisia hopwoodii* F. v. Muell. Bottomley *et al.* (*11*) examined many specimens of this solanaceous plant and found nicotine and (—), (+), and (±) forms of nornicotine present in various proportions sometimes reaching 3–5 per cent of dried material. Analyses of the alkaloid content of cured tobaccos are given by Jeffrey and Tso (*12*).

In discussing the reasons why nicotine, nornicotine and anabasine are potent insecticides, attention will be paid only to aphicidal properties. The basic structure of these compounds is of two rings, one pyridine, the other being five or six-membered. Saturation of the second ring seems requisite for Richardson and Smith found that nicotyrine, 3-(1-methyl-2-pyrryl)-pyridine (IV) was from seven to ten times less toxic to *Aphis rumicis*

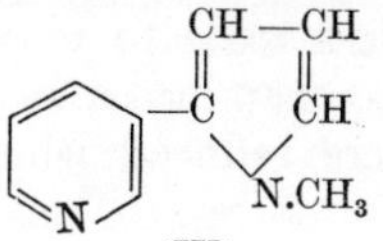

(IV)

than nicotine. Moreover the rupture of the second ring to give metanicotine (V) reduced toxicity. Secondly, the linkage between the rings must be in

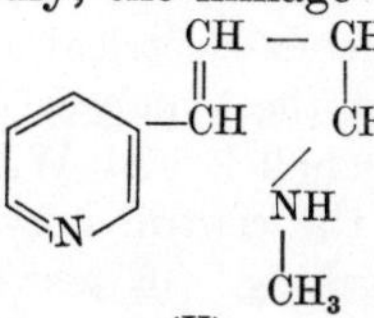

(V)

the 3-2 position for both α-nicotine and α-nornicotine, in which the linkage is 2-2, are much less toxic than the corresponding β-compounds, as shown in Table I, taken from the results of Richardson, Craig and Hansberry (*13*). Similarly, Smith, Richardson and Shepard (*14*) observed that dipyridyls and piperidyl-pyridines with the 2-3 linkage were more toxic than those with the linkage in other positions. Thirdly, it would seem of advantage if the second ring is six-membered for, as shown in Table I, anabasine is more aphicidal than nicotine. Fourthly, the spacial distribution of the rings must be of importance for it will be seen from Table I that (±)-nicotine is but half as toxic as the (—) form, from which it would appear that (+)-nicotine is non-toxic. Hansberry and Norton (*15*) in direct tests found (+)-nicotine substantially less toxic than (—)-nicotine but found, with nornicotine, that the (—), (+) and (±) forms were almost equally toxic. Finally, the presence of the *N*-methyl group is probably detrimental to the toxicity of nicotine, a result masked in Table I for a comparison of

TABLE I

Median Lethal Concentration against Aphis rumicis

Compound	MLD (mg. per 100 ml.)
(±)-β-nicotine	96
(±)-β-nornicotine	45
(±)-α-nicotine	1,496
(±)-α-nornicotine	1,514
(−)-β-nicotine	49
Anabasine	5

the inactive forms of nicotine and nornicotine is complicated by the toxicity of (+)-nornicotine. But Hansberry and Norton found (−)-nornicotine more aphicidal than (−)-nicotine, and Campbell, Sullivan and Smith (*16*) found *N*-methyl anabasine less toxic to mosquito larvae than anabasine. Turner and Saunders (*17*) who examined many nicotinium derivatives found that aphicidal properties were reduced when radicals are added to the pyrollidine nitrogen.

The insecticidal activity of nicotine and related compounds can be traced to a disruption of the nerve mechanism. Roeder and Roeder (*18*) found that treatment of the isolated roach nerve cord with nicotine at 1×10^{-3}M caused a cessation of spontaneous electrical activity due it is thought to a prevention of transmission at the synapses (the area of contact between two or more nerve cells). Schallek and Wiersma (*19*) found a similarity in the effects of nicotine on the central nervous systems of both crayfish and vertebrates in this " blocking " of nerve transmission. If, however, the blocked nerve ganglion preparation is left until recovery, a second application of nicotine or a related insecticide failed to prevent transmission of nerve impulses (*20*), a protection which Brown (*21*) suggested may be connected with the tolerance, to nicotine, of insects infesting tobacco.

REFERENCES

(*1*) Pictet, A. and Rotschy, A., *Ber. dtsch. chem. Ges.*, 1904, **37**, 1225.
(*2*) Tattersfield, F. and Gimingham, C. T., *Ann. appl. Biol.*, 1927, **14**, 217.
(*3*) Richardson, C. H. and Smith, C. R., *J. agric. Res.*, 1926, **33**, 597.
(*4*) Orékhov, A. and Menschikov, G., *Ber. dtsch. chem. Ges.*, 1931, **64**, B, 266 ; 1932, **65**, B, 232.
(*5*) Smith, C. R., *J. Amer. chem. Soc.*, 1932, **54**, 397.
(*6*) Nelson, O. A., *J. Amer. chem. Soc.*, 1934, **56**, 1989.
(*7*) Smith, C. R., *J. Amer. chem. Soc.*, 1935, **57**, 959.
(*8*) Markwood, L. N., *J. Ass. off. agric. Chem. Wash.*, 1940, **23**, 804.
(*9*) Smith, C. R., *J. econ. Ent.*, 1937, **30**, 724.
(*10*) Späth, E., Hicks, C. S. and Zajic, E., *Ber. dtsch. chem. Ges.*, 1935, **68**, B, 1388.
(*11*) Bottomley, W., Nottle, R. A. and White, D. E., *Aust. J. Sci.*, 1945, **8**, 18.

(12) Jeffrey, R. N. and Tso, T. C., *J. agric. Food Chem.*, 1955, **3**, 680.
(13) Richardson, C. H., Craig, L. C. and Hansberry, T. R., *J. econ. Ent.*, 1936, **29**, 850.
(14) Smith, C. R., Richardson, C. H. and Shepard, H. H., *J. econ. Ent.*, 1930, **23**, 862.
(15) Hansberry, R. and Norton, L. B., *J. econ. Ent.*, 1940, **33**, 734.
(16) Campbell, F. L., Sullivan, W. N. and Smith, C. R., *J. econ. Ent.*, 1933, **26**, 500.
(17) Turner, N. and Saunders, D. H., *Bull. Conn. agric. Exp. Sta.*, 512, p. 98, 1946.
(18) Roeder, K. and Roeder, S., *J. cell. comp. Physiol.*, 1939, **14**, 1.
(19) Schallek, W. and Wiersma, C. A. G., *J. cell. comp. Physiol.*, 1948, **31**, 35.
(20) Wiersma, C. A. G. and Schallek, W., *Science*, 1947, **106**, 421.
(21) Brown, A. W. A., *Insect Control by Chemicals*, London, Chapman & Hall Ltd., 1951, p. 302.

PYRETHRUM

The early history of pyrethrum as an insecticide is obscure. According to Lodeman (*1*), an Armenian named Jumtikoff discovered that an effective insect powder used by the tribes of the Caucasus was prepared from the flower-heads of certain species of pyrethrum. His son, in 1828, began the manufacture of the powder on a large scale and, in 1850, the product was introduced into France. Gnadinger (*2*) stated that the earliest sources were the species *Chrysanthemum roseum* Bieb. and *C. carneum* Bieb. and that the more effective *C. cinerariaefolium* Trev. was later discovered in Dalmatia about 1840.

Pyrethrum is today derived mainly from *C. cinerariaefolium*; other species reputed to be toxic to insects are listed by Gnadinger. The so-called African or German pyrethrum, used in pharmacy under the names "Radix pyrethri" and "Pellitory root", is derived from the roots of *Anacyclus pyrethrum* DC. and *A. officinarum* Hayne, and though of insecticidal interest (see p. 197) finds no use in horticulture. To avoid confusion the active principles of pellitory root, once named pyrethrin, have been re-named pellitorine (*3*).

Knowledge of the nature of the active constituents of *C. cinerariaefolium* is due largely to the classical researches of Staudinger and Ruzicka (*4*), but, of earlier work, that of Fujitani (*5*) and Yamamoto (*6*) is noteworthy. Fujitani isolated an effective constituent "Pyrethrone" which he showed to be a mixture of esters, the compounds of an alcohol "Pyrethrol", with various acids, of which two were later isolated by Yamamoto. Pyrethrone was critically examined by Staudinger and Ruzicka who isolated two esters to which they gave the names Pyrethrin I and Pyrethrin II which are compounds of a ketonic alcohol, pyrethrolone, with two acids which were shown to be (I), a monocarboxylic acid of structure:

$$C(CH_3)_2{:}CH.CH\left\langle \begin{array}{l} C(CH_3)_2 \\ | \\ CH.COOH \end{array} \right. \quad \text{(I)}$$

and, (II), a related dicarboxylic acid:

$$HOOC.C(CH_3){:}CH.CH\left\langle \begin{array}{l} C(CH_3)_2 \\ | \\ CH.COOH \end{array} \right. \quad \text{(II)}$$

The monocarboxylic acid is now known as chrysanthemic acid whereas the dicarboxylic acid, present in pyrethrum as the monomethyl ester, is known as pyrethric acid.

These acids may exist in *cis* and *trans* forms for the hydrogens in the 1 and 3 positions may lie on the same (*cis*) or opposite (*trans*) sides of the *cyclo*propane ring. Since neither of these positional isomers has a plane of symmetry, each has optical isomers. Campbell and Harper (*7*) established that the natural chrysanthemic acid is (±)-*trans*-2 : 2-dimethyl-3-*iso*butenyl-*cyclo*propane-1-carboxylic acid.

The alcohol, pyrethrolone, was thought by Staudinger and Ruzicka to be a single compound but in 1945 LaForge and Barthel (*8*) proved the presence of a second alcohol, cinerolone ($C_{10}H_{14}O_2$), containing one carbon atom and one double bond less than pyrethrolone ($C_{11}H_{14}O_2$). Staudinger and Ruzicka had shown that the latter on reduction gave a *cyclo*pentanone derivative proving the presence of a five-membered ring and LaForge and Haller (*9*) showed pyrethrolone to be an hydroxy-3-methyl-2-pentadienyl *cyclo*pent-2-enone. Confirmation of the suggested position of the double bond in this ring was obtained spectroscopically by Gillam and West (*10*). LaForge and Barthel (*8*) proved that cinerolone contained the same ring. Staudinger and Ruzicka considered that the hydroxyl group of pyrethrolone was in the 5-position, but LaForge and Soloway (*11*) showed that it was not adjacent to the keto group in dihydrocinerolone and suggested that, in both compounds, it is in the 4-position. Its presence confers optical activity and the natural alcohols are dextrorotatory. The elucidation of the structure of the pentadienyl side chain proved most difficult but, by spectrographic methods, West (*12*) confirmed the conclusion of LaForge and Barthel that there is a terminal $=CH_2$ group in pyrethrolone. The second double bond in the side chain of pyrethrolone and the double bond of the cinerolone side chain are in the 2 : 3 position and give rise to geometrical (*cis* and *trans*) isomers. The pyrethrolones therefore are chemically 3-methyl-2-penta-2′ : 4′-dienyl*cyclo*pent-2-en-4-olones (IIIa) and the cinerolones are 3-methyl-2-but-2′-enyl*cyclo*pent-2-en-4-olones (IIIb).

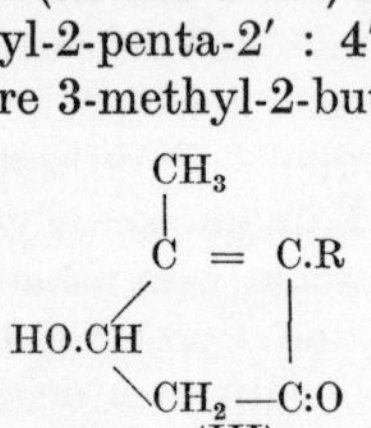

```
         CH3
          |
          C  =  C.R
         /        |
  HO.CH           |
         \CH2 —C:O
           (III)
```

(a) Pyrethrolone R= $—CH_2.CH{:}CH.CH{:}CH_2$

(b) Cinerolone R= $—CH_2.CH{:}CH.CH_3$

Confirmation, by synthesis, was obtained by Crombie, Harper and Newman (*13*) that the keto-alcohol present in the natural pyrethrins is (+)-*cis*-pyrethrolone ; a result which supports earlier opinions (see LaForge and Green, *14*) that the natural cinerins are esters of *cis*-cinerolone.

The active constituents of pyrethrum are therefore a mixture of the pyrethrins and cinerins, the former being the esters of pyrethrolone with either acid I or the monomethyl ester of acid II, the latter the corresponding esters of cinerolone. The insecticidal uses to which pyrethrum is put are largely controlled by the stability of these esters.

Inspection of the structure of the esters reveals a high degree of unsaturation indicative of instability and a ready decomposition. Staudinger and Harder (*15*), for example, pointed out that loss of toxicity might occur in " Savon-pyrèthre ", an alcoholic-soap extract of pyrethrum popular on the continent, firstly because hydrolysis of the pyrethrins would be accelerated by alkali, secondly, because of the possible replacement of pyrethrolone by ethyl alcohol. Tattersfield and Hobson (*16*) found, however, that alcoholic extracts of pyrethrum retained their toxicity in temperate climates over many months and loss of toxicity in alkaline dispersion was surprisingly small.

It has for long been known that pyrethrum flowers lose their insecticidal potency on storage and Abbott (*17*) showed that, although the whole flower-heads could be kept in sealed containers for over five years without deterioration, the ground flower-heads in open dishes practically lost their toxicity within 150 weeks. Tattersfield and Hobson (*16*) also observed a rapid loss in exposed dusts prepared by the absorption of pyrethrum extracts. Tattersfield (*18*) showed that oxidation is an important factor in the inactivation and he therefore added antioxidants, finding that many phenolic derivatives such as pyrocatechol, resorcinol, hydroquinone, pyrogallol and tannic acid, but neither phenol nor phloroglucinol, delayed loss of toxicity. Concentrated pyrethrum extracts lose toxicity even when stored in the dark. As absorption spectra show that the β-unsaturated ketonic grouping and the acidic fragments remain unaltered, West (*19*) concluded that this change, probably polymerization, involves the pentadienyl side chain.

Synthetic analogues

The successful synthesis of the chrysanthemic acids by Harper and his colleagues and of *cyclo*pentenolones of the cinerolone type by Schechter, Green and LaForge (*20*) opened up possibilities of man-made pyrethrum analogues. The first to be developed was allethrin, the commercial synthesis of which is described by Sanders and Taff (*21*), the ester of chrysanthemic acid and allethrolone, 3-methyl-2-allyl*cyclo*pent-2-en-4-olone (IIIc, $R = \text{-}CH_2.CH{:}CH_2$). It differs from cinerin I in that the terminal methyl group of the side chain of the latter is replaced by hydrogen and that the synthetic product may contain up to eight isomeric forms. Matsui, working in collaboration with LaForge and his colleagues (*22*) replaced allylacetone, used in the synthesis of allethrin, by the more available furfurylacetone producing furethronyl chrysanthemate or furethrin. The side chain of its *cyclo*pentenolone (IIId) contains the same number of carbon atoms and double bonds in the same position as pyrethrolone (IIIa). In cyclethrin (*23*)

$$\begin{array}{ll} R = -CH_2-\begin{array}{c} CH.O.CH \\ \| \quad\quad \| \\ CH-CH \end{array} & \qquad R = -\begin{array}{c} CH-CH=CH \\ | \qquad\qquad | \\ CH_2 \text{——} CH_2 \end{array} \\ \qquad\qquad \text{(IIId)} & \qquad\qquad \text{(IIIe)} \end{array}$$

the allyl side chain of allethrin is replaced by the *cyclo*pentenyl group (IIIe).

The pyrethroids, allethrin, furethrin and cyclethrin share the insecticidal properties of natural pyrethrins and cinerins though differing in relative toxicity according to the species of the test insect. The lower reactivity of the side chain renders the synthetic products more stable in storage than the pyrethrins and cinerins. Allethrin, however, appears to be less effective than the natural products in its synergism with piperonyl butoxide (*24*), a reason for the search which led to the discovery of cyclethrin which is reported to be more effective than allethrin when used with synergists.

Toxicity-structure relationships

A characteristic action of the pyrethroids is the rapidity with which the insect, particularly the housefly, is paralysed and the discussion of relationship between toxicity and molecular structure must include this " knockdown " effect. From the already mentioned loss of toxicity on storage, it is evident that the decomposition products are poor insecticides. The hydrolysis products, namely, the chrysanthemic acids and the *cyclo*pentenolones, were known from the earlier work of Staudinger and Ruzicka to be virtually non-toxic. The value of earlier comparisons of the relative potency of pyrethrins I and II, discussed by Gnadinger (*2*) is questionable now that it is known that these pyrethrins are mixtures, but the range of pyrethroids indicates that much latitude is permissible in the choice of the side chain of the *cyclo*pentenolone. Staudinger and Ruzicka concluded that at least one double bond was required in this side chain and Gersdorff (*25*) found that the saturation of both double bonds of the side chain of pyrethrin I reduced to only 6 per cent its toxicity to houseflies. Elliott (*26*) concluded, from an examination of synthetic analogues, that for highest toxicity the *cyclo*pentenolone ring should have, in the 2-position, a side chain with a double bond between its second and third carbon atoms. Similarly, saturation of the double bond of the chrysanthemic acid gives esters of lower toxicity (*25*) though the rate of " knockdown " is but little affected.

The speed of knockdown indicates an immediate penetration and a rapid spread, presumably along a nerve membrane, suggesting a high surface activity. It may be inferred that the pyrethroid molecule has a flexibility permitting a rapid and close fit to a nerve interface. The geography of the molecule and the spacial distribution of its reactive groups must then be important factors in determining toxicity.

Yet the activity of the synthetic pyrethroids indicates that this spacial distribution, for example, in the *cis-trans* optical isomers, is not as decisive as it is, for example, with nicotine. Nevertheless the *cis-trans* chrysanthemic acids yield esters differing in toxicity. Elliott found that allethrin prepared from the (+)-*trans* isomer was twice as toxic as that from (+)-*cis* isomer. Optical isomerization exerts a greater influence for Gersdorff (*27*) found that in both cinerin I and allethrin, the esters of the (+)-acid were from three to eight times as toxic to houseflies as the esters of the racemic acid. Elliott reported that the allethronyl esters of the (+)-acid were fifty times as toxic as those of the (—)-form. Gersdorff and Mitlin (*28*) concluded that against the housefly the allethronyl esters of the (+)-acid were 25 times more toxic than the esters of the (—)-acid with an even greater loss of knockdown efficiency. The esters of the (—)-acid therefore appear to contribute little to the insecticidal activity of the synthetic pyrethroids.

The effects on toxicity and knockdown of isomerization of the *cyclo*pentenolone ring are less clear cut. The data on the relative toxicities of the esters of the *cis* and *trans* isomers of the pertinent *cyclo*pentenolones are incomplete, but Metcalf (*29*) concluded that *trans*-cineronyl chrysanthemate has about the same activity as the natural *cis*-cineronyl chrysanthemate. Optical isomerization has a greater effect but the evidence is conflicting. LaForge and Green (*30*) found the (+)-*trans*-chrysanthemate of (—)-*cis*-cinerolone about twice as toxic to houseflies as the corresponding ester of (+)-*cis*-cinerolone. Elliott (*26*), however, reported that (+)-*cis*-cinerolone gave esters about four times as toxic to *Phaedon cochleariae* F. and to *Tenebrio molitor* L. as the esters of (—)-*cis*-cinerolone. He placed the toxicity of the ester of (+)-*trans*-chrysanthemic acid with (+)-allethrolone about twice that of the corresponding ester of (±)-allethrolone, a result in agreement with the conclusion of Gersdorff and Mitlin (*28*) that, with the same acid component, (+)-allethrolone yields esters about six times as toxic to houseflies as the esters of (—)-allethrolone.

The practical significance of the differences in the toxicities of the isomers, particularly of the much reduced activity of the esters of the (—)-acids, is that the chemist by using the racemic acid component is faced by the misfortune that half his synthetic product will be not only useless as an insecticide but may, by competitive action, reduce the toxicity of the active half. The theoretical significance is that the biochemical or biophysical reaction by which the pyrethroids intervene in an essential biological process is one which involves a three point attachment (see p. 103). Because the esters of the (—)-acid are non-toxic relative to those of the (+)-acid, at least two groupings of the acid component are involved. Elliott (*31*) suggested that the three groupings concerned are the *iso*butenyl group, the *cyclo*propane ring with its dimethyl group and the unsaturated side chain of the keto alcohol.

Swingle (*32*) demonstrated that pyrethrum extract is non-toxic when fed to cabbage butterfly caterpillars (*Pieris rapae*), and it is now generally accepted that the pyrethrins act solely as contact insecticides though Böttcher (*33*) reported a strong stomach poison action on honey bees. Historically it was the first protective contact insecticide for Potter (*34*) utilized, for insect control in warehouses, the treatment of surfaces with an oil solution of pyrethrum extract by which insects in subsequent contact with the surface were killed, an observation previously recorded by Tutin (*35*). The instability of the pyrethrins exposed to sunlight frustrates their effective use as protective insecticides on plants.

The rapid " knockdown " effect indicates that the primary action of the pyrethroids is on the nervous system and histological work of Hartzell (*36*), Wigglesworth (*37*), Richards and Cutkomp (*38*), and others revealed a drastic disruption of the nervous tissue which may be other than postmortem degradation. Recovery from sublethal doses, first noted by Tattersfield (*39*), indicates that the biological interaction by which the pyrethroids are toxic is reversible or that the insect is able to detoxify the insecticide. Because the pyrethroids are esters and the component acids and alcohols are of low insecticidal activity, the obvious suggestion is that detoxication is accomplished by hydrolytic enzymes. The fate of the pyrethroids in the insect body has been followed by radiotracer and chromatographic techniques by Zeid *et al.* (*40*), and by Winteringham *et al.* (*41*), but conclusive proof has not yet been obtained that hydrolysis to alcohol and acid is the mechanism of detoxication. The complexities of the problem will become more apparent in the consideration of the pyrethrum synergists.

Pyrethrum synergists. In 1940 Eagleson protected, by U.S.P. 2,202,145, the addition of sesame oil to pyrethrin insecticides on the grounds that the oil though not itself insecticidal, enabled lower concentrations of the pyrethrin to be used. Haller, LaForge and Sullivan (*42*) traced this synergistic effect to a crystalline oleoresin, sesamin (IV), present in the oil.

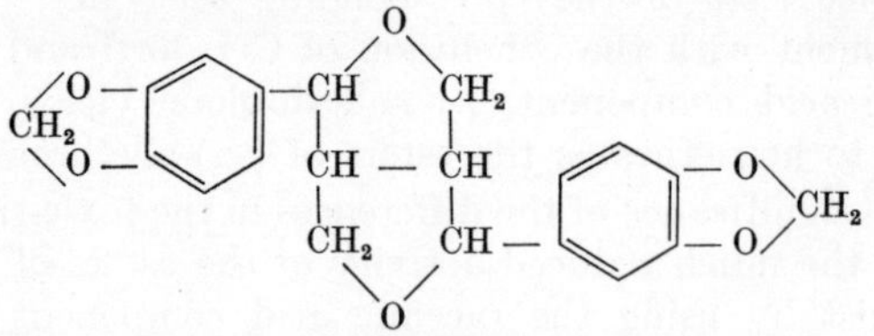

(IV)

They found that its isomers, isosesamin and asarinin, were as effective as sesamin but that pinoresinol (V) and its methyl ether were inactive. Subsequently, Beroza isolated from the non-crystalline residue, a compound sesamolin, which differs from sesamin in that one of the methylenedioxyphenyl groups is attached as an ether to the central nucleus. Gersdorff, Mitlin and Beroza (*43*) found sesamolin to be more effective than sesamin as a synergist with pyrethrins. In this series of compounds the active differ from the inactive in containing the methylenedioxy group. Many other compounds containing this group have been examined ; piperonal (VI)

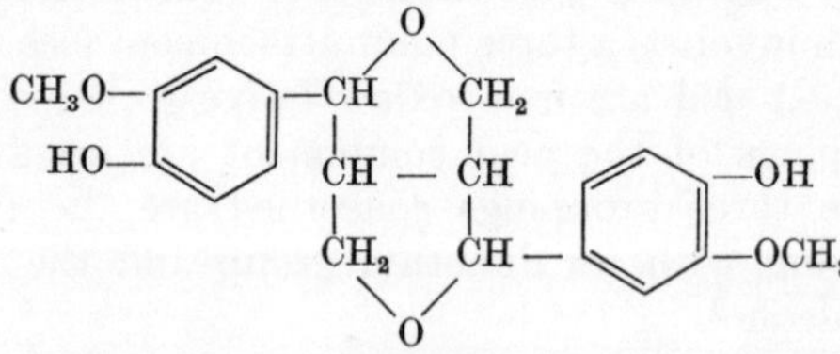

(V)

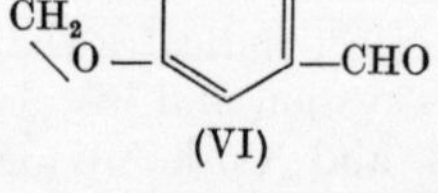

(VI)

is inactive whereas fagaramide (VII) and other related 3 : 4-methylene-

(VII)

dioxy-cinnamamides are active (U.S.P. 2,326,350). Harvill, Hartzell and Arthur (*4*) found that piperine (VIII) is active and, from this clue were

$$CH_2\langle{}^{O-}_{O-}\rangle C_6H_3-CH{:}CH.CH{:}CH.CO.N\langle{}^{CH_2-CH_2}_{CH_2-CH_2}\rangle CH_2$$

(VIII)

led to a long series of active amides and esters of piperic acid (*45*) and of condensation products of piperonal with thio ethers derived from safrole (1-allyl-3 : 4-methylenedioxybenzene) (*46*) and the corresponding sulphoxides and sulphones (*47*).

Commercial outcomes of this work were the product " Sulfoxide ", the active component of which is 1-methyl-2-(3 : 4-methylenedioxyphenyl)-ethyl *n*-octyl sulphoxide, and *n*-propyl isome, the condensation product of dipropyl maleate and *iso*safrole. From the methylenedioxyphenyl *cyclo*hexenones described by Hedenburg and Wacks (*48*), the commercial products are piperonyl cyclonene (the main active component of which is 3-*n*-hexyl-5-(3 : 4-methylenedioxyphenyl)-2-*cyclo*hexenone) and piperonyl butoxide (IX, 3 : 4-methylenedioxy-6-propylbenzyl diethyleneglycol ether)

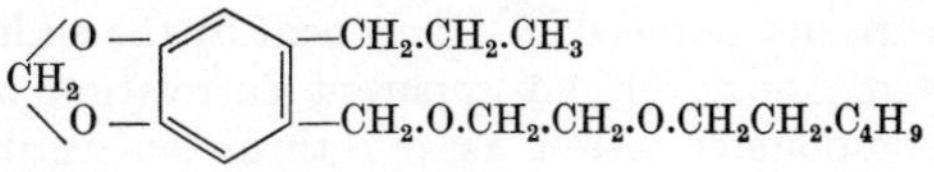

$$CH_2\langle{}^{O-}_{O-}\rangle C_6H_2\langle{}^{-CH_2.CH_2.CH_3}_{-CH_2.O.CH_2.CH_2.O.CH_2CH_2.C_4H_9}$$

(IX)

which, being of greater solubility in petroleum oils and in " Freon " than piperonyl cyclonene, is better suited for the preparation of aerosols (see p. 65).

The presence of the methylenedioxyphenyl group does not always ensure synergistic activity for piperonal (VI) is inactive. Nor is it indispensable. *N*-*iso*Butyl hendecenamide, $CH_3(CH_2)_7CH{:}CH.CO.NH.C_4H_9$, for example, was used with pyrethrum before the introduction of sesamin though Weed (*49*) regarded it as an insecticide rather than a synergist. The product now known as " MGK 264 ", the main component of which is *N*-(2-ethylhexyl)-bi*cyclo*[2,2,1]-5-heptene-2 : 3-dicarboximide, developed originally as an insecticide was found by Hartzell (*50*) to act as a synergist with pyrethrum.

Under some circumstances the improved mortality obtained by the addition of the synergist may be explained, in part at least, by an increase in the amount of toxicant received by the insect. David and Bracey (*51*) showed that the addition of high boiling solvents improved the kill of the mosquito *Aedes aegypti* L. by reducing the rate of evaporation of the mist droplets so that the insect in flying picked up larger amounts of insecticide. Parkin and Green (*52*) however, found no such increase when non-volatile oils were added to pyrethrum mists applied against the housefly. The idea that the synergist facilitates passage of the pyrethrins to their site of action within the insect was proved inadequate by Wilson (*53*) who found that the separate applications of synergist and of pyrethrins to different widely-spaced areas of the same insect led to toxic effects nearly as great as those produced when the two materials were applied to the same area. The suggestion that the synergist combines with the pyrethroid to form a more toxic compound was also questioned by Wilson, for there is little difference between the toxic effects of the mixture and of the pyrethrins alone. Page and Blackith (*54*), however, found that the toxicity of pyrethrum to *A. aegypti* was increased by the addition of sesamin to three-fold and was not increased by further addition of synergist. As this three-fold increase appeared when the molar ratio of synergist to pyrethrin was 1 : 1, they suspected the formation of a molecular complex at the peripheral nerve sheath interface. No evidence either cryoscopic or spectrometric, was obtained by Miller *et al.* (*55*) of complex formation between pyrethrin and piperonyl butoxide.

The evidence that the addition of the synergist prolongs the time of knockdown by sublethal doses indicates that the synergist has intervened in the process of recovery either, as Wilson (*53*) suggested by delaying the repair of the pyrethrin-damaged nerve or, as suggested on p. 172 by inhibiting the enzymic detoxication process, the simplest of which is hydrolysis of the esters. The latter hypothesis is supported by the demonstration by Chamberlain (*56*) that roach and housefly extracts are capable of hydrolysing the pyrethrins, a capacity reduced 25–40 per cent by the addition of piperonyl butoxide. He found, however, no apparent correlation between the efficiencies of the compounds tested as pyrethrin synergists and as lipase inhibitors. Matsubara (*57*) found the lipase activity of preparations of female flies higher than those of males yet the latter rendered the pyrethrin less toxic to mosquito larvae more rapidly than did preparations of female flies. Ester hydrolysis does not seem, therefore, to explain detoxication but the evidence, for instance, of Winteringham *et al.* (*41*) on the *in vivo* metabolism of pyrethroids and of inhibition of this metabolism by the synergist seems conclusive. Yet several experimental observations remain unexplained by the hypothesis. For instance, the pyrethrin synergists are generally less effective in augmenting the insecticidal activity of allethrin (*58*) indicating that the allethrin molecule is less susceptible to detoxication by insects. Yet Winteringham and his colleagues (*41*) found that allethrin is metabolised by the insect almost as rapidly as are the pyrethrins. Its metabolism is not as effectively inhibited by piperonyl cyclonene as that of the pyrethrins, an observation which suggests that the detoxication process may involve also the *cyclo*hexenone side chain.

REFERENCES

(*1*) Lodeman, E. G., *The Spraying of Plants*, New York, 1903, p. 78.

(*2*) Gnadinger, C. B., *Pyrethrum Flowers*, Minneapolis, 2nd Ed., 1935 ; Supplement, 1945.

(*3*) Gulland, J. M. and Hopton, C. U., *J. chem. Soc.*, 1930, 6.

(*4*) Staudinger, H. and Ruzicka, L., *Helv. chim. Acta*, 1924, **7**, 177, 201, 212, 236, 245, 377, 406, 442, 448 ; Staudinger, H., Muntwyler, C., Ruzicka, L. and Seibt, S., *ibid.*, 1924, **7**, 390.

(*5*) Fujitani, J., *Arch. exp. Path. Pharmak.*, 1909, **61**, 47.

(*6*) Yamamoto, R., *J. chem. Soc. Japan*, 1923, **44**, 311 ; *Inst. phys. chem. Res. Tokyo*, 1925, **3**, 193.

(7) Campbell, I. G. M. and Harper, S. H., *J. chem. Soc.*, 1945, p. 283.
(8) LaForge, F. B. and Barthel, W. F., *J. org. Chem.*, 1945, **10**, 106, 114, 222.
(9) LaForge, F. B. and Haller, H. L., *J. Amer. chem. Soc.*, 1936, **58**, 1777.
(10) Gillam, A. E. and West, T. F., *J. chem. Soc.*, 1942, p. 671.
(11) LaForge, F. B. and Soloway, S. B., *J. Amer. chem. Soc.*, 1947, **69**, 186.
(12) West, T. F., *J. chem. Soc.*, 1946, p. 463.
(13) Crombie, L., Harper, S. H. and Newman, F. C., *Chem. & Ind.*, 1954, 1109.
(14) LaForge, F. B. and Green, N., *J. org. Chem.*, 1952, **17**, 1635.
(15) Staudinger, H. and Harder, H., *Ann. Acad. Sci. fenn.*, 1927, A, **29**, No. 18.
(16) Tattersfield, F. and Hobson, R. P., *Ann. appl. Biol.*, 1931, **18**, 203.
(17) Abbott, W. S., *Bull. U.S. Dep. Agric.*, 771, 1919.
(18) Tattersfield, F., *J. agric. Sci.*, 1932, **22**, 396.
(19) West, T. F., *Nature, Lond.*, 1943, **152**, 660.
(20) Schechter, M. S., Green, N. and LaForge, F. B., *J. Amer. chem. Soc.*, 1949, **71**, 1517; 3165.
(21) Sanders, H. J. and Taff, A. W., *Industr. engng Chem.*, 1954, **46**, 414.
(22) Matsui, M., LaForge, F. B., Green, N. and Schechter, M. S., *J. Amer. chem. Soc.*, 1952, **74**, 2181.
(23) Haynes, H. L., Guest, H. R., Stansbury, H. A., Sousa, A. A. and Borash, A. J., *Contr. Boyce Thompson Inst.*, 1954, **18**, 1.
(24) Incho, H. H. and Greenberg, H., *J. econ. Ent.*, 1952, **45**, 794.
(25) Gersdorff, W. A., *J. econ. Ent.*, 1947, **40**, 878.
(26) Elliott, M., *J. Sci. Food Agric.*, 1954, **5**, 505.
(27) Gersdorff, W. A., *J. econ. Ent.*, 1949, **42**, 532.
(28) Gersdorff, W. A. and Mitlin, N., *J. econ. Ent.*, 1953, **46**, 999.
(29) Metcalf, R. L., *Organic Insecticides*, Interscience Publ. Ltd., New York and Lond., 1955, p. 50.
(30) LaForge, F. B. and Green, N., *J. org. Chem.*, 1952, **17**, 1635.
(31) Elliott, M., *Pyrethrum Post.*, 1951, **2**(3), 18.
(32) Swingle, M. C., *J. econ. Ent.*, 1934, **27**, 1101.
(33) Böttcher, F. K., *Z. angew. Ent.*, 1938, **25**, 419.
(34) Potter, C., *Ann. appl. Biol.*, 1938, **25**, 836.
(35) Tutin, F., *Rep. Agric. hort. Res. Sta.*, *Bristol*, 1928, p. 96.
(36) Hartzell, A., *Contr. Boyce Thompson Inst.*, 1934, **6**, 211.
(37) Wigglesworth, V. B., *Proc. R. ent. Soc. Lond.*, 1941, **16**, 11.
(38) Richards, A. G. and Cutkomp, L., *J. N. Y. ent. Soc.*, 1945, **53**, 313.
(39) Tattersfield, F., *Ann. appl. Biol.*, 1932, **19**, 28.
(40) Zeid, M. M. I., Dahm, P. A., Hein, R. E. and McFarland, R. H., *J. econ. Ent.*, 1953, **46**, 324.
(41) Winteringham. F. P. W., Harrison, A. and Bridges, P. M., *Biochem. J.*, 1955, **61**, 359.
(42) Haller, H. L., LaForge, F. B. and Sullivan, W. N., *J. org. Chem.*, 1942, **7**, 185; *J. econ. Ent.*, 1942, **35**, 247.
(43) Gersdorff, W. A., Mitlin, N. and Beroza, M., *J. econ. Ent.*, 1954, **47**, 839.
(44) Harvill, E. K., Hartzell, A. and Arthur, J. M., *Contr. Boyce Thompson Inst.*, 1943, **13**, 87.
(45) Synerholm, M. E., Hartzell, A. and Arthur, J. M., *Contr. Boyce Thompson Inst.*, 1945, **13**, 433; 1945, **14**, 79.
(46) Prill, E. A., Hartzell, A. and Arthur, J. M., *Contr. Boyce Thompson Inst.*, 1946, **14**, 127.
(47) Synerholm, M. E., Hartzell, A. and Cullmann, V., *Contr. Boyce Thompson Inst.*, 1947, **15**, 35.
(48) Hedenburg, O. F. and Wachs, H., *J. Amer. chem. Soc.*, 1948, **70**, 2216, 2695.
(49) Weed, A., *Soap, N.Y.*, 1938, **14**(6), 133.
(50) Hartzell, A., *Contr. Boyce Thompson Inst.*, 1949, **15**, 337.
(51) David, W. A. L. and Bracey, P., *Nature, Lond.*, 1944, **153**, 594.
(52) Parkin, E. A. and Green, A. A., *Nature, Lond.*, 1944, **154**, 16.
(53) Wilson, C. S., *J. econ. Ent.*, 1949, **42**, 423.
(54) Page, A. B. P. and Blackith, R. E., *Ann. appl. Biol.*, 1949, **36** 244.

(*55*) Miller, A. C., Pellegrini, J. P., Pozefsky, A. and Tomlinson, J. R., *J. econ. Ent.*, 1952, **45**, 94.
(*56*) Chamberlain, R. W., *Amer. J. Hyg.*, 1950, **52**, 153.
(*57*) Matsubara, H., *Botyu Kagaku*, 1953, **18**, 75.
(*58*) Fales, J. H. and Bodenstein, O. F., *J. econ. Ent.*, 1956, **49**, 156.

ROTENONE AND RELATED INSECTICIDES

In 1848, Oxley (*1*) suggested tuba-root, used in the Malay Archipelago as a fish and arrow poison, for the control of the insect pests of the nutmeg. Hooker (*2*) recorded that tuba-root had been used by the Chinese in Singapore for the preparation of an insecticide. The main source of tuba-root is the root of *Derris* elliptica* (Wall.) Benth. and the best known of its active principles is rotenone.

The success of tuba-root as an insecticide encouraged trials with other plants used as fish poisons. The most successful have been plants of *Derris*, *Lonchocarpus* and *Tephrosia* (*Cracca*) spp., further particulars of which are given below. The insecticidal properties of many plants reputed to possess fish-poisoning properties have been reviewed by Tattersfield and Gimingham (*4*) and by Jones (*5*). All fish poison plants having insecticidal properties are of the Leguminosae.

Derris spp. The botanical identification of the species of Derris employed in early work was made difficult by the erratic flowering habits of the plant and by the diversity of habits within the species. The sources of derris root in the Malay Peninsula have been described by Henderson (*6*), the most important commercially being *D. elliptica* and *D. malaccensis* Brain. The latter species, when grown under cultivated conditions, acquires habits different from the wild variety and is distinguished by the name *D. malaccensis* var. *sarawakensis* Hend. Cahn and Boam (*7*) found it necessary to differentiate, for chemical reasons, a Sumatra type of derris resin which they considered was probably derived from the roots of the latter variety.

Lonchocarpus spp. The fish-poison plants of South America are known collectively as " barbasco " in the Spanish-speaking countries, as " cubé " in Peru, and generally as " timbo ". They were mentioned by Geoffrey in 1892 (*8*) and have since been commercially developed. The most important of the Peruvian species was, in 1930 (*9*), identified with *L. nicou* (Aubl.) DC. but has now been given separate status, *L. utilis* A.C. Smith, by Krukoff and Smith (*10*). These investigators stated that the principal species in Brazil is *L. urucu* Killip & Smith, other species of interest being *L. martynii* A. C. Smith and *L. chrysophyllus* Kleink., which are probably the White Haiari and Black Haiari, respectively, examined by Tattersfield, Gimingham and Morris (*11*).

Tephrosia spp. The insecticidal properties of *T. vogelii* Hook, were investigated by Tattersfield, Gimingham and Morris (*12*), toxic principles being found in the leaves and seed. The following year (*11*) they showed that the roots of *T. toxicaria* Pers. possessed insecticidal properties. In

---

*The name *Derris* Lour, is antedated by *Dequelia* Aubl. (*3*), but in view of the widespread use of the former, it has been retained.

later work, Tattersfield and Gimingham (*4*) found the roots and stem of *T. macropoda* Harv. of insecticidal value. The commercial possibilities of *T. vogelii*, which is indigenous to East Africa, were explored by Worsley (*13*). The predominant insecticidal constituent isolated from these species is tephrosin, but a small content of rotenone has been found in *T. virginiana* L., which is of particular interest as it is a weed, devil's shoestring, common in southern U.S.A. (*14*).

The distribution of rotenone and allied compounds, to which Roark (*15*) gave the name "rotenoids", in other species of Papilionaceae has been examined by Worsley (*16*) who has tabulated their occurrence on a histological basis. To Worsley's list should be added the Chinese yam bean (*Pachyrrhizus erosus* Urban), the seeds of which were found of insecticidal promise by Hansberry and Lee (*17*).

The Chemistry of the Rotenoids

Extraction of derris or lonchocarpus root with organic solvents yields a resin from which, in ether or carbon tetrachloride, a white crystalline compound is obtained. That from *Lonchocarpus* was called nicouline by Geoffrey (*18*) in 1892. In 1911 Lenz (*19*), who isolated this compound from *D. elliptica*, gave it the name derrin and to it attributed the fish-poisoning properties of derris root. Independently, Ishikawa (*20*) in 1917, obtained a similar crystalline material from tuba-root and named it tubatoxin. Tubatoxin was, in 1923, shown by Kariyone and his colleagues (*21*) to be identical with a product first isolated, in 1902, by Nagai (*22*) from "Roh-ten",* which he named rotenone. The latter name has now been generally accepted for this compound, which forms colourless crystals of m.p. 163° C.

The molecular structure of rotenone was deduced almost simultaneously by LaForge, Haller and their colleagues in the United States (*23*), by Robertson (*24*) in England, by Takei and his colleagues (*25*) in Japan and by Butenandt and co-workers (*26*) in Germany. It is represented as formula I, the numbering being that of Cahn *et al.* (*27*).

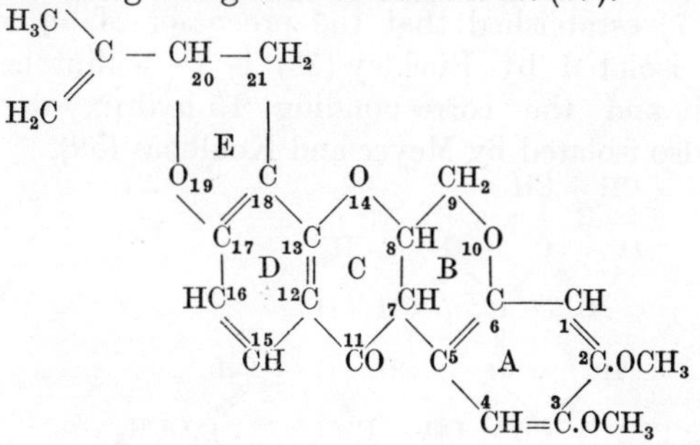

```
H3C\
     C — CH — CH2
    //   20   21
H2C         |    |
            | E  |
           O19   C      O       CH2
             \  //18\  /14\    /9\
              C17  13C     8CH   10O
              |  D  ||  C   |  B  |
             HC16  12C     7CH    C——————CH
               \\15/  \11/    \ //6     1\\
                CH     CO      C5   A     2C.OCH3
                                 \4     3/
                                  CH=C.OCH3
```

(I) Rotenone

From the resin remaining after crystallization of the rotenone from extracts of derris and related plants, a number of other compounds have been isolated. Toxicarol, first isolated by Clark (*28*), was shown by George

*"Roh-ten" has been identified as *D. chinensis* Benth. or *Millettia taiwaniana* Hayata (*22*).

et al. (*29*) to be 15-hydroxydeguelin. Deguelin, also first isolated by Clark (*30*), was found to be isomeric with rotenone and was assigned the formula II by Clark (*31*) and by Robertson (*24*).

```
              CH
             /  \\
     (H3C)2 C    CH
            |    |
            O    C     O     CH2
             \  // \  / \   /  \
              C    C     CH    O
              |    ||    |     |
             HC    C     CH    C----CH
               \\ / \   / \   //     \\
                CH   CO    C          C.OCH3
                            \        /
                             CH = C.OCH3
```

(II) Deguelin

With deguelin, Clark (*32*) also isolated tephrosin, a compound named by Hanriot (*33*) who had, in 1907, obtained it from the leaves of *T. vogelii* as white crystals of m.p. 187° C. Clark found Hanriot's compound to be a mixture of deguelin and tephrosin, which is of m.p. 197–198° C, and is a hydroxydeguelin, the hydroxyl group replacing the hydrogen atom at either carbon 7 or carbon 8.

Deguelin, toxicarol and tephrosin are optically inactive and Clark thought that they might be formed, by chemical change during the process of extraction, from precursors in the roots. The intricate changes of enolization, racemization and isomerization, which rotenone and allied compounds undergo on alkaline treatment of the type used by Clark have been discussed by Cahn, Phipers and Boam (*27*). The search for the precursors of these compounds led to the isolation of other compounds, e.g., sumatrol by Cahn and Boam (*34*), shown by Robertson and Rusby (*35*) to be 15-hydroxyrotenone, and bearing the same relationship to rotenone that toxicarol does to deguelin. Tattersfield and Martin (*36*) isolated, from *D. malaccensis*, (—)-α-toxicarol which is probably the form in which toxicarol exists in the root. Harper (*37*) established that the precursor of an optically-inactive compound first isolated by Buckley (*38*) is (—)-elliptone (formula III). This compound and the corresponding 15-hydroxy-derivative, named malaccol, were also isolated by Meyer and Koolhaas (*39*).

```
     CH = CH
     |  E  |
     O     C     O     CH2
      \  // \   / \   /  \
       C     C     CH    O
       |     ||    |     |
      HC     C     CH    C----CH
        \\  / \   / \   //     \\
         CH    CO    C          C.OCH2
                      \        /
                       CH = C.OCH2
```

(III) Elliptone

At least six rotenoids are considered to occur naturally ; rotenone, (—)-α-toxicarol, deguelin, sumatrol, elliptone and malaccol, all as the laevorotatory (—) isomers though deguelin has not yet been isolated in an optically active form. Cahn, Phipers and Boam (*40*) were able to account for over 80 per cent of the rotenoid content of typical derris resins in terms of these compounds. *D. elliptica* resin consists of about 40 per cent of rotenone and 27 per cent of (—)-deguelin, the content of sumatrol and of (—)-toxicarol being low. Resins of *D. malaccensis* of the Sumatra type, on the other hand, contain 50–60 per cent of (—)-toxicarol, 12 per cent (—)-deguelin, 5–15 per cent sumatrol and but 2–5 per cent rotenone. In plants of *Tephrosia* spp. with the exception of *T. virginiana*, tephrosin has been the predominant compound isolated but its naturally-occurring precursor is probably deguelin. Norton and Hansberry (*41*) identified rotenone and four other compounds of rotenoid structure in the extract of the seeds of the Chinese yam bean. Bickel and Schmid (*42*) concluded that one of the latter compounds is identical to pachyrrhizone, previously isolated by Meijer (*43*). Pachyrrhizone (IV) differs from other rotenoids in that the two methoxy groups of ring A are replaced by a methylenedioxy group.

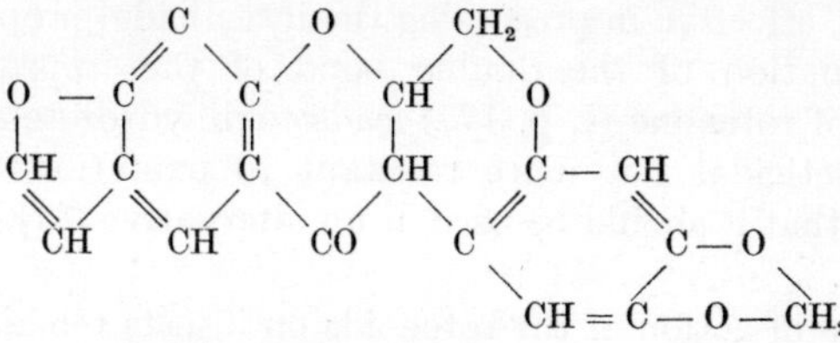

```
            C       O       CH2
O — C       C       CH      O
CH  C       C       CH      C — CH
  CH    CH      CO      C       C — O
                        CH = C — O — CH2
```

(IV) Pachyrrhizone

In structure the rotenoids have much in common; all are built up of a four ring chromonochromanone system (rings A, B, C and D) with the addition of a fifth ring (ring E) of type I in rotenone and sumatrol, of type II in elliptone and malaccol and of type III in deguelin and toxicarol. The relationship of this structure to that of other naturally-occurring plant products is of biosynthetic interest. Rotenone is converted by catalytic hydrogenation to rotenonic acid which involves the opening of ring E. Rotenonic acid may be regarded as an *iso*flavanone (V) with carbons 2 and 2′ linked by a methyleneoxy bridge. Seshadri and Varadarajan (*44*) suggested that 2-hydroxy-*iso*flavanone or its equivalent may be an intermediate in the biosynthesis of the rotenoids.

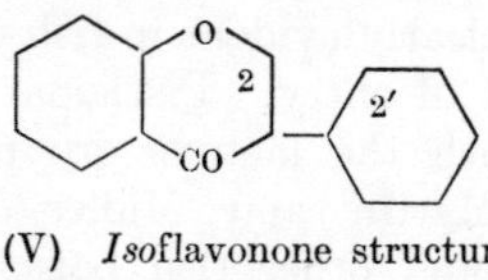

```
O
2    2′
CO
```

(V) *Iso*flavonone structure

The Toxicology of the Rotenoids

The general conclusion of a wide variety of test conditions and test insects is that rotenone is by far the most insecticidal of the naturally-occurring rotenoids. Sullivan, Goodhue and Haller (*45*) using the housefly found optically-active deguelin about half as toxic as rotenone whereas the racemic deguelin was but one tenth as toxic. Tattersfield and Martin (*46*) showed that to *Aphis rumicis* rotenone was 13–15 times more toxic than toxicarol or sumatrol. Yet the resin of *D. elliptica*, freed as far as possible from rotenone, was still one-quarter as toxic as rotenone; the rotenone-free resin of a Sumatra type root had one-sixth the toxicity of rotenone. The problem of a chemical method of assay for rotenone-containing roots has not yet been completely solved (*47*) though present day crystallization methods appear, because of their general use, to meet commercial needs.

Early in the use of derris and lonchocarpus attempts were made to extract the insecticidal principles to avoid the cost of transport of the inert material of the root. It was found that solvents used to dissolve the resinous extract hastened a degradation to products of low insecticidal properties. This degradation, studied by Jones and Haller (*48*) and by Gunther (*49*) is an oxidation accelerated by light or alkali, leading to the elimination of the hydrogens of carbons 7 and 8 (I, p. 177) giving dehydrorotenone, a compound of low insecticidal activity. Hence, attempts have been made to stabilize derris extracts, one of the most successful being due to Cahn, Phifers and Brodaty (*50*) who found the addition of traces of strong acids, e.g., phosphoric acid, effective in preserving the insecticidal properties of derris dusts.

Hydrogenation of the double bond of the *iso*propenyl side chain of carbon 20 of rotenone (I, p. 177) leads to dihydrorotenone which is not only highly insecticidal but more resistant to oxidation than rotenone. The suggestion that it should be used as an alternative (*51*) has not been adopted in practice.

The mode of action of the rotenoids on insects remains obscure but death follows a paralysis, the symptoms of which are a depressed respiration and heartbeat (*52*). So slow is this process that Fransen (*53*) suggested that the mouth parts of the silkworm fed derris dust became affected so that the insect could not feed and that death resulted from starvation, for the larva died no sooner than others deprived of food. Similarly, Speyer, Read and Orchard (*54*) concluded that the control of the cabbage moth *Mamelia brassicae* L. by derris is due to a deterrent action. Martin, Stringer and Wain (*55*) found rotenone to be virtually non-toxic to the last instar caterpillars of this moth recalling the earlier finding of Woke (*56*) that larvae of the southern army worm *Prodenia eridania* eliminated unchanged in the faeces 85 per cent of the rotenone with which they were fed.

Yet the efficiency of derris in controlling certain sensitive insects is high and its paralytic effect is clearly evident in fish of which the gills appear to provide an excellent point of entry. Piscicidal properties were associated by Priess (*57*) in 1911 with the lactone group, a suggestion which was examined by Läuger, Martin and Müller (*58*), who concluded that the toxicity of rotenone and other fish poisons is due to the grouping –CO–CL=CLOL– where L represents groups which confer lipoid solubility. This generalization, however, fails to explain the much reduced potency of other rotenoids in which this group is present.

The loss of insecticidal properties following the oxidation of rotenone may provide a clue. The reaction involved occurs at carbons 6 and 7, the sequence being an oxidation to hydroxyl and an elimination of water to form a double bond between these two carbon atoms. The latter compound, dehydrorotenone, is non-toxic and Martin (*56*) suggested that the hydrogen atoms on carbons 6 and 7 are required for toxicity perhaps because of the possibility which exists for enolization. That the methyl ether derived from the enolic form of rotenone is a poor insecticide (*27*) supports this conclusion.

Alternatively, the significance of the hydrogen atoms on carbons 6 and 7 may be due to their role in determining the shape of the rotenone molecule. Hummer and Kenaga (*60*) pointed out that the reactions of rotenone indicate that these hydrogens are in the *cis* position, causing a puckering of ring B (I, p. 177). Because of this pucker the rings of the rotenone molecule do not lie in one plane, but at an angle of inclination which corresponds to the angle between the planes of the benzene rings of DDT or of methoxychlor, in which molecules a free rotation of the benzene rings is restricted by the trichloromethyl group. Moreover, the distance between the oxygen atoms at positions 2 and 19 in the rotenone molecule is very close to that of the two methoxy oxygens of methoxychlor or of the *p-p'*-chlorine atoms of DDT. Oxidation of rotenone to dehydrorotenone removes the puckering of ring B whereas saturation in the isopropenyl side chain to give dihydrorotenone does not remove the pucker. The puckered molecule of rotenone or of dihydrorotenone can presumably come into contact with a biochemical interface close enough to be held and hence to intervene in the reactions of that interface. The shape of the dehydrorotenone molecule does not permit this close fit. It may be pointed out, however, that Hummer and Kenaga's demonstration of the similarity in molecular shape and distances between rotenone and DDT enabling the two molecules to fit to the same biological interface does not require that the biochemical lesions caused by these insecticides should be identical, nor has it yet been shown that the compounds compete in their insecticidal action. Hummer and Kenaga (*60*) did not extend their hypothesis to pachyrrhizone, the structure of which appears to meet their requirements for insecticidal activity yet it is non-insecticidal. However, Bickel and Schmid (*42*) pointed out that pachyrrhizone differs from other rotenoids, not only in the methylenedioxy group, but in that rings C, D, and E lie in one plane whereas in other rotenoids, except β-toxicarol, they do not.

No explanation has yet been attempted of the high degree of specificity shown by the rotenoids which, though highly toxic to fish and to most insect species, are almost harmless to warm-blooded animals (*61*), though recently pigs have been found to be susceptible (*62*).

REFERENCES

(*1*) Oxley, T., *J. Indian Archipelago and E. Asia*, 1848, p. 646.
(*2*) Hooker, J. D., *Rep. Progress and Condition of the Royal Gardens at Kew*, 1877, p. 43.
(*3*) Roark, R. C., *Misc. Pub. U.S. Dep. Agric.*, 120, 1932.
(*4*) Tattersfield, F. and Gimingham, C. T., *Ann. appl. Biol.*, 1932, **19**, 253.
(*5*) Jones, H. A., *U.S. Dep. Agric. Bur. Ent.*, E-571, 1942.
(*6*) Henderson, M. R., *Malayan Agric. J.*, 1934, **22**, 125.
(*7*) Cahn, R. S. and Boam, J. J., *J. Soc. chem. Ind.*, 1935, **54**, 37T, 42T.
(*8*) Geoffrey, E., *J. Pharm. Chem.*, 1892, **26**, 454.
(*9*) Killip, E. P. and Smith, A. C., *J. Wash. Acad. Sci.*, 1930, **20**, 73.
(*10*) Krukoff, B. A. and Smith, A. C., *Amer. J. Bot.*, 1937, **24**, 573.

(*11*) Tattersfield, F., Gimingham, C. T. and Morris, H. M., *Ann. appl. Biol.*, 1926, **13**, 424.
(*12*) Tattersfield, F., Gimingham, C. T. and Morris, H. M., *Ann. appl. Biol.*, 1925, **12**, 66.
(*13*) Worsley, R. R. LeG., *Ann. appl. Biol.*, 1934, **21**, 649.
(*14*) Little, V. A., *J. econ. Ent.*, 1931, **24**, 743 ; 1935, **28**, 707.
(*15*) Roark, R. C., *J. econ. Ent.*, 1940, **33**, 416.
(*16*) Worsley, R. R. LeG., *Ann. appl. Biol.*, 1939, **26**, 649.
(*17*) Hansberry, R. and Lee, C., *J. econ. Ent.*, 1943, **36**, 351.
(*18*) Geoffrey, E., *J. Pharm. Chim. Paris*, 1892, **26**, 454.
(*19*) Lenz, W., *Arch. Pharm. Berl.*, 1911, **249**, 298.
(*20*) Ishikawa, T., *J. med. Ges. Tokyo*, 1917, **31**, 187.
(*21*) Kariyone, T., Atsumi, K. and Shimada, M., *J. pharm. Soc. Japan*, 1923, **500**, 739.
(*22*) Nagai, K., *J. chem. Soc. Japan*, 1902, **23**, 744.
(*23*) LaForge, F. B. and Haller, H. L., *J. Amer. chem. Soc.*, 1932, **54**, 810.
(*24*) Robertson, A., *J. chem. Soc.*, 1932, 1380.
(*25*) Takei, S., Miyajima, S. and Ono, M., *Mem. Coll. Agric. Kyoto*, **23**, 1932 ; *Ber. dtsch. chem. Ges.*, 1932, **65**, B, 1041.
(*26*) Butenandt, A. and McCartney, W., *Liebigs Ann.*, 1932, **494**, 17.
(*27*) Cahn, R. S., Phipers, R. F. and Boam, J. J., *J. chem. Soc.*, 1938, **513**, 734.
(*28*) Clark, E. P., *J. Amer. chem. Soc.*, 1930, **52**, 2461.
(*29*) George, S. W., Robertson, A., Subramaniam, T. S. and Hilton, W., *J. chem. Soc.*, 1937, 1535.
(*30*) Clark, E. P., *J. Amer. chem. Soc.*, 1931, **53**, 313.
(*31*) Clark, E. P., *J. Amer. chem. Soc.*, 1932, **54**, 3000.
(*32*) Clark, E. P., *J. Amer. chem. Soc.*, 1931, **53**, 729.
(*33*) Hanriot, M., *C. R. Acad. Sci. Paris*, 1907, **144**, 150, 498, 651.
(*34*) Cahn, R. S. and Boam, J. J., *J. Soc. chem. Ind. Lond.*, 1935, **54**, 42T.
(*35*) Robertson, A. and Rusby, G. L., *J. chem. Soc.*, 1937, 497.
(*36*) Tattersfield, F. and Martin, J. T., *J. Soc. chem. Ind. Lond.*, 1937, **56**, 77T.
(*37*) Harper, S. H., *J. chem. Soc.*, 1939, 1099, 1424.
(*38*) Buckley, T. A., *J. Soc. chem. Ind. Lond.*, 1936, **55**, 285T.
(*39*) Meyer, T. M. and Koolhaas, D. R., *Rec, Trav. chim. Pays-Bas*, 1939, **58**, 207.
(*40*) Cahn, R. S., Phipers, R. F. and Boam, J. J., *J. Soc. chem. Ind. Lond.*, 1938, **57**, 200.
(*41*) Norton, L. B. and Hansberry, R., *J. Amer. chem. Soc.*, 1945, **67**, 1609.
(*42*) Bickel, H. and Schmid, H., *Helv. chim. Acta*, 1953, **36**, 664.
(*43*) Meijer, T. M., *Rec. Trav. chim. Pays-Bas*, 1946, **65**, 835.
(*44*) Seshadri, T. R. and Varadarajan, S., *Proc. Indian Acad. Sci.*, 1953, **37**, A, 784.
(*45*) Sullivan, W. N., Goodhue, L. D. and Haller, H. L., *Soap*, 1939, **15**(7), 107.
(*46*) Tattersfield, F. and Martin, J. T., *Ann. appl. Biol.*, 1938, **25**, 411.
(*47*) Martin, J. T., *Ann. appl. Biol.*, 1940, **27**, 274.
(*48*) Jones, H. A. and Haller, H. L., *J. Amer. chem. Soc.*, 1932, **53**, 2320.
(*49*) Gunther, F. A., *J. econ. Ent.*, 1943, **36**, 273.
(*50*) Cahn, R. S., Phipers, R. F. and Brodaty, E., *J. Soc. chem. Ind., Lond.*, 1945, **64**, 33.
(*51*) Jones, H. A., Gersdorff, W. A., Gooden, E. L., Campbell, F. L. and Sullivan, W. N., *J. econ. Ent.*, 1923, **26**, 451.
(*52*) Tischler, N., *J. econ. Ent.*, 1935, **28**, 215.
(*53*) Fransen, J. J., *Tijdschr. PfZicht.*, 1943, **49**, 126.
(*54*) Speyer, E. R., Read, W. H. and Orchard, O. B., *Rep. exp. Res. Sta. Cheshunt*, 1939, p. 39.
(*55*) Martin, H., Stringer, A. and Wain, R. L., *Rep. agric. hort. Res. Sta. Bristol*, 1943, p. 62.
(*56*) Woke, P. A., *J. agric. Res.*, 1938, **57**, 707.
(*57*) Priess, H., *Ber. dtsch. pharm. Ges.*, 1911, **21**, 267.
(*58*) Läuger, P., Martin, H. and Müller, P., *Helv. chim. Acta*, 1944, **27**, 892.
(*59*) Martin, H., *J. Soc. chem. Ind., Lond.*, 1946, **65**, 402.
(*60*) Hummer, R. W. and Kenaga, E. E., *Science*, 1951, **113**, 653.
(*61*) Ambrose, A. M. and Haag, H. B., *Industr. engng Chem.*, 1936, **28**, 815 ; 1937, **29**, 429.
(*62*) Kingscote, A. A., Baker, A. W., McGregor, J. K. and Dixon, S. E., *Rep. ent. Soc. Ont.*, 1951, p. 37.

QUASSIA

Quassia is the wood of two trees of the Simarubaceae, that known commercially as Jamaica quassia is *Picraena excelsa* (Swz.) Lindl., and that known as Surinam quassia is *Quassia amara* L. Other woods having similar bitter principles are detailed by McIndoo and Sievers (*1*). The extract of quassia wood appears to have been first used against the hop aphis (*Phorodon humuli* Schrenk.) about the year 1884 (*2*). Whitehead (*3*) recorded that the use of a quassia-soft soap extract had by 1890 become a regular feature of hop cultivation. McIndoo and Sievers found that the best extraction of the toxic principles, the quassins and the picrasmins, is secured by soap or lye solution which must be cold for a decomposition occurred on heating.

McIndoo and Sievers showed that the quassins penetrate the spiracles thereby reaching the nervous tissue, where they slowly disrupt the nerve cells. They found quassin to be effective only against certain aphides and of little use against other insects. Although Parker (*4*) found that the toxicity of quassin compares favourably with nicotine sulphate against *P. humuli*, quassia extracts have now been generally replaced by nicotine. Interest in quassia was revived by the observation of Thiem (*5*) of the potent action of the extract on certain species of sawfly (*Hoplocampa* spp.), an observation confirmed by Petherbridge and Thomas (*6*).

Clark (*7*) isolated two "amaroids" from *Q. amara* which he named quassin and neoquassin and a third from *P. excelsa* which he named picrasmin. He considered that the three amaroids were isomers, $C_{22}H_{30}O_6$, that each contained two methoxy groups and that all yielded the same isoquassin on treatment with chromic acid. Adams and Whaley (*8*) regarded picrasmin and *iso*quassin as identical, quassin to be a molecular complex of *iso*quassin and neoquassin, the latter being a hydroxy compound, $C_{20}H_{24}O_4(OCH_3)_2$, and *iso*quassin $C_{20}H_{22}O_4(OCH_3)_2$ the corresponding ketone. The presence of a hydroxyl group in *iso*quassin is doubted by Robertson and his colleagues (*9*) but the precise structure of these compounds has not yet been determined.

REFERENCES

(*1*) McIndoo, N. E. and Sievers, A. F., *J. agric. Res.*, 1917, **10**, 497.

(*2*) Ormerod, E. A., *Rep. Observ. injur. Insects*, 1884, **8**, 43.

(*3*) Whitehead, C., *J. R. agric. Soc.*, 1890, iii, **1**, 231.

(*4*) Parker, W. B., *Bull. U.S. Dep. Agric.*, 165, 1914.

(*5*) Thiem, H., *Kranke Pflanzen*, 1937, **14**, 59.

(*6*) Petherbridge, F. R. and Thomas, I., *J. Minist. Agric.*, 1937, **44**, 858.

(*7*) Clark, E. P., *J. Amer. chem. Soc.*, 1937, **59**, 927, 2511; 1938, **60**, 1146.

(*8*) Adams, R. and Whaley, W. M., *J. Amer. chem. Soc.*, 1950, **72**, 375.

(*9*) Hanson, K. R., Jaquiss, D. B., Lamberton, J. A., Robertson, A. and Savige, W. E., *J. chem. Soc.*, 1954, p. 4238.

THE HYDROCARBON OILS

As implied by their name, the hydrocarbon oils comprise those oils of which the predominant constituents are compounds solely of hydrogen and carbon. They are classified into petroleum (mineral) oils, a group subdivided by reference to origin, e.g., Pennsylvanian, Mexican, and into tar oils described by the source of the tar, whether high temperature carbonization, coke oven, etc., from which the oil was distilled.

Sufficient is known, however, of the properties of the hydrocarbon oils as spray materials to justify a physico-chemical classification. Chemically, the hydrocarbon oils are mixtures of a complexity which defies separation to individual compounds, but a chemical basis for classification is made possible by the observation that the compounds present fall into a limited number of groups. Within each group are compounds of similar molecular structure but differing in the number of carbon atoms present in the molecule and forming a series in which each member contains one $-CH_2-$ group more than its lower neighbour. In such homologous series, the individual hydrocarbons are of similar chemical properties. The different groups can be separated into the following three classes :

(1) The *saturated hydrocarbons*, which, by reason of their chemical inertness, remain unchanged when the oil is treated with sulphuric acid. The other groups of hydrocarbons are, by this process, sulphonated to acid-soluble derivatives or are polymerized to a sludge leaving the saturated oils as an unsulphonated residue. This reaction, the basis of the oldest method of refining petroleum oils, is used for the determination of unsulphonated residue, a figure which gives a measure of the degree of refinement of a particular oil.

(2) The *unsaturated hydrocarbons*, which, containing unsaturated linkages (i.e., carbon atoms united by two or more valency bonds), react with reagents such as iodine which enter the molecule at the double bond. If this reaction is carried out under standard conditions, an iodine value is obtained which gives a measure of the relative number of unsaturated linkages in the molecular structure of the compounds present in the oil.

(3) The *aromatic hydrocarbons*, which are a special group of unsaturated hydrocarbons having the aromatic properties associated with the conjugated double bonds of the benzene nucleus. Hydrocarbons of this group are distinguished by certain tests of which one of the simplest utilizes the solubility of such compounds in dimethyl sulphate, a reagent in which high-boiling oils of the first and second groups are insoluble. The percentage of neutral oil soluble in dimethyl sulphate thus affords a measure of the content of aromatic hydrocarbons.

This chemical classification, which is the simplest necessary for the explanation of the properties of the hydrocarbon oils as spray materials, is supplemented by a physical classification which serves to characterize the mean molecular weight of the mixture of hydrocarbons constituting the oil. In the homologous series the boiling-point of the hydrocarbon increases as the series is ascended. Distillation, which is the first process to which crude oils and tars are subjected, results in a fractionation according to molecular weight. The boiling range of the oil is therefore an important item in the description or specification of a particular oil, though, in the case of petroleum oils, it has been found possible to substitute, for this criterion, the viscosity of the oil. In general, the greater its molecular weight, the more viscous the hydrocarbon. Like viscosity, the specific gravity of the oil is closely correlated with and increases with boiling range, but the correlation is affected by the relative proportions of the homologous groups of which the oil is made up. As a general rule, with oils of similar boiling range, the higher the specific gravity the greater the proportion of unsaturated and, in particular, the aromatic hydrocarbons present.

Reviewing the status, under this physico-chemical classification, of the older definitions employed to characterize hydrocarbon oils, the petroleum oils consist mainly of hydrocarbons of the saturated and unsaturated groups. Petroleum oils from certain oil fields, e.g., Borneo, are especially rich in aromatic hydrocarbons and the crude oils may contain constituents such as the naphthenic acids, sulphur compounds and nitrogenous derivatives. Distillation yields fractions, the more important of which for insecticidal purposes are the kerosine and lubricating oils. The kerosines are those oils of boiling range between approximately 150–300° C, more familiar in England as paraffin used as a burning oil. Fractions of boiling range greater than 300° C are employed as lubricating oils, a group which has been subdivided, originally according to specific gravity but latterly according to viscosity, into "light", "medium" and "heavy" oils or into "thin", "medium" and "thick" lubricating oils. The refinement of these oils is effected either by sulphuric acid treatment or by solvent treatment. A highly-refined petroleum is, therefore, one of high content of saturated hydrocarbons and, as the colour of the oil disappears during the refinement processes, the terms "white", "half-white" and "red" have been employed to designate the degree of refinement of the oil.

Of the tar oils, the most important for insecticidal purposes are (1) the creosote oils, which are oils of specific gravity greater than that of water, comprising the distillate of boiling range between about 180° C and the pitching point (360–400° C, according to the type of pitch required), and (2) the anthracene or green oils which are those fractions from which crude anthracene separates on cooling and have a boiling range of above 270° C up to the pitching point. The predominant hydrocarbons present in such oils are of the aromatic group.

In addition to hydrocarbons, creosote and anthracene oils contain constituents soluble in acid and in alkali. The constituents removed by alkali treatment, which since they possess acidic properties are called tar acids or phenols, are mainly hydroxyl derivatives of the aromatic hydrocarbons. The content of phenols varies according to the source of the tar oil and its boiling range. They accumulate in the oils of intermediate boiling range (low-boiling creosotes) from which they are separated to form the cresylic acid of commerce. Correspondingly, the constituents removed by acid treatment are called tar bases and consist mainly of nitrogenous derivatives of the aromatic series, such as quinoline.

As it is now possible to produce aromatic oils from petroleum sources or to convert tar oils by hydrogenation to oils of groups 1 and 2, the older distinction between petroleum and tar oils may become useless.

Historical. It is unknown when and how petroleum oils were first employed as insecticides, but kerosine was recommended in 1865 (*1*), against scale insects of citrus, soon after it came into general use, in America, as a burning oil. As undiluted kerosine caused severe foliage and fruit injury there was need for a method whereby the material could be applied in a less concentrated state, a difficulty partly solved by its use in an emulsified form. Lodeman (*1*) was of the opinion that the employment of soap solutions for emulsifying kerosine would follow naturally, for the improvement of sprays by the addition of soap as a spreader was early recognized. Other authorities (e.g., *2*) have stated that the first kerosine emulsions were made with milk as the emulsifier.

The toxic activity of the petroleum products was at first thought to be related to their volatility. Shafer (*3*) was of the opinion that the volatile portion of kerosine and gasoline (a petroleum distillate of lower boiling point, more familiar in England as petrol), is the effective agent and concluded that after absorption into the insect body, the vapour becomes effective by preventing oxygen absorption by the tissues. Later (*4*) he showed that the vapour has an effect upon the enzymatic activity. Moore (*5*) found an excellent correlation between volatility and toxicity towards the housefly (*Musca domestica*), the more volatile gasoline being far more effective than kerosine. Later, however, Moore and Graham (*6*) found that, when used as contact insecticides in emulsified form, the higher boiling fractions of the petroleum distillates are more effective than those of lower boiling-point.

Such a result is more in accordance with practical experience. Pickering (*7*) showed that against the eggs of mussel scale (*Lepidosaphes ulmi*) the lower boiling point fractions of the petroleum distillates are ineffective because of their ready volatilization. He therefore recommended the use of a paraffin oil of which at least 40 per cent has a boiling point higher than 250° C, selecting a particular brand of paraffin oil called Solar Distillate, which distils almost entirely between 240–350° C. In the United States, crude oil emulsions were frequently recommended, but there was a tendency to replace the crude oil by distillates of relatively high gravity, as illustrated by the recommendations of Yothers and Crossman (*8*). Although lubricating oil emulsions were, according to Smith (*2*), described as early as 1907, it was not until after 1922, when Quaintance (*9*) reported their successful use against San José scale (*Aspidiotus perniciosus* Comst.), that their use became widespread.

The history of the introduction, into Great Britain, of high-boiling petroleum oils is also obscure. It would appear that the superior insecticidal properties of lubricating oils were observed, independently, about 1910 by a Tasmanian fruit-grower (Hatfield, *in litt.*) who, having no kerosine, used red engine oil instead. His success was brought to the notice of an oil company which, after experiments in Australia, introduced a miscible oil containing high-boiling petroleum oils into England in 1925. For some reason, poor emulsification according to Kent (*10*), spray damage according to Hiller (*in litt.*), the product was unsuccessful. It was not until 1928, when an American preparation bearing the name of W. H. Volck, who had begun the manufacture of this product prior to 1924 (*11*), was introduced that interest in the high-boiling petroleum oils became general.

The value of tar as a wound dressing and of the various tar oils as wood preservatives were early discoveries, but their application to crop protection dates from about 1890. Sajó (*12*) found anthracene oil effective against mussel scale. This material he painted on the dormant tree trunk, its later application causing severe foliage injury. He recorded that Robbes, in 1889, had employed tar and turpentine oils (the latter are distillates from coniferous woods) against moth eggs.

Various combinations of such tar distillates and an emulsifying agent, usually soap, were marketed under the general name of " water-soluble carbolineums ". The term " carbolineum " then referred to tar distillates employed as wood preservatives. The employment of carbolineum washes, in England, resulted from the introduction by Wiltshire (*13*) in 1921, of a Dutch product which proved so successful that other proprietary products of a similar type quickly appeared on the market.

The Phytotoxic Properties of Hydrocarbon Oils. The phytotoxic action of the hydrocarbon oils is the limiting factor determining their use. Tar oil washes were soon found unsuitable for application to foliage and so great was the injury resulting from the use of early petroleum washes that their employment was limited to dormant spraying. It is now known that tar oil washes must be applied before the buds begin to swell, if bud damage is to be avoided. Petroleum oils of unsulphonated residue as low as 60 per cent (*14*) have been found safe, under certain conditions, for application at periods intermediate between dormancy and blossoming.

The phytotoxic properties of petroleum oils to citrus foliage were examined by Gray and de Ong (*15*), who concluded that the constituents responsible for acute injury are removed by sulphuric acid treatment. Spuler, Overley and Green (*16*) showed that oils refined by either the sulphuric acid or sulphur dioxide treatments can be safely applied to deciduous foliage. These studies were extended by Green (*17*), by Young and Morris (*18*), and by Kearns, Marsh and Martin (*19*), and confirmation was obtained of Gray and de Ong's observation that acute oil injury is correlated with the percentage of oil removed by sulphonation, though certain oils behaved exceptionally.

De Ong (*20*) suggested that, in addition to unsaturated and aromatic hydrocarbons, compounds likely to oxidize on exposure to acidic derivatives might prove phytotoxic. Tucker (*21*) supported this idea and attributed the damage to oil-soluble asphaltogenic acids formed by the oxidation of the unsaturated hydrocarbons themselves. Further, de Ong (*20*) suggested that certain types of sulphur compounds would prove to be phytotoxic; the application of petroleum oil emulsion to foliage already bearing residues of a sulphur spray is apt to cause foliage damage.

Chronic injury by petroleum oils takes the form of a yellowing and early shedding of leaves and a delayed ripening and premature drop of fruit. With saturated hydrocarbons, de Ong, Knight and Chamberlin (*11*) found that oils of low viscosity are apparently safer to citrus than those of high viscosity. They attributed this difference to the more rapid disappearance of oils of low viscosity from the foliage, a process which they thought due primarily to absorption rather than to volatilization. Knight, Chamberlin and Samuels (*22*) observed the penetration of the oil into the vascular system and showed that the recovery of citrus from the effects of oil sprays was more rapid, the lighter the oil applied. On deciduous trees, Kelley (*23*), who investigated the effect of hydrocarbon oils on transpiration rate, found that reduction of this rate was apparently due to the physical properties of the oil and concluded that high viscosity is associated with greater interference with metabolic processes. Spuler, Overley and Green (*24*), who employed the starch content of apple leaves as a measure of the effects of oil sprays on the metabolic processes, also found that the accumulation of starch in sprayed foliage increased with the viscosity of the oil applied. It may therefore be concluded that the higher the viscosity of the oil, and consequently the higher its boiling range, the more liable it is to produce injury of the chronic type. On certain herbaceous plants, unsaturated hydrocarbons of the correct viscosity may cause chronic injury which appears to be associated with the interference of transpiration. Young tomato plants, especially if recently watered, assume a characteristic appearance termed oedema.

The extent of foliage damage will be determined not only by the quality of the oil applied but also by the quantity retained upon the sprayed surface. This amount will depend on the oil concentration of the spray applied and on the stability of the emulsion. De Ong, Knight and Chamberlin (*11*), and Griffin, Richardson and Burdette (*25*) independently showed that the amount of oil deposited on a standard surface was greater with an unstable emulsion than with a stable emulsion. The term instability is here used not to imply instability in the spray tank but the rapid breaking of the emulsion after application. De Ong, Knight and Chamberlin decreased the stability of their emulsions by reducing the content of emulsifier, but the type of emulsifier used also affects the rate of breaking of the emulsion. De Ong and his co-workers recommended lime casein or soda casein as more suitable emulsifiers than soap for quick-breaking emulsions. Ebeling (*26*) concluded that liability to injury was inversely proportional to the concentration of emulsifier used.

To summarize, it has been shown that, for application to foliage, only highly refined hydrocarbon oils are suitable and that these oils should be of the minimum boiling range or viscosity and applied as emulsions of the greatest stability consistent with insecticidal efficiency.

For use upon dormant trees, the success of tar oil washes indicates that oils in which aromatic hydrocarbons predominate may be applied with safety. Bud damage is usually associated with the quantity of oil applied and factors which tend to a disproportionate concentration of oil, e.g., the presence of solid matter such as crude anthracene, the application of washes inadequately emulsified, or application in strong wind causing the repeated deposition of spray by drift, may be the primary reason for spray damage.

The Insecticidal Properties of Hydrocarbon Oils. Phytotoxicity limits the type of hydrocarbon oil suitable for application to foliage to oils of high unsulphonated residue and not too high a viscosity. The influence of the chemical properties of the oil upon its insecticidal efficiency is therefore of little practical importance, though it is of interest to note that de Ong's (*27*) suggestion that the removal of the more reactive constituents of the oil by refinement may decrease its toxicity was not confirmed by Griffin, Richardson and Burdette (*29*). Spuler, Overley and Green (*16*) reported that, in tests upon codling moth eggs, oils of the same viscosity but varying 48 per cent in unsulphonated residue were equally toxic. Ebeling (*26*) concluded that, on citrus, the qualities of the spray making for greater safety to the tree also reduce insecticidal efficiency with the exception of degree of refinement. He found no correlation between degree of refinement and insecticidal properties.

The increase of insecticidal efficiency with boiling range of the oil has already been mentioned and the correlation is well shown in the results of de Ong, Knight and Chamberlin (*11*) upon citrus scale insects. They concluded that oils, with a rather wide range of high viscosity will give complete control but that below the minimum of this range, the lighter the oil, the less certain will be the kill. Against pests of deciduous and herbaceous plants, the action of hydrocarbon oils was examined by Griffin, Richardson and Burdette, employing *Aphis rumicis* as the test organism. They showed that the insecticidal efficiency of oils was independent of viscosity when the latter was above a certain minimum figure.

It is difficult to associate a chemical toxicity with the bland saturated hydrocarbons and it seems, as suggested by de Ong, Knight and Chamberlin (*11*), that their insecticidal action is dependent upon the permanence of the oil deposit and, with scale insects, upon the inability of the insect to expel oils of viscosity above a certain minimum from its tracheal system. The efficiency of the spray will therefore be dependent on the amount of oil retained upon the insect after spraying. The relationship between insecticidal efficiency and the stability of the emulsion was shown by de Ong and his colleagues. Using a high-refined lubricating oil at 2 per cent and lime casein at various concentrations as the emulsifier, they found that approximately 40 per cent of the red scale *Aonidiella aurantii* Maskell survived when the emulsifier concentration exceeded 0.5 per cent, whereas all were killed at emulsifier concentrations below 0.125 per cent. This increased efficiency was found to run parallel, firstly, to the greater amount of oil deposited from the emulsions of low emulsifier content and, secondly, to the average size of the oil droplets in the emulsion which was greater the less stable the emulsion. This observation was confirmed against aphides by Griffin, Richardson and Burdette (*25*) and by English (*28*), and is in accord with the results of later work on the factors affecting the spray retention of emulsions (*29*). The relationship between concentration of emulsifier and insecticidal efficiency was confirmed by Smith (*2*) and by Ebeling (*26*) with blood albumin as the emulsifier and by Cressman and Dawsey (*30*) with soap emulsions.

The excellent penetrating properties of the oils not only permit their entry into the scale insect through the tracheae and armour but also into the bark (*31*). Ebeling (*32*) suggested that the insecticidal efficiency might be improved if the rapid loss of oil from the plant surface by penetration into the bark could be inhibited. For this purpose he experimented with the addition, to the oil, of paraffin wax, and later (*33*), with oil-soluble amphipathic (see p. 73) compounds such as the triethanolamine soaps. Knight and Cleveland (*34*) added compounds such as glyceryl oleate and aluminium naphthenate to increase the surface persistence of the oil and reported an improved insecticidal action. The physico-chemical basis of the use of anti-spreaders is obscure but the related problem of the spread of oils, used as mosquito larvicides, on water has been studied by Murray (*35*).

For application to foliage it may therefore be concluded that the saturated hydrocarbon oil should have a boiling range, or to follow American usage, a viscosity, above a certain minimum and that the emulsion should have a stability sufficient to ensure the application of a spray of uniform oil concentration yet sufficiently quick-breaking to furnish an oil deposit of the maximum insecticidal efficiency.

For dormant application, a wider range of hydrocarbon oils is available because phytotoxic properties become less limiting. The main purpose of the dormant wash is to kill hibernating insects and insect eggs. Against the former, Swingle and Snapp (*36*) carried out trials with petroleum oils of different characteristics and found all equally effective provided that their viscosity was above a certain minimum. The importance of viscosity and boiling range was also shown in work upon the physical and chemical characteristics which determine the ovicidal efficiency of hydrocarbon oils. To the eggs of the capsid bug *Lygus pabulinus* L., Austin, Jary and Martin (*37*) showed that ovicidal efficiency was independent of the boiling range and viscosity of the oil provided, firstly, that these values were above a certain minimum and, secondly, that the content of saturated hydrocarbons was greater than a certain percentage. The second of these provisos is discussed below. Their conclusions are in agreement with the general experience that ovicidal properties are distributed over a wide range of high-boiling or high-viscosity petroleum oils, shown, for example, in the conclusions of Melander, Spuler and Green (*38*) upon toxicity to eggs of the fruit-tree leaf-roller (*Archips argyrospila* Wlk.) and of Hamilton (*39*) concerning the control of red spider mite (*Metatetranychus ulmi*).

Employing more refined methods, including Waterman analysis to such properties as refractive index and " aniline point ", for the deduction of the chemical character and molecular weight of the hydrocarbons present, Pearce, Chapman and Avens (*40*) concluded that the best dormant oil against eggs of *A. argyrospila* is of high paraffinic and low aromatic character. Similarly for summer use against codling moth, the same workers (*41*) found that high " paraffinicity " gave the most ovicidal oil and, of such saturated oils, Pearce, Chapman and Frear (*42*) found the paraffinic oils superior to naphthenic oils.

In the case of tar oils, the insecticidal properties of neutral oils towards the grain weevil *Calandra granaria*, were found by Goetze (*43*) to increase with boiling range, a conclusion also reached by Beran and Watzl (*44*) who, in addition to the grain weevil, employed larvae of the scale *Lecanium corni*

Bch. The ovicidal properties of tar oils were examined by Tutin (*45*) upon the eggs of *Aphis pomi* DeG. and of the winter moth *Operophthera brumata* L. He found that the most toxic fraction was the high-boiling neutral oil.

That the chemical character of the hydrocarbons influences insecticidal and ovicidal efficiency may be illustrated from Austin, Jary and Martin's results. The control of eggs of *L. pabulinus* by petroleum oils consisting mainly of saturated hydrocarbons was significantly better than that by tar oils consisting mainly of aromatic hydrocarbons. Allowance for the somewhat higher boiling range of the former and for the presence of constituents of a non-hydrocarbon character in the latter did not provide a complete explanation of the differences and it was suggested that oils of a low content of saturated hydrocarbons have inferior ovicidal properties towards these eggs, though this inferiority may not become of practical importance until the unsulphonated residue of the oil falls below 60 per cent by volume. The conclusion that a certain content of unsaturated and aromatic hydrocarbons may not affect the efficiency of an oil is in agreement with the results of Spuler, Overley and Green (*16*), who showed that refinement to a stage beyond that represented by 50 per cent unsulphonated residue was not necessary with petroleum oils for use against the eggs of *A. argyrospila*, and of Swingle and Snapp (*36*), who found unsulphonated residue of no effect upon the toxicity of the petroleum oils they used against *Aspidiotus perniciosus*. The reduction of ovicidal efficiency when the character of the hydrocarbons is changed from saturated to aromatic is most strikingly shown by practical experience of red spider mite (*Metatetranychus ulmi*) control. Whereas petroleum oils exert a degree of control when applied as winter washes against this pest, the tar oils are useless (see e.g., *46*).

Generalizations on the relative efficiency of saturated and aromatic hydrocarbons of similar boiling range are complicated by the fact that, against eggs of Aphididae and Psyllidae, the saturated hydrocarbons are devoid of ovicidal properties. To explain this apparent reversal of the conclusions reached in the preceding paragraph, Staniland, Tutin and Walton (*47*) suggested that the ovicidal properties of hydrocarbon oils are exerted in two ways: firstly, a physical "stifling" action associated with the permanence of the oil film over the insect egg and to which is due the ovicidal properties of hydrocarbon oils upon geometrid and capsid eggs but which has no effect upon aphid and psyllid eggs; secondly, a chemical toxic action peculiar to tar oils and to which is due their action upon aphid and psyllid eggs.

The first hypothesis, of a "stifling" action, is not only in accord with the conclusion that the hydrocarbon should be liquid, high-boiling and non-volatile but has been experimentally verified by Powers and Headlee (*48*) on eggs of the mosquito *Aedes aegypti*. They found that the main cause of death was oxygen starvation rather than the accumulation of carbon dioxide within the oil-covered egg, and concluded that the variations in ovicidal properties of the petroleum oils examined were probably factors such as too short a period of oil retention or the failure of the oil to make a complete seal. Further the hypothesis may explain the superiority of the saturated hydrocarbons over the unsaturated and aromatic hydrocarbons for the latter being more chemically reactive, may form water-soluble derivatives removed by rain. The effect of water on the weathering of creosote was found, by Gillander *et al.* (*49*) to be confined mainly to the removal of the more polar compounds present and to hydrocarbons distilling below 270° C.

The predominant group of polar compounds present in aromatic oils is the phenols and it was observed by Tutin (*45*) that their removal increases the toxicity of tar oils to eggs of *O. brumata*. His conclusion was confirmed by the field work of Staniland, Tutin and Walton (*53*) upon *Plesiocoris rugicollis* and by the results of the laboratory trials of Austin, Jary and Martin (*37*) upon eggs of *L. pabulinus*.

The second hypothesis of a chemical toxic action to which aphid and psyllid eggs are peculiarly susceptible was examined by Austin, Jary and Martin (*37*) on *Psyllia mali* and by Kearns, Martin and Wilkins (*50*) on *A. pomi*. The relationship between volatility and ovicidal properties does not follow as a corollary to this hypothesis as it does to the first but the earlier observation of Tutin, that ovicidal properties to *A. pomi* ascends with distillation range, was confirmed. Further, it was shown that, in the oils examined, toxicity was associated with the content of aromatic hydrocarbons as assessed by solubility in dimethyl sulphate. This criterion has provided a satisfactory test in the standardization of tar-oil products (see e.g., *51*) but, theoretically, it implies an absence of marked differences in ovicidal action among the compounds which constitute high-boiling hydrocarbons soluble in dimethyl sulphate. Such differences were not found by Kearns, Martin and Wilkins but every new source of hydrocarbons requires investigation for the discovery of differences would greatly facilitate research on the isolation of the compounds responsible for ovicidal action. Shaw, Steer and Davies (*52*), for example, found that in low temperature tar oils ovicidal properties to *A. pomi* were lower than indicated by their content of high-boiling, dimethyl sulphate soluble, neutral oils.

The influence of the polar compounds present in tar oils on ovicidal action of the chemical type appears to be slight. The tar bases examined by Shaw and Steer (*53*) exhibited ovicidal properties, outstanding in nicotine, but not of an order which makes them of importance at the low concentrations present in tar oils. Though Tutin (*45*) stated that the presence of phenols reduced the toxicity of tar oils to *A. pomi*, no confirmatory evidence of this observation is available.

Contrary to the results obtained by the application of petroleum oils to foliage, the influence of the stability of the emulsion does not appear to affect markedly the ovicidal efficiency of winter washes. Austin, Jary and Martin (*37*) and Steer (*54*) obtained a similar control of *L. pabulinus* by a tar-petroleum oil mixture whether emulsified by Bordeaux mixture or by the two-solution oleic acid method. Speyer (*55*), who examined the action of a number of proprietary preparations upon the eggs of *P. mali*, found no relationship between toxicity and the stability or surface tension of the diluted washes.

On the other hand, Beran's (*56*) successful use of low oil concentrations in sprays applied at sub-zero temperatures in Austria may be at least partly explained by instability of the emulsion induced by the freezing of the droplets. Beran found that the oil concentration could be halved without loss of ovicidal effect and attributed the saving to the lower loss of oil in the " run-off ".

Summarizing, it may be concluded that the insecticidal efficiency of hydrocarbon oils applied as dormant washes is determined by two sets of factors. When employed against insects or eggs susceptible to a " stifling " action, the oil should have a boiling range or viscosity above a certain minimum and should contain the maximum practicable of saturated hydrocarbons and the minimum of phenols. When employed against insect eggs susceptible to the chemical toxic action, the content of aromatic hydrocarbons should be above a certain minimum. By the use of a suitable combination of petroleum and tar oils it becomes possible to control adequately both types of insect eggs, provided that the total oil concentration of the wash is not large enough to cause spray damage.

REFERENCES

(*1*) Lodeman, E. G., *The Spraying of Plants*, New York, 1903, p. 79.
(*2*) Smith, R. H., *Bull. Calif. agric. Exp. Sta.*, 527, 1932.
(*3*) Shafer, G. D., *Tech. Bull. Mich. agric. Exp. Sta.*, 11, 1911.
(*4*) Shafer, G. D., *Tech. Bull. Mich. agric. Exp. Sta.*, 21, 1915.
(*5*) Moore, W., *J. agric. Res.*, 1917, **10**, 365.
(*6*) Moore, W. and Graham, S. A., *J. econ. Ent.*, 1918, **11**, 70.
(*7*) Bedford, Duke of, and Pickering, S. U., *6th Rep. Woburn exp. Fruit Farm*, 1906, p. 75.
(*8*) Yothers, W. W. and Crossman, S. S., abstr. in *Exp. Sta. Rec.*, 1911, **25**, 153.
(*9*) Quaintance, A. L., *Clip Sheet U.S. Dep. Agric.*, 193, 1922.
(*10*) Kent, W. G., *Rep. Wye Provincial Conf.*, 11/4/35, p. 5.
(*11*) de Ong, E. R., Knight, H. and Chamberlin, J. C., *Hilgardia*, 1927, **2**, 351.
(*12*) Sajó, K., *Z. PflKrankh.*, 1894, **4**, 4.
(*13*) See Lees, A. H., *Rep. agric. hort. Res. Sta., Bristol*, 1924, p. 51.
(*14*) Martin, H., *Ann. appl. Biol.*, 1935, **22**, 334.
(*15*) Gray, G. P. and de Ong, E. R., *Industr. engng Chem.*, 1926, **18**, 175.
(*16*) Spuler, A., Overley, F. L. and Green, E. L., *Bull. Wash. St. agric. Exp. Sta.*, 247, 1931.
(*17*) Green, J. R., *J. agric. Res.*, 1932, **44**, 773.
(*18*) Young, P. A. and Morris, H. E., *J. agric. Res.*, 1933, **47**, 505.
(*19*) Kearns, H. G. H., Marsh, R. W. and Martin, H., *Rep. agric. hort. Res. Sta., Bristol*, 1937, p. 65.
(*20*) de Ong, E. R., *IV Int. Congr. Ent., Ithaca*, 1928, **2**, 145.
(*21*) Tucker, R. P., *Industr. engng Chem.*, 1936, **28**, 458.
(*22*) Knight, H., Chamberlin, J. C. and Samuels, C. D., *Plant Physiol.*, 1929, **4**, 299.
(*23*) Kelley, V. W., *Bull. Ill. agric. Exp. Sta.*, 353, 1930, p. 579.
(*24*) Spuler, A., Overley, F. L. and Green, E. L., *Bull. Wash. St. agric. Exp. Sta.*, 252, 1931.
(*25*) Griffin, E. L., Richardson, C. H. and Burdette, R. C., *J. agric. Res.*, 1927, **34**, 727.
(*26*) Ebeling, W., *J. econ. Ent.*, 1932, **25**, 1007.
(*27*) de Ong, E. R., *Industr. engng Chem.*, 1928, **20**, 826.
(*28*) English, L. L., *Bull. Ill. nat. Hist. Surv.*, 1928, **17**, 235.
(*29*) Fajans, E. and Martin, H., *J. Pomol.*, 1938, **16**, 14 ; Ben-Amotz, Y. and Hoskins, W. M., *J. econ. Ent.*, 1937, **30**, 879 ; Hoskins, W. M. and Ben-Amotz, Y., *Hilgardia*, 1938, **12**, 83 ; Brown, G. T. and Hoskins, W. M., *J. econ. Ent.*, 1939, **32**, 57.
(*30*) Cressman, A. W. and Dawsey, L. H., *J. agric. Res.*, 1934, **49**, 1.
(*31*) Hoskins, W. M., *Hilgardia*, 1933, **8**, 49.
(*32*) Ebeling, W., *Hilgardia*, 1936, **10**, 95.

(*33*) Ebeling, W., *Verh. VII Int. Congr. Ent. Berlin*, 1938, **4**, 2966.
(*34*) Knight, H. and Cleveland, C. R., *J. econ. Ent.*, 1934, **27**, 269 ; Cleveland, C. R., *J. econ. Ent.*, 1935, **28**, 715.
(*35*) Murray, D. R. P., *Bull. ent. Res.*, 1936, **27**, 289 ; 1938, **29**, 11 ; 1939, **30**, 211.
(*36*) Swingle, H. S. and Snapp, O. I., *Tech. Bull. U.S. Dep. Agric.*, 253, 1931.
(*37*) Austin, M. D., Jary, S. G. and Martin, H., *J. S. E. agric. Coll. Wye*, 1932, **30**, 63 ; 1933, **32**, 63 ; 1934, **34**, 114 ; 1935, **36**, 86.
(*38*) Melander, A. L., Spuler, A. and Green, E. L., *Bull. Wash. St. agric. Exp. Sta.*, 197, 1926.
(*39*) Hamilton, C. C., *Circ. N. J. agric. Exp. Sta.*, 187, 1926.
(*40*) Pearce, G. W., Chapman, P. J. and Avens. A. W., *J. econ. Ent.*, 1942, **35**, 211.
(*41*) Chapman, P. J., Pearce, G. W. and Avens, A. W., *J. econ. Ent.*, 1943, **36**, 241.
(*42*) Pearce, G. W., Chapman, P. J. and Frear, D. E. H., *Industr. engng Chem.*, 1948, **40**, 284.
(*43*) Goetze, G., *Zbl. Bakt.*, 1931, **83**, ii, 136.
(*44*) Beran, F. and Watzl, O., *Z. angew. Ent.*, 1933, **20**, 382.
(*45*) Tutin, F., *Rep. agric. hort. Res. Sta., Bristol*, 1927, p. 81.
(*46*) Carroll, J., *J. Dep. Agric.* (*Ireland*), 1929, **29**, 86.
(*47*) Staniland, L. N., Tutin, F. and Walton, C. L., *J. Pomol.*, 1930, **8**, 129.
(*48*) Powers, G. E. and Headlee, T. J., *J. econ. Ent.*, 1939, **32**, 219.
(*49*) Gillander, H. E., King, C. G., Rhodes, E. O., and Roche, J. N., *Industr. engng Chem.*, 1934, **26**, 175.
(*50*) Kearns, H. G. H., Martin, H. and Wilkins, A., *J. Pomol.*, 1937, **15**, 56.
(*51*) *Minist. Agric. Tech. Bull.*, 1, 1951.
(*52*) Shaw, H., Steer, W. and Davies, R. G., *J. hort. Sci.*, 1950, **25**, 190.
(*53*) Shaw, H., and Steer, W., *J. Pomol.*, 1939, **16**, 364.
(*54*) Steer, W., *Rep. E. Malling Res. Sta.*, 1932, p. 132.
(*55*) Speyer, W., *Z. angew. Ent.*, 1934, **20**, 565.
(*56*) Beran, F., *PflnSchBer.*, 1948, 2, 161.

THE GLYCERIDE OILS AND SOAPS

The use, as spray supplements, of glyceride oils, known more familiarly as animal and vegetable oils, has already been reviewed (p. 161). The oils possess also insecticidal properties. Staniland (*1*) searching for a cheap substitute for nicotine, selected rape and linseed oil on account of their easier emulsification with soft soap (a property now known to be due to their content of free fatty acid) but found the latter unsuitable owing to a rapid " varnishing " after application. Vegetable oil emulsions were found by Austin, Jary and Martin (*2*) to be intermediate in efficiency against the eggs of *Lygus pabulinus* between the high-boiling petroleum oils and anthracene oils. On the score of cost, the glyceride oils cannot compete with petroleum oils as winter washes, whilst for summer use, hydrocarbon oils are more effective against red spider mites.

The soaps, which are the alkali salts of the higher fatty acids, have long been known to possess contact insecticidal properties. Thus Lodeman (*3*) recorded that D. Haggerston, in 1842, employed a whale-oil soap against aphis, red spider mites, and thrips. This action was at first attributed to the caustic action of the alkali constituent of the soap, but Siegler and Popenoe (*4*) have shown that the acid constituent is insecticidal. They observed a striking increase in toxicity towards certain aphides as the series of normal saturated monocarboxylic fatty acids is ascended, maximum toxicity being reached at capric acid, $C_9H_{19}COOH$. Their results were confirmed by Tattersfield and Gimingham (*5*), who showed that the relative toxicities to *Aphis rumicis* of the fatty acids increase with molecular weight as the series is ascended from acetic to undecylic acid, $C_{10}H_{21}COOH$, and that beyond this point there is a fall in toxicity. The sodium salts were in most cases less toxic than the corresponding acids, a difference not so marked with the ammonium soaps, due probably to their readier hydrolysis, which results in the liberation of the fatty acid.

Similar investigations were made by Dills and Menusan (*6*), who found capric acid and lauric acid, $C_{11}H_{23}COOH$, the most toxic of the saturated fatty acids towards *A. rumicis* and *Macrosiphum rosae* L. Of the potassium soaps, the laurate was the most toxic of the saturated derivatives but was surpassed by the oleate. Sodium oleate was found by Tattersfield and Gimingham to be as toxic as oleic acid, an unsaturated acid which was slightly less effective than the more toxic of the saturated acids.

This example of the dependence of insecticidal properties upon molecular structure has attracted theoretical consideration. By plotting the logarithm of the reciprocal of the molecular concentration necessary to give 100 per cent mortality, against the molecular weight of the acid, O'Kane *et al.* (*7*) showed that Tattersfield and Gimingham's results gave points lying on a straight line. An absorption process is evidently involved in the gradual increase of toxicity with the length of the carbon chain, until the "cut-off" point is reached (see p. 98). Tattersfield and Gimingham traced a relationship between toxicity and the partition coefficient (see p. 97) of the acid between olive oil and water. O'Kane and his colleagues showed that the surface activity of the fatty acid, which they employed as a measure of surface adsorption, when plotted on a semi-logarithmic scale against molecular weight also gives a straight line. In a later paper, O'Kane and Westgate (*8*) established a relationship between surface activity (as measured by contact angle, see p. 69) and molecular weight of the sodium soaps of the saturated acids similar to that between insecticidal properties and molecular weight, the most active soap being sodium laurate. This relationship between toxicity and molecular weight in an homologous series has been given in detail for it is an early example of the general rule which is further discussed on p. 98.

Of attempts to extend the use of soaps and fatty acids as practical spray materials, mention may be made of the recommendation by Siegler and Popenoe (*4*) of the product known as "Doubly-distilled coconut fatty acids", which contains a high percentage of capric and lauric acids. The most serious limitation to the employment of the fatty acids is their high phytotoxicity, observed by Martin and Salmon (*9*). Dills and Menusan found that the most phytotoxic of the fatty acids were capric and lauric acids and that an increase of molecular weight beyond this point was accompanied by a decrease in phytotoxic properties. In the case of the potash soaps, however, they found no correlation between insecticidal and phytotoxic properties, plant injury decreasing with molecular weight.

REFERENCES

(*1*) Staniland, L. N., *Rep. agric. hort. Res. Sta. Bristol*, 1926, p. 78.

(*2*) Austin, M. D., Jary, S. G. and Martin, H., *J. S. E. agric. Coll. Wye*, 1933, **32**, 63; 1934, **34**, 114.

(3) Lodeman, E. G., *The Spraying of Plants*, New York, 1903, p. 14.
(4) Siegler, E. H. and Popenoe, C. H., *J. agric. Res.*, 1924, **29**, 259.
(5) Tattersfield, F. and Gimingham, C. T., *Ann. appl. Biol.*, 1927, **14**, 331.
(6) Dills, L. E. and Menusan, H., *Contr. Boyce Thompson Inst.*, 1935, **7**, 63.
(7) O'Kane, W. C., Westgate, W. A., Glover, L. C., and Lowry, P. R., *Tech. Bull. N. H. agric. Exp. Sta.*, 39, 1930.
(8) O'Kane, W. C. and Westgate, W. A., *Tech. Bull. N. H. agric. Exp. Sta.*, 48, 1932.
(9) Martin, H. and Salmon, E. S., *J. agric. Sci.*, 1933, **23**, 228.

MISCELLANEOUS CONTACT INSECTICIDES OF PLANT ORIGIN

Empirical selection, which revealed tobacco, derris, pyrethrum and quassia, has left a long list of plants reputed to possess insecticidal properties, of which Roark (*1*) cited no less than 180 species. Many of the materials he mentioned, however, are used more to prevent insect attack as, for example, Rosemary, a herb used to repel clothes moth, and Eucalyptus, the scent of which is said to keep away mosquitoes. Of those reputed to possess actual insecticidal properties, McIndoo and Sievers (*2*) tested the action of various preparations derived from 54 species upon various insects. Included in their trials were certain of the plant species already discussed, but of the others, **Stavesacre** (Lousewort) and **Tomato** warrant mention. Davidson (*3*) found the oil and alkaloids from the seeds of Stavesacre (*Delphinium staphysagria* L.) and **Larkspur** (*D. consolida* L.) insecticidal. Decoctions of tomato leaves and stems (*Lycopersicum esculentum* L.) were used in Russia (*4*) against caterpillars of the large cabbage butterfly (*Pieris brassicae*), and Blin (*5*) found an extract of tomato stems in potash lye a substitute for nicotine. McIndoo and Sievers found the alcoholic and ether extracts of tomato vines effective against flies and bees.

Other surveys of plant products for insecticidal properties were made at the Boyce Thompson Institute (*6*), (*7*), by Lee and Hansberry (*8*) on Chinese plants and by Tattersfield and his co-workers (*9*). Of non-rotenone-containing plants, the latter workers found *Annona* spp. of interest and, although alkaloids are present, Harper, Potter and Gillham (*10*) concluded that the insecticidal constituents of the seeds and roots of *A. reticula* L. and *A. squamosa* L. were glycerides of hydroxylated unsaturated acids.

Acetone extracts of the fruits of the **Amur corktree** (*Phellodendron amurense* Rupr.) were found to be highly toxic to houseflies by Haller (*11*). The insecticidal components were examined by Schechter and Haller (*12*), but on finding the extracts of little toxicity to houseflies when dissolved in kerosine it was decided that the product was unsuitable for commercial development (*13*). Of the five plants from tropical America selected by Sievers *et al.* (*14*) as worthy of further examination, Mamey (*Mammea americana* L.) had previously been reported on by Plank (*15*). Jones and Plank (*16*) isolated from the seed of this tree an insecticidal oil containing pyrethrin-like compounds.

From the roots of a Mexican plant, thought to be *Erigeron affinis* DC., Acree, Jacobson and Haller (*17*) isolated an insecticidal amide, **affinin**, identified as *N-iso*butyl-2 : 6 : 8-decatrienamide : $CH_3CH{=}CH.CH{=}CH(CH_2)_2CH{=}CH.CO.NH.CH_2CH(CH_3)_2$. In 1947 this plant was identified as *Heliopsis longipes* (Gray) Blake and, in an investigation of the related species, Gersdorff and Mitlin (*18*) found the roots of *H. scabra* Dunal. highly toxic to houseflies. Jacobson (*19*) isolated from the roots of this plant, **scabrin**, an *iso*butylamide of an unsaturated C_{18} acid appreciably more toxic to houseflies than the pyrethrins. Several other aliphatic *iso*butylamides of vegetable origin are known, including **pellitorine** from the roots of *Anacyclus pyrethrum* (see p. 167), thought by Jacobson (*20*) to be *N*-*iso*butyl-2 : 6-decadienamide, but Crombie (*21*) prepared the four *cis-trans* isomers and found none identical to pellitorine. Moreover the latter compound is about half as toxic to houseflies as the pyrethrin tested by Jacobson but Crombie's compounds were of low insecticidal activity. Crombie (*21*) continued his investigation for pellitorine is responsible for only part of the insecticidal amide fraction of the roots. He isolated a new unsaturated *iso*butylamide, **anacyclin**, of low insecticidal activity yet which, when hydrogenated, is converted to a highly potent insecticide shown to be *N*-*iso*butyltetradeca-*trans*-2-*trans*-4-*cis*-8-*cis*-10-tetraenamide.

REFERENCES

(*1*) Roark, R. C., *Amer. J. Pharm.*, 1919, **91**, 25, 91.
(*2*) McIndoo, N. E. and Sievers, A. F., *Bull. U.S. Dep. Agric.*, 1201, 1924.
(*3*) Davidson, W. M., *J. econ. Ent.*, 1929, **22**, 226.
(*4*) See Schreiber, A. F., abstr. in *Rev. appl. Ent.*, 1916, A, **4**, 59.
(*5*) Blin, H., *J. Agric. prat.*, 1920, **34**, 17.
(*6*) Hartzell, A. and Wilcoxon, F., *Contr. Boyce Thompson Inst.*, 1941, **12**, 127.
(*7*) Hartzell, A., *Contr. Boyce Thompson Inst.*, 1944, **13**, 243 ; 1947, **15**, 21.
(*8*) Lee, C. S. and Hansberry, R., *J. econ. Ent.*, 1943, **36**, 915.
(*9*) Tattersfield, F., Potter, C., Lord, K. A., Gillham, E. M., Way, M. J. and Stoker, R. I., *Kew Bull.*, 1948, **3**, 329.
(*10*) Harper, S. H., Potter, C. and Gillham, E. M., *Ann. appl. Biol.*, 1947, **34**, 104.
(*11*) Haller, H. L., *J. econ. Ent.*, 1940, **33**, 941.
(*12*) Schechter, M. S. and Haller, H. L., *J. org. Chem.*, 1943, **8**, 194.
(*13*) Sullivan, W. N., Schechter, M. S. and Haller, H. L., *J. econ. Ent.*, 1943, **36**, 937.
(*14*) Sievers, A. F., Archer, W. A., Moore, R. H. and McGovran, E. R., *J. econ. Ent.*, 1949, **42**, 549.
(*15*) Plank, H. K., *J. econ. Ent.*, 1944, **37**, 737 ; *Trop. Agric.*, 1950, **27**, 38.
(*16*) Jones, M. A. and Plank, H. K., *J. Amer. chem. Soc.*, 1945, **67**, 2266.
(*17*) Acree, F., Jacobson, M. and Haller, H. L., *J. org. Chem.*, 1945, **10**, 236.
(*18*) Gersdorff, W. A. and Mitlin, N., *J. econ. Ent.*, 1950, **43**, 554.
(*19*) Jacobson, M., *J. Amer. chem. Soc.*, 1951, **73**, 100 ; 1952, **74**, 3423.
(*20*) Jacobson, M., *J. Amer. chem. Soc.*, 1949, **71**, 366.
(*21*) Crombie, L., *J. chem. Soc.*, 1955, p, 999.

CHAPTER X

INSECTICIDES (III)

SYNTHETIC CONTACT INSECTICIDES

A RIGID classification of the many synthetic insecticides and acaricides, most of which have emerged since 1940, is impossible but, by permitting some elasticity in definition, most can be assigned to certain broad groups such as the chlorinated hydrocarbons, the organic phosphates, the organic thiocyanates and the dinitrophenols. Historically the latter group comes first:

THE DINITROPHENOLS

In their pioneer systematic work on insecticidal properties, Tattersfield and his colleagues (*1*) included derivatives of the aromatic hydrocarbons. They found but little difference in the toxicities of phenol and the three isomeric cresols to eggs of the purple thorn moth, *Selenia tetralunaria* Hufn., but observed a great increase when nitro-groups are introduced. The addition of a second nitro-group gave a further increase in toxicity, but the relative position of the two nitro-groups had some effect. Further nitration led to a decrease in toxicity.

3 : 5-Dinitro-*o*-cresol was found to possess so high an ovicidal value that field trials (*2*) were carried out with this compound and its salts as dormant spray materials. The potassium salt had been introduced some time before as an active constituent of "Antinonnin" (see p. 158) but its use was limited by its phytocidal properties which are, however, not objectionable for winter wash purposes. Successful trials with compounded winter washes based on dinitrocresol were reported by Blijdorp (*3*), Hey (*4*), Shaw and Steer (*5*), and Kearns and Martin (*6*). In these products the acid is dissolved in petroleum oil whereby the ovicidal properties of the latter are combined with the toxic properties of the cresol against eggs of Aphididae and Psyllidae and against moss and lichen.

3 : 5-Dinitro-*o*-cresol, which in stricter chemical nomenclature should be 2-methyl-4 : 6-dinitrophenol (I, p. 233), is usually abbreviated to DNC now adopted as its common name. The pure cresol is a yellow solid of m.p. 85.8° C, almost insoluble in water though the alkali and ammonium cresylates are freely soluble. Nor is the cresol of high solubility in petroleum oils, one reason for the further survey of related phenols by Kagy and Richardson (*7*). These workers reported that the compound which showed consistently the highest toxicity was 2-*cyclo*hexyl-4 : 6-dinitrophenol (U.S.P. 1,880,404), a compound which differs in constitution from the cresol only in the substitution of the methyl ($-CH_3$) group of the latter by

the *cyclo*hexyl ($-C_6H_{11}$) group. The introduction of this heavier hydrocarbon group increases the oil solubility of the compound. The results of spray trials of washes containing this compound, at one time designated by the symbol DNOCHP but since named dinex, have been consistently successful (see *8*) and the observation of synergism (see p. 94) between the phenol and the oil by Kagy and Richardson was verified by Bliss (*9*).

Boyce *et al.* (*10*), in studying the distribution of dinex between the continuous and disperse phases of petroleum oil emulsions, observed that, under acid conditions when the phenol is present mainly in the oil phase, the emulsion was more toxic to eggs of *Lygaeus kalmii* Stål. than alkaline emulsions of similar oil and phenol contents. A similar result was obtained by Barker *et al.* (*11*) with DNC-petroleum oil emulsions against eggs of *Aphis fabae* and the effect was ascribed to the greater toxicity of the undissociated phenol compared to the phenate ion, for Krahl and Clowes (*12*) showed that the nitrophenols penetrate the eggs of *Arbacia* only in unionized form. Bennett *et al.* (*13*) could trace no difference between phenol and phenate at concentrations usually employed against eggs of *Psyllia mali* and winter moth (*Operophtera brumata*) but against the eggs of the red spider mite *Metatetranychus ulmi* Koch the phenol was the more toxic. Chapman and Avens (*14*) found no difference, at equal DNC content, in the ovicidal action of solutions of the cresol or of cresylates on various aphid eggs.

Of attempts to prepare dinitrophenols safe for application to foliage, the most promising is that of Kagy and McCall (*15*) who selected the di*cyclo*hexylamine salt of dinex on the score of its low water-solubility and low vapour pressure (m.p. 197° C). Successful use of this salt on both hop and tomato foliage was reported by Simpson (*16*).

The mammalian toxicity of the dinitrophenols is so high that their one-time use for the treatment of obesity has been discontinued (*17*), while in Great Britain legislation requires that precautionary measures be taken by those engaged in their use for crop protection. The predominant effect is a stimulation of metabolism ; an increased oxygen consumption by honey bees treated with sub-lethal amounts of DNC was measured by Goble and Patton (*18*). Biochemical investigations have shown that the dinitrophenols interfere in the enzymic processes by which the animal stores the energy it derives from the oxidation of carbohydrates. This energy is converted to chemical energy in the form of " high-energy " phosphate bonds, e.g., by the phosphorylation of adenosine diphosphate to adenosine triphosphate. Loomis and Lipman (*19*) first reported the prevention of this phosphorylation by traces of dinitrophenol and their observation has been repeatedly confirmed and extended to other dinitrophenols. The application of the hypothesis to the insecticidal action of the dinitrophenols is facilitated by the observation that oxidative phosphorylation occurs in insects (*20*). Moreover, in the rabbit detoxication of DNC is by reduction to 2-amino-4-nitro-6-methyl phenol (*21*), a compound formed under certain conditions in the storage of DNC formulations and which is devoid of ovicidal properties (*13*).

REFERENCES

(*1*) Tattersfield, F., Gimingham, C. T. and Morris, H. M., *Ann. appl. Biol.*, 1925, **12**, 218; Tattersfield, F., *J. agric. Sci.*, 1927, **17**, 181.
(*2*) Gimingham, C. T. and Tattersfield, F., *J. agric. Sci.*, 1927, **17**, 162.
(*3*) Blijdorp, P. A., *Verh. VIII Int. Congr. Ent. Berlin*, 1938, **4**, 2941.
(*4*) Hey, G. L., *J. Min. Agric.*, 1938, **45**, 932.
(*5*) Shaw, H. and Steer, W., *J. Pomol.*, 1939, **16**, 364.
(*6*) Kearns, H. G. H. and Martin, H., *Rep. agric. hort. Res. Sta. Bristol*, 1938, p. 66.
(*7*) Kagy, J. F. and Richardson, C. H., *J. econ. Ent.*, 1936, **29**, 52; Kagy, J. F., *J. econ. Ent.*, 1941, **34**, 660.
(*8*) Dutton, W. C., *J. econ. Ent.*, 1936, **29**, 62.
(*9*) Bliss, C. I., *Ann. appl. Biol.*, 1939, **26**, 585.
(*10*) Boyce, A. M., Kagy, J. F., Persing, C. O. and Hansen, J. W., *J. econ. Ent.*, 1939, **32**, 432.
(*11*) Barker, C. H., Ripper, W. E. and Warburg, J. W., *J. Soc. chem. Ind.*, 1945, **64**, 187.
(*12*) Krahl, M. E. and Clowes, G. H. A., *J. cell. comp. Physiol.*, 1938, **11**, 1, 21.
(*13*) Bennett, S. H., Kearns, H. G. H., Martin, H. and Wain, R. L., *Ann. appl. Biol.*, 1946, **33**, 396.
(*14*) Chapman, P. J. and Avens, A. W., *J. econ. Ent.*, 1948, **41**, 190.
(*15*) Kagy, J. F. and McCall, G. L., *J. econ. Ent.*, 1941, **34**, 119.
(*16*) Simpson, A. C., *Nature, Lond.*, 1945, **155**, 241.
(*17*) Bidstrup, P. L. and Payne, D. J. H., *Brit. med. J.*, 1951, **2**, 16.
(*18*) Goble, G. J. and Patton, R. L., *J. econ. Ent.*, 1946, **39**, 177.
(*19*) Loomis, W. F. and Lipmann, F., *J. biol. Chem.*, 1948, **173**, 807.
(*20*) Lewis, S. E. and Slater, E. C., *Biochem. J.*, 1954, **58**, 207.
(*21*) Smith, J. N., Smithies, R. H. and Williams, R. T., *Biochem. J.*, 1953, **54**, 225.

THE ORGANIC THIOCYANATES

Murphy and Peet (*1*), in a study of the insecticidal properties of various organic radicals, found the most promising to be the thiocyanate group, –S–C≡N. They did not specify the actual compound used but reported that an aliphatic thiocyanate was highly toxic to *Aphis rumicis*, and, in a second paper (*2*), successful trials of the same compound upon the mealy bug *Pseudococcus citri* Risso and its eggs are recorded. Compounded products based on this thiocyanate were placed on the market under the name " Lethane " and, in 1936, Murphy (*3*) described the active constituent as β-butoxy-β′-thiocyanodiethyl ether, $C_4H_9OCH_2.CH_2.OCH_2.CH_2.SCN$ (*n*-butyl " Carbitol " thiocyanate).

Bousquet, Salzberg and Dietz (*4*) examined the insecticidal properties of a number of aliphatic thiocyanates, selecting a series of the thiocyanates of the saturated even-carbon fatty alcohols from ethyl to cetyl alcohol. They established the marked influence of the alkyl radical upon the toxicity of these compounds to various aphides, thrips and red spider. In all cases the C_{12} member, *n*-dodecyl (=*n*-lauryl) thiocyanate, $CH_3(CH_2)_{10}CH_2.SCN$ was found to be the most toxic and this compound appeared in the proprietary product " Loro ".

The success of these organic thiocyanates prompted wide investigation among this group of which over three hundred were included by Frear (*5*) in his catalogue of insecticides. Among those developed commercially may be mentioned 2-thiocyanoethyl laurate, $C_{11}H_{25}.CO.OCH_2CH_2SCN$,*

*"Lethane 60" is a 50% solution, in kerosine, of the 2-thiocyanoethyl ester of C_{10-18} aliphatic acids.

(U.S.P. 2,220,521), *iso*bornyl thiocyanoacetate (I) (" Thanite "), and fenchyl thiocyanoacetate (II) (U.S.P. 2,209,184 and 2,217,611). The insecticidal properties of the two latter compounds are described by Pierpont (*6*).

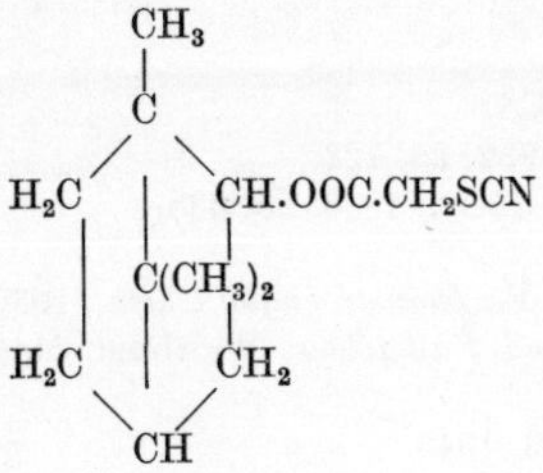

```
            CH3                                   CH3
             |                                     |
             C                                     C
           /   \                                 /   \
      H2C   |   CH.OOC.CH2SCN               H2C   |   CH.OOC.CH2SCN
       |    |    |                           |    |    |
       |  C(CH3)2                            |   CH2   |
       |    |    |                           |    |    |
      H2C   |   CH2                         H2C   |   C(CH3)2
           \   /                                 \   /
            CH                                    CH
            (I)                                  (II)
```

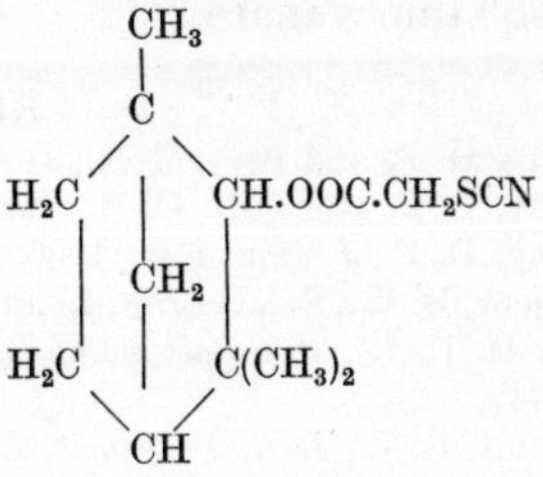

Work of more academic interest includes that at the Boyce Thompson Institute where Hartzell and Wilcoxon (*7*) showed that γ-thiocyanopropyl phenyl ether, $SCN.CH_2CH_2CH_2OC_6H_5$, was highly toxic to *A. rumicis* and better than the corresponding halogen derivative ; subsequently they found trimethylene dithiocyanate, $SCN.CH_2CH_2CH_2SCN$ more effective and less phytotoxic. Harvill and Arthur (*8*) having ascribed the toxicity of certain essential oils to the presence of a nuclear allyl group, tried the effects of the incorporation of the –SCN group. The resulting γ-thiocyanopropyl and β-thiocyanoethyl esters of the allyl phenols were more toxic and caused a rapid paralysis of the treated flies. The latter property, in which the pyrethrins excel, is of special value in fly sprays giving a rapid " knock-down ". Grove and Bovington (*9*) suggested that, as the knockdown activity of the thiocyanoacetates is greater than that of the corresponding alkyl thiocyanates, the linkage of the thiocyano group through a keto-methylene group to a lipoid-soluble hydrocarbon radical should be sought. The α-thiocyanoketones $R.COCH_2.SCN$ and thiocyanoacetates $RO.CO.CH_2SCN$ which they examined proved too irritant to mucous membranes for use in fly sprays. In these series the length of the alkyl R is again critical though in the alkyl thiocyanoacetates " cut-off " occurs earlier than in the alkyl thiocyanates ; dodecyl thiocyanoacetate proved inactive in contrast to dodecyl thiocyanate and 2-thiocyanoethyl laurate.

Grove and Bovington pointed out that the thiocyanoketones are readily converted to hydroxythiazoles which are structurally related to the thiazole moiety of the vitamin thiamin. They suggested that activity might arise through an inhibition of an enzyme process involving thiamin, but as the two hydroxythiazoles they synthesized were inactive they reverted to the observation of von Oettingen and his colleagues (*10*) of the liberation of hydrogen cyanide from the lower alkyl thiocyanates by rabbit liver. The suggestion that the *in vivo* production of hydrogen cyanide is responsible for the biological activity of the organic thiocyanates was also supported by Coon (*11*) and by Gustafson *et al.* (*12*).

Though this group of insecticides has been largely utilized as fly sprays, it finds special uses in crop protection. Kearns and Martin (*13*) found *n*-dodecyl thiocyanate toxic to the eggs of *A. pomi* and *Pysllia mali*, a result confirmed by Shaw and Steer (*14*) who also found the thiocyanates of the corresponding secondary alcohols less ovicidal. Hey (*15*), Shaw and Steer (*14*) and Kearns and Martin (*16*) reported successful trials of butyl " Carbitol " thiocyanate.

REFERENCES

(*1*) Murphy, D. F. and Peet, C. H., *J. econ. Ent.*, 1932, **25**, 123.
(*2*) Murphy, D. F. and Peet, C. H., *Industr. engng Chem.*, 1933, **25**, 638.
(*3*) Murphy, D. F., *J. econ. Ent.*, 1936, **29**, 606.
(*4*) Bousquet, E. W., Salzberg, P. L. and Dietz, H. F., *Industr. engng Chem.*, 1935, **27**, 1342.
(*5*) Frear, D. E. H., *A Catalogue of Insecticides and Fungicides*, Waltham, Mass., U.S.A., 1947.
(*6*) Pierpoint, R. L., *Bull. Del. agric. Exp. Sta.*, 253, 1945.
(*7*) Hartzell, A. and Wilcoxon, F., *Contr. Boyce Thompson Inst.*, 1934, **6**, 269 ; 1935, **7**, 29.
(*8*) Harvill, E. K. and Arthur, J. M., *Contr. Boyce Thompson Inst.*, 1943, **13**, 79.
(*9*) Grove, J. F. and Bovington, H. H. S., *Ann. appl. Biol.*, 1947, **34**, 113.
(*10*) von Oettingen, W. F., Hueper, W. C. and Deichmann-Gruebler, W., *J. industr. Hyg.*, 1936, **18**, 310.
(*11*) Coon, B. F., *J. econ. Ent.*, 1944, **37**, 785.
(*12*) Gustafson, C., Lies, T. and Wagner-Jauregg, T., *J. econ. Ent.*, 1953, **46**, 620.
(*13*) Kearns, H. G. H. and Martin, H., *Rep. agric. hort. Res. Sta. Bristol*, 1935, p. 49.
(*14*) Shaw, H. and Steer, W., *J. Pomol.*, 1939, **16**, 364.
(*15*) Hey, G. L., *J. Minist. Agric.*, 1938, **45**, 932.
(*16*) Kearns, H. G. H. and Martin, H., *Rep. agric. hort. Res. Sta. Bristol*, 1938, p. 66 ; Bennett, S. H., Kearns, H. G. H. and Martin, H., *J. Pomol.*, 1947, **23**, 38.

THE CHLORINATED HYDROCARBONS

(i) **DDT and Related Compounds**

By B.P. 547,871 (Sept. 15, 1942) the Swiss firm of J. R. Geigy A.G. protected, for insecticidal purposes, compounds of the general formula RR′.CH–CX_3, where X is chlorine or bromine, R and R′ organic radicals of at least three and five carbon atoms respectively. In their products " Neocid " and " Gesarol " (" Guesarol " in Great Britain), the main active component was of this formula with R = R′ = *p*-chlorophenyl and X = chlorine, i.e., 1 : 1 : 1-trichloro-2 : 2-di-(*p*-chlorophenyl)ethane (I), known also as dichlorodiphenyltrichloroethane, abbreviated to DDT. So effective was this compound against the body louse and other disease-carrying insect pests of man, of particular menace in wartime, that by 1945 the annual production of DDT had reached 30,000,000 lb.

DDT is made by the condensation of chloral with chlorobenzene and the composition of the crude product varies slightly with conditions of manufacture. Under suitable conditions the *p,p*′-isomer (I) constitutes some 80 per cent of the product, most of the remainder being the *o,p*′-isomer (II) ; typical analyses of early DDT samples are given by Haller *et al.* (*1*).

Cl–C_6H_4–CH(CCl_3)–C_6H_4–Cl
(I)

o-Cl–C_6H_4–CH(CCl_3)–C_6H_4–Cl
(II)

p,p′-DDT is a white crystalline compound of m.p. 108.5° C, readily soluble in many organic solvents (*2*) but almost insoluble in water, a property which, coupled with high stability and low vapour pressure (1.5×10^{-7} mm. Hg at 20° C (*3*)), renders spray residues of DDT on inert surfaces of extreme persistence. On leaf surfaces loss of insecticidal properties may follow through solution of the DDT in the cuticle waxes. Gunther *et al.* (*4*) found, however, a loss of 71–97 per cent of DDT in 86 days from spray residues in citrus foliage, a loss which was ascribed by Gunther, Carman and Elliott (*5*) to decomposition, the initial reaction being one of dehydrochlorination to 1 : 1-dichloro-2 : 2-di-(*p*-chlorophenyl)ethylene (III) :—

$$\underset{\text{(I)}}{(p\text{-Cl.C}_6\text{H}_4)_2\text{CH.CCl}_3} \rightarrow \underset{\text{(III)}}{(p\text{-Cl.C}_6\text{H}_4)_2\text{C:CCl}_2}$$

Fleck (*6*) assumed that this reaction also proceeded under the action of ultraviolet light in the presence of air, for he isolated 4 : 4′-dichlorobenzophenone, the oxidation product of III, from irradiated DDT deposits. Dehydrochlorination also proceeds readily when DDT is heated to 195° C or is treated, in solution, with alkalies.

The success of DDT rests largely on its extreme toxicity to insects, especially to those walking on the spray deposit. Läuger, Martin and Müller (*7*) estimated that a surface deposit of 10^{-12} g. per cm.^{2} was toxic to flies and clothes moth larvae. Only a few of the compounds of the general formula $RR'CH.CX_3$ approach this high toxicity but some have achieved commercial development. Replacement of one of the aliphatic chlorines of DDT by hydrogen gives 1 : 1-dichloro-2 : 2-di-(*p*-chlorophenyl)ethane (IV), abbreviated to DDD or TDE (tetrachlorodiphenylethane).

$$\underset{\text{(IV)}}{(p\text{-Cl.C}_6\text{H}_4)_2\text{CH.CHCl}_2} \qquad \underset{\text{(V)}}{(p\text{-F.C}_6\text{H}_4)_2\text{CH.CCl}_3}$$

Replacement of the nuclear chlorines by fluorine gives 1 : 1 : 1-trichloro-2 : 2-di-(*p*-fluorophenyl)ethane (V), at one time produced in Germany as the active component of " Gix " (*8*). The condensation of chloral with anisole or toluene in place of chlorobenzene gives 1 : 1 : 1-trichloro-2 : 2-di-(*p*-methoxyphenyl)ethane (VI) and 1 : 1 : 1-trichloro-2 : 2-di-(*p*-tolyl)-ethane (VII) respectively :

$$\underset{\text{(VI)}}{(p\text{-CH}_3\text{O.C}_6\text{H}_4)_2\text{CH.CCl}_3} \qquad \underset{\text{(VII)}}{(p\text{-CH}_3\text{.C}_6\text{H}_4)_2\text{CH.CCl}_3}$$

the former being called methoxychlor. The practical efficiency of these four insecticides (IV–VII) relative to DDT varies with insect species and method of use, but none has shown the same order of toxicity as DDT except against specific pests. On the other hand, it is claimed that both IV and VI are of lower mammalian toxicity than DDT. This topic has been of wide interest since the discovery that DDT used to control flies in the milking shed, can appear in the milk. When methoxychlor was used instead, Carter *et al.* (*9*) were unable to detect the insecticide in the milk.

Relationships between Structure and Insecticidal Properties

Laüger, Martin and Müller (*7*) traced the evolution of DDT from their work on moth-proofing agents which led to the discovery of the Geigy product " Mitin FF ". In this product, toxicity to moth larvae was attributed to the *p,p*′-dichlorophenyl ether fragment of the molecule. They showed that other dichlorophenyl derivatives, for example, 4 : 4′-dichlorophenyl sulphone $(p\text{-Cl.}C_6H_4)_2SO_2$, were effective stomach poisons and Müller suggested that by the inclusion of a fat-solubilizing group, such as the trichloromethyl fragment of chloroform, the molecule would become an effective contact insecticide. Hence the Swiss workers regarded the dichlorophenyl-methylene group of DDT as the toxophore and the trichloromethyl group as the conductophoric group imparting to the molecule those physical properties required for the penetration of DDT to its site of action within the insect. It was on this hypothesis and by the substitution of other fat-solubilizing groups for the trichloromethyl group that Läuger, Martin and Müller discovered a wide range of insecticidal compounds (see *10*) including 1 : 1-di-(*p*-chlorophenyl)-2-nitroethane (Swiss P. 236,227).

An alternative suggestion by Martin and Wain (*11*) arose from the observation that neither the ethylene (III) formed by the dehydrochlorination of DDT nor the tetrachloroethane (VIII), produced by the addition of chlorine to III, is insecticidal. VIII has the structure required by the Swiss hypothesis, but, like III, it is not susceptible to dehydrochlorination.

$$\underset{\text{(III)}}{(p\text{-Cl.}C_6H_4)_2C{:}CCl_2} \xrightarrow{Cl_2} \underset{\text{(VIII)}}{(p\text{-Cl.}C_6H_4)_2CCl.CCl_3}$$

For this and other reasons, Martin and Wain suggested that easy dehydrochlorination is requisite for insecticidal properties in this series of compounds. Indeed, insecticidal properties can be correlated with the ease of dehydrochlorination in closely related series such as :

DDT (I) > DDD (IV) > 1-chloro-2 : 2-di-(*p*-chlorophenyl)ethane (IX)
> 2 : 2-di-(*p*-chlorophenyl)ethane (X)

$$\underset{\text{(IX)}}{(p\text{-Cl.}C_6H_4)_2CH.CH_2Cl} \qquad \underset{\text{(X)}}{(p\text{-Cl.}C_6H_4)_2CH.CH_3}$$

or the *p,p′*-(I) > *o,p′*-(II) > *o,o′*-(XI) isomers of DDT. In both series the final member is non-insecticidal and inert to alcoholic potash (*12*). However, outside these limited series ease of dehydrochlorination was shown by

$$\underset{\text{(XI)}}{(o\text{-Cl.}C_6H_4)_2CH.CCl_3}$$

Busvine (*13*), Müller (*14*) and Cristol (*15*) to be an unreliable index of insecticidal potency.

Busvine emphasized that the most insecticidal analogues of DDT were those of similar molecular weight and structure. Picard and Kearns (*16*) considered that *p,p′* substitution was requisite and that the substituents should be of limited size. The comparative insecticidal effect of the *p*-substituent is instructive : chlorine, the methyl and methoxy groups yield powerful insecticides. The volumes occupied by the chlorine atoms and the methyl and methoxy groups are almost the same ; the compounds I, VI and VII are therefore of the same shape and size, i.e., they are isosteric.

Incidentally, the hydroxy derivative, 1 : 1 : 1-trichloro-2 : 2-di-(*p*-hydroxyphenyl)ethane, is also an isoster of DDT but is non-insecticidal for, being water soluble, it lacks penetrative properties.

Whenever it is found that the biological activity of a molecule is profoundly modified by a change in its shape or size and that isosters of similar lipoid solubility are of similar activity, it may be inferred that activity is dependent upon the ability of the molecule to fit so closely to some interface within the molecule that London forces come into play (see p. 100). Several hypotheses have been put forward relating structure and activity in the DDT analogues which directly or indirectly involves consideration of complementariness or, as expressed by Gunther and his colleagues (*17*), structural topography.

The first of these hypotheses is due to Brown and his co-workers (*18*). The DDT molecule contains a carbon atom linked to two planar groups (i.e., phenyl groups) and a group $-CCl_3$ which is just large enough to hinder a free rotation of the phenyl groups. The latter will, therefore, tend to occupy positions of maximum clearance imparting to the molecule what Brown *et al.* described as a trihedralized configuration, a configuration which they considered to be linked to insecticidal activity. To test this idea they examined a series of DDT analogues in which the $-CCl_3$ group is replaced by a group of similar or larger dimensions and found that the group $-C(CH_3)_3$ gave substituted diphenyldimethylpropanes of good insecticidal activity, including the chlorine-free isoster of DDT, 1 : 1-dianisyl neopentane (XII). Support for this hypothesis is also derived from the

$CH_3O-C_6H_4-CH(C(CH_3)_3)-C_6H_4-OCH_3$

(XII)

$Cl-C_6H_4-CH(CH(NO_2)CH_3)-C_6H_4-Cl$

(XIII)

insecticidal activities of 1 : 1-dichlorophenyl-2-nitropropane (XIII) and the corresponding butane, which had previously been introduced in admixture under the name " Dilan " (U.S.P. 2,516,186). In these compounds the CCl_3 group of DDT is replaced by the nitro alkyl groups $-CH(NO_2)CH_3$ and $-CH(NO_2)C_2H_5$, respectively; the homologous ethane and pentane derivatives have little insecticidal activity. But, contrary to the hypothesis, in the nitro alkanes studied by Skerrett and Woodcock (*19*) in which the $-CCl_3$ group of DDT is replaced by the groups $-CH(NO_2)Cl$ or $-CCl_2(NO_2)$, insecticidal activity is absent. Moreover, these workers (*20*) found that replacement by the group $-CCl(CH_3)_2$ removed insecticidal properties yet trihedralization is present in the resultant compound 2-chloro-1 : 1-di-(*p*-chlorophenyl)-2-methylpropane.

Rotation of the planar phenyl groups of DDT is also a decisive factor in the hypothesis put forward by Riemschneider and Otto (*21*), for they considered that if rotation was impossible as in *o,o′*-DDT, the analogue lacked insecticidal properties. They successfully extended this idea to the tetramethyl analogues examined by Müller (*14*). The 3 : 3′-4 : 4′-isomer was insecticidal whereas the 2 : 2′-4 : 4′- and 2 : 2′-5 : 5′-isomers were non-insecticidal; rotation of the benzene rings is possible in the former but is impossible in the two latter compounds. In the 2 : 3 : 3-trichloro-1 : 1-diphenylpropylenes, however, rotation of the phenyl groups is not hindered by the trichloropropylene group (–CH–CCl=CCl_2) and mild insecticidal properties are shown by these compounds even when the *o*-positions are occupied by methyl groups. The least satisfactory of the correlations derived by Riemschneider and Otto concerned the positional isomers of DDT. Free rotation of the phenyl groups was found impossible by Wild (*22*) in *o,p′*-DDT yet its insecticidal activity is not greatly below that of 1-(*p*-chlorophenyl)-1-phenyl-2 : 2 : 2-trichloroethane in which rotation is not restricted.

When regarded as an application of structural topography these two hypotheses come together, for that of Riemschneider and Otto requires that the molecules must be capable of assuming the trihedral configuration which Brown *et al.* regarded as necessary for insecticidal activity.

The third hypothesis was deduced by Mullins (*23*) on the basis of Ferguson's concept of thermodynamic activity (see p. 97) to explain narcotic action. Narcosis appears to follow when a constant volume of some non-aqueous phase within the organism is occupied by molecules of the narcotic. Mullins pictured this phase as a membrane lattice of cylindrical lipoprotein molecules orientated to allow the passing of ions and small molecules. He assigned dimensions to the interstices of the lattice such that molecules of DDT could penetrate whereas analogues of larger dimensions are excluded. He suggested that if the penetrating molecule has regions of strong attractive forces, e.g., chlorine atoms, their interaction with the membrane molecules might distort the membrane producing regions of instability affecting its permeability.

Finally, Gunther and his colleagues have attempted a quantitative test of structural topography. In a series of compounds obtained by replacing one or more of the five chlorine atoms of DDT by hydrogen or methyl groups, they found a direct relationship between the sum of the logarithms of the van der Waal's attractive forces of the substituent groups and the negative logarithm of the LD50 of the compound.

Action on the Insect

Dresden and Krijgsman (*24*) showed that the toxicities of DDT by intravenous injection to vertebrates and by intra-abdominal injection of insects are of the same order in mg./kg., yet when DDT is applied externally, the dose lethal to vertebrates is far greater. Moreover, Tobias *et al.* (*25*) found that the toxicity of acetone solutions of DDT to the cockroach is similar whether applied externally or intra-abdominally. The specific toxicity of DDT to insects is therefore a result of the high permeativity of the insect cuticle to DDT to which the vertebrate skin is an effective barrier.

The symptons of DDT poisoning are aptly described by the word "jitters"; violent tremors followed by a stage in which the insect is incapable of coordinated movement. Sooner or later, according to the insect species, death follows accompanied, in the case of larvae, by an extreme loss of weight and size (*26*). Neither histological studies nor the examination of metabolic effects, reviewed by Dresden (*27*), by Metcalf (*28*) and by Wigglesworth (*29*) have provided clues of the cause of death. Biochemical studies have given no evidence that DDT interferes with any specific enzyme system. But investigations of the electrical activity of the DDT-treated nerve leads to the general conclusion that death follows a disorganization of the nervous mechanism. According to Roeder and Weiant (*30*), DDT caused intense trains of impulses travelling towards the central nervous system which stimulate the motor nerve cells to general incoordinated activity; according to Dresden (*27*) DDT causes an irreversible block of the synapses. Welsh and Gordon (*31*) suggested that DDT and its insecticidal analogues are absorbed in the surface lipins of the nerve axon producing a distortion of the membrane which results in spontaneous activity. They suggested, for example, that the distortion of the membrane might reduce the affinity of its phosphoryl groups for calcium ions and hence delay the restoration of the chelate linkages broken during the passage of the nerve impulses. Welsh and Gordon thus provided a mechanism of action which is a direct example of that deduced by Mullins (see p. 206) and which, in principle, is derived from the structural topography theories based on physico-chemical evidence. It is strange to think of DDT as a narcotic, but, in its early history, Gavaudan and Poussel (*32*) deduced, from solubility data, that DDT acts in a purely physical manner as an "indifferent narcotic" (*33*).

Reports in 1947 from Italy (*34*) and from Sweden (*35*) that DDT was failing to control houseflies, were followed by frequent demonstrations that, by breeding from the survivors of sublethal doses, it was possible to select flies tolerant to high and impractical amounts of DDT. Since 1947 the problems raised by the appearance of DDT-resistant strains, particularly of the housefly, have become increasingly serious as will be seen from the reviews of insecticide resistance by Metcalf (*36*) and by Hoskins and Gordon (*37*). The process is one of selection for DDT as ordinarily used has no mutagenic activity (*38, 39*), nor did two inbred populations of *Drosophila melanogaster* become more resistant in the experiments described by Merrell and Underhill (*40*). Genetical studies have so far yielded confused results. Resistance in the housefly to knockdown, as distinct from mortality, was shown by Harrison (*41*) to be associated, in a resistant Italian strain with which she worked, with a single Mendelian factor, non-resistance being dominant. Resistance to mortality, however, appears to be polygenic though Crow (*42*) pointed out that the methods used might not have disclosed a monofactorial resistance. It would appear that DDT resistance in houseflies may arise through several mechanisms combined in various ways in the different strains examined.

The various mechanisms invoked to explain the increased tolerance have been classified by Metcalf (*36*), Hoskins and Gordon (*37*), and by Chadwick (*43*). There is first the selection of individuals better able to withstand the insecticide by reason of morphological differences such as Wiesmann (*35*) found in the thick cuticle of the Swedish strain of resistant houseflies, or a more robust constitution enabling the individual to withstand stress. Hoskins and Gordon described this as " vigor tolerance " and pointed out that in highly resistant strains the vigour effect is likely to be relatively small. A second group of hypotheses invoked a change of habit or behaviour resulting in a lessened chance that the insect will receive a toxic dose of the insecticide, e.g., the resting of flies on untreated feed troughs and floors rather than the treated walls, a behaviour not due to the repellent action (*44*). The third group comprise biochemical mechanisms of which that which has been verified by experiment is the detoxication of DDT to the non-insecticidal ethylene (III, p. 203), first demonstrated by Sternburg, Kearns and Bruce (*45*), and by Perry and Hoskins (*46*). In the five fly strains examined by Perry and his colleagues (*47*), there is a striking correlation between the degree of resistance of the strain and its ability to metabolise DDT expressed as the percentage of absorbed DDT converted to the ethylene. The conversion is enzymic and the enzyme, DDT-dehydrochlorinase, has been extracted from all resistant strains examined by Sternburg and his colleagues (*48*) who could not find it in susceptible strains. The ability of the enzyme to dehydrochlorinate analogues of DDT is closely parallel to the insecticidal activity of the analogue ; for example, *o,p'*-DDT is not attacked. The enzyme was extracted from the cuticle and hypoderma of resistant flies and may be localised for, as Winteringham *et al.* (*49*) pointed out, resistant flies have been found to contain unchanged DDT in amount sufficient to kill susceptible flies. It may be that in the resistant fly the DDT has accumulated in tissues where it is unable to exert a toxic effect and to encounter the dehydrochlorinase. Such storage would be in fat and Wiesmann (*50*) found more lipoid in the cuticle and in the epidermal cells at the base of the retae in resistant flies than in susceptible flies. Langenbuch (*51*) also ascribed the greater tolerance to DDT of the last instar of larvae of the Colorado beetle over earlier instars to their greater content of lipoids ; the fact that insects held at high temperatures withstand more DDT than those at lower temperatures has also been explained by the increased lipoid solubility.

If enzymic dehydrochlorination is an important detoxication mechanism in resistant flies, the competitive inhibition of the enzyme should render the flies again susceptible to DDT. Sumerford and his colleagues (*52*) surveyed the effects of compounds structurally related to DDT as synergists for use with DDT on resistant flies. This list was extended by March, Metcalf and Lewallen (*53*) ; the compounds most effective in rendering DDT toxic to the resistant strains examined were those of similar structure to DDT, in particular, the carbinol (1 : 1-di-(*p*-chlorophenyl)ethanol), di-(*p*-chlorophenyl)ethane, di-(*p*-chlorophenyl)chloromethane and di-(*p*-chlorophenyl)ethynyl carbinol. Moorefield and Kearns (*54*) showed that these compounds effectively inhibit the *in vitro* activity of dehydrochlorinase. These results support the suggestion of Speroni (*55*) that the synergist competes with DDT for its site on the dehydrochlorinase and confirm that this detoxication mechanism is a reason for DDT resistance. The ability of the pyrethrum synergist piperonyl cyclonone (*56*) to render DDT toxic to resistant flies is also related to a slower rate of dehydrochlorination, but the mechanism is still obscure. Nor should it be inferred that detoxication is the sole reason for resistance or that it is the effective mechanism in DDT resistant insects other than the housefly. Winteringham (*57*), for instance, could not detect dehydrochlorination in larvae of *Trogoderma* which are naturally resistant to DDT largely because their hairs prevent contact.

(ii) **Lindane**

In 1942, during a search for possible wireworm repellents among the foul-smelling monochloro- and *o*-dichloro-benzene hexachlorides and their less objectionable parent benzene hexachloride by workers of I.C.I. Ltd. (*58*), the latter compound was found to be highly insecticidal. For reasons of wartime secrecy this information was not published until 1945 when it was found that the same discovery had been made in France by Dupire and Raucourt (*59*).

Benzene hexachloride, more specifically 1 : 2 : 3 : 4 : 5 : 6-hexachloro-*cyclo*hexane, is produced by the chlorination of benzene and is a mixture of isomers, of which four were isolated by van der Linden in 1912. The carbon atoms of *cyclo*hexane, unlike those of benzene, do not lie in one plane, but in the more stable form, lie three in one plane and three in a parallel plane, giving a side elevation resembling (a) below and known as the chair form. Each carbon of benzene hexachloride has bonded to it a chlorine and a hydrogen atom but these bonds differ in their direction relative to the trigonal axis of the carbon ring. One bond will lie approximately parallel to the axis and is designated a polar (p) bond ; the other bond will be approximately at right angles to the axis and is designated an equatorial (e) bond. Hassel (see *60*) deduced that the normal oscillation of the carbon atoms may cause the ring conversion of the type represented by (a) to that represented by (b)

(a) ⇌ (b)

by which each p bond becomes an e bond. Of the eight theoretically-possible isomers, five are known and may be represented by the disposition of the chlorine atoms around the ring as follows :

α	p p e e e e	65–70%
β	e e e e e e	5– 6%
γ	p p p e e e	13%
δ	p e e e e e	6%
ε	p e e p e e	—

Of these isomers the α form exists as optical isomers. These isomers are the least strained forms, and are present together with small amounts of hepta- and octa-chlorocyclohexane, in technical benzene hexachloride roughly in the amounts indicated above.

The I.C.I. workers quickly found, because samples of crude BHC differed in insecticidal activity, that the γ isomer is a far more potent insecticide than the other isomers. Metcalf (*61*) showed that the presence of the other isomers did not affect the toxicity of the γ isomer to the thrips *Heliothrips haemorrhoidalis* Bouché. Against roaches, van Asperen (*62*) found an antagonistic effect of the α and δ isomers on the γ isomer, though these effects were weaker than those shown in mice. The content of the γ isomer is an adequate criterion of the insecticidal potency of crude BHC. The musty odour of crude BHC precludes its domestic use and its propensity for imparting off-flavours to potato and fruit limits its agricultural use. Methods have, therefore, been derived for the separation of the γ isomer which, if over 99 per cent pure, is known as lindane.

Lindane has an appreciable vapour pressure, given by Balsom (*3*) as 9.4×10^{-6} mm. Hg at 20° C. It is surprisingly stable to heat and can be volatilized. But in the presence of alkali, even of lime water at 60° C, it is dehydrochlorinated. The β isomer on the other hand is resistant to dehydrochlorination.

Action on Insects

Like DDT, benzene hexachloride is capable of a rapid penetration of the insect cuticle and of the isomers lindane is the most efficient (*63*). The symptoms produced are similar to those of DDT poisoning (*64*), rapid tremors, followed by ataxia, convulsions and prostration, but a few perhaps important differences have been recorded. Whereas with DDT an increase of temperature from 14.5 to 32° C was found by Guthrie (*65*) to increase the LD_{50} to roaches of DDT from 2.1 to 40.8 μg./roach, the LD_{50} for lindane rose from 0.071 to only 0.18 μg./roach. Busvine (*66*) referred to the " fanning " of the wings of lindane-treated houseflies which is a characteristic of the action of the cyclodiene insecticides but not of DDT. He coupled this difference with the finding that, in houseflies, tolerance to lindane is coupled with resistance to the cyclodiene insecticides but is independent of DDT resistance. For these reasons, it would appear that the biochemical reactions by which lindane is toxic differ from those exerted by DDT.

The critical dependence of structural topography which renders isomers of BHC other than the γ of low insecticidal activity coupled with the absence of any evidence that lindane inhibits a specific enzyme, renders it probable that lindane is toxic because, like DDT, it is adsorbed on an interface and there disrupts permeability. Mullins (*23*) employed the same model of a membrane lattice which he used to explain the differing insecticidal properties of the positional isomers of DDT (see p. 206). He concluded that of the isomers of BHC only lindane can orientate within the interstices of the lattice, and suggested that only when orientation brings into play the attractive forces of all the chlorine atoms of BHC can the membrane lattice be distorted enough to lead to a fatal lesion.

(iii) Chlorinated Terpenes

The success of DDT and BHC prompted a search for insecticidal properties among other chlorinated compounds. Desalbres and Labutut (*67*) found the chlorination of terpenes a promising route but, though 2 : 6 : 7-trichlorocamphane was highly insecticidal (*68*), no commercial development seems to have followed their work.

In 1947 Parker and Beacher (*69*) reported successful trials of toxaphene, made by the chlorination of camphene to a chlorine content of 67–69 per cent (U.S.P. 2,565,471). Though the product moved as if homogeneous on the chromatogram described by O'Calla (*70*) the precise nature of the active components is unknown. The empirical formula is $C_{10}H_{10}Cl_8$ and the product loses hydrogen chloride to alkali, on heating or exposure to light. An apparently similar product was introduced in 1951 under the name " Strobane ".

(iv) The Cyclodiene Group

In 1946, Kearns, Ingle and Metcalf (*71*) reported successful tests, as an insecticide, of a new product of empirical formula $C_{10}H_6Cl_8$ and known in its early days as 1068. This product was the first of a remarkable series of insecticides, a common feature of which is the use of hexachloro*cyclo*pentadiene in Diels–Alder reactions.

The condensation of hexachloro*cyclo*pentadiene with the *cyclo*pentadiene itself gives a hexachlorodi*cyclo*pentadiene (XIV), known commonly as chlordene.

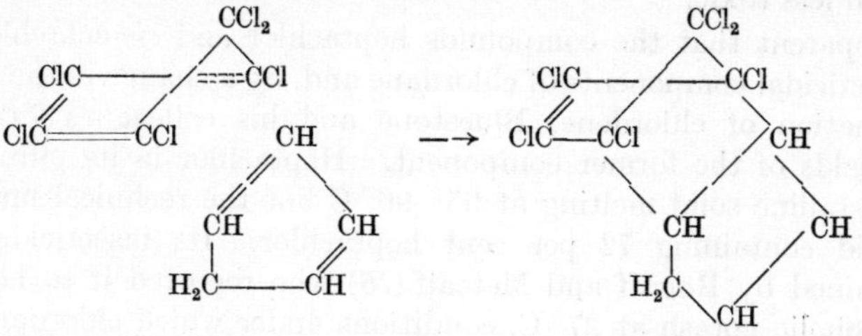

CCl_2 ClC CCl ClC CCl + CH CH CH H_2C CH → CCl_2 ClC CCl ClC CCl CH CH CH H_2C CH

In theory this condensation could give two stereoisomers which, if the methylene bridge be ignored, could be represented in side elevation by (a) or (b) but by analogy (the second Alder " rule ") it is thought that only the (a) or " *endo* " form is produced.

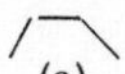

(a) 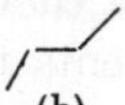 (b)

Chlordene is not highly insecticidal but if chlorinated it gives rise to octachlorodi*cyclo*pentadiene, which is the main component of the insecticide chlordane. This compound is in full 1 : 2 : 4 : 5 : 6 : 7 : 10 : 10-octachloro-4 : 7 : 8 : 9-tetrahydro-4 : 7-*endo*methyleneindane, but may conveniently be called octachlor.

In theory and neglecting optical isomers, four stereoisomers are possible ; the additional chlorine atoms to the *cyclo*pentene ring may be disposed in the four arrangements :

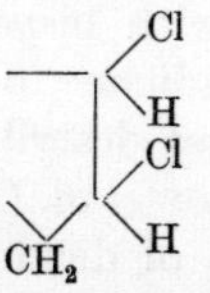

Cl H Cl H CH_2 (c) · H Cl H Cl CH_2 (d) · Cl H H Cl CH_2 (e) · H Cl Cl H CH_2 (f)

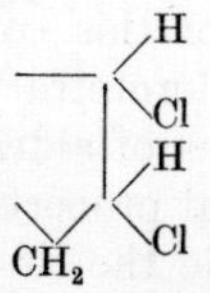

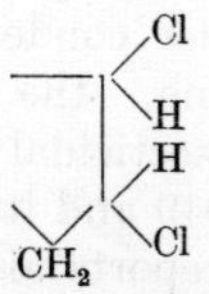

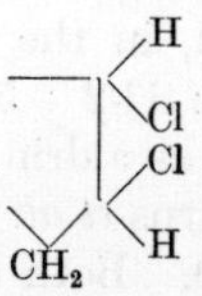

two of *cis* (c and d) and two of *trans* (e and f). But not all four have yet been isolated. Chromatographic analyses of technical chlordane by March (*72*) yielded, in addition to unreacted hexachloro*cyclo*pentadiene and chlordene, two isomers of octachlor of m.p. 106.5°–108° and 104.5°–106°, and a compound of formula $C_{10}H_5Cl_7$ (m.p. 92–93°). Vogelbach (*73*) isolated a third compound of formula $C_{10}H_5Cl_8$ of m.p. 141°–141.5°, three compounds of formula $C_{10}H_5Cl_7$ and one of formula $C_{10}H_5Cl_9$. Bluestone *et al.* (*74*) referred, however, only to the heptachlor ($C_{10}H_5Cl_7$) of m.p. 95°–96°, the two isomers of octachlor and nonachlor ($C_{10}H_5Cl_9$, XVI). Of the isomers of octachlor, that of lower melting point loses hydrogen chloride to alcoholic potash more readily than that of higher melting point and it is therefore thought to be the *cis*-isomer. The insecticidal dosages to houseflies of these components relative to that of technical chlordane (= 1) quoted by Cristol (*75*) are heptachlor (0.43), *cis*-octachlor (0.56), *trans*-octachlor (1.7) and nonachlor (2.0). March (*72*) found that toward the milkweed bug *Oncopeltus fasciatus*, heptachlor and *cis*-octachlor were three to four times more toxic than technical chlordane whereas *trans*-octachlor and nonachlor were much less toxic.

It is apparent that the compounds heptachlor and *cis*-octachlor are the main insecticidal components of chlordane and, by a change in the conditions of chlorination of chlordene, Bluestone and his colleagues (*74*) obtained greater yields of the former component. Heptachlor in its pure form is a white crystalline solid melting at 95°–96° C but the technical material is a waxy solid containing 72 per cent heptachlor. Its insecticidal activity was examined by Rogoff and Metcalf (*76*) who reported it to be stable to dilute alcoholic potash at 27° C, conditions under which chlordane released one half a mole and DDT just over one mole of hydrogen chloride per mole of compound. Bluestone and his colleagues found that silver acetate in boiling acetic acid removed exactly one gram atom of chlorine per mole heptachlor whereas none was removed from octachlor. The constituents of heptachlor are therefore thought to be represented by formula XV and in which case XVI represents nonachlor.

```
      CCl     CH              CCl     CH              CCl     CHCl
ClC      CH       CH     ClC      CH       CH    ClC      CH       CHCl
||   CCl2                ||   CCl2               ||   CCl2
ClC      CH              ClC      CH             ClC      CH
      CCl     CH2             CCl     CHCl            CCl     CHCl
      (XIV)                   (XV)                    (XVI)
```

Pursuing the hexachloro*cyclo*pentadiene route, Lidov (*77*) in 1947 discovered, in the Diels–Alder condensation of this compound with bi*cyclo*-[2 : 2 : 1]-2 : 5-heptadiene, the hexachlorotetra*cyclo*dodecadiene now known as aldrin. The insecticidal properties of aldrin were first described by Kearns *et al.* (*78*) in 1949 and its chemical properties by Lidov *et al.* (*79*) in 1950. Both of these reports also provide the first mention of dieldrin, the epoxide obtained by the oxidation of aldrin. In Great Britain and the

United States the name aldrin may be applied to a product containing at least 95 per cent of 1 : 2 : 3 : 4 : 10 : 10-hexachloro-1 : 4 : 4a : 5 : 8 : 8a-hexahydro-1 : 4-5 : 8-dimethanonaphthalene (HHDN), the planar formula of which is represented in XVII. Similarly the name dieldrin is applicable to products containing at least 85 per cent 1 : 2 : 3 : 4 : 10 : 10-hexachloro-6 : 7-epoxy-1 : 4 : 4a : 5 : 6 : 7 : 8 : 8a-octahydro-1 : 4-5 : 8-dimethanonaphthalene (HEOD), of planar structure XVIII.

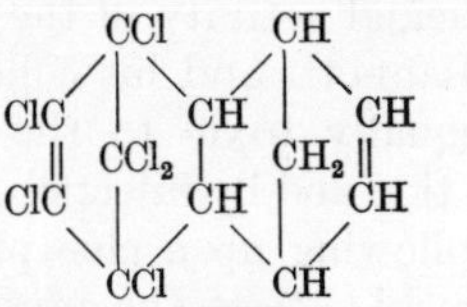

(XVII) 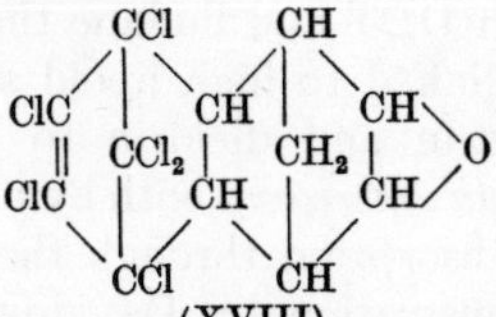 (XVIII)

The bi*cyclo*heptadiene referred to above is itself a Diels–Alder condensation product prepared either by the addition of ethylene to *cyclo*pentadiene or by the dehydrochlorination of the condensation product of vinyl chloride and *cyclo*pentadiene. If now the procedure, which led to the preparation of aldrin, be reversed and *cyclo*pentadiene is condensed with hexachlorobi*cyclo*[2 : 2 : 1]-2 : 5-heptadiene the resultant hexachlorotetra*cyclo*dodecadiene although of the same planar formula as aldrin is a different compound. Both the new compound isodrin and its epoxide endrin were introduced as experimental insecticides in 1951 but only endrin has been introduced commercially.

The spacial configuration of the isomers is not known with certainty but Lidov (*80*) considered that the reaction of hexachloro*cyclo*pentadiene with the bi*cyclo*heptadiene leads to either the *exo-exo* or the *endo-exo* structure whereas that of *cyclo*pentadiene with the hexachlorobi*cyclo*heptadiene leads to either the *endo-endo* or *exo-endo* structure.* The evidence that heptachlor and its relatives are of the *endo* structure leads to the conclusion that aldrin and dieldrin have the *endo-exo* structure depicted in XIX and that isodrin and endrin have the *endo-endo* structure of XX.

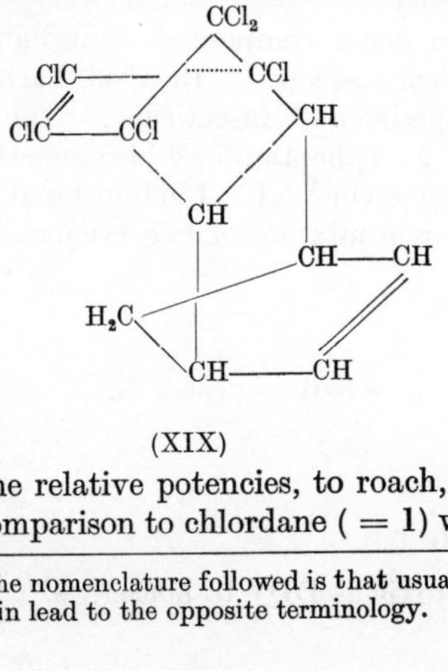

(XIX) 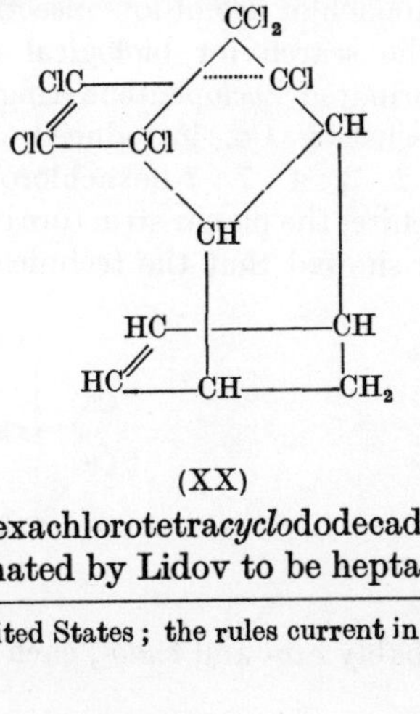 (XX)

The relative potencies, to roach, of the hexachlorotetra*cyclo*dodecadienes in comparison to chlordane (= 1) were estimated by Lidov to be heptachlor

*The nomenclature followed is that usual in the United States ; the rules current in Great Britain lead to the opposite terminology.

3.5, aldrin 3.5 and isodrin 5.45, whereas, to the milkweed bug, aldrin and isodrin were 16 and 26 times as potent as chlordane. The comparative toxicities of the stereoisomers aldrin and isodrin, dieldrin and endrin are of particular interest in their bearing on the mode of action of these compounds, but appear to be dependent on the type of test insect. To aphids isodrin and endrin are much more toxic than are the isomers whereas to houseflies the order is reversed.

As with DDT and lindane the high insecticidal activity of the cyclodiene group is linked to high lipoid solubility. Giannotti and his colleagues (*81*) found aldrin and dieldrin to be almost equally toxic to the cockroach *Periplaneta americana* both by topical application and by injection, indicating a rapid absorption through the cuticle. Following up a clue provided by the demonstration by Davidow and Radomski (*82*) of the presence of an epoxide of heptachlor in the liver of dogs fed with the insecticide, they found evidence that aldrin administered to roaches is converted to its epoxide dieldrin. Busvine (*83*) observed that the relative resistance to cyclodiene insecticides and to lindane of various strains of flies followed the same order which was different to that observed with DDT. He therefore argued that the defence mechanisms of the fly against the former insecticides were similar but different to DDT detoxication. Moreover, he pointed out that the symptoms of poisoning in houseflies differ from those of DDT-poisoning. He therefore concluded that there is a common factor in the mode of action of the cyclodiene group and lindane. The cyclodienes are alike in the chlorinated *cyclo*pentene ring, which in end view of the molecular model appears as a pentagon of chlorine atoms with the sixth protruding from the centre. Of the isomers of benzene hexachloride only lindane has a similar configuration. Busvine suggested that the chlorine pentagon is the toxophore responsible for the insecticidal activity of lindane and the cyclodienes, but it is more probably the area of attachment to the biological interface requisite for insecticidal activity. Clearly, that part of the molecule other than the chlorine pentagon has a critical influence, for chlordene or nonachlor are of low insecticidal activity.

The search for biological activity in other compounds containing the chlorinated *cyclo*pentene ring is a logical sequence. In 1956 Farbwerke Hoechst & Co. introduced, as an experimental insecticide, " Thiodan " (1 : 2 : 3 : 4 : 7 : 7-hexachlorobi*cyclo*[2 : 2 : 1]-heptan-5 : 6-bisoxymethylene sulphite) the planar structure of which is given in XXI. Lindquist and Dahm (*84*) showed that the technical product is a mixture of two isomers,

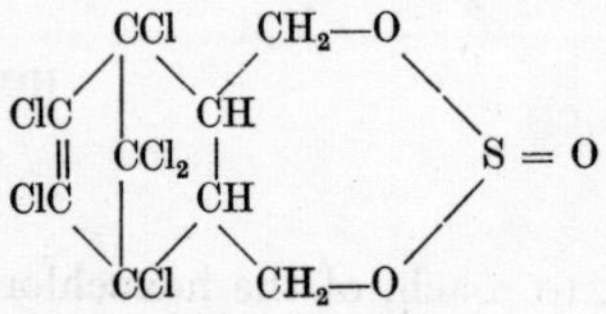

```
         CCl    CH2—O
       /  |  \  /       \
   ClC    |   CH          \
    ||  CCl2 |             S = O
   ClC    |   CH          /
       \  |  /  \       /
         CCl    CH2—O
```

(XXI)

probably *exo*- and *endo*-, each about as toxic as DDT to houseflies.

REFERENCES

(*1*) Haller, H. L., Bartlett, P. D., Drake, N. L., Newman, M. S., Cristol, S. J., Eaker, C. M., Hayes, R. A., Kilmer, G. W., Magerlein, B., Mueller, G. P., Schneider, A. and Wheatley, W., *J. Amer. chem. Soc.*, 1945, **67**, 1591.

(*2*) Gunther, F. A., *J. Amer. chem. Soc.*, 1945, **67**, 189.

(*3*) Balsom, E. W., *Trans. Faraday Soc.*, 1947, **43**, 54.

(*4*) Gunther, F. A., Lindgren, D. L., Elliot, M. I. and LaDue, J. P., *J. econ. Ent.*, 1946, **39**, 624.

(*5*) Gunther, F. A., Carman, G. E. and Elliot, M. I., *J. econ. Ent.*, 1948, **41**, 895.

(*6*) Fleck, E. E., *J. Amer. chem. Soc.*, 1949, **71**, 1034.

(*7*) Läuger, P., Martin, H. and Müller, P., *Helv. chim. Acta*, 1944, **27**, 892.

(*8*) Tanner, C. C., Greaves, W. S., Orell, W. R., Smith, N. K. and Wood, R. E. G., *B.I.O.S. Report* 1480, 1947.

(*9*) Carter, R. H., Wells, R. W., Radeleff, R. D., Smith, C. L., Hubanks, P. E. and Mann, H. D., *J. econ. Ent.*, 1949, **42**, 116.

(*10*) Mylius, A. and Koechlin, H., *Helv. chim. Acta*, 1946, **29**, 405.

(*11*) Martin, H. and Wain, R. L., *Nature, Lond.*, 1944, **154**, 512 ; *Rep. agric. hort. Res. Sta., Long Ashton*, 1944, p. 121.

(*12*) Gatai, K., *Helv. chim. Acta*, 1946, **65**, 356.

(*13*) Busvine, J. R., *J. Soc. chem. Ind.*, 1946, **65**, 356.

(*14*) Müller, P., *Helv. chim. Acta*, 1946, **29**, 1560.

(*15*) Cristol, S. J., *J. Amer. chem. Soc.*, 1945, **67**, 1494.

(*16*) Picard, J. P. and Kearns, C. W., *Canad. J. Res.*, 1949, **27**(D), 59.

(*17*) Gunther, F. A., Blinn, R. C., Carman, G. E. and Metcalf, R. L., *Arch. Biochem.*, 1954, **50**, 504.

(*18*) Brown, H. D. and Rogers, E. F., *J. Amer. chem. Soc.*, 1950, **72**, 1864 ; Rogers, E. F., Brown, H. D., Rasmussen, I. M. and Heal, R. E., *J. Amer. chem. Soc.*, 1953, **75**, 2991.

(*19*) Skerrett, E. J. and Woodcock, D., *J. chem. Soc.*, 1952, **3308**.

(*20*) Stringer, A., Woodcock, D. and Skerrett, E. J., *Ann. appl. Biol.*, 1955, **43**, 366.

(*21*) Riemschneider, R. and Otto, H. D., *Z. Naturf.*, 1954, **9** b, 95.

(*22*) Wild, H., *Helv. chim. Acta*, 1946, **29**, 497.

(*23*) Mullins, L. J., *Chem. Rev.*, 1954, **54**, 289 ; *Science*, 1955, **122**, 118.

(*24*) Dresden, D. and Krijgsman, B. J., *Bull. ent. Res.*, 1948, **38**, 575.

(*25*) Tobias, J. M., Kollros, J. J. and Savit, J., *J. Pharmacol. exp. Therap.*, 1946, **86**, 287.

(*26*) Martin, H., Stringer, A. and Wain, R. L., *Rep. agric. hort. Res. Sta. Bristol*, 1943, p. 62.

(*27*) Dresden, D., *Physiological Investigations into the action of DDT*, Arnheim, 1949.

(*28*) Metcalf, R. L., *Organic Insecticides*, Interscience Publ., New York, 1955, p. 169.

(*29*) Wigglesworth, V. B., In *DDT insektizid dichlordiphenyltrichloräthan und seine bedeutung*, ed. by P. Müller, Birkhäuser Verlag, Basel, 1955, p. 93.

(*30*) Roeder, K. D. and Weiant, E. A., *Science*, 1946, **103**, 304.

(*31*) Welsh, J. H. and Gordon, H. T., *J. cell. comp. Physiol.*, 1947, **30**, 147.

(*32*) Gavaudan, P. and Poussel, H., *C. R. Acad. Sci. Paris*, 1947, **224**, 683.

(*33*) Meyer, K. H. and Hemmi, H., *Biochem. Z.*, 1935, **277**, 39.

(*34*) Missiroli, A., *Riv. Parassit.*, 1947, **8**, 141.

(*35*) Wiesmann, R., *Mitt. schweiz. ent. Ges.*, 1947, **20**, 484.

(*36*) Metcalf, R. L., *Physiol. Rev.*, 1955, **35**, 197.

(*37*) Hoskins, W. M. and Gordon, H. T., *Annu. Rev. Ent.*, 1956, **1**, 89.

(*38*) Luers, H., *Naturwissenschaften*, 1953, **10**, 293.

(*39*) Pielou, D. P., *Canad. J. Zool.*, 1952, **30**, 375.

(*40*) Merrell, D. J. and Underhill, J. C., *J. econ. Ent.*, 1956, **49**, 300.

(*41*) Harrison, C. M., *J. econ. Ent.*, 1953, **46**, 528.

(*42*) Crow, J. F., *Annu. Rep. Ent.*, 1957, **2**, 227.

(*43*) Chadwick, L. E., *In Origins of Resistance to Toxic Agents*, ed. by Sevag, M. G., Reid, R. D. and Reynolds, O. E. Academic Press Inc., New York, 1955, p. 133.

(*44*) King, W. V. and Gahan, J. B., *J. econ. Ent.*, 1949, **42**, 405.

(*45*) Sternburg, J., Kearns, C. W. and Bruce, W. N., *J. econ. Ent.*, 1950, **43**, 214.

(*46*) Perry, A. S. and Hoskins, W. M., *Science*, 1950, **111**, 600.
(*47*) Perry, A. S., Fay, R. W. and Buckner, A. J., *J. econ. Ent.*, 1953, **46**, 972.
(*48*) Sternburg, J., Kearns, C. W. and Moorefield, H., *J. Sci. Food Agr.*, 1954, **2**, 1125.
(*49*) Winteringham, F. P. W., Loveday, P. M. and Harrison, A., *Nature, Lond.*, 1951, **167**, 106.
(*50*) Wiesmann, R., *J. Insect Physiol.*, 1957, **1**, 187.
(*51*) Langenbuch, R., *Naturwissenschaften*, 1954, **41**, 70.
(*52*) Sumerford, W. T., Goette, M. B., Quarterman, K. D. and Schenck, S. L., *Science*, 1951, **114**, 6.
(*53*) March, R. B., Metcalf, R. L. and Lewallen, L. L., *J. econ. Ent.*, 1952, **45**, 851.
(*54*) Moorefield, H. H. and Kearns, C. W., *J. econ. Ent.*, 1955, **48**, 403.
(*55*) Speroni, G., *Chim. e. Industr.*, 1952, **43**, 391.
(*56*) Perry, A. S. and Hoskins, W. M., *J. econ. Ent.*, 1951, **44**, 839.
(*57*) Winteringham, F. P. W., *Proc. nat. Res. Coun., Wash.*, 1952, Pub. 219, p. 61.
(*58*) Slade, R. E., *Chem. & Ind.*, 1945, p. 314.
(*59*) Dupire, A. and Raucourt, M., *C. R. Acad. Agric. Fr.*, 1942, **29**, 470.
(*60*) Hassel, O., *Quart. Rev. chem. Soc. Lond.*, 1953, **7**, 221.
(*61*) Metcalf, R. L., *J. econ. Ent.*, 1947, **40**, 522.
(*62*) van Asperen, K., *Bull. ent. Res.*. 1956, **46**, 837.
(*63*) Armstrong, G., Bradbury, F. R. and Standen, H., *Ann. appl. Biol.*, 1951, **38**, 555.
(*64*) Savit, J., Kollros, J. J. and Tobias, J. M., *Proc. Soc. exp. Biol. N.Y.*, 1946, **62**, 44.
(*65*) Guthrie, F. E., *J. econ. Ent.*, 1950, **43**, 559.
(*66*) Busvine, J. R., *Nature. Lond.*, 1954, **174**, 783.
(*67*) Desalbres, L. and Labatut, R., *Chim. et Industr.*, 1947, **58**, 443.
(*68*) Desalbres, L. and Rache, J., *Chim. et Industr.*, 1948, **59**, 236.
(*69*) Parker, W. L. and Beacher, J. H., *Bull. Del. agric. Exp. Sta.*, 264, 1947.
(*70*) O'Colla, P., *J. Sci. Food Agric.*, 1952, **3**, 130.
(*71*) Kearns, C. W., Ingle, L., and Metcalf, R. L., *J. econ. Ent.*, 1945, **38**, 661.
(*72*) March, R. B., *J. econ. Ent.*, 1952, **45**, 452.
(*73*) Vogelbach, C., *Angew. Chem.*, 1951, **63**, 378.
(*74*) Bluestone, H., Lidov, R. E., Knaus, J. H. and Howerton, P. W., U.S.P. 2,576,666 (Nov. 27, 1951).
(*75*) Cristol, S. J., In *Advanc. Chem. Ser. I*, p. 184, *Amer. Chem. Soc.*, Washington, 1950.
(*76*) Rogoff, W. M. and Metcalf, R. L., *J. econ. Ent.*, 1951, **44**, 910.
(*77*) Lidov, R. E., U.S.P. 2,635,977, Apr. 21, 1953.
(*78*) Kearns, C. W., Weinman, C. J. and Decker, G. C., *J. econ. Ent.*, 1949, **42**, 127.
(*79*) Lidov, R. E., Bluestone, H. and Soloway, S. B., In *Advanc. Chem. Ser. I*, p. 175, *Amer. Chem. Soc.*, Washington, 1950.
(*80*) Lidov, R. E., U.S.P. 2,717,851, Sept. 13, 1955.
(*81*) Giannotti, O., Metcalf, R. L. and March, R. P., *Ann. ent. Soc. Amer.*, 1956, **49**, 588.
(*82*) Davidow, B. and Radomski, J. L., *J. Pharmacol.*, 1953, **107**, 259.
(*83*) Busvine, J. R., *Nature, Lond.*, 1954, **174**, 783.
(*84*) Lindquist, D. A. and Dahm, P. A., *J. econ. Ent.*, 1957, **50**, 483.

ORGANOPHOSPHORUS COMPOUNDS

Since the introduction in Germany, in 1944, of the nicotine substitute "Bladan", the range of organophosphorus compounds found promising for crop protection has extended from aphicides to the full range of insecticides, from fumigants to systemic and long-lasting protective insecticides. To a large extent this development has been achieved through the work of Gerhard Schrader who, at the request of allied administration in West Germany at the close of the 1939–45 war, prepared a report (*1*) on his earlier work, which later formed the basis of his monograph (*2*). Many industrial firms have since become interested but of them only the American Cyanamid Company in the United States and Albright and Wilson in

Great Britain appear to have worked in this field prior to the publication of Schrader's report.

On the basis of his work prior to 1947 Schrader (*3*) concluded that the insecticidal properties of the organophosphates are associated with the structure $(RO)_2PX.Z$ or $(R_2N)_2PX.Z$ where R is alkyl, X is oxygen or sulphur and Z is a grouping such that HZ has acidic properties. Though exceptions are now known to this hypothesis it has experimental support and serves as a pattern by which to classify the many insecticidal organophosphates.

Speaking generally, the alkoxy group (RO) of the formula is limited to methoxy or ethoxy, as in tetraethylpyrophosphate, known as TEPP, $(C_2H_5O)_2PO.O.PO(OC_2H_5)_2$, a colourless hygroscopic liquid miscible with water but rapidly hydrolysed in aqueous solution to a non-insecticidal diethylphosphoric acid $(C_2H_5O)_2PO.OH$. Because of this rapid decomposition, TEPP can function only as a direct contact insecticide and acaricide which, in spite of its high mammalian toxicity, can be applied shortly before harvest without risk of poisonous hazards. " Bladan ", the nicotine substitute mentioned above, was thought by Schrader to contain hexaethyl tetraphosphate but its main component is now known to be TEPP.

Having found that the esters of thiophosphoric acids are less rapidly hydrolysed by water than the corresponding phosphoric acid esters, Schrader prepared the tetraethyl dithionopyrophosphate $(C_2H_5O)_2PS.O.PS(OC_2H_5)_2$. This compound, in the now accepted nomenclature, bis-*OO*-diethylphosphorothionic anhydride but christened sulfotep, is a potent contact insecticide less soluble but more stable in water than TEPP but of low persistence because of its volatility. It survives mainly for use as a " smoke " for greenhouses. The related tetra-*n*-propyl dithionopyrophosphate appeared on the market under the name " NPD " (du Pont).

An amido analogue of TEPP is bis-*NNN'N'*-tetramethylphosphorodiamidic anhydride, also known as octamethylpyrophosphoramide $[(CH_3)_2N]_2PO.O.PO[N(CH_3)_2]_2$, abbreviated to OMPA but since christened schradan. The pure compound is miscible with water to form stable solutions, but, as with TEPP, the technical product contains higher analogues which are active insecticides. Schradan is of historic interest for it was the first organophosphorus compound to be studied as a systemic insecticide. The ability of the plant to translocate this compound throughout its tissues rendering it toxic to sucking insects was first discovered by Kükenthal and Schrader in the early 1940's. The same property was later discovered in other organophosphates, in particular, the esters of diethylphosphoric acid and glycol ethers or thioethers. Demeton, chosen for commercial development around 1950 by Farbenfabriken Bayer A.G., to whom the Schrader patents are assigned, is such an ester of 2-ethylthioethanol. This compound, which chemically is *OO*-diethyl *O*-2-ethylthioethyl phosphorothionate (XXII) spontaneously isomerises to the thiol *OO*-diethyl *S*-ethylthioethyl phosphorothiolate (XXIII).

$$\underset{\text{(XXII)}}{(C_2H_5O)_2PS.OCH_2CH_2.SCH_2CH_3} \rightarrow \underset{\text{(XXIII)}}{(C_2H_5O)_2PO.S.CH_2.CH_2SCH_2CH_3}$$

The commercial product, with the trade name " Systox ", is a mixture of the two compounds named demeton-O and demeton-S, respectively. Its dimethyl homologue, *OO*-dimethyl-*O*-2-ethylthioethyl phosphorothionate (demeton-O-methyl), again in mixture with its thiol isomer (demeton-S-methyl), appeared as " Metasystox ", intended also for systemic use. More recently in 1956 the firm of Bayer has introduced, under the name " Disyston ", the dithioate corresponding to demeton, namely, *OO*-diethyl *S*-2-ethylthioethyl phosphorodithioate (XXIV), in which spontaneous

$$(C_2H_5O)_2PS.SCH_2CH_2.SCH_2CH_3$$
(XXIV)

isomerization has no effect. This compound is suggested for the treatment of seed affording protection of the seedlings from sucking insects and mites. Such a seed treatment had been found successful by the American Cyanamid Company with their insecticide " Thimet ", *OO*-diethyl *S*-ethylthiomethyl phosphorodithioate (XXV), chosen from the homologous series examined by Clark *et al.* (*4*). The replacement of the terminal

$$(C_2H_5O)_2PS.SCH_2.SCH_2CH_3$$
(XXV)

thioethyl group of XXII by the diethylamino group gives *OO*-diethyl *S*-2-diethylaminoethyl phosphorothiolate (XXVI), christened amiton and introduced in 1955 by I.C.I. Ltd. in the form of the oxalate salt. The insecticidal properties are comparatively insignificant but it is a potent systemic acaricide.

$$(C_2H_5O)_2PO.SCH_2CH_2N(C_2H_5)_2$$
(XXVI)

Having noted that the more insecticidal of his earlier organophosphates were acid anhydrides, Schrader explored other substituents strongly acid in character such as the nitro and chlorophenols. Parathion, *OO*-diethyl *O*-*p*-nitrophenyl phosphorothioate (XXVII) first introduced in 1944,

$$(C_2H_5O)_2PS.O-C_6H_4-NO_2$$
(XXVII)

rapidly became a widely-used insecticide, but, largely because of its high toxicity to mammals, it has been superseded by less hazardous phosphates. The dimethyl analogue, methyl parathion, for example, is less toxic than parathion to mice by subcutaneous injection. The addition of a nuclear chlorine markedly reduces mammalian toxicity, and, in 1952, Bayer introduced *OO*-dimethyl *O*-3-chloro-4-nitrophenyl phosphorothioate (XXVIII) often called " Chlorthion ". At about the same time, American Cyanamid

$$(CH_3O)_2PS.O-C_6H_3(Cl)-NO_2 \qquad (CH_3O)_2PS.O-C_6H_3(Cl)-NO_2$$
(XXVIII) (XXIX)

Company introduced for experimental purposes the related *OO*-dimethyl *O*-2-chloro-4-nitrophenyl phosphorothioate (XXIX). Schrader (*5*) placed the toxic oral dose to rats of the 3-chloro isomer at 500 mg./kg. and of the 2-chloro isomer at 200 mg./kg., but DuBois and his colleagues (*6*) estimated the acute oral LD50 to rats of chlorthion at about 1500 mg./kg., a remarkable elimination of the poisonous hazards of methyl parathion by the insertion of a single chlorine atom. Still more surprising is the discovery by the U.S. Department of Agriculture (*7*) during a search for insecticides suitable for internal administration to cattle against parasitic insects, that a compound from the Dow Chemical Company under the code number ET-57 was effective for use as a systemic insecticide against the maggots of the warble flies *Hypoderma lineatum* (De Vill) and *H. bovis* L. This compound, the technical grade of which is known as korlan, is *OO*-dimethyl *O*-2 : 4 : 5-trichlorophenyl phosphorothioate (XXX) and it is of so low a mammalian toxicity that it can be fed to cattle at rates of 100 mg./kg.

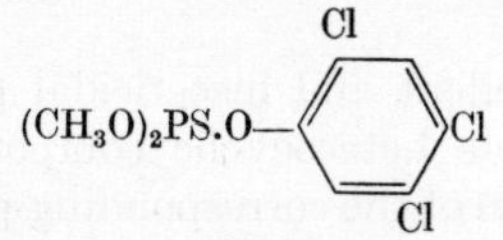

$(CH_3O)_2PS.O$— (2,4,5-trichlorophenyl: Cl, Cl, Cl)

(XXX)

$(C_2H_5O)_2PS.O$ (2,4-dichlorophenyl: Cl, Cl)

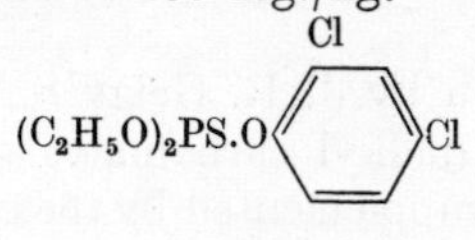

(XXXI)

The diethyl ester of the dichlorophenol, *OO*-diethyl *O*-2 : 4-dichlorophenyl phosphorothioate (XXXI), introduced in 1956 as an experimental nematicide (U.S.P. 2,761,806) is likewise of low mammalian toxicity for the acute oral LD50 to male albino rats is 270 mg./kg. An extension of the series is illustrated by the introduction, in 1955 under the name " Trithion ", of *OO*-diethyl *S*-*p*-chlorophenylthiomethyl phosphorodithioate (XXXII), which being stable to water and of low volatility, has been suggested as a non-systemic insecticide and acaricide with residual activity.

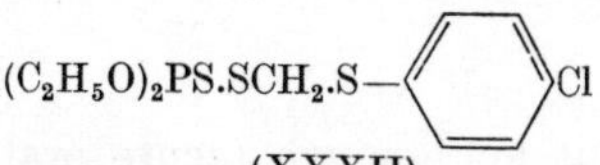

$(C_2H_5O)_2PS.SCH_2.S$— (p-chlorophenyl: Cl)

(XXXII)

Continuing his search for insecticidal organophosphates, Schrader in 1947 replaced the phenols by heterocyclic compounds having enolic hydroxyl groups. Using methyl umbelliferone he prepared *OO*-diethyl *O*-4-methyl-7-coumarinyl phosphorothioate (XXXIII) which was found particularly effective against the Colorado potato beetle, then a plague in Germany.

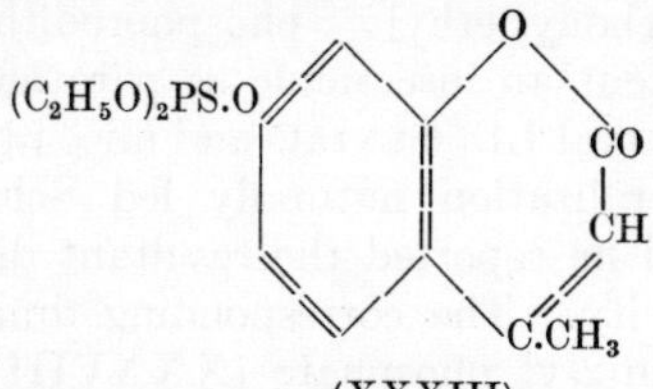

$(C_2H_5O)_2PS.O$ — O, CO, CH, $C.CH_3$

(XXXIII)

$(C_2H_5O)_2PS.O$— O, CO, CCl, $C.CH_3$

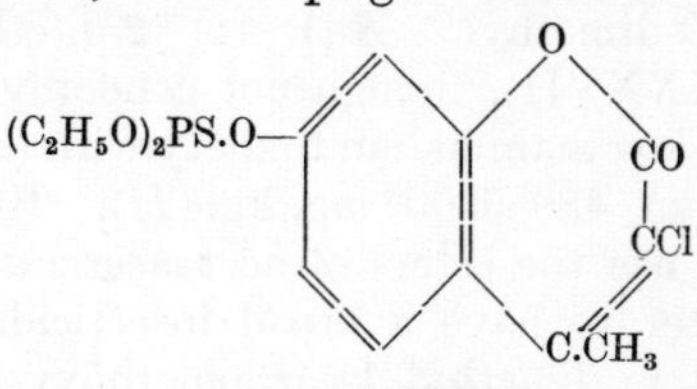

(XXXIV)

Introduced under the trade name " Potosan " it lacks systemic properties and is not a potent aphicide. Frawley *et al.* (*8*) calculated its acute oral LD50 to male and female rats to be 42 and 19 mg./kg. respectively, and because of this high toxicity, Schrader again tried the effect of chlorine substitution, which gives *OO*-diethyl *O*-3-chloro-4-methylcoumarinyl phosphorothioate (XXXIV), a good fly larvicide of low mammalian toxicity.

The wide range of heterocyclic radicals available is illustrated by the compound of code number Bayer 17147 introduced in 1954 and first described by Ivy *et al.* (*9*) who found it particularly effective against cotton pests. It is a benzotriazine derivative, *OO*-dimethyl *S*-4-oxobenzotriazine-3-methyl phosphorodithioate (XXXV), and being of low volatility and solubility, has a long residual activity. In N. America its trade name is " Guthion ".

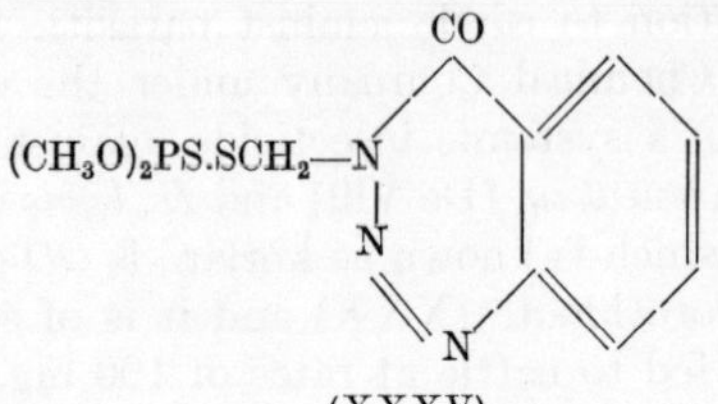

CO

$(CH_3O)_2PS.SCH_2$—N

N

N

(XXXV)

The search by J. R. Geigy S.A. for repellent and insecticidal properties among the dialkyl carbamates of enolisable heterocyclic compounds (see p. 227) was supplemented by the examination of the corresponding phosphate esters and led to the discovery of a number of promising compounds including diazinon. The insecticidal properties of this compound, *OO*-diethyl *O*-(2-*iso*propyl-4-methyl-6-pyrimidyl) phosphorothioate (XXXVI), were first described by Gasser (*10*).

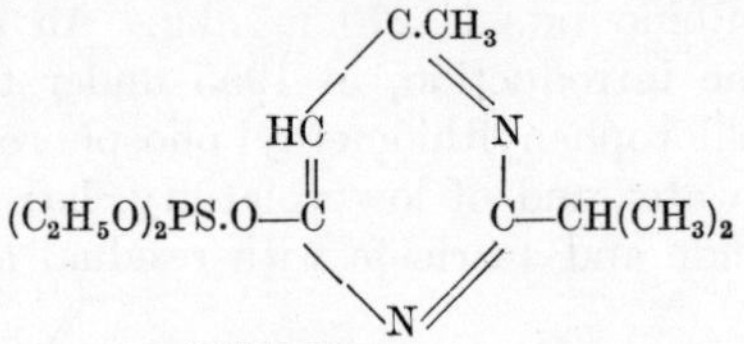

$C.CH_3$

HC N

$(C_2H_5O)_2PS.O$—C C—$CH(CH_3)_2$

N

(XXXVI)

Its mammalian toxicity is low enough (acute oral LD50 to rats 100–150 mg./kg.) to permit its use for fly control, a purpose for which its moderate volatility is an advantage.

Schrader's generalization, that in the insecticidal compound $(RO)_2PX.Z$, the group Z is such that HZ has acidic properties, was violated in 1950, when the American Cyanamid Company introduced malathion. This compound, *OO*-dimethyl *S*-(1 : 2-di(ethoxycarbonyl)ethyl phosphorodithioate (XXXVII), though not generally so potent an insecticide as parathion, is of low mammalian toxicity with an acute oral LD50 to rats and mice ranging from 480–5800 mg./kg. (*11*). His generalization naturally led Schrader to use the esters of acetoacetic acid and he reported the resultant diethyl ester to have a broad insecticidal activity. The corresponding dimethyl ester, dimethyl 1-carbomethoxy-1-propen-2-yl phosphate (XXXVIII) was introduced as a systemic insecticide and acaricide by the Shell Chemical

$(CH_3O)_2PS.SCH.COOC_2H_5$ $(CH_3O)_2PO.OC(CH_3):CH.COOCH_3$

$CH_2.COOC_2H_5$

(XXXVII) (XXXVIII)

Corporation in 1953 (*12*) under the name " Phosdrin ". This compound exists in *cis* and *trans* isomers and Casida *et al.* (*13*) found that one, thought to be the *cis* isomer, is about a hundred times as toxic both to insects and to mammals as the other form. When applied to plants the technical mixture, which contains about twice as much of the supposed *cis* as of the *trans* isomer, is effective for only a few days being lost both by evaporation and decomposition. In the study of the possible decomposition products Casida and his colleagues examined the activity of dimethyl 1-methylvinyl phosphate but found it of low toxicity to houseflies. The related diethyl 2-chlorovinyl phosphate (XXXIX) was reported by Corey *et al.* (*12*) to be of promise as a systemic insecticide and acaricide, and dimethyl 2 : 2-dichlorovinyl phosphate (XL) is powerfully toxic to flies and to mammals. This compound was recognized by Mattson *et al.* (*14*) as a volatile impurity of the

$$(C_2H_5O)_2PO.OCH{:}CHCl \qquad (CH_3O)_2PO.OCH{:}CCl_2$$

(XXXIX) (XL)

technical grade of " Dipterex ", a product introduced in 1952 by Farbenfabriken Bayer for use against flies. The main component, dimethyl 2 : 2 : 2-trichloro-1-hydroxyethyl phosphonate (XLI) was synthesized by Schrader's colleague Lorenz (*15*), and is relatively stable in aqueous solution but is dehydrochlorinated in alkali, the resultant product undergoing rearrangement to the phosphate XL. The phosphonate XLI is highly insecticidal especially to houseflies yet is of low toxicity to warm-blooded animals, the intraperitoneal LD50 to rats being 225 mg./kg. (*16*).

$$(CH_3O)_2PO.CH(OH).CCl_3$$

(XLI)

It will be noted that compound XLI is a phosphonate and therefore contains, unlike all the other organophosphates mentioned, a direct carbon phosphorus linkage. In his early discussions Schrader (*17*) expressed his view that the phosphates were a more promising field for exploration because his experience was that the compounds which contain the C–P linkage are highly poisonous yet not more insecticidal than those containing the C–O–P linkage. Apart from " Dipterex " the only phosphonate to be marketed for insecticidal purposes was *O*-ethyl *O*-*p*-nitrophenyl phenylphosphonothioate (XLII) introduced in 1949 under the name " EPN ".

$$C_6H_5{-}PS(OC_2H_5){-}O{-}C_6H_4{-}NO_2$$

(XLII)

Action on the Insect : The observations of Lange and Krueger (*18*) in 1932 on the physiological action of diethyl phosphorofluoridate $(C_2H_5O)_2POF$, led to an examination of such compounds as possible chemical warfare agents. Early in 1941 at Cambridge, Adrian and his colleagues (*19*) showed that this action was associated with an inhibition of cholinesterase, to which Dixon (see *20*) found the di*iso*propyl homologue, called DFP from the older nomenclature di*iso*propyl fluorophosphonate, of exceptional potency being active at concentrations of the order of 10^{-10} molar, though without effect on the other enzymes he tested.

The physiological function of cholinesterase is to accelerate the hydrolysis of acetylcholine which is produced at the nerve endings and is responsible for the transmission of nerve impulses across the synapses. If the acetylcholine is allowed to remain, derangement of the nervous mechanism results in death. The mammalian toxicity of anticholinesterases such as DFP is therefore thought to be due to the accumulation of acetylcholine. Accordingly the insecticidal activity of the organophosphates is likewise attributed to an inhibition of insect cholinesterase, the presence of which in insect nerve tissue has been repeatedly demonstrated since Means (*21*), in 1942, found it in the grasshopper. Moreover, the presence in insects of relatively large amounts of a substrate with the pharmacological reactions of acetylcholine was shown by Corteggiani and Serfaty (*22*) in 1939 though final biochemical proof of its identity with acetylcholine came later (*23*, *24*, *25*). Finally the presence in insects, again in large amounts in comparison with mammals, of the enzymic machinery for the biosynthesis of acetylcholine was established by Smallman (*26*).

For these reasons, it seemed reasonable to adopt the mammalian hypothesis of cholinesterase inhibition to the insecticidal action of the organophosphates, but there was evidence to indicate that the hypothesis may be inadequate. Lord and Potter (*27*) pointed out that TEPP kills the eggs of the tomato moth and of the Mediterranean flour moth, apparently acting at a stage before the nervous system is fully differentiated. Mehrotra and Smallman (*28*), however, found in fly eggs a coincidence between the appearance and rise in the acetylcholine content of the eggs and the lethal action of parathion.

The hypothesis requires that acetylcholine be insecticidal yet Hopf (*29*) found that locusts survived the injection of acetylcholine and other choline esters. In earlier work Tobias *et al.* (*30*) had found acetylcholine toxic when injected into cockroach but only at the excessive dose of 7–10 g./kg. The explanation of this anomaly is now thought to rest in the differences in structure between the insect and the mammal nerve. The haemolymph of herbivorous insects contains concentrations of potassium salts which would be lethal in the blood of mammals. Indeed, by feeding the blood-sucking bug *Rhodnius prolixus* blood enriched with potassium chloride, Ramsay (*31*) secured a tenfold increase above the normal potassium level which was tolerated for several hours without loss of nervous coordination.

Hoyle (*32*) established that the whole nervous system of the locust is protected by a continuous sheath which is an effective barrier to potassium ions. Twarog and Roeder (*33*) found that when this sheath is removed the cockroach nerve responds to acetylcholine at high concentrations (10^{-2} M) which had no effect on the intact nerve. O'Brien (*34*) showed that when the tissue of a cockroach opened by a dorsal cut in such a way that the membranes surrounding the nerves and other important organs are undamaged, is bathed with acetylcholine, the latter is not hydrolysed indicating that it fails to reach the site of the cholinesterase. The non-toxicity of acetylcholine noted above may therefore be explained by its failure to reach an effective site through the agency of this barrier to ionised compounds. O'Brien and

Fisher (*35*) extended this explanation to account for the non-insecticidal activity of other ionisable compounds which function as anticholinesterases in mammals.

It may therefore be concluded that the main reason for the insecticidal activity of the organophosphates is their capacity for inhibiting cholinesterase. Satisfactory correlations between insecticidal activity and *in vitro* inhibition were found, in limited series, by Chadwick and Hill (*36*) and, in a more extensive series, by Metcalf and March (*37*). A quantitative study of the nature of the reactions between the enzyme and its inhibitor is difficult because cholinesterase has not yet been obtained in a pure condition. But esterases other than cholinesterase are in general inhibited by active organophosphates (*38*) and of these enzymes chymotrypsin, a proteolytic enzyme with some esterase activity, can be crystallized. Jansen *et al.* (*39*) showed that its inhibition of DFP was the result of phosphorylation, hydrogen fluoride being released at a rate of approximately one mole per mole of α-chymotrypsin. Burgen (*40*) had suggested that the inhibition of cholinesterase was likewise the result of phosphorylation, a view supported by Boursnell and Webb's results (*41*) with horse serum cholinesterase and radioactive DFP.

The stoichiometric reaction of enzyme (EnzOH) and phosphate suggests that the reaction may be represented :—

$$\text{EnzOH} + (\text{RO})_2\text{PX.Z} \rightarrow (\text{RO})_2\text{PX.OEnz} + \text{HZ}$$

and the basis for Schrader's generalization on the nature of the insecticidal phosphates becomes apparent.

If enzyme inhibition is due to phosphorylation the anti-enzyme activity of the phosphate should be related to its ability to phosphorylate water, or, in simpler terms, its rate of hydrolysis. Aldridge and Davison (*42*) examined a series of diethyl phenyl phosphates substituted in the aromatic ring and established a linear relationship between the stability of the inhibitor to aqueous hydrolysis at physiological *p*H and temperatures, expressed as the negative logarithm of the hydrolysis constant, and the inhibitory power of the phosphate, expressed as the logarithm of the rate of reaction of the inhibitor with cholinesterase.

It is tempting now to apply the hypothesis to insecticidal activity which would be expected to parallel ease of hydrolysis ; but many factors intervene between the site of application of the phosphate insecticide and its arrival at the enzyme. Clearly the phosphate must not be so readily hydrolysable that it is attacked by water before it can react with the enzyme. Di*iso*propyl phosphorochloridate, as Burgen (*40*) pointed out, is more readily hydrolysed than DFP yet is far less active as an anticholinesterase. Another obvious requirement is that of appropriate lipoid solubility and Ketelaar (*43*) suggested that Schrader's generalization that the thionophosphate is usually more insecticidal yet of lower mammalian toxicity than the corresponding phosphate is due to the high polarity of the P = O bond which induces in the phosphate a greater water solubility and poorer lipoid solubility than in the thionophosphate. Finally, the intervention of metabolic processes may profoundly change the anti-enzyme activity of the compound applied.

Parathion (XXVII), for example, produces no measurable inhibition of cholinesterase in *in vitro* tests, as first shown by Diggle and Gage (*44*) for rat brain cholinesterase. It is now considered that insecticidal thionophosphates in general are subject to an *in vivo* conversion to an active anticholinesterase (*45*). Gage showed that the livers of rats fed with parathion contained an anticholinesterase which he identified as paraoxon, diethyl *p*-nitrophenyl phosphate, in which the sulphur atom of parathion is replaced by oxygen. Evidence that malathion is oxidised to malaoxon in the cockroach and the mouse was obtained by O'Brien (*46*) who also found that both malathion and malaoxon are, in the mouse, rapidly hydrolysed at the carboxylic ester linkages of the side chains to far less potent antiesterases (*47*). He suggested that because in the mouse hydrolysis destroys the malaoxon more rapidly than in the cockroach, the balance of the oxidative (activation) and hydrolytic (detoxication) activities accounts for the much greater toxicity of malathion for the cockroach than for the mouse.

The metabolism of demeton XXII in both plant (*48*) and in mouse and cockroach (*49*), studied by Metcalf and his colleagues, is again primarily an oxidation but the importance of the oxidation of the thionosulphur is overshadowed by the oxidation of the mercapto sulphur to sulphoxide and ultimately to sulphone. The increase, in this series, of anticholinesterase activity with the degree of oxidation is strikingly shown in the results given by Fukuto *et al.* (*50*).

Schradan itself is a poor cholinesterase inhibitor yet is a systemic insecticide effective against sap-feeding but not against leaf-eating insects. Larvae of *Pieris brassicae* for instance are not affected by feeding on treated plants toxic to *Brevicoryne brassicae* (*51*). An active anticholinesterase arises by the oxidation of schradan in the test tube (*52*), the plant (*53*), the mammal (*54*) and the insect (*55*), the oxidation process being similar in each (*56*). The process, as first suggested by Hartley (*57*), is an oxidation of one of the dimethylamido groups of schradan to a phosphoramide *N*-oxide (XLIV) which may rearrange to the methyl ether (XLV) or to the methylol derivative (XLVI) which will release formaldehyde to form heptamethylpyrophosphoramide.

$$—PO\begin{cases}N(CH_3)_2\\N(CH_3)_2\end{cases} \longrightarrow —PO\begin{cases}N(O)(CH_3)_2\\N(CH_3)_2\end{cases}$$

(XLIII) (XLIV)

$$\swarrow \qquad \searrow$$

$$—PO\begin{cases}N(CH_3)(CH_2OH)\\N(CH_3)_2\end{cases} \qquad —PO\begin{cases}N(OCH_3)(CH_3)\\N(CH_3)_2\end{cases}$$

(XLVI) (XLV)

Spencer, O'Brien and White (*58*) synthesized the methyl ether and showed that it was not a metabolic product of schradan. They eliminated the

N-oxide for infrared spectrographic reasons and thus demonstrated that the likely active anticholinesterase is the methylol XLVI. This compound is readily hydrolysed but is non-toxic to the roach because it fails to penetrate to the nerve cord (*59*).

The history of the organophosphates so far has been one of exploration by which the first hazardous members of the series have been replaced by insecticides of lower mammalian toxicity. This development has revealed the importance of a fuller knowledge of the fate of the chemical within the organism whereby advantage may be taken of differences between its metabolism by insect and by mammal.

REFERENCES

(*1*) Schrader, G., *B.I.O.S. Final Report*, 714, 1946.
(*2*) Schrader, G., *Die Entwicklung neuer Insketizide auf Grundlage organischer Fluor- und Phosphor-Verbindungen*, Monogr. *Angew. Chemie* No. 62, Verlag Chemie, GMBH, Weinheim, 2nd ed., 1952.
(*3*) Schrader, G., *B.I.O.S. Final Report*, 1808, 1947.
(*4*) Clark, E. L., Johnson, G. A. and Mattson, E. L., *J. Agric. Food Chem.*, 1955, **3**, 834.
(*5*) Schrader, G., *Angew. Chem.*, 1954, **66**, 265.
(*6*) DuBois, K. P., Doull, J., Deroin, J. and Cummings, O. K., *Arch. industr. Hyg.*, 1953, **8**, 350.
(*7*) McGregor, W. S. and Bushland, R. C., *J. econ. Ent.*, 1957, **50**, 246.
(*8*) Frawley, J. P., Hagan, E. C. and Fitzhugh, O. G., *J. Pharmacol.*, 1952, **105**, 156.
(*9*) Ivy, E. E., Brazzel, J. H., Scales, A. L. and Martin, D. F., *J. econ. Ent.*, 1955, **48**, 293.
(*10*) Gasser, R., *Z. Naturf.*, 1953, **8**b, 225.
(*11*) Hazelton, L. W. and Holland, E. G., *Arch. industr. Hyg.*, 1953, **8**, 399.
(*12*) Corey, R. A., Dorman, S. C., Hall, W. E., Glover, L. C. and Whetstone, R. R., *Science*, 1953, **118**, 28.
(*13*) Casida, J. E., Gatterdam, P. E., Getzin, L. W. and Chapman, R. K., *J. agric. Food Chem.*, 1956, **4**, 236.
(*14*) Mattson, A. M., Spillane, J. T. and Pearce, G. W., *J. agric. Food Chem.*, 1955, **3**, 319.
(*15*) Lorenz, W., Henglein, A. and Schrader, G., *J. Amer. chem. Soc.*, 1955, **77**, 2554.
(*16*) DuBois, K. P. and Cotter, G. J., *Arch. industr. Hyg.*, 1955, **11**, 53.
(*17*) Martin, H. and Shaw, H., *B.I.O.S. Final Report*, 1095, 1945.
(*18*) Lange, W. and Krueger, G. von, *Ber. dtsch. chem. Ges.*, 1932, **65**, 1598.
(*19*) Adrian, E. D., Feldberg, W. and Kilby, B. A., *Brit. J. Pharmacol.*, 1947, **2**, 56.
(*20*) McCombie, H. and Saunders, B. C., *Nature, Lond.*, 1946, **157**, 287, 776.
(*21*) Means, O. W., *J. cell. comp. Physiol.*, 1942, **20**, 319.
(*22*) Corteggiani, E. and Serfaty, A., *C. R. Soc. Biol. Paris*, 1939, **131**, 1124.
(*23*) Lewis, S. E., *Nature, Lond.*, 1953, **172**, 1004.
(*24*) Augustinsson, K. B., *Acta physiol. scand.*, 1954, **32**, 174.
(*25*) Chefurka, W. and Smallman, B. N., *Nature, Lond.*, 1955, **175**, 946.
(*26*) Smallman, B. N., *J. Physiol.*, 1956, **132**, 343.
(*27*) Lord, K. A. and Potter, C., *Ann. appl. Biol.*, 1951, **38**, 495.
(*28*) Mehrotra, K. N. and Smallman, B. N., *Nature, Lond.*, 1957, **180**, 97.
(*29*) Hopf, H. S., *Ann. appl. Biol.*, 1952, **39**, 193.
(*30*) Tobias, J. M., Kollross, J. J. and Savit, J., *J. cell. comp. Physiol.*, 1946, **28**, 159.
(*31*) Ramsay, J. A., *J. exp. Biol.*, 1952, **29**, 110.
(*32*) Hoyle, G., *J. exp. Biol.*, 1953, **30**, 121.
(*33*) Twarog, B. M. and Roeder, K. D., *Biol. Bull. Wood's Hole*, 1956, **111**, 278.
(*34*) O'Brien, R. D., *Ann. ent. Soc. Amer.*, 1957, **50**, 223.
(*35*) O'Brien, R. D. and Fisher, R. W., *J. econ. Ent.*, 1958, **51**, 169.
(*36*) Chadwick, L. E. and Hill, D. L., *J. Neurophysiol.*, 1947, **10**, 235.
(*37*) Metcalf, R. L. and March, R. B., *J. econ. Ent.*, 1949, **42**, 721.
(*38*) Webb, E. C., *Biochem. J.*, 1948, **42**, 96.

(*39*) Jansen, E. F., Nutting, M. D. F.,Jang, R. and Balls, A. K., *J. biol. Chem.*, 1950, **185**, 209.
(*40*) Burgen, A. S. V., *Brit. J. Pharmacol.*, 1949, **4**, 219.
(*41*) Boursnell, J. C. and Webb, E. C., *Nature, Lond.*, 1949, **164**, 875.
(*42*) Aldridge, W. N. and Davison, A. N., *Biochem. J.*, 1952, **51**, 62.
(*43*) Ketelaar, J. A. A., *Trans. IX Int. Congr. Ent.*, 1953, **2**, 318.
(*44*) Diggle, W. M. and Gage, J. C., *Biochem. J.*, 1951, **49**, 491.
(*45*) Metcalf, R. L. and March, R. B., *Ann. ent. Soc. Amer.*, 1953, **46**, 63.
(*46*) O'Brien, R. D., *J. econ. Ent.*, 1957, **50**, 159.
(*47*) March, R. B., Fukuto, T. R., Metcalf, R. L. and Maxon, M. G., *J. econ. Ent.*, 1956, **49**, 185.
(*48*) Metcalf, R. L., March, R. B., Fukuto, T. R. and Maxon, M. G., *J. econ. Ent.*, 1954, **47**, 1045.
(*49*) March, R. B., Metcalf, R. L., Fukuto, T. R. and Maxon, M. G., *J. econ. Ent.*, 1955, **48**, 355.
(*50*) Fukuto, T. R., Metcalf, R. L., March, R. B. and Maxon, M. G., *J. econ. Ent.*, 1955, **48**, 347.
(*51*) Ripper, W. E., Greenslade, R. M. and Hartley, G. S., *Bull. ent. Res.*, 1950, **40**, 481.
(*52*) Spencer, E. Y. and O'Brien, R. D., *J. agric. Food Chem.*, 1953, **1**, 716.
(*53*) Heath, D. F., Lane, W. J. and Llewellyn, M., *J. Sci. Food agric.*, 1952, **3**, 60.
(*54*) Casida, J. E., Allen, T. C. and Stahmann, M. A., *J. biol. Chem.*, 1954, **210**, 607.
(*55*) O'Brien, R. D. and Spencer, E. Y., *J. agric. Food Chem.*, 1953, **1**, 946; 1955, **3**, 56.
(*56*) Casida, J. E., Chapman, R. K., Stahmann, M. A. and Allen, T. C., *J. econ. Ent.*, 1954, **47**, 64.
(*57*) Hartley, G. S. at *XV Int. Chem. Congr.*, 11 Sept., 1951.
(*58*) Spencer, E. Y., O'Brien, R. D. and White, R. W., *J. agric. Food Chem.*, 1957, **5**, 123.
(*59*) O'Brien, R. D. and Spencer, E. Y., *Nature, Lond.*, 1957, **179**, 52.

THE CARBAMATE GROUP

The successful development of the organophosphates as insecticides directed attention to other compounds which act as anticholinesterases, among which physostigmine or eserine is prominent. The suggestion that the physiological properties of this alkaloid arise because it is a phenyl ester of methylcarbamic acid led Stedman (*1*) to examine the aminophenyl esters which he found shared cholinergic activity. From this work arose a wide range of parasympathomimetic drugs such as neostigmine (*m*-hydroxyphenyl trimethylammonium methylsulphate dimethylcarbamate, I).

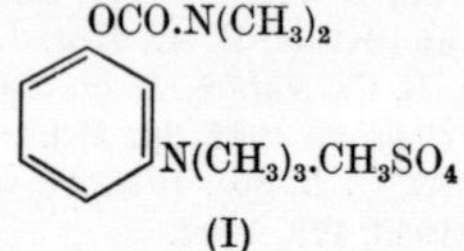

(I)

These drugs, purposely made quaternary salts to obtain water solubility, are not effective against insects, a result which Kolbezen, Metcalf and Fukuto (*2*) attributed to their low lipoid solubility. O'Brien and Fisher (*3*), however, also considered that the ion-impermeable sheath of the insect nerve (see p. 222) would also contribute, for these drugs, being largely ionized at physiological *p*H, would be excluded from their site of action. Of lipoid-soluble derivatives, Kolbezen and his colleagues (*2*) found that the *N*-methylcarbamates of certain phenols were highly toxic to houseflies and greenhouse thrips.

Another line of development was pursued by Gysin (*4*) whose colleagues with J. R. Geigy S.A. had found that certain diethylcarbamates, including 5 : 5-dimethyldihydroresorcinol diethylcarbamate, were insect repellents.

Gysin found that the corresponding dimethylcarbamate was not repellent but had promising insecticidal properties, first described by Wiesmann *et al.* (*5*). This compound (II) introduced, under the name " Dimetan ", is moderately water-soluble.

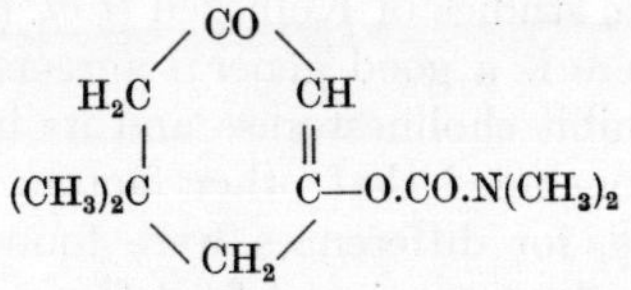

CO
H_2C CH
$(CH_3)_2C$ C—O.CO.N$(CH_3)_2$
CH_2
(II)

Following his discovery through other carbamates of the enolizable heterocyclic compounds, Gysin found the dimethylcarbamates of 1-phenyl-3-methyl-5-pyrazole and of 1-*iso*propyl-3-methyl-5-pyrazole of potential insecticidal value and these compounds were introduced for experimental use under the name " Pyrolan " (III) and " Isolan " (IV).

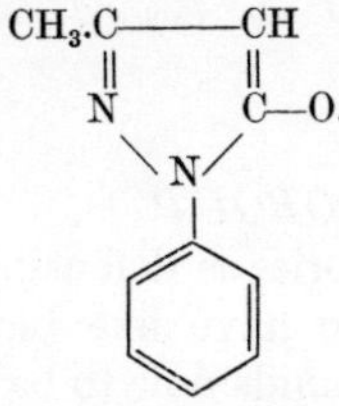

CH_3.C——CH
N C—O.CO.N$(CH_3)_2$
N
(III)

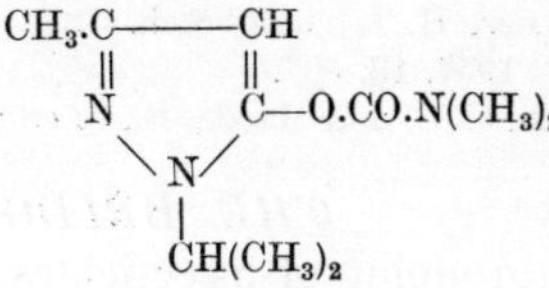

CH_3.C——CH
N C—O.CO.N$(CH_3)_2$
N
CH$(CH_3)_2$
(IV)

The replacement of the phenyl group of III by the *iso*propyl group to form IV increases mammalian toxicity but Gysin reported an improvement in systemic activity. In the pyrimidine series the most promising compound found by Gysin and his colleagues was the 2-*n*-propyl-4-methylpyrimidyl-(6)-dimethylcarbamate (V). This compound named " Pyramat " is of mammalian toxicity low enough to be considered for household use against flies to which it is highly toxic. Gysin also examined the effect of the substitution of the dimethylcarbamate group by the diethyl phosphate group and was so led, as mentioned on p. 220, to diazinon, the analogue of V below but with the *iso*propyl group replacing the *n*-propyl group.

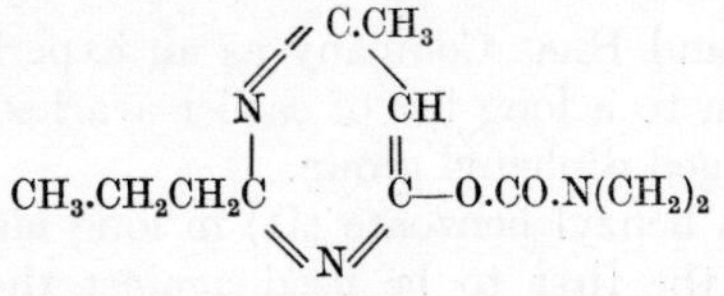

C.CH_3
N CH
$CH_3.CH_2CH_2C$ C—O.CO.N$(CH_2)_2$
N
(V)

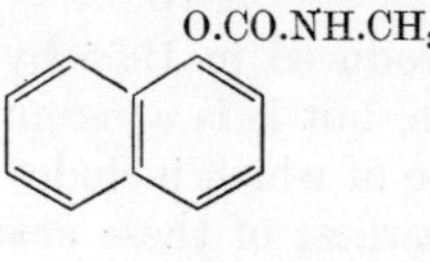

O.CO.NH.CH_3
(VI)

The long series synthesized by Gysin and examined by his colleagues were mainly dimethylcarbamates for the substitution of the dimethylamine in the carbamate radical by other amines led to a decrease in insecticidal activity. The methylcarbamates have, however, yielded a promising insecticide of broad activity in 1-naphthyl *N*-methylcarbamate (VI) introduced in 1956 by Union Carbide Chemicals Co. under the name " Sevin ". Its insecticidal properties were first described by Haynes *et al.* (*6*). Data on the mammalian toxicity of this compound indicate that it is without undue hazard.

The hypothesis that the action of the carbamates on mammal and on insect is an inhibition of the cholinesterase appears to answer all requirements. The enzymic studies of Kolbezen *et al.* (*2*) and of Roan and Maeda (*7*) revealed that there is a good general agreement between the ability of the compound to inhibit cholinesterase and its insecticidal activity, though the latter authors concluded that other factors might be of importance in the *in vivo* reactions, for differences were found in the relative toxicities of the compounds to three species of fruit fly.

REFERENCES

(*1*) Stedman, E., *Biochem. J.*, 1926, **20**, 719.
(*2*) Kolbezen, M. J., Metcalf, R. L., Fukuto, T. R., *J. agric. Food Chem.*, 1954, **2**, 864.
(*3*) O'Brien, R. D. and Fisher, R. W., *J. econ. Ent.*, 1958, **51**, 169.
(*4*) Gysin, H., *Chimia*, 1954, **8**, 205, 221.
(*5*) Wiesmann, R., Gasser, R. and Grob, H., *Experientia*, 1951, **7**, 117.
(*6*) Haynes, H. L., Lambrech, J. A. and Moorefield, H. H., *Contr. Boyce Thompson Inst.*, 1957, **18**, 507.
(*7*) Roan, C. C. and Maeda, S., *J. econ. Ent.*, 1954, **47**, 507.

THE BRIDGED DIPHENYL GROUP

The grouping of insecticides into chemical categories is defensible on the grounds that like chemicals might be expected to have like biochemical reactions. But the rule breaks down, for the compounds now to be reviewed differ widely in activity from DDT and its analogues which logically would be the first members of the group. Indeed, the insecticidal activity of DDT has been related to the trihedral configuration of its two phenyl groups (see p. 205). Yet the substitution of the hydroxyl group for the hydrogen of the carbon atoms between these groups giving 1 : 1-dichlorophenyl-2 : 2 : 2-trichloroethanol (I) virtually eliminates toxicity to true insects but introduces toxicity to phytophagous mites. This compound

OH, Cl, C, Cl, CCl_3
(I)

$CH_2.O.CO$
(II)

was introduced in 1955 by Rohm and Haas Company as an experimental acaricide, but it is a recent addition to a long list of earlier acaricides, the structure of which includes the bridged diphenyl group.

The earliest of these acaricides is benzyl benzoate (II) in long medicinal use for the treatment of scabies; the first to be used against the mites infesting crop plants was azobenzene (III) developed by Blauvelt (*1*) mainly for greenhouse use.

N:N
(III)

In the early days of DDT, its use on fruit was often followed by serious attacks of phytophagous mites (see *2*) usually attributed to destruction of

DDT-sensitive predators of the mites. Not only is DDT non-toxic to the mites but there is evidence (*3*) that egg production is stimulated. The resultant search for effective acaricides for fruit tree use was helped by the discovery that two DDT-like compounds, bis-(*p*-chlorophenoxy)methane (IV) and 1 : 1-di-(*p*-chlorophenyl)ethanol (V), are powerful acaricides. The former compound " Neotran ", developed by the Dow Chemical Company, was described by Jeppson (*4*) ; the latter, " DMC " (dichlorophenyl)methyl carbinol by Grummitt (*5*). DMC is again so similar in struc-

$Cl\text{-}C_6H_4\text{-}OCH_2.O\text{-}C_6H_4\text{-}Cl$ (IV)

$Cl\text{-}C_6H_4\text{-}C(OH)(CH_3)\text{-}C_6H_4\text{-}Cl$ (V)

ture to DDT that it is able to render DDT toxic to resistant flies, presumably by competing for the DDT-dehydrochlorinase (see p. 208). Metcalf (*6*) found these two compounds IV and V the most acaricidal of the series of DDT analogues he tested and observed that the acaricidal properties of V are lost by the removal of the methyl group. Its replacement by the carbethoxy or the ethoxymethyl groups to give compounds VI and VII restores acaricidal properties as first described by Gasser (*7*). The former

$Cl\text{-}C_6H_4\text{-}C(OH)(CO.OC_2H_5)\text{-}C_6H_4\text{-}Cl$ (VI)

$Cl\text{-}C_6H_4\text{-}C(OH)(CH_2.OC_2H_5)\text{-}C_6H_4\text{-}Cl$ (VII)

of these compounds (VI), ethyl 4 : 4′-dichlorobenzilate, was introduced by J. R. Geigy A.G. under the name " Chlorobenzilate ".

Among the developments directly or indirectly inspired by the acaricidal action of azobenzene is that of diphenyl sulphone (VIII), which Eaton and Davies (*8*) found particularly effective against summer eggs of the fruit tree spider mite *Metatetranychus ulmi*. They found that, as with azobenzene, the introduction of one *p*-chlorine did not reduce ovicidal activity but that both 4 : 4′-dichloroazobenzene and 4 : 4′-dichlorophenyl sulphone were inferior ovicides ; *p*-chlorophenyl phenyl sulphone (IX) is an active

$C_6H_5\text{-}SO_2\text{-}C_6H_5$ (VIII)

$Cl\text{-}C_6H_4\text{-}SO_2\text{-}C_6H_5$ (IX)

component of the product " Sulphenone " reported on by Barnes (*9*). Unexpectedly a further chlorination restored acaricidal properties for Huisman, van der Veen and Meltzer (*10*) reported that both 2 : 4 : 5-trichlorodiphenyl sulphone and 2 : 4 : 5 : 4′-tetrachlorodiphenyl sulphone (X) are effective against all stages, except adults, of *Tetranychus urticea* Koch (= *T. telarius* L.) and lack the phytotoxic peculiarities of diphenylsulphone. The tetrachloro derivative is marketed under the name " Tedion " and Meltzer and Dietvorst (*11*) attributed its long residual effect to stability

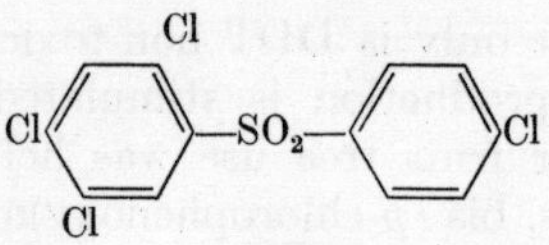

$$\text{Cl}-\text{C}_6\text{H}_2\text{Cl}_2-\text{SO}_2-\text{C}_6\text{H}_4-\text{Cl}$$

(X)

$$\text{Cl}-\text{C}_6\text{H}_4-\text{CH}_2.\text{S}-\text{C}_6\text{H}_4-\text{Cl}$$

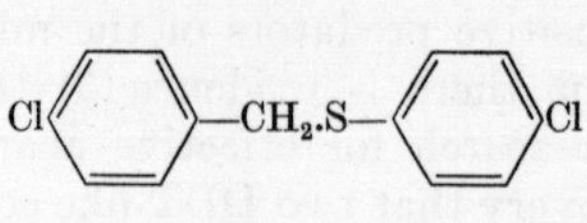

(XI)

and a slower penetration into the sprayed foliage than occurs with the corresponding sulphide or sulphoxide. Of the sulphides, *p*-chlorobenzyl *p*-chlorophenyl sulphide (XI) was introduced in 1953 by Boots Pure Drug Co. (*12*) under the common name chlorbenside. This compound is effective against all stages except adults of the Tetranycid mites, is somewhat volatile and lipoid-soluble enabling it to penetrate into leaves without, however, imparting true systemic acaricidal properties to the treated plant. After application oxidation proceeds, certainly to the sulphoxide and possibly to the sulphone, and it is to this compound that the high persistence of chlorbenside is attributed. The corresponding fluorine compound, fluorbenside, *p*-chlorobenzyl *p*-fluorophenyl sulphide, was introduced in 1955 for, being more volatile than chlorbenside, it is more suitable for greenhouse use.

The acaricidal and insecticidal properties of the substituted phenyl benzenesulphonates were explored by Kenaga and Hummer (*13*) and, of the series tested, they found *p*-chlorophenyl *p*-chlorobenzenesulphonate (XII)

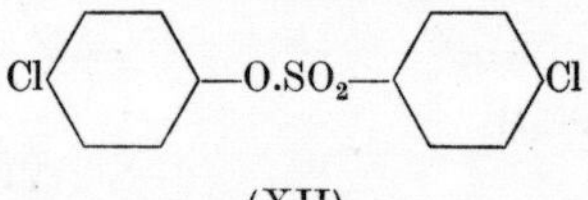

$$\text{Cl}-\text{C}_6\text{H}_4-\text{O.SO}_2-\text{C}_6\text{H}_4-\text{Cl}$$

(XII)

the most effective against eggs of *Tetranychus bimaculatus*. This compound has been given the common names chlorfenson in Great Britain and ovex in the United States. The corresponding *p*-chlorophenyl benzenesulphonate, under the name fenson, has been widely used as an acaricide in Great Britain for it is not inferior to chlorfenson as an acaricide towards *M. ulmi* and *T. telarius* (*14*).

It is permissible to include in this category of acaricidal compounds " Aramite ", though this compound is not of bridged diphenyl structure. Nevertheless, its discovery arose from the observation that 2-chloroethyl *p*-chlorobenzenesulphonate is toxic to citrus mites. This exploratory work was described by Harris and Zukel (*15*) and led to the introduction of 2(*p*-*tert*.-butylphenoxy)*iso*propyl 2′-chloroethyl sulphite (XIII) under the name " Aramite " by the U.S. Rubber Company.

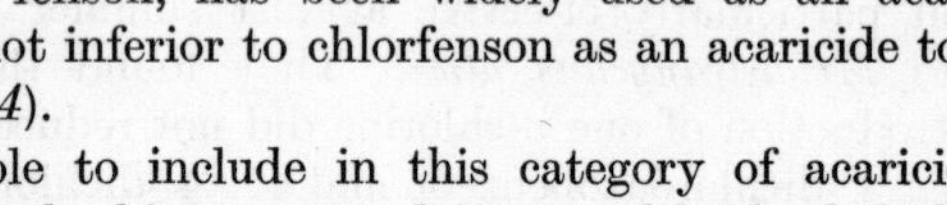

$$(\text{CH}_3)_3\text{C}-\text{C}_6\text{H}_4-\text{OCH}_2.\underset{\displaystyle\text{CH}_3}{\underset{|}{\text{CH}}}-\text{O.SO.O.CH}_2\text{CH}_2\text{Cl}$$

(XIII)

Little can yet be said of the reasons why this wide group of compounds should be acaricidal. Physiological study of the mites is still in its infancy, an exception being the morphological examination by Beament (*16*) of the summer and winter eggs of *M. ulmi*. The chemist is embarrassed by the variety of bivalent bridging groups which render the diphenyl structure acaricidal, though it is evident that this structure is of high biochemical significance in acaricidal as well as insecticidal activity. The selective toxicity of these compounds, not only between mites and insects, but among the mites themselves, adds interest to the study of the reasons for their biological activity.

REFERENCES

(*1*) Blauvelt, W. E., *N.Y. St. Flower Growers Inc. Bull.*, **2**, 1945.

(*2*) Ripper, W. E., *Annu. Rev. Ent.*, 1956, **1**, 403.

(*3*) Hueck, H. J., Kuenen, D. J., den Boer, P. J. and Jaeger-Draafsel, E., *Physiol. comp.*, 1952, **2**, 371.

(*4*) Jeppson, L. R., *J. econ. Ent.*, 1946, **39**, 813.

(*5*) Grummitt, O., *Science*, 1950, **111**, 361.

(*6*) Metcalf, R. L., *J. econ. Ent.*, 1948, **41**, 875.

(*7*) Gasser, R., *Experientia*, 1952, **8**, 65.

(*8*) Eaton, J. K. and Davies, R. G., *Nature, Lond.*, 1948, **161**, 644 ; *Ann. appl. Biol.*, 1950, **37**, 471.

(*9*) Barnes, M. M., *J. econ. Ent.*, 1951, **44**, 672.

(*10*) Huisman, H. O., van der Veen, R. and Meltzer, J., *Nature, Lond.*, 1955, **176**, 515.

(*11*) Meltzer, J. and Dietvorst, F. C., *IV Intern. Cong. Crop Prot.*, 1957.

(*12*) Cranham, J. E., Higgons, D. J. and Stevenson, H. A., *Chem. & Ind.*, 1953, p. 1206.

(*13*) Kenaga, E. E. and Hummer, R. W., *J. econ. Ent.*, 1949, **42**, 996.

(*14*) Kirby, A. H. M. and McKinlay, K. S., *Rep. E. Malling Res. Sta.*, 1950, p. 164.

(*15*) Harris, W. D. and Zukel, J. W., *J. agric. Food Chem.*, 1954, **2**, 140.

(*16*) Beament, J. W. L., *Ann. appl. Biol.*, 1951, **38**, 1.

CHAPTER XI

WEEDKILLERS

INTRODUCTION

If dirt may be defined as " matter in the wrong place ", weeds become " plants in the wrong place ". Attention will be given not so much to the control of weeds growing as a nuisance, as upon paths, or as a hazard, as below hydroelectric power lines, but to that of weeds growing as pests among crops. In the former cases it is possible to apply herbicides toxic to all vegetation ; in the latter case the control measure must cause no undue injury to the crop. There are, however, occasions when the defoliation of the crop plant is sought. The destruction of potato haulm infected with blight *Phytophthora infestans* prior to harvesting will not only facilitate lifting the tubers but will reduce their chance infestation by spores from the blighted haulm (*1*). The defoliation of cotton is widely practiced to permit mechanical picking and the practice reduces boll worm infection (*2*).

Korsmo (*3*) demonstrated that weed eradication gave an increase of 20–25 per cent in the cereal yield, a figure which substantiates the seemingly large estimates of crop loss, for example, an annual loss of $3,000,000,000 in the United States alone (*4*). Such figures usually include loss through interference with cultivation and harvesting, and the reduction of market quality. Less tangible are the losses due to the poisonous or noxious weeds such as corn cockle (*Agrostemma githago* L.) with its poisonous glucosides or wild onion (*Allium* spp.) which imparts a disagreeable flavour to the milk of cows feeding in an infested pasture. Other weeds are host plants for the pests and pathogens of crop plants ; others may produce root diffusates directly or indirectly inimical to other plants. Pickering (*5*), for instance, was of the opinion that the dire effects of grass growing around young fruit trees were explainable only because of such toxic diffusates, but his views were not at that time generally accepted. However, more recent evidence reviewed by Martin (*6*) indicates that many species of higher plants are capable of producing root diffusates of intense biological activity.

Suppression of weed growth would clearly by a factor operating in the evolution of crop husbandry and, particularly, of crop rotation. With root crops, hoeing is a commonly-used method and the good effects of inter-row cultivation on water retention are due primarily to weed control (*7*). The destruction of wild seeds is often the chief object of partial sterilization of the soil and, in other cases of seed-propagated weeds, the harvesting of the crop prior to the ripening of the weed provides control. But too often cultural methods are inadequate and use is made of the selective herbicidal action of certain chemicals.

This selective action was first put to practical use around 1895 when Bonnet in France, Bolley in America and Schultz in Germany used solutions of copper sulphate for the destruction of charlock, *Sinapsis arvensis* L., in cereal crops. As an alternative to copper sulphate, ferrous sulphate was recommended by Bolley (*8*) and is still popular as a component of lawn sand. Following the developmental work of Rabaté (*9*), sulphuric acid became widely used in France after 1911 and, despite its corrosive action on clothing and machinery, is still used for the destruction of potato haulm. Rabaté considered that the acid would also destroy certain fungal pathogens of cereal straw, notably *Leptosphaeria herpotrichoides* de Not and *Ophiobolus graminis*. The effect of the acid on the soil is not unduly serious for, as Rabaté pointed out, ammonium sulphate in equivalent amount would decalcify the soil to the same degree. Ammonium sulphate has long been used for the treatment of turf and Blackman (*10*) showed that the previously-held view that its success in eradicating plant species other than grasses is due to an increase in soil acidity is untenable. He suggested that the ammonium ions are selectively toxic and that plants with a high carbohydrate and organic acid content withstand the treatment better than those of low content. Sulphamic acid, H_2NSO_2OH, a crystalline solid, stable when dry but hydrolysing slowly when in solution to form ammonium hydrogen sulphate, was introduced in 1942 (U.S.P. 2,277,744) for herbicidal use and like ammonium sulphamate, provides an alternative method of using ammonium sulphate.

In 1932 Truffaut and Pastac (*11*) patented the use of nitrophenols as selective herbicides and their product " Sinox ", containing the sodium salt of 3 : 5-dinitro-*o*-cresol (I), became widely used in Europe and in the

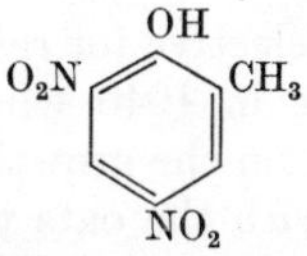

OH
O_2N CH_3
NO_2

(I)

United States. The abbreviation DNC was adopted as its common name in Great Britain and the United States, displacing the older DNOC frequently used in the earlier literature. The effect of substitution of the methyl group by higher alkyl groups was studied by Crafts (*12*) who found that the 2 : 4-dinitro-*o*-*sec.*-butylphenol (II) (2-(1-methyl-*n*-propyl)-4 : 6-dinitrophenol) is more effective than DNC on certain weeds and its greater oil

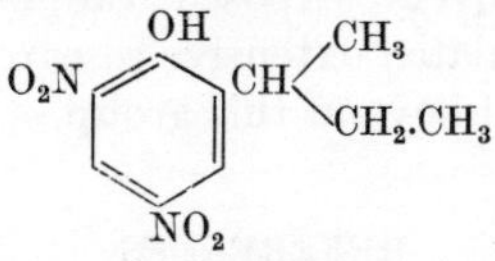

OH
O_2N $CH\langle^{CH_3}_{CH_2.CH_3}$
NO_2

(II)

solubility is often advantageous. Blackman and his colleagues (*13*) showed that generalizations on relative toxicity in this series are subject to modification according to test plant and its stage of growth. The butyl phenol has proved less injurious than DNC to certain plants such as peas and is widely used for the chemical weeding of pulse crops. The common name dinoseb is used in Great Britain but the abbreviation DNBP is approved by the Weed Society of America.

Harris and Hyslop (*14*), examining the effect of adding nitrogenous fertilizers to DNC weedkillers, observed an increase in herbicidal activity with ammonium sulphate in amounts of negligible nutrient value. The "activating" effect of acid salts was confirmed and led to the use of the ammonium dinitrocresylate (*15*). "Activation" is usually regarded (*16*) as a consequence of the effect of acidity on the rate of penetration. The reduction of the extent of dissociation and, as in ovicidal action (see p. 199), increase of the proportion of undissociated cresol which, being more lipoid soluble than its ions, facilitates penetration of the cuticle of the plant. Blackman and his colleagues (*13*), however, considered that the effect cannot be explained solely on the basis of *p*H.

As in the case of their insecticidal action, the dinitrophenols prevent the formation of the phosphate bonds by which the plant conserves its metabolic energy, a process which Bonner and Millerd (*17*) showed proceeds in the mitochondria.

The selective action of the herbicides so far mentioned is mainly the result of differential wetting. The herbicidal spray readily runs off the long narrow and erect cereal leaves which are difficult to wet. The leaves of charlock and most other dicotyledonous weeds are more easily wetted and because of their more horizontal spread, retain more of the spray. Moreover, in cereals and plants such as onion, the growing point is protected by leaf sheaths whereas in most annual weeds it is exposed at the apex of the shoot.

The discovery of herbicides selective for reasons not wholly based on such morphological differences came in 1940 when Templeman examining the effects of α-naphthylacetic acid on the growth of oats, observed that chance seedlings of charlock growing with the oats were alone killed. His interest in this chemical arose because of its auxin-like action on plant growth and his observation led to the testing of a number of such growth regulators (*18*). Among the most active compounds were 2-methyl-4-chlorophenoxyacetic and 2 : 4-dichlorophenoxyacetic acids. The herbicidal activity of the latter compound, now known as 2,4-D, was independently discovered by Zimmerman and Hitchcock (*19*) and by Quastel and his colleagues (*20*) but their work was not known publicly until the end of the 1939–45 war. The success of their discovery exposed the commercial possibilities of chemical weeding and stimulated extensive research work which has brought in a range of herbicides well beyond this group.

REFERENCES

(*1*) Wilson, A. R., Boyd, A. E. W., Mitchell, J. G. and Greaves, W. S., *Ann. appl. Biol.*, 1947, **34**, 1.

(*2*) Dunnam, E. W. and Calhoun, S. L., *J. econ. Ent.*, 1950, **43**, 488.

(*3*) Korsmo, E., *Meld. Norg. LandbrHøisk.*, 1932, **12**, 305.

(*4*) Robbins, W. W., Crafts, A. S. and Raynor, R. N., *Weed Control*, McGraw-Hill Book Co. Inc., New York, 2nd ed., 1952.
(*5*) Pickering, S. U. and Duke of Bedford, *Science and Fruit Growing*, MacMillan & Co., London, 1919.
(*6*) Martin, H., *Chemical Aspects of Ecology in Relation to Agriculture*, Canad. Dep. Agric., 1958.
(*7*) Russell, E. W. and Keen, B. A., *J. agric. Sci.*, 1941, **31**, 326.
(*8*) Bolley, H. L., *Rep. N. Dak. agric. Exp. Sta.*, 1901, **11**, 48.
(*9*) Rabaté, E., *La destruction des mauvaises herbes*, Nat. Acad. Agric., Paris, 1927.
(*10*) Blackman, G. E., *Ann. appl. Biol.*, 1932, **19**, 204, 443.
(*11*) Truffaut, G. and Pastac, I., B.P. 425,295, May 29, 1933.
(*12*) Crafts, A. S., *Science*, 1945, **101**, 417 ; *Plant Physiol.*, 1946, **21**, 345.
(*13*) Blackman, G. E., Holly, K. and Roberts, H. A., *Symp. Soc. exp. Biol.*, 1949, **3**, 283.
(*14*) Harris, L. E. and Hyslop, G. R., *Bull. Ore. agric. Exp. Sta.*, 403, 1942.
(*15*) Crafts, A. S. and Reiber, H. G., *Hilgardia*, 1945, **16**, 487.
(*16*) Fogg, G. E., *Ann. appl. Biol.*, 1948, **35**, 315.
(*17*) Bonner, J. and Millerd, A., *Arch. Biochem.*, 1953, **42**, 135.
(*18*) Slade, R. E., Templeman, W. G. and Sexton, W. A., *Nature, Lond.*, 1945, **155**, 497.
(*19*) Zimmerman, P. W. and Hitchcock, A. E., *Contr. Boyce Thompson Inst.*, 1942, **12**, 321.
(*20*) Nutman, P. S., Thornton, H. G. and Quastel, J. H., *Nature, Lond.*, 1945, **155**, 498.

THE 2,4-D GROUP

The extensive examination of this group which, less conveniently but more explicitly, has been called the aryloxyalkylcarboxylate group, has not yielded compounds of greater general herbicidal action than 2,4-D. But others of the group have properties which permit their use for specific purposes and they must accordingly be reviewed.

2 : 4-Dichlorophenoxyacetic acid (III), being practically insoluble in water, was formerly used either as its sodium salt or in admixture with sodium bicarbonate, precautions being taken to prevent precipitation of the calcium salt. Alternatively the salts of organic amines such as triethanolamine were used when high water solubility was required, or the esters if oil solubility was needed. Of the wide variety of such salts and esters which have been tested (*1*), the *iso*propyl ester is the most popular for emulsion formulation. Unfortunately, the indiscriminate use of 2,4-D led to the damage of crops neighbouring those sprayed and because the *iso*propyl ester is somewhat volatile, the less volatile polypropylene glycol esters have been used (*2*) to lessen this drift hazard. Spray drift, however, is responsible more often than is vapour drift.

4-Chloro-2-methylphenoxyacetic acid (MCPA, IV) was more widely used than 2,4-D in Great Britain, largely because of the availability of the cresol required as intermediate in its manufacture. Because the chlorination of *o*-cresol gives two isomers, crude MCPA may contain up to 35 per cent of the isomeric 6-chloro-2-methylphenoxyacetic acid (*3*). Although Hansen found the isomer less potent in its action on plant growth, he considered that its presence somewhat enhances the herbicidal action of the 4-chloro-2-methyl isomer. It will be noted that 2,4-D and MCPA are isosters and the replacement of the chlorine of MCPA by a second methyl group yields 2 : 4-dimethylphenoxyacetic acid, a compound likewise of high herbicidal activity (*4*). Further chlorination of 2,4-D yields 2 : 4 : 5-trichlorophenoxyacetic acid (2, 4, 5-T) (V) which was found by Hamner and Tukey

(*5*) more effective in field use against woody plants such as mesquite and brushwood, whereas the corresponding propionic acid (VI), for which the awkward abbreviation 2-(2, 4, 5-TP) was adopted by the Weed Society of America but now replaced by the common name silvex, is peculiarly effective against oak species.

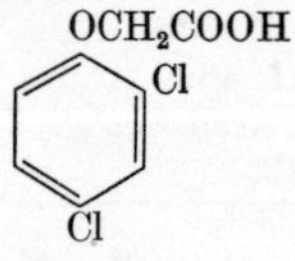

OCH_2COOH Cl Cl (III)

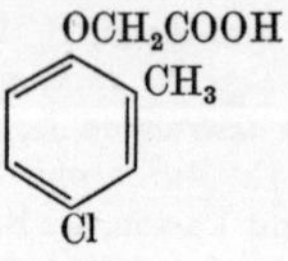

OCH_2COOH CH_3 Cl (IV)

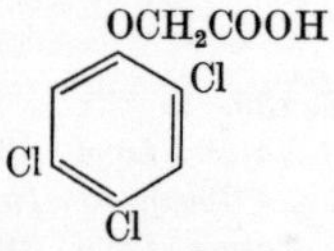

OCH_2COOH Cl Cl Cl (V)

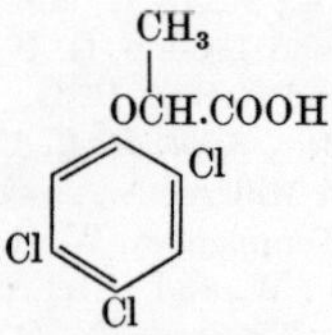

CH_3 $OCH.COOH$ Cl Cl Cl (VI)

The compounds so far described in this section are most effective when applied direct to the foliage for they are absorbed and translocated being systemic herbicides. They are also effective, when applied to soil, in preventing the growth of germinating weed seeds. In order to utilize this latter action without damaging established plants, King and his colleagues (*6*) at the Boyce Thompson Institute employed certain esters of the chlorophenoxy alcohols such as the sodium salt of 2 : 4-dichlorophenoxyethyl sulphate (VII). These esters have little effect when applied to foliage but in the moist unsterilized soil are converted to active compounds.

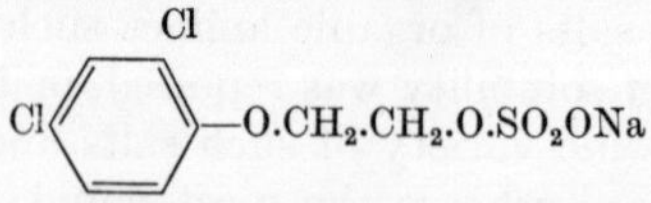

Cl Cl $—O.CH_2.CH_2.O.SO_2ONa$ (VII)

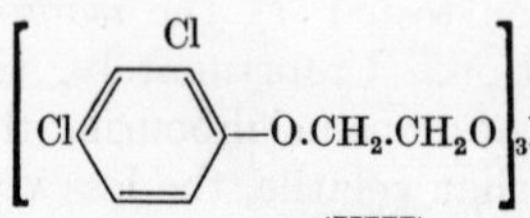

[Cl Cl $—O.CH_2.CH_2O$]$_3$P (VIII)

Sesone which has been proposed as the common name for this sulphate and the corresponding benzoate (" Sesin ") which has a longer residual action, are marketed by the Union Carbide Chemicals Co. The active breakdown product of sesone is almost certainly 2,4-D but, as Vlitos (*7*) pointed out, the process is not fully understood. The common soil organism *Bacillus cereus* var. *mycoides* converts sesone to the ethanol which is oxidised to 2,4-D but this organism had no effect on the benzoate nor could arylsulphatase activity be detected. Audus (*8*), being able to detect only 2,4-D as the active herbicidal product of the ester, concluded that the oxidation of the ethanol is too rapid to permit its accumulation in detectable amounts. To illustrate the range of esters of promise, mention may be made of tris-(2 : 4-dichlorophenoxyethyl)phosphite (VIII) recently introduced by the U.S. Rubber Co. as an experimental herbicide.

An ingenious selective action, attained by another biological agency, was put to use by Wain and his colleagues. Grace (*9*) in examining the root-promoting activity of an homologous series from 1-naphthylacetic to ϵ-(1-naphthyl)-hexoic acids, observed that only those acids with an odd number of methylene groups in the side chain were active. Synerholm and Zimmerman (*10*) found a similar alternation in the homologous series of 2 : 4-dichlorophenoxyalkylcarboxylic acids ; those compounds with an odd number of methylene groups in the side chain produced epinasty, those with an even number did not. They suggested that, by β-oxidation the side chain lost two carbon atoms at each stage, whereby the final product was 2,4-D in the odd-numbered and the inactive 2 : 4-dichlorophenol in the even-numbered members. Wain *et al.* (*11*) established the existence of the expected breakdown products in the plant and found that the β-oxidation systems of plants differ in their capacity to degrade the side chain. For example, wheat, pea and tomato plants can degrade the alternate homologues of 2,4-D but, in the ω-(2 : 4 : 5-trichlorophenoxy) series, only wheat gave growth response in alternate members. But if these members were pretreated with wheat coleoptile tissue, they provoked growth response in the pea. Similarly, γ-(2-methyl-4-chlorophenoxy)butyric acid (IX) applied to celery growing with nettles killed the nettles only whereas the corresponding acetic acid killed both plants (*12*). Herbicidal selectivity is thus obtained through the specific action of the plants on the compound applied. A commercial outcome of this work is " Tropotox " of which the active component compound is IX which in sensitive plants is oxidised to MCPA.

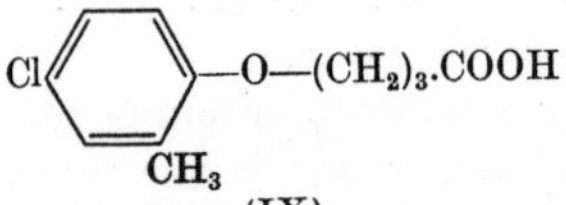

Cl—⟨ring⟩—O—$(CH_2)_3$.COOH
CH_3

(IX)

Action on Plant

By the use of radio-2,4-D it has been shown, particularly by Crafts (*13*), that the herbicide passes through the phloem from regions of food synthesis to those of food use, moving only when photosynthesis or carbohydrate movement is proceeding. If applied to roots it moves in the transpiration stream to the leaves where it is transferred to the phloem.

When in responsive plant tissue, 2,4-D exerts an action apparently similar to that of the natural growth hormone, 3-indoleacetic acid, which it is generally thought, so affects the cell wall that water is taken up with the resultant elongation of the cell. In the normal plant such processes are controlled by various mechanisms, including the enzymic oxidation of indoleacetic acid to inactive products. 2,4-D and its relatives induce the growth response of indoleacetic acid but the plant is apparently unable to oxidise or otherwise inactivate the herbicide.

To the biochemist few subjects are as attractive as the correlation of auxin-like activity and chemical structure, but this topic belongs to a broader field of plant biochemistry than herbicidal action. Moreover the subject has been frequently reviewed, that of Muir and Hansch (*14*) dealing specifically with structural relationships.

REFERENCES

(*1*) Thompson, H. E., Swanson, C. P. and Norman, A. G., *Bot. Gaz.*, 1946, **107**, 476.
(*2*) King, L. J. and Kramer, J. A., *Contr. Boyce Thompson Inst.*, 1951, **16**, 267.
(*3*) Hansen, B., *Physiol. plant.*, 1951, **4**, 667.
(*4*) Slade, R. E., Templeman, W. G. and Sexton, W. A., *Nature, Lond.*, 1945, **155**, 497.
(*5*) Hamner, C. L. and Tukey, H. B., *Science*, 1944, **100**, 154.
(*6*) King, L. J., Lambrech, J. A. and Finn, T. P., *Contr. Boyce Thompson Inst.*, 1951, **16**, 191.
(*7*) Vlitos, A. J., *Contr. Boyce Thompson Inst.*, 1953, **17**, 127.
(*8*) Audus, L. J., *Nature, Lond.*, 1953, **171**, 523.
(*9*) Grace, N. H., *Canad. J. Res.*, 1939, **17**, C, 247.
(*10*) Synerholm, M. E. and Zimmerman, P. W., *Contr. Boyce Thompson Inst.*, 1947, **14**, 369.
(*11*) Fawcett, C. H., Ingram, J. M. A. and Wain, R. L., *Proc. roy. soc.*, 1954, **142**, B, 60; Wain, R. L. and Wightman, F., *Proc. roy. Soc.*, 1954, **142**, B, 525.
(*12*) Wain, R. L., *J. agric. Food Chem.*, 1955, **3**, 128.
(*13*) Crafts, A. S., *Hilgardia*, 1956, **26**, 287.
(*14*) Muir, R. M. and Hansch, C., *Annu. Rev. Plant Physiol.*, 1955, **6**, 157.

THE CARBAMATE GROUP

In following up their work on plant growth substances as herbicides, Templeman and Sexton (*1*) examined chemicals known to affect cell division as distinct from cell extension. One such chemical is ethyl phenylcarbamate which they found to inhibit the growth of oats without serious effect on interplanted charlock. Of related compounds they found *iso*propyl *N*-phenylcarbamate (X) about thrice as effective. This compound has been named propham in Great Britain, though the abbreviation IPC is recommended for use by the Weed Society of America. The older chemical name of *iso*propyl carbanilate is rarely used.

The sensitivity of many monocotyledonous plants and the resistance of most dicotyledons to propham, which is the reverse of the order shown to 2,4-D, attracted wide attention but, in practice, the results of its use are disappointing. It was quickly found that soil applications gave better results than foliage application (*2*) and to lessen the danger of a rapid decomposition by soil organisms, the chloro-derivatives, which are generally more resistant to microbial attack, were tried. *Iso*propyl 3-chlorophenylcarbamate (XI), known as CIPC, was found to be effective against grasses

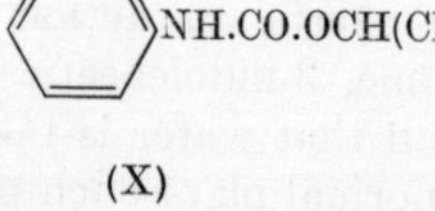

$NH.CO.OCH(CH_3)_2$ (X)

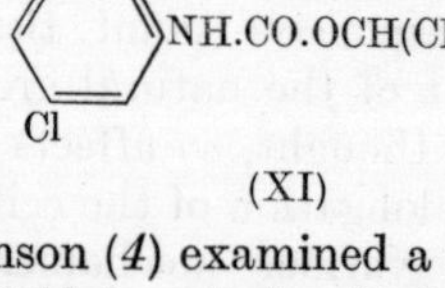

$NH.CO.OCH(CH_3)_2$, Cl (XI)

resistant to propham (*3*). Shaw and Swanson (*4*) examined a wide range of substituted carbamates and found herbicidal activity highly correlated with substitution by chlorine, methyl and methoxy groups, particularly in the 3 and 6 positions of the benzene ring.

Although the discovery of their herbicidal action does not seem to have been inspired by the earlier work on phenylcarbamates, the phthalamates fall into this group. This discovery was due to Hoffmann and Smith (*5*) and *N*-1-naphthylphthalamic acid (XII) has been marketed under the name " Alanap-1 " (U.S. Rubber Co.), its sodium salt as " Alanap-3 " and the phthalimide as " Alanap-2 ".

COOH

—NH.CO—

(XII)

The phthalamates are of more general phytotoxicity than the carbanilates but are used at pre-emergent stages for the control of grasses and broad-leaved weeds, being less effective when applied to the foliage. Little is yet known of the reactions by which these compounds are herbicidal ; nor indeed may they be common to all members of the group. The carbanilates interfere with cell division producing nuclear abnormalities (*6*) ; the phthalamates produce epinastic, parthenocarpic and geotropic effects reminiscent of an interference in the auxin mechanisms.

The failure of the roots of seedlings treated with 1-naphthyl phthalamic acid to grow downwards was first observed by Mentzer *et al.* (*7*). In the comprehensive tests of Jones, Metcalfe and Sexton (*8*) this property of abolishing the geotropic and phototropic responses of seedlings was found, not only in this compound, but in a wide range of derivatives of 2-benzoylbenzoic acids and in certain nitramines, of which 2-bromo-4 : 6-dimethylphenylnitramine was the most potent. The use of the latter compounds as selective herbicides was discussed by Templeman (*9*).

REFERENCES

(*1*) Templeman, W. C. and Sexton, W. A., *Nature, Lond.*, 1945, **156**, 630 ; *Proc. roy. Soc.*, 1946, **133**, B, 480.

(*2*) Ennis, W. B., *Science*, 1947, **105**, 95.

(*3*) DeRose, H. R., *Agron. J.*, 1951, **43**, 139.

(*4*) Shaw, W. S. and Swanson, C. R., *Weeds*, 1953, **2**, 43.

(*5*) Hoffmann, O. L. and Smith, A. E., *Science*, 1949, **109**, 588.

(*6*) Ivens, G. W. and Blackman, G. E., *Symp. Soc. exp. Biol.*, 1949, **3**, 266.

(*7*) Mentzer, C., Molho, D. and Pachéco, H., *Bull. Soc. Chim. biol., Paris*, 1950, **32**, 572.

(*8*) Jones, R. L., Metcalfe, T. P. and Sexton, W. A., *J. Sci. Food agric.*, 1954, **5**, 32, 38, 44.

(*9*) Templeman, W. C., *Proc. Brit. Weed Control. Conf.*, 1954, **1**, 3.

THE HETEROCYCLIC NITROGEN GROUP

The first compound of this group to find practical use as a herbicide was maleic hydrazide, the potent phytotoxicity of which was described by Schoene and Hoffmann (*1*). Maleic hydrazide is somewhat soluble in water behaving as a weak monobasic acid and is generally applied to the plant as the monoethanolamine salt. Although chemically it is usually defined as 1 : 2-dihydroxypyridazine-3 : 6-dione, its properties are better represented by structure XIII which is 6-hydroxy-3(2H)pyridazinone (*2*).

Its effect on plant growth is unique for it inhibits growth, particularly of the internodes, in striking contrast to the effect of the gibberellins which promote internodal growth. Its action in preventing the sprouting of potato tubers and the suckering of tobacco may render its practical use that of inhibitor rather than herbicide, though its promise as a selective herbicide against certain grasses was reported (*3*) early in its history. Leopold and

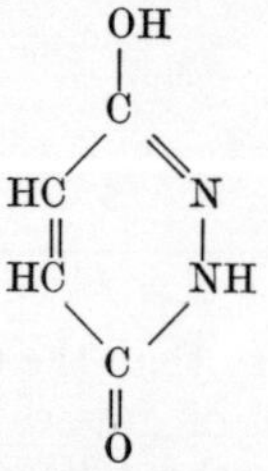

OH
C
HC N
HC NH
C
O
(XIII)

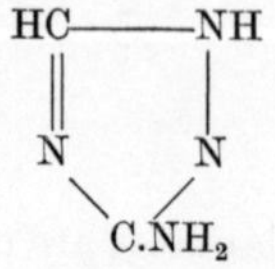

HC—NH
N N
C.NH$_2$
(XIV)

Klein (*4*) showed that maleic hydrazide will counter the action of 3-indoleacetic acid in the standard split pea curvature test for auxin activity and hence describe it as an " anti-auxin ". Andreae and Andreae (*5*) finding that maleic hydrazide accelerates the oxidation of indoleacetic acid suggested that it may inhibit growth by causing an excessive destruction of auxin, a suggestion supported by the results of Waygood *et al.* (*6*).

The possible herbicidal use of 3-amino-1 : 2 : 4-triazole (XIV) was first described in 1954 by Allen (*7*) when it was introduced by the American Chemical Paint Company under the names of " Weedazol " and " Amizol ". This compound is fairly soluble in water and is freely translocated by the plant (*8*) both from root and from leaves. Apart from its trial as a selective herbicide it has proved excellent for the defoliation of cotton for which both the base and its salts have been tried (*9*). A first symptom of its phytotoxic action is a chlorosis though whether due to an interruption of chlorophyll synthesis or a destruction of chlorophyll is unknown. This property is shared by a curious assemblage of other compounds; Hamner and his colleagues (*10*) found that certain tetronic acid derivatives could cause treated plants to become chlorotic. Cucumber seed treated with 3-(α-iminoethyl)-5-methyltetronic acid (XV) produced yellow cotyledons and white seedlings which died in five or six days. Ready *et al.* (*11*) recorded that, among the many compounds tested for phytotoxicity at Camp Detrick, a similar " albinism " was produced by certain 3-nitro substituted benzoic acids such as 3-nitro-4-hydroxybenzoic acid (XVI).

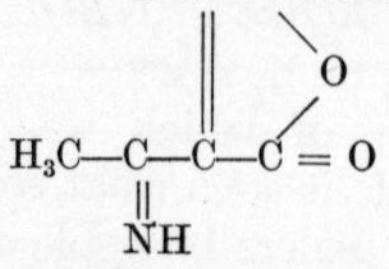

HO.C—CH.CH$_3$
O
H$_3$C—C—C—C = O
NH
(XV)

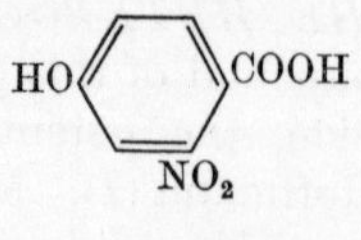

HO COOH
NO$_2$
(XVI)

In 1955 Gast and his colleagues reported the discovery in the laboratories of J. R. Geigy A.G. of the phytotoxic and plant growth regulating properties of a series of aminotriazines, from which group 2-chloro-4 : 6-bis(ethylamino)-*s*-triazine (XVII) was chosen for experimental study (*12*). This compound, introduced into the United States under the trade name "Simazin", has given promising results as a pre-emergent herbicide.

$$\begin{array}{c} CCl \\ / \quad \backslash\!\backslash \\ N \qquad N \\ \| \qquad | \\ H_5C_2.NH.C \diagdown_{N} \!/\!\!/ C.NH.C_2H_5 \end{array}$$

(XVII)

Little is known of the reactions underlying its phytotoxicity but the property is not confined to the group examined by the Geigy workers for Koopman and his colleagues (*13*) of Philips-Roxane have reported many other *s*-triazole derivatives with strong phytotoxic properties.

While on the subject of the herbicidal properties of heterocyclic nitrogen compounds, mention is permissible of the suggested use as a pre-emergent weedkiller of the thiadiazine referred to on p. 133 as a potential fungicide.

REFERENCES

(*1*) Schoene, D. L. and Hoffmann, O. L., *Science*, 1949, **109**, 588.
(*2*) Miller, D. M. and White, R. W., *Can. J. Chem.*, 1956, **34**, 1510.
(*3*) Currier, H. B. and Crafts, A. S., *Science*, 1950, **111**, 152.
(*4*) Leopold, A. C. and Klein, W. H., *Physiol. Plant.*, 1952, **5**, 91.
(*5*) Andreae, W. A. and Andreae, S. R., *Can. J. Bot.*, 1953, **31**, 426.
(*6*) Waygood, E. R., Oaks, A. and Maclachlan, G. A., *Can. J. Bot.*, 1956, **34**, 905.
(*7*) Allen, W. W., U.S.P. 2,670,282, Feb. 23, 1954.
(*8*) Rogers, B. J., *Weeds*, 1957, **5**, 5.
(*9*) Miller, C. S. and Hall, W. C., *Weeds*, 1957, **5**, 218.
(*10*) Alamercery, J., Hamner, C. L. and Latus, M., *Nature, Lond.*, 1951, **168**, 85.
(*11*) Ready, D., Minarik, C. E., Bradbury, D., Thompson, H. E. and Owings, J. F., *Plant Physiol.*, 1952, **27**, 210.
(*12*) Gast, A., Knüsli, E. and Gysin, H., *Experientia*, 1955, **11**, 107; 1956, **12**, 146.
(*13*) Koopman, H., Uhlenbroek, J. H. and Daams, J., *Nature, Lond.*, 1957, **180**, 147.

THE SUBSTITUTED UREA GROUP

The first of the main group, the phenyl ureas, to be discovered as a herbicide was 3-(*p*-chlorophenyl)-1 : 1-dimethylurea (XIX) by Bucha and Todd (*1*) in 1951. This compound, christened monuron is, in the British nomenclature, *N*-*p*-chlorophenyl-*N'N'*-dimethyl urea. Although it is possible to use the compound at carefully controlled rates for selective weeding of crops, its widest use is to render soil barren of plants. It is the first of a series selected by the du Pont organization in which the requirements of water solubility and resistance to microbial destruction are purposely regulated by attention to the substituents present. Thus for greater soil stability, the degree of chlorination was increased; for greater water insolubility, that is to say resistance to leaching, both chlorination and larger alkyl groups were used. Fenuron (3-phenyl-1 : 1-dimethylurea, XVIII) is water soluble to the extent of about 3000 ppm. at ordinary temperatures and is suggested for use against deep-rooted perennial weeds; monuron (XIX) has a solubility of about 230 ppm. and is recommended for use in regions of average rainfall; diuron (XX) (3-(3 : 4-dichlorophenyl)-1 : 1-dimethylurea) with solubility of 42 ppm. is used in regions of heavy rainfall; neburon (3-(3 : 4-dichlorophenyl)-1-methyl-1-*n*-butylurea, XXI) is only soluble to the extent of 4.8 ppm. and is suggested for use where

$C_6H_5NH.CO.N(CH_3)_2$

(XVIII)

$Cl C_6H_4 NH.CO.N(CH_3)_2$

(XIX)

$Cl_2C_6H_3 NH.CO.N(CH_3)_2$

(XX)

$Cl_2C_6H_3 NH.CO.N<\begin{matrix} CH_3 \\ CH_2(CH_2)_2CH_3 \end{matrix}$

(XXI)

long persistence is needed. Abel (*2*) analysed the practical use of these herbicides on the basis of solubility, inherent toxicity, soil adsorption and stability in soil.

The phenyldimethylureas move freely in the transpiration stream of the plant and accumulate in the leaves causing a collapse of the parenchyma and, in corn, a collapse of the stem at ground level. Minshall (*3*) finding that the symptoms are aggravated by light suggested that monuron affects photosynthesis in which connection the discovery by Wessels and van der Veen (*4*) and by Cooke (*5*) that the compound inhibits the Hill reaction is of importance.

Of the disubstituted ureas the only representative is the dichloral urea (1 : 3-bis(3-hydroxy-2 : 2 : 2-trichloroethyl) urea (XXII) introduced in 1950 by the Union Carbide Chemicals Co. under the name " Crag Herbicide 2 ", though the abbreviation DCU has been accepted by the Weed Society of America. Its main use is as a pre-emergent herbicide for the control

$$Cl_3C.C(OH)H.NH.CO.NH.CH(OH).CCl_3$$

(XXII)

of annual grasses in certain crops. Little is known of its action but it seems likely in view of the herbicidal action of trichloroacetic acid, that the trichlorochloral moieties are involved.

REFERENCES

(*1*) Bucha, H. C. and Todd, C. W., *Science*, 1951, **114**, 493.
(*2*) Abel, A. L., *Chem. & Ind.*, 1957, p. 1106.
(*3*) Minshall, W. H., *Can. J. Plant Sci.*, 1957, **37**, 157; *Weeds*, 1957, **5**, 29.
(*4*) Wessels, J. S. C. and van der Veen, R., *Biochim. biophys. Acta*, 1956, **19**, 548.
(*5*) Cooke, A. R., *Weeds*, 1956, **4**, 397.

THE TRICHLOROACETIC ACID GROUP

Trichloroacetic acid ($CCl_3.COOH$), TCA (XXIII) was introduced by the du Pont organization around 1947 for weed-killing purposes and it rapidly became popular, in the form of its sodium salt, for use against grasses, either as a pre-emergent spray, or, at higher rates, for the control of annual weeds on non-agricultural land. Replacement of one of the chlorine atoms by the methyl group gives rise to 2 : 2-dichloropropionic acid (XXIV), named dalapon by the Dow Chemical Co. who introduced its sodium salt as the experimental herbicide in 1953 (*1*).

$$CH_3.CCl_2.COOH$$

(XXIV)

The sodium salts of both acids are effective grass herbicides, absorbed by the roots and translocated by the plant. The salts of TCA are taken up by the roots but not by the leaves (*2*), and have little effect on perennial grasses. But the salts of dalapon are taken up by both roots and leaves and hence are better suited for use against perennial grasses.

The substitution of other organic bases for the metal in the salts proved an easy way of combining two herbicidal materials. Examples are erbon, the common name proposed by the Dow Chemical Co. for 2-(2 : 4 : 5-trichlorophenoxy)ethyl 2 : 2-dichloropropionate, and " Urox " the trade name of Allied Chemical & Dye Corp. for chlorophenyl dimethylurea trichloroacetate.

REFERENCES

(*1*) Barrons, K. C. and Hummer, R. W., *Agric. Chemic.*, 1951, **6**(6), 48.
(*2*) Barrons, K. C., U.S.P. 2,642,354, June 16, 1953.

MISCELLANEOUS GROUP

Although little is known of the reasons why the compounds are phytotoxic, it is necessary to review briefly the remaining list of available experimental and established herbicides.

The plant growth regulating properties of the salts of 3 : 6-endoxohexahydrophthalic acid were first described in 1951 by Tischler and Bell (*1*) and the sodium salt (XXV) was introduced that year as an experimental herbicide under the trade name of " Endothal ". Because the molecule

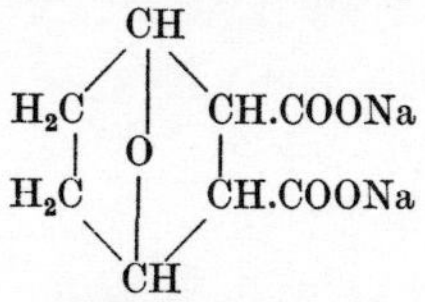

```
              CH
            / | \
        H2C   |   CH.COONa
         |    O   |
        H2C   |   CH.COONa
            \ | /
              CH
            (XXV)
```

is not planar in structure its three isomers are not represented in the two-dimensional formula XXV. Tischler and Bell concluded that the *exo-cis* isomer is generally of greater activity and is easier to prepare. The sodium salt is readily soluble in water and is taken up by the roots acting as a pre-emergent herbicide. Applied to foliage the acid and its salt act as defoliants and are translocated, for the treatment of the stem will cause the primary leaves to fall. Susceptible and resistant plants are to be bound both among grasses and among dicotyledonous species.

The survey by Hamm and Speziale (*2*) of the herbicidal activity of the *N*-substituted α-chloroacetamides led to the introduction in 1956 of a number of these compounds as pre-emergent grass killers. Aromatic substituents gave almost inactive compounds but with aliphatic substituents, disubstitution with three carbon groups gave greatest activity. The allyl derivative, α-chloro-*NN*-diallylacetamide (XXVI), was introduced by Monsanto Chemical Co. under the trade name " Randox ". This development recalls the suggested use of allyl alcohol for the weeding of forest nursery beds.

$$CH_2Cl.CO.N(CH_2.CH{:}CH_2)_2$$
(XXVI)

Hydrocarbon oils, usually in the form of discarded lubricating oils (sump oil), kerosine or diesel oils, have long been used to clear ground of unwanted vegetation and it was noticed that umbelliferous weeds are resistant. In California (*3*) stove oil, a petroleum oil of distillation range 330–570° F, was successfully used for weeding carrots and, in the Eastern States where such an oil was less available, Lackman (*4*) tried paint thinners and cleaning solvents such as " Stoddard Solvent ". Crafts and Reiber (*3*) confirmed that selective action is greatest in oils of medium distillation range containing about 25 per cent of unsaturated and aromatic hydrocarbons. The physiological response of plants to oils has been examined by van Overbeek and Blondeau (*5*), by Currier and Peoples (*6*) and by Helson and Minshall (*7*), but the reasons for the lower susceptibility of umbelliferous plants are still obscure.

REFERENCES

(*1*) Tischler, N. and Bell, E. P., U.S.P. 2,576,081, Nov. 20, 1951.
(*2*) Hamm, P. C. and Speziale, A. J., *J. agric. Food Chem.*, 1956, **4**, 518.
(*3*) Crafts, A. S. and Reiber, H. G., *Hilgardia*, 1948, **18**, 77.
(*4*) Lackman, W. H., *Spec. Circ. Mass. St. Coll.*, 120, 1945.
(*5*) van Overbeek, J. and Blondeau, R., *Weeds*, 1954, **3**, 55.
(*6*) Currier, H. B. and Peoples, S. A., *Hilgardia*, 1954, **23**, 155.
(*7*) Helson, V. A. and Minshall, W. H., *Plant Physiol.*, 1956, **31**, 5.

CHAPTER XII
FUMIGANTS

FUMIGATION is undoubtedly of great antiquity; the burning of aromatic herbs, of resins and of incense, though probably of little real effect upon microorganisms, was adopted as a sanitary measure, for the strong odour would effectively mask the smell of putrefaction which was considered the cause of disease. Originally applied for the treatment of houses, the method was extended for use in glasshouses and for the treatment of citrus trees. For the latter purpose, the tree is covered by an air-tight tent into which is introduced the requisite amount of fumigant. After exposure for the necessary time, the tent is removed and re-erected over the next tree. The material of the tent must be light for easy manipulation, must be pliable to lie close to the ground, of sufficient strength to prevent tearing and as gas-tight as possible. Tarpaulins were widely used but may be replaced by plastic and rubber-coated fabrics (*1*).

Fumigation is, in principle, the use of the toxic chemical in volatile form. It can be practised therefore only in an enclosed space and, because the concentration of the fumigant must be controlled, a knowledge is required not only of the volume of the space to be fumigated but of the amount of fumigant evolved in unit time. The determination of the volume of the greenhouse is usually a matter of simple arithmetic and, for tent fumigation, formulae based on the resemblance of the tent to a cylinder surmounted by a hemisphere or "dosage tables" based on the circumference at the base of the tent and the longest distance over its top may be used. To maintain the required concentration for the correct time, freedom from leaks is necessary and fumigation is therefore carried out in cooler weather.

The early fumigants were chosen because of their ready volatility. But methods of atomization (see p. 65) or of volatilization from thermostatically controlled heaters or by means of pyrotechnic mixtures (see p. 64) permit the use of compounds, as fumigants, of comparatively low vapour pressure at ordinary temperatures. Among such compounds the widest used are lindane, DDT, sulfotep and azobenzene. But many of the older true fumigants survive, the most important being:

Hydrocyanic Acid

Hydrocyanic acid was first employed by D. W. Coquillett (*2*), in 1886, against the scale insects of citrus in California. He introduced the method more particularly against the cottony cushion scale (*Icerya purchasi*), then a serious menace to citrus-growers. With improvements in technique, the method has now become an important routine not only in citrus culture but in general glasshouse management.

Hydrocyanic acid was, at first, generated by the action of an acid upon a cyanide, the materials employed by Coquillett being potassium cyanide and sulphuric acid. As the amount of fumigant used has to be carefully controlled, rigid directions of use must be followed. The potassium cyanide should be of not less than 93 per cent purity (*3*) and the acid of sp. gr. 1.84. The acid should not be diluted to an extent sufficient to prevent the complete evolution of the hydrocyanic acid which is readily soluble in excess of water. Yet enough water should be added to prevent secondary reactions and the separation of solid potassium sulphate which, by occluding the cyanide, prevents its complete decomposition; the solidification of the reacting mixture is known popularly as "freezing". A typical formula is 1–1–3; i.e. one ounce 98 per cent potassium cyanide added to one fluid ounce of strong acid diluted with three fluid ounces of water. The cyanide should be added to the diluted acid to prevent a polymerization of the reaction products.

The more expensive potassium cyanide was later replaced by sodium cyanide, used in a similar manner. As high-grade (97 per cent purity) sodium cyanide liberates a greater relative amount of hydrocyanic acid than potassium cyanide, the proportions of sodium cyanide, sulphuric acid and water usually employed are 1–1½–2 to 1–1½–3. The 97–98 per cent sodium cyanide was by virtue of this higher proportion of cyanide, sometimes described in terms of potassium cyanide as of 129–130 per cent cyanide.

Although with cautious management there is ample time for the operator to move away to safety after dropping the loose cyanide into the prepared "pots" of acid, there have appeared on the market "Safety Cyanide Packages" obviating danger (*4*). These packages are of metal with thin zinc foil sides containing weighed amounts of sodium cyanide. The acid first dissolves away the zinc before coming into contact with the cyanide. Standardization of the amount of zinc used is necessary and allowance must be made for the extra amount of sulphuric acid neutralized by this reaction and for the water required for the solution of the zinc sulphate formed.

The ready liquefaction of hydrocyanic acid at 26.5° C led to the use by Mally (*5*) in 1915, of liquid hydrocyanic acid for fumigation purposes. The anhydrous liquid may be pumped into the enclosed space, enabling an easier and more rapid application and, since it is introduced as a fine spray, a more thorough distribution of the gas. Further, in acid-cyanide methods there is a danger of the burning of the fabric of the tent by the acid carried up by the gas, a risk obviated by the use of liquid hydrocyanic acid. This method became widely recommended in the United States and the technique of the process was worked out by Quayle (*6*) and by Woglum (*7*).

A more convenient method is the use of solids, such as calcium cyanide, which are decomposed by moisture to liberate hydrogen cyanide. Calcium cyanide is made by the fusion of salt with calcium cyanamide, the "Cyanogas" process, or by the combination of liquid hydrogen cyanide with calcium carbide ("Citrofume", *8*). According to Metzger (*9*) the latter reaction yields a product of composition more akin to the acid cyanide, $CaH_2(CN)_4$ than to the normal cyanide $Ca(CN)_2$. In the presence of atmospheric moisture, these products decompose to form calcium hydroxide and hydrocyanic acid. Their application therefore merely involves the discharge of the requisite amount of the dust into the enclosed space containing air of a sufficient humidity to permit its decomposition. The first trials with calcium cyanide were reported in 1923 by Quayle (*10*) for tent fumigation, and the method was extended to glasshouse work with excellent results (*11*). The relationships between humidity and the decomposition of calcium cyanide dusts were examined by Quayle (*12*).

Other products which evolve hydrocyanic acid gas when exposed to moisture are the sodium cyanide-anhydrous magnesium sulphate mixture used by Ladell (*13*) and the bicarbonate-cyanide mixture devised by Speyer and Owen (*14*).

Action upon the Insect : Shafer (*15*) suggested that insects are killed by hydrogen cyanide through a disturbance of the normal balance of the reductases, catalase and oxidases. This early view was made more specific by Warburg's work (*16*) on the inhibition of his " Atmungsferment " by cyanide which he suggested combined with the iron present in this enzyme since identified with cytochrome oxidase (*17*). Peroxidase and catalase are likewise iron-containing enzymes and each is inhibited by cyanide. The prosthetic groups of most oxidases are now known to be iron (present usually as a porphyrin complex) or copper compounds, the activity of the enzyme being associated with a valency change of the metal. This valency change is prevented by cyanide through the formation of a non-dissociable compound.

The concentration (c) of hydrogen cyanide required for the treatment of a particular insect is dependent, among other factors, on time of exposure (t) for the applicability of Haber's rule (ct = constant) was confirmed by Quayle and Knight (*18*). Not only does the relative susceptibility to hydrogen cyanide differ between insect species but Knight (*19*) found the effect of temperature on susceptibility differs. The friendly ladybird beetle (*Hippodamia convergens* Guér.) is less readily killed at low temperatures when the beetle is less active, whereas the effectiveness of cyanide fumigation against the red scale (*Aonidiella aurantii*) is not greatly affected by change of temperature. Knight therefore recommended that fumigation against this pest should be carried out only when the citrus trees are partially dormant for then the effects of the fumigant on tree and on predator are likely to be less.

It has also been stated that the insect exposed to sub-lethal cyanide concentration becomes more difficult to kill, a phenomenon termed " protective stupefaction " by Gray and Kirkpatrick (*20*). Hence suggestions of the addition of irritant or other gaseous substances to provoke the insect to greater activity during fumigation, though the observation of Pratt *et al.* (*21*) that the addition of 0.1 per cent trichloroethylene or 1 per cent carbon dioxide to 0.2 per cent hydrogen cyanide caused a decreased kill of *H, convergens*, is contrary to this idea.

In California, it was found that the effectiveness of the standard fumigation methods suffered because of the selection of strains of the scale insects of increased resistance to the poison. Quayle (*22*) reported, in 1922, that in at least two localities in California, the red scale had become so resistant that the dosages of hydrocyanic acid found, in earlier fumigation practice, to be fatal were tolerated. The increased dosage necessary for a satisfactory kill had become too great for the safety of the tree. This development (*23*) confirmed by Woglum (*24*), is due to the selection of resistant strains, an hypothesis verified by Boyce (*25*) and by Gough (*26*) who showed that resistance to hydrogen cyanide is an hereditary property.

Action on the Plant : Hydrocyanic acid is toxic to plant growth, but de Ong (*27*) showed that there is a margin of safety between the strength necessary to kill an insect such as *Phylloxera* and that causing injury to hardy plants. Owing to the differences in the susceptibility of different species, there exist, however, certain plants on which it is unwise to employ hydrocyanic acid as a fumigant. Weigel (*28*) mentioned, among the ornamental greenhouse plants, sweet pea, chrysanthemum and rose.

The injurious action of hydrocyanic acid upon plant foliage proved the most severe difficulty met in the early days of cyanide fumigation. Coquillett reduced injury by permitting only a slow evolution of the gas, and F. W. Morse (*29*) introduced the " soda " process in which sodium bicarbonate was mixed with the cyanide, the hydrocyanic acid thereby being diluted by the carbon dioxide simultaneously evolved. Later, however, Coquillett found that the extent of the foliage injury was dependent upon the moisture present and the action of sunlight during fumigation.

Quayle (*30*) suggested that the lessened danger of plant injury in night fumigation was due to the differences of light, temperature and moisture affecting the stomatal opening and the chemical relations governing gas absorption and cuticular defence. Clayton (*31*) found that the conditions which bring about an opening of the stomata were those under which the injury was most severe. Further, he showed a positive correlation of the sugar content of the plant tissue and resistance to hydrocyanic acid. The lower the temperature, the less active the plant and the less the injury caused.

For these reasons, hydrogen cyanide fumigation is conducted at night time and watering is withheld, though if calcium cyanide is used the relative humidity should not be below 79 per cent.

It was at one time thought that the hydrochloric acid generated from chloride impurities of the commercial cyanide was responsible for injury (*32*). Moore and Willaman (*33*), however, established the injurious nature of hydrocyanic acid and found that the immediate effect was a reduction in respiration due to a disturbance of the activity of the oxidases and catalase. Shill (*34*) was unable to confirm this conclusion with citrus for he showed that the hydrocyanic acid treatment led to an increase in respiration which returned to normal some thirty-five hours after fumigation. Moore and Willaman also showed that there was an increase in the permeability of the leaf septa (presumably, the plasma membranes) causing a less rapid intake of water from the stem and a more rapid cuticular transpiration. Thus, in mild fumigation, the plant may wilt temporarily.

Woglum (*35*), in 1918, noticed a greater injury caused by cyanide fumigation following the application of Bordeaux mixture. It was found also that citrus, treated with copper salts for the control of exanthema, a copper-deficiency disease, were more susceptible to hydrogen cyanide (*36*). Butler and Jenkins (*37*) ascribed the injury which follows Bordeaux spraying to the formation of a soluble phytotoxic complex copper cyanide, a conclusion confirmed by Guba and Holland (*38*) who regarded the cuprocyanide $CaCu_2(CN)_4,5H_2O$ or a similar soluble compound as responsible.

On the other hand, there is evidence to show that the presence of traces of hydrocyanic acid brings about a stimulation of plant growth. Moore and Willaman (*33*) found that the increase of permeability, if not too severe, was followed by a rate of growth and of fruit production in excess of the normal. Clayton found that the maximum beneficial results from stimulation occurred at a dosage of hydrocyanic acid just below that of injury to the plant. Gassner (*39*) utilized this stimulation of plant growth as a method of " forcing " plants by regular treatment in dilute atmospheres of hydrocyanic acid.

REFERENCES

(*1*) Phillips, G. L. and Nelson, H. D., *J. econ. Ent.*, 1957, **50**, 452.
(*2*) See Howard, L. O., *Yearb. U.S. Dep. Agric.*, 1899, p. 150, and, for the early history of the use of HCN, Johnson, W. G., *Fumigation Methods*, New York, 1902.
(*3*) See *Bull. Minist. Agric.*, 82, 1934.
(*4*) See Lloyd, L., *Ann. appl. Biol.*, 1922, **9**, 1.
(*5*) Mally, C. W., *S. Afr. J. Sci.*, 1915, **12**(3), 95.
(*6*) Quayle, H. J., *Bull. Calif. agric. Exp. Sta.*, 308, p. 393, 1919.
(*7*) Woglum, R. S., *J. econ. Ent.*, 1919, **12**, 117.
(*8*) Quayle, H. J., *J. econ. Ent.*, 1927, **20**, 200.
(*9*) Metzger, F. J., *Industr. engng Chem.*, 1926, **18**, 161.
(*10*) Quayle, H. J., *J. econ. Ent.*, 1923, **16**, 327.
(*11*) See Weigel, C. A., *J. econ. Ent.*, 1925, **18**, 161.
(*12*) Quayle, H. J., *Hilgardia*, 1928, **3**, 207.
(*13*) Ladell, W. R. S., *Ann. appl. Biol.*, 1928, **25**, 341.
(*14*) Speyer, E. R. and Owen, O., *Ann. appl. Biol.*, 1926, **13**, 144.
(*15*) Shafer, G. D., *Tech. Bull. Mich. agric. Exp. Sta.*, 21, 1915.
(*16*) Warburg, O., *Biochem. Z.*, 1925, **165**, 196 ; 1926, **177**, 471.
(*17*) Keilin, D., *Proc. roy. Soc.*, 1925, **98**, B, 312.
(*18*) Quayle, H. J. and Knight, H., *Calif. Citrograph.*, 1921, **6**, 196.
(*19*) Knight, H., *Hilgardia*, 1925, **1**, 35.
(*20*) Gray, G. P. and Kirkpatrick, A. F., *J. econ. Ent.*, 1929, **22**, 878.
(*21*) Pratt, F. S., Swain, A. F. and Eldred, D. N., *J. econ. Ent.*, 1933, **26**, 1031.
(*22*) Quayle, H. J., *J. econ. Ent.*, 1922, **15**, 400.
(*23*) Quayle, H. J., *Hilgardia*, 1938, **11**, 183.
(*24*) Woglum, R. S., *J. econ. Ent.*, 1925, **18**, 593.
(*25*) Boyce, A. M., *J. econ. Ent.*, 1928, **21**, 715.
(*26*) Gough, H. C., *Ann. appl. Biol.*, 1939, **26**, 533.
(*27*) de Ong, E. R., *J. agric. Res.*, 1917, **11**, 421.
(*28*) Weigel, C. A., *Dept. Circ. U.S. Dep. Agric.*, 380, 1926.
(*29*) See Johnson, W. G., *Fumigation Methods*, New York, 1902, p. 3.
(*30*) Quayle, H. J., *Calif. Citrograph*, 1919, **4**, 292.
(*31*) Clayton, E. E., *Bot. Gaz.*, 1919, **67**, 483.

(*32*) Woglum, R. S., *Bull. U.S. Dep. Agric. Bur. Entom.*, 90, p. 83, 1911.
(*33*) Moore, W. and Willaman, J. J., *J. agric. Res.*, 1917, **11**, 319.
(*34*) Shill, A. C., *Univ. Calif. Publ. agric. Sci.*, 1931, **5**, 167.
(*35*) Woglum, R. S., *Fmrs'. Bull. U.S. Dep. Agric.*, 923, 1918.
(*36*) Haas, A. R. C. and Quayle, H. J., *Hilgardia*, 1935, **9**, 143.
(*37*) Butler, O. and Jenkins, R. R., *Phytopathology*, 1930, **20**, 419.
(*38*) Guba, E. F. and Holland, E. B., *Bull. Mass. agric. Exp. Sta.*, 303, 1933.
(*39*) Gassner, G., *Ber. dtsch. bot. Ges.*, 1925, **43**, 132.

Naphthalene

This substance is obtained from coal-tar and is marketed in various stages of purity. The least pure form, the Undrained Salts or Unwhizzed Naphthalene, contains relatively large amounts of creosote and phenols (see p. 272). Part of the creosote is removed by allowing the " salts " to drain, a process yielding the " Drained Salts ", and the mother liquor may be more thoroughly separated by pressure or by the centrifuge processes which yield the " hot-pressed " and " Whizzed " Naphthalene. Finally, the purest commercial form, the " Pure Flake " Naphthalene is obtained free from carbonaceous matter by sublimation.

Pure naphthalene is a white crystalline substance of m.p. 80° C with a characteristic odour. The material has long been known to possess insecticidal properties and was employed, under the popular name of " Carbon " or " Camphor " balls, as a repellent for clothes moths (Tineidae), and as a fumigant in museum cases, etc.

Only in a few cases has naphthalene proved a successful competitor to hydrocyanic acid in crop protection. One case is the red spider *Tetranychus telarius* which Speyer (*1*) found was killed after forty hours' exposure to naphthalene vapour. He recommended for fumigation, " Pure commercial White Flake Naphthalene, Grade 16 ", i.e., passing a sieve of 1/16th inch mesh. This method, although successful in cucumber houses, proved less so in tomato houses (*2*), one factor being that, owing to the relatively low temperature, effective fumigation cannot be completed in a single night. To shorten the time required, Parker (*3*) introduced a vaporizing lamp. This is frequently used in carnation houses, for the control of concentration is not as rigid as is required to avoid injury to other plants. Hartzell and Wilcoxon (*4*) devised an apparatus whereby air, drawn from outside, was saturated with naphthalene vapour and discharged into the house. Later, with Youden (*5*), they adopted the method to a recirculatory process, controlling the concentration of naphthalene vapour by the use of solutions of naphthalene in inert solvents such as light lubricating (hydrocarbon) oil.

Naphthalene vapour has, under certain conditions, a pronounced scorching action upon the cucumber plant, an action which may be prevented by the presence of sufficient moisture. It would appear that the naphthalene vapour enters the plant tissue via the stomata and, being insoluble in water, the amount of naphthalene which enters is reduced if the plants are giving out and not taking in moisture. Speyer found that cucumber fruit absorbed the vapour, but if exposed to fresh air after the treatment, the taste imparted to it soon disappears.

The tolerance of ornamental plants to naphthalene fumigation differs greatly (*6*) but the susceptible varieties and species have been tabulated in most practical hand-books (see *7*).

REFERENCES

(*1*) Speyer, E. R., *Rep. exp. Res. Sta. Cheshunt*, 1923, p. 69.
(*2*) Speyer, E. R., *Rep. exp. Res. Sta. Cheshunt*, 1925, p. 89.
(*3*) Parker, T., *Ann. appl. Biol.*, 1928, **15**, 81.
(*4*) Hartzell, A. and Wilcoxon, F., *J. econ. Ent.*, 1930, **23**, 608.
(*5*) Wilcoxon, F., Hartzell, A. and Youden, W. J., *Contr. Boyce Thompson Inst.*, 1933, **5**, 461.
(*6*) Hartzell, A., *J. econ. Ent.*, 1926, **19**, 780.
(*7*) Whitcomb, W. D., *Bull. Mass. agric. Exp. Sta.*, 326, 1935.

Tetrachloroethane

This liquid is the active agent of certain proprietary fumigants for use against the greenhouse whitefly (*Trialeurodes vaporariorum*). It is produced commercially as a fat solvent, and has the advantage of non-inflammability. Tetrachloroethane, $C_2H_2Cl_4$, was first used experimentally upon whitefly by Lefroy (see *1*) in 1915. Lloyd (*2*) considered that it killed the scale through its effect upon the waxy covering of the larval stages. The flies died in the attempt to emerge from this covering. This hypothesis is in agreement with the observation that tetrachloroethane has little effect upon aphides but that mealy bugs are affected. Fumigation may easily be carried out by suspending sacks soaked with the required amount of the fumigant in the closed glasshouse at dusk, the temperature of the house being maintained at 65–70° F.

Tetrachloroethane is phytotoxic to many plant species. Plants resistant and susceptible to the fumigant are quoted by Parker (*3*) and by Fox Wilson (*1*). Speyer (*4*) found a remarkable difference even in the varietal susceptibilities of various ornamental plants. It was, in the case of the chrysanthemum, almost possible to name the variety from the nature of the injury to the foliage.

REFERENCES

(*1*) Fox Wilson, G., *Gdnrs'. Chron.*, 1926, **79**, 138.
(*2*) Lloyd, L., *Ann. appl. Biol.*, 1922, **9**, 1.
(*3*) Parker, T., *Ann. appl. Biol.*, 1928, **15**, 251.
(*4*) Speyer, E. R., *Rep. exp. Res. Sta. Cheshunt*, 1925, p. 107.

Nicotine

Nicotine is still used as a glasshouse fumigant, mainly for aphis control (*1*). The alkaloid may be heated on an iron plate over a small charcoal fire or spirit lamp. Alternatively, " tobacco shreds " may be used, consisting of paper soaked in a solution of saltpetre, dried and impregnated with a mixture of linseed oil and nicotine (*2*). A proportion of the nicotine is thereby evolved as vapour which acts in the same way as nicotine from dusts or sprays.

REFERENCES

(*1*) Herrick, G. W. and Griswold, G. H., *Bull. Cornell agric. Exp. Sta.*, 474, 1929.
(*2*) Reid, J. H., *Chem. & Ind.*, 1931, **50**, 954.

Methyl Bromide

An important application of fumigants for plant protection purposes is in the treatment of " seed " and nursery stock. Under legal requirements, enforced in most countries, the importation of such plant products is permissible only after a satisfactory fumigation or sterilization. Hydrogen cyanide is the usual fumigant employed for such purposes, but methyl bromide is an important competitor (*1*).

Methyl bromide, or bromomethane CH_3Br, is a colourless liquid boiling at 4.5° C; its insecticidal properties were first reported by Le Goupil (*2*) in 1932, and its early use reviewed by Busbey (*3*). In addition to its use for the fumigation of dormant nursery stock and for food protection in storage (*4*), it is effective as a soil fumigant (*5*) being used at concentrations which kill weed seed. Rigid precautions are made necessary by its insidious toxicity to man (*6*).

The biochemical interactions of methyl bromide were shown by Winteringham and his colleagues (*7*) to involve methylation and the methylation of the sulphydryl enzymes was regarded by Lewis (*8*) as a probable cause of the toxicity of methyl bromide.

Methanesulphonyl fluoride CH_3SO_2F, has insecticidal properties which directed Schrader's attention in the late 1930's (*9*) to the possible use of this non-inflammable low-boiling liquid as a fumigant, but the compound has only recently been used. Its mammalian toxicity, high when injected subcutaneously, is low enough to permit its use for the control of animal ectoparasites (*10*). It is reported to be harmless to plants at insecticidal concentrations. Sulphuryl fluoride SO_2F_2 was recommended in 1957 as an insecticidal fumigant by Kenaga (*11*) but little information has yet appeared on the results of its use.

REFERENCES

(*1*) Richardson, H. H., Johnson, A. C., Bulger, J. W., Casanges, A. H., and Johnson, G. V., *Tech. Bull. U.S. Dep. Agric.*, 853, 1943.
(*2*) Le Goupil, —., *Rev. Path. vég.*, 1932, **19**, 169.
(*3*) Busbey, R. L., *U.S. Dep. Agric. Bur. Ent. Plant Quar.*, E-612, 1944.
(*4*) Page, A. B. P., Lubatti, O. F. and Russell, J., *J. Soc. chem. Ind., Lond.*, 1949, **68**, 102, 151.
(*5*) Newhall, A. G. and Lear, B., *Phytopathology*, 1948, **38**, 38.
(*6*) Fumigation with Methyl Bromide. Precautionary Measures. *H.M. Stationery Office, London*, 1947 see also von Oettingen, W. F., *U.S. Dep. Health Pub.* 414, Washington, 1955.
(*7*) Winteringham, F. P. W., *J. Sci. Food Agric.*, 1955, **6**, 269.
(*8*) Lewis, S. E., *Nature, Lond.*, 1948, **161**, 692.
(*9*) Schrader, G., Die Entwicklung neuer Insektizide auf Grundlage organischer Fluor- und Phosphor-Verbindungen. Monogr. *Agnew. Chem.*, No. 62, Verlag Chemie, GMBH, Weinheim, 2nd Ed., 1952.
(*10*) Feils, G., *Tierarztl. Umsch.*, 1954.
(*11*) Kenaga, E. E., *J. econ. Ent.*, 1957, **50**, 1.

Miscellaneous Fumigants

A brief reference may be made to insecticidal fumigants which, for reasons such as their high phytotoxic properties, are unsuitable for use in the presence of growing plants, but which may be safe for use in empty glasshouse or infested stores. Moreover, the literature of the control of pests of stored products contains valuable studies of the factors of chemical structure and of insect physiology influencing toxic action.

The pioneer work of Moore (*1*) indicated a correlation between volatility and toxicity over a wide range of organic compounds, a discovery extended by Ferguson and Pirie (*2*) to the action of the chlorinated ethylenes on the grain weevil *Calandra granaria*. A comprehensive search for stored produce fumigants undertaken by Neifert and his colleagues (*3*) led to the introduction of a mixture of ethyl acetate (40 volumes) and carbon tetrachloride (60 volumes). The addition of carbon tetrachloride, which was necessary to lower the inflammability of the ethyl acetate vapour, reduced the already low toxicity of the latter. Roark and Cotton (*4*) therefore examined a wider range of aliphatic compounds, reviewing their results in relation to chemical structure. The most promising of the compounds tested were ethylene dichloride, ethylene oxide, certain alkyl formates and esters of the halogenated fatty acids. The halogen- and sulphur-substituted organic compounds were also surveyed by Shepard, Lindgren and Thomas (*5*). Of the physiological studies, mention may be made of Busvine's (*6*) investigation of the insecticidal action of ethylene oxide which is generally used in admixture with carbon dioxide to reduce inflammability, e.g., " Carboxide " of the Union Carbide Chemicals Co.

Of the halogenated compounds ethylene dibromide and ethylene dichloride found use against stored product insects and for soil treatment (see p. 273). β,β′-Dichloroethyl ether, $ClCH_2.CH_2O.CH_2.CH_2Cl$, selected by Roark and Cotton (*4*), has proved too phytotoxic for general crop protection but is used for the sterilization of bare soil. 1 : 1-Dichloro-1-nitroethane, $CCl_2(NO_2)CH_3$, first described as a fumigant by O'Kane and Smith (*7*), is nowadays mainly used for the fumigation of stored products where long exposure times are practicable.

A special example of fumigation for purposes of crop protection is provided by the use of benzene C_6H_6 for the control of blue mould of tobacco (*Peronospora tabacina* Adam) in tobacco seed beds, a practice introduced by Angell, Hill and Allen (*8*) and developed in the U.S.A. by Wolf and co-workers (*9*). Clayton (*10*) reported the successful substitution of benzene by *p*-dichlorobenzene for the control of this disease.

REFERENCES

(*1*) Moore, W., *J. agric. Res.*, 1917, **9**, 371 ; **10**, 365.

(*2*) Ferguson, J. and Pirie, H., *Ann. appl. Biol.*, 1948, **35**, 532.

(*3*) Neifert, I. E., Cook, F. C., Roark, R. C., Tonkin, W. H., Back, E. A. and Cotton, R. T. *Bull. U.S. Dep. Agric.* 1313, 1925.

(*4*) Roark, R. C. and Cotton, R. T., *Tech. Bull. U.S. Dep. Agric.*, 162, 1929.

(*5*) Shepard, H. H., Lindgren, D. L. and Thomas, E. L., *Tech. Bull. Minnesota agric. Exp. Sta.* 120, 1937.

(*6*) Busvine, J. R., *Ann. appl. Biol.*, 1938, **25**, 605.

(*7*) O'Kane, W. C. and Smith, H. W., *J. econ. Ent.*, 1941, **34**, 438.

(*8*) Angell, H. R., Hill, A. V. and Allan, J. M., *J. Coun. sci. industr. Res. Aust.*, 1935, **8**, 203.

(*9*) Wolf, F. A., Pinckard, J. A., Darkis, F. R., McLean, R. and Gross, P. M., *Phytopathology*, 1939, **29**, 103.

(*10*) Clayton, E. E., *Science*, 1938, **88**, 56.

CHAPTER XIII

SEED TREATMENT

To ensure the continuation of the species, it is necessary that the pathogen or pest, during the absence or dormancy of its host, should itself either pass into a resting stage or find an alternate host. For example, it may be attached, in a like dormant condition, on or within the seed of its host. The term " seed " is here used in its widest sense to embrace fruit, tuber, bulb, true seed, or other form in which the plant over-winters and which is sown or planted to yield the crop. It is during this stage of seed transmission that the pest is attacked in seed disinfection. Alternatively, it may be possible by treatment of the seed to protect either the seed or seedling from attack by soil-borne fungi, a principle termed seed protection.

The various methods of seed treatment employed fall conveniently under the three heads : (i) Mechanical ; (ii) Chemical ; (iii) Physical.

MECHANICAL METHODS

Such methods are only available when the presence of the resting stage of the pest causes some abnormality in size, weight or appearance, by which it is possible to separate good from bad seed. Such methods include the examination of seed samples and a discarding of those which show an abundance of weed seeds. A satisfactory cleaning of seed is now enforced by legislation in most countries.

A frequent source of trouble is seed of Dodder (*Cuscuta* spp.), parasitic plants, in seed of the clover group. Screening and sieving may not give complete separation and there have been several ingenious suggestions, such as the Dossier machine with velvet linings on which the rougher-coated dodder seeds are retained. Foy (*1*) employed a magnetic device in which the rough-coated dodder in the impure seed, when mixed with a powder containing iron remains covered, and is therefore removed by passing under a magnet. An ingenious demonstration, due to von Degen and recorded by Saunders (*2*), depends on the difference in elasticity of the seed coats of dodder and clover. If the seeds are dropped from a height of about eighteen inches into a saucer, the clover seeds bounce out while the dodder seeds remain.

The elimination by seedsmen of misshapen and damaged beans and peas removes many infected with various fungal and bacterial diseases, e.g., *Mycosphaerella pinodes* (Berk. & Blox.) Vestergr., *Ascochyta pisi* Lib., *Pseudomonas medicaginis* var. *phaseolicola* (Burk.) Dows (*3*). Hennig (*4*) proposed an interesting method for the mechanical treatment of the loose smut of barley (*Ustilago nuda* (Jens.) Rostr.). With the two-rowed barleys of the nodding type, *Hordeum distichum nutans*, only those flowers at the apex and base of the ear usually open during fertilization. It is at this stage that infection occurs and the closed cleistogamous flowers of the middle of the ear may escape. As the flowers at the base and apex of the ear yield smaller grain, Hennig concluded and showed that the small grain bears a higher infection than the larger grain from the middle of the ear. Sifting to remove the smaller grain was also found to give partial control of the stripe disease of barley (*Helminthosporium gramineum* Rabenk.) and there were indications that oats from small seed were more liable to attack by stem rust.

In the " Purples " disease of wheat, caused by the nematode *Anguillulina tritici* (Steinb.) Gerv. & v. Ben., the normal wheat grain is replaced by a gall enclosing numerous larvae. These galls resemble the normal grain in appearance, but have a lower specific gravity and can therefore be separated by flotation in a 20 per cent sodium chloride solution. The good grain is rinsed in clean water and, to prevent poor germination, Leukel (*5*) recommended sowing immediately after rinsing. Chester (*6*) suggested the flotation of delinted cotton seed in water, for he found that the fraction sinking gave a germination and emergence twice as good as that of the lighter seed.

REFERENCES

(*1*) Foy, N. R., *N.Z. J. Agric.*, 1924, **29**, 44.
(*2*) Saunders, C. B., *J. Minist. Agric.*, 1921, **28**, 551.
(*3*) Ogilvie, L., *Bull. Minist. Agric.*, 123, 1941.
(*4*) Hennig, E., *K. Landtbr Akad. Handl. Stockh.*, 1916, **50**, 282.
(*5*) Leukel, R. W., *J. agric. Res.*, 1924, **27**, 925.
(*6*) Chester, K. S., *Phytopathology*, 1938, **28**, 745.

CHEMICAL METHODS

The practice of steeping seed in liquids such as wine or urine was employed from the earliest times (see *1*), though perhaps more for the purpose of destroying the insect pests present. In the seventeenth century, in England, it was usual to " brine " seed corn against " smuttiness ". Tull (*2*) in 1733, recorded how, seventy years previously, a ship laden with corn had gone ashore near Bristol. Local farmers sowed salvaged wheat, and the crop produced was free from smut, in strong contrast to that of neighbouring fields sown with normal wheat. Even in 1733, however, it was found that sodium chloride gave but an incomplete control of the disease, for Tull himself had the suspicion that the " drowned " wheat at Bristol, being foreign, might have been free from smut and " might not have been smutty the next year though it had not been soaked in sea water ". In 1761, Schulthuss (*3*) suggested the use of copper sulphate in place of salt.

The scientific foundations of chemical seed disinfection were laid by Prévost (*4*), who, in 1807, discovered the nature of the " smut " disease, against which copper sulphate had been used with success. He observed the germination of the smut spores in water and the prevention of this germination by the presence of a trace of copper sulphate. His work was repeated and extended by Kühn (*5*), who, in 1859, popularized the copper sulphate treatment of cereal grain.

The copper sulphate treatment was modified, in 1873, by Dreisch (see *6*), who suggested an after treatment of the soaked seed by immersion for five minutes in milk of lime. What reason prompted such a proposal is not given, but it had been the practice, according to Tull, to sift quicklime over the heap of " brined " seed. The object here was evidently to hasten the drying of the seed, though Tull stated that the quicklime confined the brine to the surface of the grains and " suffers none of it to be exhaled by the air ". In this fashion were the basic copper seed disinfectants introduced, materials which survive today in the copper carbonate dusts.

From 1860 to 1895, the search for a suitable substitute for copper sulphate, which was found to injure the seed, continued. Hollrung (*6*) has given the history of this period in detail. The most promising of the materials tried appears to have been salicylic acid, suggested by Schröder in 1892, which however is too expensive. Kellerman and Swingle, in 1891, found in liver of sulphur (potassium polysulphide) a satisfactory disinfectant. This material formed the active agent of " Ceres " powder, a seed fungicide placed on the market by Jensen, the originator of the hot-water treatment (see p. 267).

About 1890, a new field was opened by Bolley (*7*) who, for the control of potato skin diseases, immersed the tubers in mercuric chloride. At the beginning of the century, Hiltner employed this compound for the control of cereal diseases. From his success and from experience with salicylic acid may be traced the development of the organomercury seed disinfectants. In 1895, the then new disinfectant, formaldehyde, was introduced for the treatment of cereals. For some time, formaldehyde proved one of the most popular of the seed disinfectants until the development of the organomercury seed treatments.

REFERENCES

(*1*) Buttress, F. A. and Dennis, R. W. G., *Agric. Hist.*, 1947, **21**, 93.
(*2*) Tull, J., *The Horse-hoing husbandry*, London, 1733, p. 66.
(*3*) Schulthuss, H. H., *Abh. natur. Ges. Zurich*, 1761, **1**, 497.
(*4*) Prévost, B., *Mémoire sur la cause immédiate de la Carie ou Charbon des Blés*, Montauban, 1807.
(*5*) Kühn, J., *Die Krankheiten der Kulturgewachse, ihre Ursachen und ihre Verhütung*, Berlin, 1859, p. 88.
(*6*) Hollrung, M., *Landw. Jb.*, 1897, **26**, 145.
(*7*) Bolley, H. L., *Bull. N. Dakota agric. Exp. Sta.*, 4, 1891.

The Copper Group

The older treatment with copper sulphate solutions required submersion of the seed, which especially when it entailed after-treatment with milk of lime, was too cumbersome for the average farmer. Special vats and apparatus for hoisting the sacks of grain were needed. The grain took up large amounts of water and needed a long time to re-dry. To ease the labour involved, the " heap " method was introduced in which the heap of grain was sprinkled with a solution of 1 to 10 per cent bluestone at the rate of one gallon per sack of wheat. The heap was then shovelled until all the rain was wetted, afterwards being spread out to dry. The method has the advantages that no special apparatus (apart from a suitable floor) is required and that less time is taken for the grain to dry.

The heap method was, until the introduction of formaldehyde, widely used for the control of bunt of wheat (*Tilletia caries* and *T. foetida* (Wallr.) Liro) but was unsatisfactory because of seed injury. That the extent of seed injury was correlated with the extent of seed damage during threshing was established by Nobbe (*1*). This relationship was explained by the work of Brown (*2*), who showed that the uninjured seed coat, though permitting water to pass into the seed, excluded salts such as copper sulphate.

Dry treatment with copper fungicides was not generally adopted until after 1917, when Darnell Smith (*3*), in Australia, recommended basic copper carbonates for the control of bunt. The basic carbonates of commerce are generally assumed to approximate in composition to malachite, $Cu(OH)_2.CuCO_3$, which corresponds to the standard proposed by Mackie and Briggs (*4*). Leukel (*5*) stated that the better grades of copper carbonate marketed in the U.S.A. were of this formula but that cheaper grades contained but 18–25 per cent copper with some sulphate. Southern (*6*) also found that the products used in Australia included basic sulphates and basic chlorides, but he could find no better test than copper content to distinguish the inferior products.

The action of copper salts upon the bunt spore is generally held to follow an adsorption of cupric ions on the spore membrane. Bodnár and Terényi (*7*), in addition to showing that the amounts of copper taken up by bunt spores accorded with the adsorption equation, confirmed earlier reports that leaching of the treated spores with hydrochloric acid restored their ability to germinate. Terényi (*8*) confirmed these results with *Ustilago avenae* but found that, after prolonged steepage, ability to germinate was no longer restored by acid treatment, indicative of a penetration of copper into the spore. The mode of formation of cupric ions from the basic " carbonates " was investigated by Southern who was unable to confirm Pickering's hypothesis that carbon dioxide is the primary agent causing the solution of copper.

The dry copper carbonate treatment was found by Dillon Weston (*9*) to control bunt on slightly contaminated seed and, following a comprehensive test against the older " pickling " or heap methods, Pethybridge and Moore (*10*) reported that the dry method was worth comparison by growers with the older methods. Against other cereal diseases, Tisdale, Taylor, Leukel and Griffiths (*11*) found copper carbonate less effective against covered smut of barley, *U. hordei* (Pers.) Lagerh., and loose smut of oats (*U. avenae*). Heald, Zundel and Boyle (*12*) reported that copper carbonate was effective both for bunt of wheat and smut of oats (*U. levis* (K. & S.) Magn.) (= *U. kolleri* Wille), being more effective on the hull-less varieties of oats than on the commoner varieties.

The use of copper derivatives as seed protectants against *Pythium ultimum* and other soil fungi, was developed by Horsfall (*13*) who first experimented with monohydrated copper sulphate (*14*). Later (*15*) he selected cuprous oxide which, because of its better adherence and high copper content, enabled the seed to retain a greater amount of fungicide. Horsfall, Newhall and Guterman (*16*) enumerated many varieties of plant which benefited by the treatment, but recorded damage to the cabbage family, for which zinc oxide treatment was recommended (*17*). Frictional difficulties met in drilling treated seed were overcome by the addition of powdered graphite as a lubricant (*18*).

REFERENCES

(*1*) Nobbe, F., *Landw. Vers. Sta.*, 1872, **15**, 252.
(*2*) Brown, A. J., *Ann. Bot.*, 1907, **21**, 79.
(*3*) Darnell Smith, G. P., *Agric. Gaz. N.S.W.*, 1915, **26**, 242 ; 1917, **28**, 185.
(*4*) Mackie, W. W. and Briggs, F. N., *Bull. Calif. agric. Exp. Sta.*, 364, 1924, p. 533.
(*5*) Leukel, R. W., *Bot. Rev.*, 1936, **2**, 498.
(*6*) Southern, B. L., *J. roy. Soc. W. Aust.*, 1931–32, **18**, 85.
(*7*) Bodnár, J. and Terényi, A., *Z. physiol. Chem.*, 1930, **186**, 157.
(*8*) Terényi, A., *Z. physiol. Chem.*, 1930, **192**, 274.
(*9*) Dillon Weston, W. A. R., *Ann. appl. Biol.*, 1929, **16**, 86.
(*10*) Pethybridge, G. H. and Moore, W. C., *J. Minist. Agric.*, 1930, **37**, 429.
(*11*) Tisdale, W. H., Taylor, J. W., Leukel, R. W. and Griffiths, M. A., *Phytopathology*, 1925, **15**, 651.
(*12*) Heald, F. D., Zundel, G. L. and Boyle, L. W., *Phytopathology*, 1923, **13**, 169.
(*13*) Horsfall, J. G., *Bull. N.Y. St. agric. Exp. Sta.*, 683, 1938.
(*14*) Horsfall, J. G., *Tech. Bull. N.Y. St. agric. Exp. Sta.* 198, 1932.
(*15*) Horsfall, J. G., *Bull. N.Y. St. agric. Exp. Sta.* 615, 1932.
(*16*) Horsfall, J. G., Newhall, A. G. and Guterman, C. E. F., *Bull. N.Y. St. agric. Exp. Sta.* 643, 1934.
(*17*) Horsfall, J. G., *Bull. N.Y. St. agric. Exp. Sta.* 650, 1934.
(*18*) Arnold, E. L. and Horsfall, J. G., *Bull. N.Y. St. agric. Exp. Sta.* 660, 1936.

Formaldehyde

Formaldehyde, first employed as a disinfectant by Loew in 1888, was introduced as a seed disinfectant by Geuther (*1*) in 1895. Its use for the purposes of seed treatment extended rapidly in N. America following Bolley's (*2*) recommendation in 1897, but in Europe its use did not become widespread until it was employed in Germany, because of the shortage of copper in 1914–18, as a substitute for copper sulphate. In England, its use followed mainly, despite the earlier recommendation of Johnson (*3*), from the work of Salmon and Wormald (*4*). These workers found that the germination injury caused by formaldehyde is less serious than that caused by copper sulphate.

As to the action of formaldehyde upon the fungus spore, little is known. Hailer (*5*) showed that the inhibition of the germination could be prevented by the timely addition of sulphite (the condensation with a sulphite is a characteristic reaction of formaldehyde). With a prolonged action, however, Hailer concluded that an irreversible complex of formaldehyde and the amino groups of the spore protoplasm is formed.

Further, McAlpine (*6*) found that unless the grain was sown immediately after treatment, germination injury became severe, which he attributed to the hardening of the seed coat. As injury is less severe if the grain is dried rapidly, the heap of treated grain should be spread out to dry after treatment.

Formaldehyde appears to be useless as a seed protectant; indeed, Machacek and Greaney (*7*) found that it increased attack of cereals by "Foot-rot" (due to soil fungi such as *Fusarium culmorum* and *Helminthosporium sativum*) probably by retarding the growth of the seedlings.

The treatment of seed potatoes for two hours in a 1 : 240 dilution of 40 per cent formalin was suggested by Arthur (*8*) for the control of scab, *Streptomyces scabies*. To avoid this lengthy immersion, Melhus, Gilman and Kendrick (*9*) tried the effect of increasing temperature and found that a five-minute immersion at 48–50° C was as effective as the cold treatment against both scab and Rhizoctonia disease (*Corticium vagum* B. & C. = *Pellicularia filamentosa* (Pat.) Rogers). Injury to the sprouting of the seed tubers resulted only if the temperature was raised above 55° C. It must be remembered, however, that these potato diseases can also be soil-borne; and that tuber treatment is of secondary importance until it is established that there is no danger of soil infection.

REFERENCES

(*1*) Geuther, T., abstr. in *Bied. Zbl.*, 1896, 25, 879.
(*2*) Bolley, H. L., *Bull. N. Dak. agric. Exp. Sta.* 27, 1897, p. 109.
(*3*) Johnson, J. C., *J. Bd. Agric.*, 1913, **20**, 120.
(*4*) Salmon, E. S. and Wormald, H., *J. Minist. Agric.*, 1921, **27**, 1013.
(*5*) Hailer, E., *Biochem. Z.*, 1921, **125**, 69.
(*6*) McAlpine, D., *Agric. Gaz. N.S.W.*, 1906, **17**, 423.
(*7*) Machacek, J. E. and Greaney, F. J., *Sci. Agric.*, 1935, **15**, 607.
(*8*) Arthur, J. C., *Bull. Ind. agric. Exp. Sta.* 65, 1897, p. 19.
(*9*) Melhus, I. E., Gilman, J. C. and Kendrick, J. B., *Res. Bull. Ia. agric. Exp. Sta.* 59, 1920

The Sulphur Group

Apart from the polysulphides, recommended by Kellerman and Swingle (*1*) and the "Ceres" powder of Jensen, little use has been made of sulphur and its compounds as fungicides for seed treatment. Mackie and Briggs (*2*) found dry treatment with flowers of sulphur effective for the control of bunt and, in 1923, Uppal and Malelu (*3*) successfully used sulphur for the control of grain smut of millet, *Sphacelotheca sorghi* (Link) Clinton, a success which led Howard Jones (*4*) to include sulphur in his trials upon the control of covered smut of barley (*U. hordei*). The simpleness and foolproof character of the process rendered it suitable for recommendation to native Egyptian farmers following but primitive agricultural methods. Howard Jones recommended sulphur treatment of the grain as soon as possible after harvest, for the sulphur afforded a useful measure of protection against insect pests during storage.

REFERENCES

(*1*) Kellerman, W. A. and Swingle, W. T., *Bull. Kans. agric. Exp. Sta.* 12, 1890, p. 27.
(*2*) Mackie, W. W. and Briggs, F. N., *Bull. Calif. agric. Exp. Sta.* 364, 1923, p. 533.
(*3*) Uppal, B. N. and Malelu, J. S., *Agric. J. India*, 1928, **23**, 471.
(*4*) Howard Jones, G., *Bull. Minist. Agric. Egypt., Tech. Sci. Ser.* 142, 1934.

The Mercury Group

The powerful bactericidal action of mercuric chloride (corrosive sublimate, $HgCl_2$) naturally led to its trial as a seed disinfectant. For the treatment of cereal grain it was tested by Kellerman and Swingle (*1*) but without great success. Hiltner (*2*) later found it valuable for the control of Fusarium disease of rye (*Calonectria graminicola* (Berk. & Br.) Wr.). The significance of Hiltner's discovery is that *C. graminicola* is not carried over as spores on the exterior of the grain but as a dormant mycelium within the seed. The fungicide must remain on the seed to prevent infection of the coleoptile. Hiltner's observation thus indicated the possibility of utilizing a protective fungicide for the control of diseases carried over within the seed, of which examples are the Helminthosporia and the loose smuts of wheat, *Ustilago tritici* (Pers.) Rostr. and of barley, *U. nuda*.

For pharmaceutical purposes, mercuric chloride was, on account of its intense poisonous properties, replaced by organomercury derivatives. According to Bonrath (*3*), G. Wesenberg, of the I.G. Farbenindustrie A.-G., first suggested the use of such materials as seed disinfectants. The first record of their use is by Riehm (*4*), who, in 1914, reported " Chlorphenol-mercury " successful for the control of bunt. The development of the organomercury seed disinfectants has been due entirely to the interested commercial firms and, as little of this work has been published, this development can be traced only by a review of the materials which have been placed on the market and which presumably mark definite advances in the investigations of the manufacturers.

The true organomercury derivatives are those in which the mercury atom is attached direct, by one or both valency bonds, to carbon atoms. The general structure of those derivatives found as the active constituents of seed disinfectants is R.Hg.X, where R represents a hydrocarbon with or without substituent groups, and X represents an acidic radical.

Reviewing the more important of the proprietary seed disinfectants, the first was " Uspulun ", placed on the market in 1915 by Fr. Bayer in Germany. The active constituent was described as " chlorphenol-mercury " probably of the structure $Cl(OH)C_6H_3.Hg.OSO_3Na$, and the product, which contained 18.8 per cent metallic mercury, was intended for the steeping or pickling method. As alkali is required for its solution, Uspulun contained sodium hydroxide and a colouring matter was added to conform to poison regulations. Its success led to the introduction of a number of similar products, in particular, " Semesan " of E. I. du Pont de Nemours in the United States in 1924. " Germisan ", introduced about 1920 by the Saccharin Fabrik A.-G., contained 16.1 per cent. metallic mercury in the form of cresylmercuric cyanide, $(HO)(CH_3)C_6H_3.Hg.CN$. Bonrath (*3*) stated that the employment of cyanide, iodide or complex cyanides to satisfy the second mercury valency added to the efficiency of the mercurated cresols.

The adoption of the dry treatment with basic copper carbonate resulted in the introduction, by Fr. Bayer in 1924, of organomercury dusts, of which the first was " Uspulun Trockenbeize ". The active constituent of this product was stated to be an *ortho*-nitrophenol mercury derivative, probably represented by $(HO)(NO_2)C_6H_3.Hg.OH$, the dust containing 3.4 per cent metallic mercury.

The next important group of products are the mercurated hydrocarbons, noteworthy in that their organic radicals, instead of being phenolic, were unsubstituted hydrocarbon groups. " Ceresan ", introduced by the I.G. for the European market, was reported to have as the active ingredient phenylmercuric acetate, $C_6H_5.Hg.O.CO.CH_3$, and to contain 1.5 per cent metallic mercury. " Agrosan G.", introduced by I.C.I. Ltd., contained a like amount of mercury in the form of tolylmercuric acetate, $CH_3.C_6H_4.Hg.O.CO.CH_3$. " Ceresan ", introduced by Bayer-Semesan Co. Inc., for the American market, contained 2.0 per cent ethylmercuric chloride $C_2H_5.Hg.Cl$, though the figure 2.0 may refer to the content of metallic mercury. This product was later replaced, in America, by " New improved Ceresan ", containing 1.3 per cent mercury in the form of ethylmercuric phosphate.

Around 1930, the active component of European " Ceresan ", was altered to what Bonrath termed " an ether of a mercurated alcohol ". The new compound of the general formula, $CH_3O.C_2H_4.HgX$, was methoxyethylmercuric silicate, which has survived as the active component of " Ceresan Universal Trockenbeize " (*6*), the corresponding chloride being used in the product intended for slurry treatments. Substitution of the silicate group by other acidic radicals did not greatly affect fungicidal properties except against *U. avenae* (*5*). Unfortunately, the trade name " Ceresan " is also used by Du Pont in the United States and since 1948 their main product, " Ceresan M ", has contained 3.2 per cent mercury as *N*-(ethylmercuri)-*p*-toluenesulphonamide (I) compounded for use either as a dust or slurry.

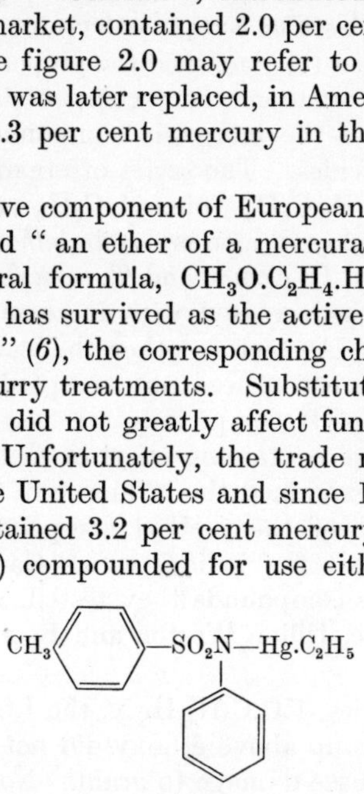

CH_3—C_6H_4—SO_2N—$Hg.C_2H_5$ (with C_6H_5 on N)

(I)

Slurry treatments became popular by reason of the reduced hazards and inconvenience to operators from the dust inevitable in dust treatments, and were re-introduced with " Panogen ", a product originating in Sweden, the active component of which is methylmercuric dicyandiamide $CH_3Hg.NHC(:NH).NHCN$. The related phenylmercury urea is used in the British product " Leytosan " (F. W. Berk & Co. Ltd.).

The dependence of fungicidal efficiency on the molecular structure of the organomercury compounds was first studied by Riehm (*7*), who applied Ehrlich's chemotherapeutic techniques to seed disinfectants. Riehm determined the minimum concentration necessary to inhibit the germination, under standard conditions, of bunt spores, which he termed the *dosis curativa*, by which Ehrlich referred to the minimum concentration which will effect a cure. The maximum concentration without deleterious effect upon germination, Riehm called the *dosis tolerata*, a term by which Ehrlich described the maximum dose which the patient can tolerate. Continuing this work Gassner and Esdorn (*8*) found that the *dosis curativa* of mercuric chloride was 0.025, that of mercuric cyanide was greater than 10 ; that of compounds similar to those present in " Uspulun ", " Germisan " and " Tillantin-R " was between 0.07 and 0.12, whereas that of methyl mercuric iodide was as low as 0.001. The last-named compound, although the most active of the compounds tested, was discarded by Gassner and Esdorn on the score of its highly poisonous character. It will be noted, however, that in this compound, R is a hydrocarbon without substituent groupings. To this observation and to the work of Klages (see *3*), the introduction of the mercurated hydrocarbons as seed disinfectants is due.

Dillon Weston and Booer (*9*) examined the properties of inorganic mercury compounds and of those organomercury compounds in which R is a hydrocarbon. The majority of the inorganic compounds were found to be of little value as seed fungicides. The series of organoderivatives in which R was limited to the tolyl, $CH_3.C_6H_4$–, phenyl, C_6H_5–, ethyl, C_2H_5– and methyl, CH_3– groups, were employed against *Helminthosporium avenae* (Bri. & Cav.) Eid., *H. gramineum*, *U. avenae* and *T. caries*, and the results indicated that fungicidal properties decreased with increase of molecular weight of the hydrocarbon group. As factors other than actual toxicity determine practical value, this conclusion was not accepted as meaning that dusts containing phenyl (or tolyl) compounds are necessarily less effective in disease control than those containing ethyl or methyl compounds. Dillon Weston and Booer also examined the influence of the acidic group X upon the fungicidal efficiency of the methyl compounds and, although their results would suggest that the iodide, which Gassner and Esdorn found the most effective of the compounds they tested, was relatively inferior to the chloride or phosphate, Dillon Weston and Booer derived no conclusions upon the point.

In the alkoxyethyl series, $RO.C_2H_4.HgX$, the I.G. workers (*5*) found that increase of the alkoxy group above ethoxy did not greatly affect fungicidal value but tended to increase damage to grain. Nor did the acidic group X have great effect when X was an inorganic or carboxyacid radical, with the exception of the silicate already mentioned. But if, in X, the second valency of the mercury is attached to nitrogen or sulphur, efficiency is reduced though the resulting compounds are effective for dry seed treatment at 2 per cent mercury. If the mercury is attached to a second carbon atom, as in methoxyethylmercuricarbide, $CH_3O.C_2H_4.HgC{:}CHg.C_2H_4.OCH_3$, activity is further depressed and this compound, employed by one firm for the recovery of mercury from the mother liquors (*10*), becomes effective for dry treatment only at a content of 3 per cent mercury. The I.G. workers considered that the discovery of compounds of greater practical value than the methoxyethyl series was unlikely. In reaching this decision, they bore in mind factors other than fungicidal value, e.g., harmlessness to seed and man, all round efficiency against the controllable seed-borne cereal diseases.

The general chemical properties of these two groups of organomercury compounds, R.Hg.X and $R'O.C_2H_4.Hg.X$, are summarized by Booer (*11*).

Ethyl- and methoxyethyl-mercuric hydroxides are strongly alkaline and give, with strong acids, highly ionized salts of water-solubility similar to the corresponding sodium salts except that ethylmercury iodide is almost insoluble in water. All these compounds are appreciably volatile at ordinary temperatures. The phenyl- and tolyl-mercury radicals resemble silver and their chlorides are almost water-insoluble. In the ethyl-, phenyl- and tolyl-mercury compounds the R–Hg linkage is stable and most of the pure compounds melt or sublime without decomposition. The alkoxy-ethylmercury compounds are not so stable and are decomposed by strong acids to ethylene, alcohol and the mercury salt.

Against diseases other than the seed-borne cereal diseases, mercuric chloride has been mainly used for the treatment of "seed" potatoes. Bolley (*12*) found it a most promising agent against potato scab and it has since been used in the United States for the control of scab, Rhizoctonia disease and powdery scab (*Spongospora subterranea* (Wallr.) Lagerh.). For these purposes it has been found, on the whole, to be more effective than formaldehyde. Leach, Johnson and Parson (*13*) recommended an acid-mercuric chloride treatment, the tubers being immersed for five minutes in a solution of 1 in 500 mercuric chloride containing 1 per cent commercial hydrochloric acid. They pointed out that this treatment will not always give perfect control as the diseases are not exclusively seed-borne. The extension to organomercury disinfectants followed and a short immersion ("instant dip") with suspensions or solutions of these compounds was found by Clayton (*14*) more satisfactory than mercuric chloride treatment. The greater convenience of the instant dip method led to its recommendation by Cairns, Greeves and Muskett (*15*) for the control of *Streptomyces scabies* and by Greeves and Muskett (*16*) against skin spot, *Oospora pustulans* Owen & Wakef. Foister (*17*) used a dip in an organomercury solution for the control of dry rot of potato due to *Fusarium caeruleum* (Lib.) Sacc.

The use of organomercury compounds for seed protection has not extended beyond the recommendations of Brett, Dillon Weston and Booer (*18*) for the treatment of early-sown peas. An example of the protection of cuttings is the use of organomercury dips in the control of pineapple disease of sugar cane (*Ceratostomella paradoxa* Dade = *Endoconidiophora paradoxa* (Dade) Davidson) (*19*, *20*).

Action on the Fungus Spore. As with other heavy metals, the toxic action of mercury is generally ascribed to an adsorption of mercuric ions which, in the case of spores of *T. caries*, was shown by Bodnár and Terényi (*21*) to conform to the general adsorption equation. Upon this hypothesis, the non-ionized mercuric compounds, such as mercuric cyanide, should be deficient in fungicidal properties, a conclusion confirmed by Gassner and Esdorn (*8*) and by Bodnár and Terényi. The latter workers showed, however, that spores treated with mercuric chloride or bromide did not germinate when sown upon moist soil, under which conditions those treated with mercuric acetate germinated. As the amount of adsorbed mercury on the latter spores was greater, the quantity of adsorbed mercury is not the sole criterion of fungicidal efficiency. Bodnár and Terényi suggested that the difference is associated with the adsorption, not of mercuric ions, but of mercuric chloride or bromide molecules which, by virtue of their lipoid solubility, can penetrate within the spore membrane. A similar explanation was put forward by Walker (*22*), who showed that phenylmercuric chloride and allylmercuric chloride were more toxic than mercuric chloride to the protozoon *Colpidium colpoda*. He associated this greater toxicity with the greater lipoid solubility of the organomercury compounds which facilitates diffusion to the site of action within the spore.

An alternative hypothesis, applied by Daines (*23*), suggests that metallic mercury, formed from mercury compounds by reduction in the soil, is the active fungicide. Booer (*11*) examined the reaction of mercury compounds in soil. He found that the organomercury compounds are rapidly decomposed and suggested that the initial reaction is the formation of an organomercury clay by base exchange. The diethylmercury clay then decomposes to mercury diethyl, lost by evaporation, and mercury clay ; the diphenylmercury clay to mercury diphenyl and mercury clay ; the methoxyethylmercury clay, in the presence of acid, to mercury clay and the corresponding mercuric salt. The mercury clay so formed gives rise, by base exchange, to mercuric salts. The mercuric salts, probably as chloride, are reduced to mercurous chloride and metallic mercury ; the mercurous chloride dissociates to mercuric chloride and mercury thus ultimately liberating most of the mercury in the metallic form. A detoxication can be effected by the addition of sulphur which converts the metal to the innocuous and insoluble mercuric sulphide.

It is generally accepted that the site of action of the fungicidal activity of mercury, either as vapour or ion, is at the sulphydryl (–SH) groups of the susceptible enzymes.

Action on the Seed. Apart from one effect, the organomercury disinfectants appear to be harmless to the seed. The exception is a characteristic hypertrophy of cereal seedlings induced by disinfectants containing compounds of the R.Hg.X type (*9*, *24*) when these are overdosed or the treated seeds incorrectly stored. The plumule and radical of the injured seed fail to elongate but remain swollen, a malformation attributed by Sass (*15*) to incomplete mitosis.

It has been frequently claimed (see *26*), as with the basic copper carbonates, that the treatment with organomercury disinfectants has a stimulating effect not only upon germination but upon the whole development of the plant. The proof of this stimulation is difficult, for it must be shown that the improved growth is not due to the destruction of soil fungi and bacteria which are normally present and might cause direct or indirect injury to the germinating seed of the control plot. Sampson and Davies (*27*), by trials with bunt-free seed treated with either " Uspulun ", " Tillantin " or basic copper carbonate, concluded that stimulation was attributable solely to the fungicidal properties of the disinfectant which inhibited the growth of competing organisms.

In efforts to hasten germination and the establishment of the seedling, whereby the risks of infection would be reduced, it has been suggested that growth-promoting substances should be added to seed dressings. Grace (*28*) used them to overcome the ill effects of formaldehyde on germination; Croxall and Ogilvie (*29*) used them with promising results with cuprous oxide and with organomercury seed dressings. Templeman and Marmoy (*30*), however, found no effect of the presence of the phytohormone on germination, tillering or growth of cereals; their use for this purpose has not become general.

Fillers for Mercury Dusts. The inert carrier or filler used for the preparation of organomercury seed dressings should be one giving maximum retention and tenacity of the dust on the grain yet should not interfere with drilling operations. Brett and Dillon Weston (*31*) noted rather wide variations in the amounts retained of the different proprietary dressings they examined, a point emphasized by Fitzgibbon (*32*) who concluded that the best filler was a crystalline powder of an average particle size of 10 microns diameter with an upper limit of 30 microns.

To reduce dustiness, in view of the noxious character of the organomercury compounds, most proprietary seed disinfectants contain an anti-dust, usually of an oil character (e.g., B.P. 241,568). Dillon Weston and Booer (*9*) found that the addition of oil had no deleterious action in the case of the compound and dust they tested.

REFERENCES

(*1*) Kellerman, W. A. and Swingle, W. T., *Bull. Kans. agric. Exp. Sta.* 12, p. 27, 1890.
(*2*) Hiltner, L., *Prat. Bl. PflBau.*, 1915, **18**, 65.
(*3*) Bonrath, W., *Nachr. SchadlBekampf., Leverkusen*, 1935, **10**, 23.
(*4*) Riehm, E., *Zbl. Bakt.*, 1914, ii, **40**, 424.
(*5*) Martin, H. and Shaw, H., *Brit. Intell. Obj. Sub-comm.*, Final Report 1095, 1947.
(*6*) *Bayer Pflanzenschutz Compendium*, Farbenfabriken Bayer A.G., Leverkusen (1954).
(*7*) Riehm, E., *Z. angew. Chem.*, 1923, **36**, 3.
(*8*) Gassner, G. and Esdorn, I., *Arb. biol. Abt.* (*Anst. Reichsanst.*) *Berl.*, 1923, **11**, 373.
(*9*) Dillon Weston, W. A. R. and Booer, J. R., *J. agric. Sci.*, 1935, **25**, 628.
(*10*) Tanner, C. C., Greaves, W. S., Orrell, W. R., Smith, N. K. and Wood, R. E. G., *Brit. Intell. Obj. Sub-comm.*, Final Report 1480, 1947.
(*11*) Booer, J. R., *Ann. appl. Biol.*, 1944, **31**, 340.
(*12*) Bolley, H. L., *Bull. N. Dak. agric. Exp. Sta.* 9, 1893.
(*13*) Leach, J. G., Johnson, H. W. and Parson, H. E., *Phytopathology*, 1929, **19**, 713.
(*14*) Clayton, E. E., *Bull. N.Y. St. agric. Exp. Sta.* 564, 1929.
(*15*) Cairns, H., Greeves, T. N. and Muskett, A. E., *Ann. appl. Biol.*, 1936, **23**, 718.
(*16*) Greeves, T. N. and Muskett, A. E., *Ann. appl. Biol.*, 1939, **26**, 481.
(*17*) Foister, C. E., *Scot. J. Agric.*, 1940, **23**, 60.
(*18*) Brett, C. C., Dillon Weston, W. A. R. and Booer, J. R., *J. agric. Sci.*, 1937, **27**, 53.
(*19*) McMartin, A., *S. Afr. Sug. J.*, 1944, **28**, 509.
(*20*) Wiehe, P. O., *Rev. agric. Maurice*, 1947, **26**, 138.
(*21*) Bodnár, J. and Terényi, A., *Z. physiol. Chem.*, 1932, **207**, 78.
(*22*) Walker, E., *Biochem. J.*, 1928, **22**, 292.
(*23*) Daines, R. H., *Phytopathology*, 1936, **26**, 90.
(*24*) Crosier, W., *Phytopathology*, 1934, **24**, 544.
(*25*) Sass, J. E., *Phytopathology*, 1937, **27**, 95.
(*26*) Stützer, A., *Dtsch. Landw. Presse*, 1918, **45**, 361.
(*27*) Sampson, K. and Davies, D. W., *Ann. appl. Biol.*, 1928, **15**, 408.
(*28*) Grace, N. H., *Canad. J. Res.*, Sect. C., 1938, **16**, 313.
(*29*) Croxall, H. E. and Ogilvie, L., *J. Pomol.*, 1940, **17**, 362.

(*30*) Templeman, W. G. and Marmoy, C. J., *Ann. appl. Biol.*, 1940, **27**, 453.
(*31*) Brett, C. C. and Dillon Weston, W. A. R., *J. agric. Sci.*, 1941, **31**, 500.
(*32*) Fitzgibbon, M., *J. Soc. chem. Ind. Lond.*, 1943, **62**, 8.

Miscellaneous Seed Disinfectants

No organic compound has yet been found able to compete with the organomercury compounds for the seed treatment of cereals, mainly because the competitors have been found effective only against a limited range of the diseases against which the mercury compounds are effective. The poisonous properties of the mercury compounds is a defect which makes the prospect of a non-poisonous seed disinfectant attractive but, speaking generally, the organic fungicides of Chapter VII lack the qualities required for cereal seed treatment though some have wide use for the treatment of vegetable seed. In most cases the function of the fungicide here becomes not to disinfect the seed of seed-borne pathogens but to protect the germinating seed and seedling from soil-borne pathogens. **Thiram** (tetramethylthiuram disulphide, see p. 129), for example, gave a satisfactory control of *Colletotrichum lini* (Westerd.) Toch on flax seed, but it was less effective than the organomercury dressings against *Polyspora lini* Laff. In proprietary products such as " Nomersan ", it is used primarily for seed protection where its action may involve the general soil microflora. Richardson (*1*), for example, found a reduction in fungal species and the survivors were predominantly species of *Trichoderma* and *Penicillium*. As protection of pea seed from *Pythium ultimum* was achieved for periods long after the content of thiram had fallen below levels toxic to the pathogen, it would appear that this lasting protection is due to the activities of the fungicide-resistant soil fungi. These survivors are known to produce, at least in culture media, metabolic products, such as gliotoxin, powerfully toxic to the pathogenic fungi.

Chloranil (tetrachloro-*p*-benzoquinone) was selected in trials at the Crop Protection Institute (*2*) for the protection of lima beans from damping-off and has since been used, under U.S.P. 2,349,771, for the treatment of the seed of many other vegetables. It is the active component of " Spergon " (*3*). The related dichlone is described on p. 136 for its main use is for foliage protection. Of other quinones described for seed treatment is **quinone oxime benzoylhydrazine** (II) introduced in 1954 by Farbenfabriken Bayer as the active component of " Cerenox ".

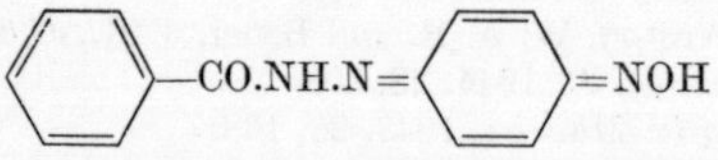

C_6H_5—CO.NH.N=C_6H_4=NOH

(II)

Urbschat and his colleagues (*4*) had found that quinone acyl hydrazones containing the grouping –CO–NH–N = had promising fungicidal properties and that the mono-oximes, though useless against cereal pathogens, were highly effective against soil borne diseases such as *Phoma betae* (Oudem.) Frank. The wide range of the biological activity of II, examined by Frohberger (*5*), prompted its introduction for vegetable seed protection.

In 1945 the use of hexachlorobenzene (III) for the control of bunt of wheat (*Tilletia caries*) was reported in France (*6*). The compound had little effect on other cereal born pathogens but it had the merit of controlling not only the seed-borne spores of common bunt but also the soil-borne spores

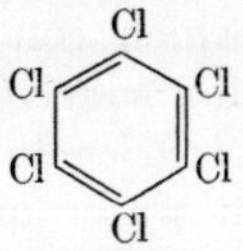

$$C_6Cl_6$$

(III)

of this bunt and of dwarf bunt (*Tilletia contraversa* Kühn) (*7*). Because of the importance of the soil phase the organomercury compounds are ineffective against dwarf bunt (*8*).

REFERENCES

(*1*) Richardson, L. T., *Canad. J. Bot.*, 1954, **32**, 335.
(*2*) Cunningham, H. S. and Sharvelle, E. G., *Phytopathology*, 1940, **30**, 4.
(*3*) Felix, E. L., *Phytopathology*, 1942, **32**, 4.
(*4*) Petersen, S., Gauss, W. and Urbschat, E., *Angew. Chem.*, 1955, **67**, 217.
(*5*) Frohberger, P. E., *Phytopath. Z.*, 1956, **27**, 427.
(*6*) Yersin, H., Chomette, A., Baumann, G. and Lhoste, J., *C. R. Acad. Agric. Fr.*, 1945, **31**, 24.
(*7*) Siang, W. N. and Holton, C. S., *Plant. Dis. Reptr.*, 1953, **37**, 63.
(*8*) Purdy, L. H., *Plant. Dis. Reptr.*, 1957, **41**, 916.

PHYSICAL METHODS

The development of the use of heat as a seed disinfectant was due to Jensen (*1*). In 1882, he demonstrated that the hyphae and spores of *Phytophthora infestans* on diseased potato tubers could be killed by a four-hour treatment in a current of hot air at 40° C. His method was used against potato blight until 1886, the date of the introduction of Bordeaux mixture.

In 1887, for the prevention of smut in oats and barley, Jensen (*2*) found that the chemical steeps then known were ineffective, but that treatment with hot water gave, at least with oats, a measure of control. Prior to this, Sinclair (*3*) had suggested treatment with hot air from an oven, a process which Jensen found ineffective even after an exposure for seven hours at 51.5–54° C. Jensen, noting that the control of the loose smut of barley was not so complete as that secured by his method with oats and that hot-air treatment was ineffective, concluded that moisture plays some role other than that of heat transfer. He considered that the smut " germ " when moistened was more vulnerable and so proposed his " modified hot-water treatment " in which the seed is immersed in cold water before treatment with hot water. This pre-treatment was placed on a scientific basis by Appel and Riehm (*4*), who showed that, by a four-hour treatment in water at 20–30° C, the dormant mycelium develops activity and is more easily killed by a short exposure at 50–52° C. The modified hot-water treatment is, despite its cumbersomeness, still used against the loose smut of wheat, *Ustilago tritici*, and of barley, *U. nuda*, in both of which diseases the fungus is carried over as dormant mycelium within the seed.

Hot-water treatment has also been found effective for the control of certain nematodes. Ramsbottom (*5*) showed that the immersion of narcissus bulbs for one hour in water at 110° F reduced attack by *Anguillulina dipsaci* (Kuhn) Gerv. & v. Ben. In its present form (*6*) treatment is for three hours at 110° F, under which conditions the eelworms within the bulbs are killed whilst the bulbs survive. Similarly, for the control of eelworm of chrysanthemum, *Aphelenchoides ritzema-bosi* Schwartz, the infested stocks are immersed in water maintained at 110° F for twenty minutes (*7*, *8*, *9*). Another application of the method (*10*) is for the control of tarsonemid mite of strawberry (*Tarsonemus fragariae* Zimm.), by the immersion of the runners in water held at 110° F for twenty minutes (*11*). In view of the high phytotoxic properties of temperature above 110° F, Massee preferred to call the method a warm-water treatment.

The possibility that heat treatment might free plant stock material from virus was suggested in 1923 by Wilbrink (*12*) who immersed sugar cane cuttings in water at 52° C for 20–30 minutes for the control of Sereh disease. Various treatments have been used in other virus-infected " seed " such as potato but without success though Kunkel (*13*) obtained healthy stock from virus-infected peach budsticks by warm-water treatment. He also noted that trees kept at 35° C for periods of a fortnight or longer were free of peach yellows. He suggested that the absence of peach yellows in lowlands south of Washington, D.C. may be because of this heat inactivation. In most cases, however, the practical use of heat for virus control is frustrated by the ill effects of the treatment on the host plant. Miller (*14*), for instance, recorded a fatality rate of 70 per cent which would rule out the commercial use of the method for the treatment of strawberry runners though it might prove a possible means of obtaining disease-free clones.

REFERENCES

(*1*) Jensen, J. L., *Mem. Soc. nat. Agric. France*, 1887, **131**, 31.
(*2*) Jensen, J. L., *J. R. agric. Soc.*, 1888, **24**, 397.
(*3*) Sinclair, J., *Code Book of Agriculture*, 5th Ed., London, 1832, p. 58.
(*4*) Appel, O. and Riehm, E., *Arb. biol. Abt.* (*Anst. Reichsanst.*) *Berl.*, 1911, **8**, **343**.
(*5*) Ramsbottom, J. K., *J. R. hort. Soc.*, 1918, **43**, 51, 65.
(*6*) Staniland, L. N. and Barber, D. R., *Bull. Minist. Agric.* 105, 1937.
(*7*) Hodson, W. E. H., *Hort. Educ. Assoc. Year Book*, 1933, **2**, 85.
(*8*) Edwards, E. E., *J. Helm.*, 1934, **12**, 23.
(*9*) Kearns, H. G. H. and Walton, C. L., *Rep. agric. hort. Res. Sta. Bristol.* 1933, p. 66.
(*10*) Hodson, W. E. H., *Pamph. Seale Hayne Coll.* 36, 1930, p. 9.
(*11*) Massee, A. M., *Rep. E. Malling Res. Sta.*, 1933, p. 256.
(*12*) Wilbrink, G., *Arch. Suikerind. Ned.-Ind.*, 1923, **31**(3), 1.
(*13*) Kunkel, L. O., *Phytopathology*, 1936, **26**, 809.
(*14*) Miller, P. W., *Plant Dis. Reptr.*, 1953, **37**, 609.

CHAPTER XIV

SOIL TREATMENT

UNDER " Soil Treatment " are classed those methods of pest control which aim at the destruction of the pest in the soil itself. The pests include root-eating insects such as wireworm (the larvae of click beetles, Elateridae), parasitic nematodes, soil fungi and those " biological factors " at one time regarded as responsible for soil sickness. The latter condition, described in more detail below, is now corrected by the partial sterilization of the soil by physical and chemical methods and to this subject first attention is given. There follow chemical control of specific soil pests and pathogenic fungi and, finally, those mechanical methods of which the most important is " Crop Rotation " which, by withholding host plants, hinders the multiplication of the harmful organism.

PARTIAL STERILIZATION

Partial Sterilization has assumed great importance, especially to the horticulturist, for not only are weed seeds and other harmful organisms destroyed, but there is a great increase in the fertility of the treated soil. Indeed in certain cases, partial sterilization of the soil has now become, not so much a measure of pest control, but a cultural operation profitable because of its relationship to the nutrient supply. It would be impossible to enter at all fully into this " manurial " aspect of Partial Sterilization.

" Soil Sickness " is a popular term applied to the condition of a soil which although showing no lack of nutrient material fails to yield a satisfactory crop. It generally appears in soils which under an intensive system of cultivation have grown one crop continually, conditions which encourage the growth of the disease organisms and pests of that crop. The condition may be due, in part, to the deterioration of the physical properties of the soil by heavy dressings of artificial manures and watering and may then be alleviated by the incorporation of straw in the soil by the method described by Bewley (*1*). In other cases, soil sickness has been traced to a superabundance in the soil of definite organisms, for example, root knot due to the eelworm *Heterodera marioni*, flax sickness due to the fungus *Fusarium lini*, clover sickness, one type of which is due to the eelworm *Anguillulina dipsaci* and another type to the fungus *Sclerotinia trifoliorum* Erikss. The failure of lucerne (" alfalfa fatigue ") has been traced to the presence in the soil of a bacteriophage which, by attacking the lucerne strain of *Rhizobium* prevents nitrogen fixation (*2*). By controlling these biological " limiting factors ", partial sterilization becomes a true method of pest control.

The first observations of the increase of yield following partial sterilization were made independently by Girard (*3*) and by Oberlin (*4*) in 1894. These workers had employed carbon disulphide as their sterilizing agent. Similar results were found after the application of other volatile antiseptics, and in 1907, Darbishire and Russell (*5*) showed that partial sterilization by heat produced a like result.

Heat

It has been suggested that the practice of exposure of the soil to the heat of the sun, still carried out in India, has as its basis partial sterilization. Prescott (*6*) suggested that the " sheraqi " fallow, a feature of Egyptian cultivation during which the soil is left fallow at the hot period, may secure a partial sterilization of the soil. Drying out of the soil may bring about similar results and Lebediantzeff (*7*) showed that the fertility of the Russian soils with which he worked was increased by air-drying.

Practical difficulties render the partial sterilization of large areas of soil by heat impossible. Fortunately those areas which show the greatest need for treatment are sufficiently small for the purpose. It is mainly in glasshouses, seed and frame beds, especially where the same crop is grown continually, that the climatic and nutritive conditions are so favourable for the rapid multiplication of soil pests and fungi. It is just these soils which show the greatest need for partial sterilization.

To secure the beneficial effects of partial sterilization the soil should be maintained at a temperature of 97° C (207° F) for one hour. Various methods are now available for the economical heating of the soil to the required temperature (*8*). For the small grower the baking of small quantities of soil in a suitable oven is satisfactory, provided that a careful control of temperature is possible. It is of course necessary to guard the soil from re-infection after treatment. The labour of moving the soil is saved by the use of steam as the heat-carrying agent. The general procedure of such methods is to conduct steam, at high or low pressure, into the soil until the temperature required is reached and maintained. In the United States, the " inverted steam pan " method, introduced in 1909 by Gilbert, is a popular method of soil treatment by heat (*9*). Electrical methods of partial sterilization or pasteurization of batches of soil are available (*10*).

Chemical Methods

Theoretically the application of a toxic chemical to the soil should be a far simpler matter than its application to foliage. It is possible to apply the material when the soil is fallow and provided that it disappears by volatilization or by decomposition before planting there should be no damage to the plant. The greatest difficulty is to secure the penetration of the toxic agent to the greatest depth occupied by the organisms against which it is to act. The penetration of water-soluble agents may be hastened by watering but normally diffusion in solution or in vapour phase is the only route. The physics of diffusion in soil were examined by Penman (*11*) and, applying his techniques, van Bavel (*12*) found that diffusivity of a vapour through a porous medium such as soil is only about one half as fast as in a gaseous medium. The diffusion of a water-insoluble vapour through soil is of course reduced in wet soil and Flegg (*13*) showed that this reduction is approximately proportional to the reduction in porosity due to the volume of water present.

In the treatment of soil sickness it is usual to describe the chemical employed as an " antiseptic ", whereas if volatile and against a definite pest, the name " soil fumigant " is more frequent. The requirements of such antiseptics and soil fumigants may be summarized briefly thus: (i) cheapness ; (ii) good penetration into soil ; (iii) non-injuriousness to the plant either direct or through a deleterious action upon the physical properties of the soil. The literature and uses of early soil insecticides have been reviewed by Gough (*14*).

Dealing with those chemicals of practical use for partial soil sterilization in its broadest sense, one of the most interesting historically is :

Carbon Disulphide. This substance was originally employed in 1872 by Thenard, against phylloxera of the vine (*Phylloxera vitifoliae* Fitch) by injection of the required amount of carbon disulphide into the soil at distances around the infected vine. The treatment was so successful that by 1887 over 66,000 ha. of land were so treated in France alone (*15*).

Carbon disulphide has the advantages of cheapness, volatility and good insecticidal properties. It was, at one time, thought that the density of its vapour would permit it to sink into the soil but Higgins and Pollard (*16*) concluded that the movement of the vapour was largely a simple diffusion process obtaining no evidence of marked gravitational flow. The diffusion was also studied by O'Kane (*17*) and by Fleming (*18*), the general conclusion being that carbon disulphide is rapidly lost by evaporation, a loss not adequately prevented by covering the surface with sacking. Further, it was found that though penetration is good in sandy or humus soils of not too great a moisture content, it was inefficient in moist, or heavy clay soils. Fleming deduced the interesting point that part of the original carbon disulphide is taken up by the soil and that by a second treatment, the persistence of the vapour in the soil is greatly enhanced.

To prevent rapid loss of the carbon disulphide by evaporation, a number of compounded products have been marketed notably in Europe. " Sulphoergethan ", a preparation of carbon disulphide and tetrachloroethane in the form of a cake which, when buried in the soil, slowly evolves carbon disulphide was found unsatisfactory by Börner and Thiem (*19*). The latter workers also found that a trade preparation of carbon disulphide and nitrobenzene under the name " Horlin " was no more efficient than carbon disulphide alone. To lessen the fire hazard of carbon disulphide, which adds greatly to transit costs and risks of use, Truffaut (*20*) proposed the employment of an emulsion of carbon disulphide. Leach, Fleming and Johnson (*21*) recommended the use of an emulsion of carbon disulphide and wormseed oil for the control of the soil stages of the Japanese beetle.

A modification of the carbon disulphide treatment was suggested in 1874 by Dumas (*22*), who proposed the use of potassium sulphocarbonate (thiocarbonate), K_2CS_3. This material was marketed in France as a dark-red aqueous solution containing 18–20 per cent potash. Penetration into the soil is good provided that the soil be sufficiently watered. Under the action of carbon dioxide, the compound decomposes with the evolution of carbon disulphide. The material is therefore slower in action than carbon disulphide but is more persistent. Despite a plea by Molinas (*23*) for its continued use, its employment has not spread beyond the vine-growing centres of France. The cost of the treatment is offset by the manurial value of the potash present. Indeed as the effects of any form of partial sterilization of the soil on the succeeding crop are similar to those of a nitrogenous manuring, the need for potash becomes greater.

Coal-tar Antiseptics. The commercial coal-tar antiseptics, in use as disinfectants, suggested themselves for partial sterilization of soil. Hiltner (*24*) proposed the use of carbolineums (see p. 187) and Molz (*25*) found them excellent for soil treatment against nematodes. An objection to their use, pointed out by Nostitz (*26*) is damage to plant growth unless applied some five months before sowing.

Russell (*27*) and his colleagues found a cheap and effective antiseptic in cresylic acid (liquid carbolic acid, pale straw coloured, 97–99 per cent purity), a product of the crude tar acids consisting mainly of phenol and the three isomeric cresols. These compounds are decomposed with sufficient rapidity in the soil. The diluted cresylic acid is either " watered " on or is mixed with dry soil containing gypsum and then dug in. Of the constituents of cresylic acid, phenol and *o*-cresol have about the same toxic value (*28*) but the *m*- and *p*-cresols do not appear to be so effective (*29*).

A systematic examination was carried out at Rothamsted of the pure aromatic derivatives which are derived from the tar oils. It was found that the introduction of chlorine into the aromatic compound led to an increased toxicity. Dichlorocresylic acid, for example, was more effective than cresylic acid. Extending this work, Matthews (*28*) found that the chlorine group by itself or with a nitro group produced high stability but that one or two nitro groups and one chlorine group gave an unstable product. Thus, chlorodinitrobenzene was superior in toxicity towards eelworm, fungi and protozoa to nitrodichlorobenzene and was more easily and rapidly decomposed in the soil. According to Russell (*30*), however, this decomposition is not sufficiently rapid to prevent injury to plants. The stability of highly chlorinated hydrocarbons such as DDT and BHC has already been noted (see pp. 202, 210).

p-Dichlorobenzene was recommended in 1919 by Blakeslee (*31*) for the control of the peach-tree borer *Sanninoidea exitiosa* Say. He found other soil fumigants unreliable against this pest under field conditions but this compound, though sufficiently volatile to prove toxic to the larvae, did not injure trees over six years of age. According to Essig (*22*), the best results are at soil temperatures of 75–83° F, and in regions where such soil temperatures are usual.

Halogenated hydrocarbons. The increased activity due to the introduction of chlorine into the molecule is again well shown in the case of the chlorinated hydrocarbons. Schwaebel (*33*) showed that, whereas an increase in yield follows treatment of the soil with both dichloroethylene, CHCl:CHCl, and trichloroethylene $CHCl{:}CCl_2$, tetrachloroethylene $CCl_2{:}CCl_2$ is so toxic that it brings about a reduction of fertility. 1 : 3-dichloropropene (dichloropropylene), $CHCl{:}CH.CH_2Cl$, constitutes about one third of the proprietary product " D–D Soil Fumigant ", first reported by Carter (*34*) as an effective soil fumigant. A typical analysis of this product, introduced by the Shell Chemical Corp. and a by-product of plastic manufacture, was given by Chisholm (*35*). Carter used it on pineapple soils infested by nematodes, larvae of *Anomala orientalis* Waterh. and fungi (*Pythium* spp.). He found (*36*) that the propene is more insecticidal than the other main component of the mixture, 1 : 2-dichloropropane, but that the two mixed were better than either alone. " D–D Soil Fumigant " has been tried under English conditions for the control of the eelworm *Heterodera rostochiensis* Woll., but though giving higher potato yields left a higher eelworm population than before treatment (*37*). Peters (*38*) obtained improved growth of potatoes in eelworm-free soil following treatment with D–D Mixture, an effect he attributed to partial sterilization. 1 : 3-Dichloropropene itself was introduced in 1956 as a nematicide by the Dow Chemical Co. under the trade name " Telone ". The technical product contains both the α and β isomers and, admixed with ethylene dibromide in the proportions 5 : 1 by volume, forms the Dow product " Dorlone ".

The use of methyl bromide as a soil fumigant has already been mentioned (p. 252). Its fungicidal properties are seemingly unique among soil fungicides for it is reported (*39*) to control soil fungi such as *Sclerotinia sclerotiorum* (Lib.) de Bary, the sclerotia of which are resistant to most chemicals. 1 : 2-Dibromo-3-chloropropane was first described as a nematicide by McBeth and Bergeson (*40*) in 1955. It is an active component of the trade products " Fumazone " (The Dow Chemical Co.) and " Nemagon " (Shell Development Co.) and is sparingly soluble in water. Because of its relatively high boiling point (196° C) its nematicidal effects are slow in appearing and Ichikawa *et al.* (*41*), in studying the diffusion pattern through soil, reported that the larvae of the root knot nematode (*Meloidogyne* sp.) though apparently unaffected, were unable to infest tomato roots.

The association of chlorine and the nitro group in chloropicrin (trichloronitromethane, CCl_3NO_2) gives a highly efficient soil fumigant (*42*) but this compound is too hazardous for ordinary use.

Formaldehyde. Formaldehyde was apparently first recommended for soil treatment by Gifford (*43*), who found it more effective, for his particular purpose, than steam. It is volatile and soluble in water, appearing on the market in aqueous solution under the name Formalin. The solution should contain not less than 37.5 and not more than 40.5 per cent weight in volume of formaldehyde (*44*) and up to 14 per cent methyl alcohol is sometimes added to delay polymerization to paraformaldehyde.

A special advantage of formaldehyde was revealed by Hunt, O'Donnell and Marshall (*45*), who showed that its penetration is apparently equal to that of the water carrying it in solution. This result would indicate that, with formaldehyde, there is no adsorption of the toxic material on the soil, the cause of the poor penetration of most of the soil disinfectants.

The disadvantages of formaldehyde, of which usually 2–3 quarts of a 0.5 per cent solution are added per square foot of soil, are the general objection that the soil takes too long to re-dry ; that the fumes of formaldehyde may cause a retardation or reduction of germination ; but more important that formaldehyde, although an efficient soil fungicide, has but slight insecticidal properties.

Miscellany. The utilization of the wide variety of insecticides and fungicides which have been met in previous pages for soil treatment has proved successful in limited cases. Hydrocyanic acid, for example, clearly had to be tried as a soil fumigant and potassium cyanide was indeed suggested by Mamelle (*36*) as a substitute for carbon disulphide. De Ong (*47*) found, however, that, because of the variable manner in which the gas was adsorbed by the soil, it was impossible to estimate the dosage with sufficient accuracy to avoid damage to the host plant. Calcium cyanide was found successful against pear midge larvae in New Zealand (*49*) but its widest use has been for the control of wireworms. To reduce the cost of treatment to workable levels, Campbell (*49*) introduced pre-baiting, a process in which the area is sowed in spaced rows with seeds which by their germination attract the wireworms so that the pest is destroyed with lower amounts of cyanide.

The introduction of BHC naturally led to tests of its use against wireworm with outstanding success (*50*) due not only to a reduction in wireworm number but to the effect of the insecticide in rendering the wireworms incapable of attacking the crop plant (*51*). Because it taints certain food crops BHC was replaced by lindane which, with the cyclodiene group, is extremely effective against many soil insects at rates as low as a few ounces per acre.

Mercuric chloride was found effective by Preston (*52*) for the control of " clubroot " of brassicas, caused by *Plasmodiophora brassicae*, a dilute solution being used to water-in the young cabbage plants at transplanting from the seed bed. Protection from the cabbage root fly *Hylermia brassicae* Bch. is also obtained (*53, 54*) if the treatment is continued at intervals after transplanting. The intensely poisonous mercuric chloride was replaced by mercurous chloride (calomel) which Glasgow (*55*) found effective against diptera infesting vegetable crops. The reactions of both mercuric and mercurous chlorides in soil were examined by Booer (*56*) (see p. 264) and with both the active insecticide or fungicide is probably metallic mercury. The final immobilization of mercury as a biologically inert mercuric sulphide removes the hazard that the continued use of calomel will render the soil infertile.

Effects of Partial Sterilization on Plant Growth

Apart from the direct effect of heat and of chemicals upon soil pests and pathogens, there results from partial sterilization certain changes in the microflora and fauna and in the chemical and physical properties of the soil which exert that extraordinary influence upon plant growth which has rendered partial sterilization of such importance. An early theory, due to Koch (*57*), suggested that a small amount of toxicant acts as a stimulant to the plant. Such a theory fails, however, to account for the lasting effect of partial sterilization.

In the study of the increased fertility of partially-sterilized soils, attention has been mainly directed to the rapid multiplication of bacteria in the treated soil, first noted by Hiltner and Störmer (*58*). The general course of events following treatment is a temporary reduction in bacterial numbers for the first few days with a subsequent rise dependent on the treatment. Hiltner and Störmer explained the bacterial increase by the theory that carbon disulphide and other poisons upset the equilibrium of the soil bacteria to the benefit of certain groups. It is probable that partial sterilization by heat has a similar result for the non-sporing bacteria will be killed by heat while the spores survive—the ammonifying bacteria are spore formers. The initial fall of bacterial numbers therefore follows from the simplification of the bacterial flora which leaves the survivors free of the competition which, in the untreated soil, limits their numbers.

Russell and Hutchinson (*59*) suggested that the limiting factor was the protozoa of the soil. The status of protozoa as soil organisms had till that time received but little attention, and, as a result of this suggestion, extended observations were made at Rothamsted upon the part played by these organisms. It was found by Cutler (*60*) that, owing to the adherence of these organisms to the soil particles, the number present in normal soil had previously been underestimated. Daily counts of soil bacteria and soil protozoa revealed an inverse relationship between the numbers ; when the bacterial numbers were low the number of trophic amoebae was high and vice versa. Cutler and Crump (*61*) regarded this observation as a clear demonstration of the relationship of protozoa and bacteria put forward by Russell and Hutchinson.

Waksman and Starkey (*62*) were unable to support the protozoa theory except in the case of certain abnormal soils of high moisture and organic matter content. They held that account should be taken of the part played by soil fungi and actinomycetes and they suggested that in normal soils much of the decomposition of organic matter is carried out by fungi. After partial sterilization, the bacteria are left to carry on this decomposition and as they assimilate less of the carbon and nitrogen, a larger amount of ammonia is produced. They traced a relationship between soil fungi and bacteria similar to that observed by Cutler and Crump between soil protozoa and bacteria.

Other workers have found the assumption of the destruction of a biological limiting factor not necessary to account for the increase of bacterial numbers. Not only is there a simplification of the soil flora and fauna to the advantage of the survivors but the material of the organisms killed by the treatment becomes available as food supply for the bacteria. Further, Richter (*63*) observed in soils, partially sterilized by heat, that though the total nitrogen content was unaltered, there was an increase of readily available nitrogen and a decomposition of the organic matter. Greig Smith (*64*) suggested that carbon disulphide, when applied to soil, acts as a solvent for the soil wax or " agricere " which normally " waterproofs " the decomposable organic matter. The treatment therefore renders this supply of nutrient available to the surviving bacteria. A new concept was introduced by Matthews (*65*), who showed that the relative extent of the increase of bacterial numbers in the early days following treatment with aromatic hydrocarbons, is governed by the toxic chemical employed. The extent of increase was proportional to the heat of combustion of the chemical used. Matthews therefore concluded that the increase of bacterial numbers is largely due to the nutrient value (as given by the heat of combustion) of the antiseptic.

Secondary effects of partial sterilization upon the plant are of importance. Schulze (*65*) observed that the development of the plant in soils, partially sterilized by heat, is, in the initial stages of growth, slower than that of the untreated control. Later, however, the plant on the treated soil showed the more luxuriant growth. The temperatures Schulze employed were higher than usual for partial sterilization but Pickering (*67*) showed that working with temperatures from 60–150° C, there is a retardation of germination in the treated soil which he attributed to the formation of a toxic material, probably a nitrogenous compound, which slowly loses some of its inhibitory properties. Russell and Petherbridge (*68*) considered the retardation, which they also found in partial sterilization by means of chemicals, due to changes in the soluble soil constituents which, though useful to older plants, are detrimental to the delicate processes of germination. The precautions necessary to reduce the toxicity of steam-sterilized soil used for the preparation of composts for nursery work were studied by Lawrence and Newell (*69*). Johnson (*70*) attributed the damage to the formation and accumulation of ammonia, Pickering's nitrogenous compound, which is now generally accepted as the main cause, subsidiary causes being the increase in water-soluble organic matter and of available manganese (*71*). Damage may be reduced by suitable pre-sterilization treatment, the avoidance of liming and the use of superphosphate (*69*).

Finally, in the case of leguminous plants (which differ from others in that to take up their nitrogen they enter into a symbiotic relationship with certain nitrogen-fixing bacteria), to secure a satisfactory crop after partial sterilization it is generally necessary to re-inoculate the treated soil with the proper bacteria.

REFERENCES

(*1*) Bewley, W. F., *Rep. exp. Sta. Cheshunt*, 1933, p. 27.
(*2*) Demolon, A. and Dunez, A., *C. R. Acad. Sci. Paris*, 1933, **197**, 1344.
(*3*) Girard, A., *Bull. Soc. nat. Agric. France*, 1894, **54**, 356.
(*4*) Oberlin, C., *Bodenmüdigkeit und Schwefelkohlenstoff*, Mainz, 1894.
(*5*) Darbishire, F. V. and Russell, E. J., *J. agric. Sci.*, 1907, **2**, 305.
(*6*) Prescott, J. A., *J. agric. Sci.*, 1919, **9**, 216.
(*7*) Lebediantzeff, A., *C. R. Acad. Sci. Paris*, 1924, **178**, 793.
(*8*) For detailed methods see *Bull. Minist. Agric.* 22, 1939 ; Lawrence, W. J. C. and Newell, J., *Seeds and Potting Composts*, 2nd Ed. Allen and Unwin Ltd., London, 1941.
(*9*) See Hunt, N. R., O'Donnell, F. G. and Marshall, R. P., *J. agric. Res.*, 1925, **31**, 301.
(*10*) Newhall, A. G. and Nixon, M. W., *Bull. Cornell agric. Exp. Sta.* 636, 1935.
(*11*) Penman, H. L., *J. agric. Sci.*, 1940, **30**, 437, 570.
(*12*) van Bavel, C. H. M., *Soil Sci.*, 1952, **73**, 91.

(*13*) Flegg, P. B., *J. Sci. Food Agric.*, 1953, **4**, 104.
(*14*) Gough, H. C., *A Review of the Literature on Soil Insecticides*, Imp. Inst. Entom., London, 1945.
(*15*) Vogt, E., *Zbl. Bakt.*, 1924, ii, **61**, 323.
(*16*) Higgins, J. C. and Pollard, A. G., *Ann. appl. Biol.*, 1937, **24**, 895.
(*17*) O'Kane, W. C., *Tech. Bull. N.H. agric. Exp. Sta.* 20, 1922.
(*18*) Fleming, W. E., *Bull. N. J. agric. Exp. Sta.* 380, 1923.
(*19*) Börner, C. and Thiem, H., *Mitt. biol. Abt.* (*Anst. Reichsanst.*) *Berl.*, 1921, **21**, 167.
(*20*) See Russell, E. J., *J. R. hort. Soc.*, 1920, **45**, 237.
(*21*) Leach, B. R., Fleming, W. E. and Johnson, J. P., *J. econ. Ent.*, 1924, **17**, 361.
(*22*) Dumas, J. B. A., *C. R. Acad. Sci. Paris*, 1874, **79**, 645.
(*23*) Molinas, E., *Progrès agric. vitic.*, 1914, **31**, 374.
(*24*) Hiltner, L., *Jb. Ver. angew. Bot.*, 1907–08, p. 200.
(*25*) Molz, E., *Zbl. Bakt.*, 1911, ii, **30**, 181.
(*26*) Nostitz, A. von, *Landw. Jb.*, 1915, **48**, 587.
(*27*) See Russell, E. J. and Petherbridge, F. R., *J. Bd. Agric.*, 1913, **19**, 809.
(*28*) Matthews, A., *J. agric. Sci.*, 1924, **14**, 1.
(*29*) Buddin, W., *J. agric. Sci.*, 1914, **6**, 417.
(*30*) Russell, E. J., *J. R. hort. Soc.*, 1920, **45**, 237.
(*31*) Blakeslee, E. B., *Bull. U.S. Dept. Agric.*, 796, 1919.
(*32*) Essig, E. O., *Bull. Calif. agric. Exp. Sta.* 411, 1926.
(*33*) Schwaebel, —. *Zbl., Bakt.*, 1923, ii, **60**, 316.
(*34*) Carter, W., *Science*, 1943, **97**, 383.
(*35*) Chisholm, R. D., *U.S. Dept. Agric.*, Yearb. 1952, p. 331.
(*36*) Carter, W., *J. econ. Ent.*, 1945, **38**, 35.
(*37*) Peters, B. G. and Fenwick, D. W., *Ann. appl. Biol.*, 1949, **36**, 364.
(*38*) Peters, B. G., *J. Helminth.*, 1948, **22**, 128.
(*39*) Newhall, A. G. and Lear, B., *Phytopathology*, 1948, **38**, 38.
(*40*) McBeth, C. W. and Bergeson, G. B., *Plant Dis. Reptr.*, 1955, **39**, 223.
(*41*) Ichikawa, S. T., Gilpatrick, J. D. and McBeth, C. W., *Phytopathology*, 1955, **45**, 576.
(*42*) Godfrey, G. H., *Phytopathology*, 1936, **26**, 246.
(*43*) Gifford, C. M., *Bull. Vermont agric. Exp. Sta.* 157, 1911, p. 143.
(*44*) *Tech. Bull. Minist. Agric.* 1, 1951.
(*45*) Hunt, N. R., O'Donnell, F. G. and Marshall, R. P., *J. agric. Res.*, 1925, **31**, 301.
(*46*) Mamelle, T., *C. R. Acad. Sci., Paris*, 1910, **150**, 50.
(*47*) de Ong, E. R., *J. agric. Res.*, 1917, **11**, 421.
(*48*) Miller, D., *N.Z. J. Agric.*, 1925, **30**, 220.
(*49*) Campbell, R. E., *J. econ. Ent.*, 1926, **19**, 636.
(*50*) Jameson, H. R., Thomas, F. J. D. and Woodward, R. C., *Ann. appl. Biol.*, 1947, **34**, 346.
(*51*) Dunn, E., Henderson, V. E. and Stapley, J. H., *Nature, Lond.*, 1946, **158**, 587.
(*52*) Preston, N. C., *Welsh J. Agric.*, 1928, **4**, 280 ; *J. Minist. Agric.*, 1931, **38**, 272.
(*53*) Brittain, W. H., *Bull. Dept. nat. Resources, Nova Scotia*, 11, 1927.
(*54*) Edwards, E. E., *J. Minist. Agric.*, 1932, **38**, 1230.
(*55*) Glasgow, H., *J. econ. Ent.*, 1929, **22**, 335.
(*56*) Booer, J. R., *Ann. appl. Biol.*, 1944, **31**, 340.
(*57*) Koch, A., *Arb. dtsch. landw. Ges.*, 1899, **40**, 44.
(*58*) Hiltner, L. and Störmer, K., *Arb. biol. Abt.* (*Anst. Reichsanst.*) *Berl.*, 1903, **3**, 479.
(*59*) Russell, E. J. and Hutchinson, H. B., *J. agric. Sci.*, 1909, **3**, 111 ; 1913, **5**, 152.
(*60*) Cutler, D. W., *J. agric. Sci.*, 1920, **10**, 135.
(*61*) Cutler, D. W. and Crump, L. M., *Ann. appl. Biol.*, 1920, **7**, 11.
(*62*) Waksman, S. A. and Starkey, R. L., *Soil Sci.*, 1923, **16**, 137, 247, 343.
(*63*) Richter, L., *Landw. Vers. Sta.* 1896, **47**, 269.
(*64*) Greig Smith, R., *Zbl. Bakt.*, 1911, ii, **30**, 154.
(*65*) Matthews, A., *J. agric. Sci.*, 1924, **14**, 1.
(*66*) Schulze, C., *Landw. Vers. Sta.* 1906, **65**, 137.
(*67*) Pickering, S. U., *J. agric. Sci.*, 1908, **2**, 411.

(*68*) Russell, E. J. and Petherbridge, F. R., *J. Bd. Agric.*, 1912, **18**, 809.
(*69*) Lawrence, W. J. C. and Newell, J., *Sci. Hort.*, 1936, **4**, 165.
(*70*) Johnson, J., *Soil Sci.*, 1919, **7**, 1.
(*71*) Walker, T. W. and Thompson, R., *J. hort. Sci.*, 1949, **25**, 19.

SOIL CONDITIONS AND THE PEST

Next for consideration come the methods by which the soil conditions are rendered less favourable for the pest (see p. 53). This question has been regarded mainly, so far, from the aspect of soil acidity or, better, the hydrogen ion concentration of the soil. Plants are tolerant to certain ranges of hydrogen ion concentration of the soil. In a similar manner, pathogens and pests have their characteristic tolerated ranges of *p*H. If the tolerated range of host plant and pest do not coincide, it is possible to adjust the hydrogen ion concentration of the soil, rendering it favourable for the plant and unfavourable for the pest (*1*).

One of the earliest examples of this type of control is due to Halstead (*2*), who found that liming eradicated clubroot of turnip. It is now known that the organism concerned, *Plasmodiophora brassicae*, producing clubroot of many brassicas, thrives in a soil more acid than is most suitable for this type of plant. By liming the soil the acidity is reduced, improving the soil for the crop and rendering it less suitable for the pathogen.

To take an example at the other end of the scale, Halstead found that the application of sulphur as a fertilizer reduced the amount of scab on potatoes. Gillespie and Hurst (*3*) found that the distribution of scab, due to *Streptomyces scabies*, was closely related to the reaction of the soil. The disease was rare on acid soils of *p*H below 5.2, but was common on soils of *p*H above 5.2. It is for this reason that liming is regarded with caution among potato growers for the decrease of soil acidity which it brings about will favour scab. The success of sulphur, used as suggested by Halstead, is now thought to depend upon the increase of soil acidity arising from the oxidation of the sulphur to sulphuric acid, for not only are the more finely-divided sulphurs more effective but improvements were obtained by inoculating the soil with sulphur bacteria (e.g., *Thiobacillus thiooxydans*), thereby aiding the oxidation processes.

A second method of rapidly increasing soil acidity is by green manuring, the ploughing in or application of green vegetable matter, a practice found by Millard (*4*) to inhibit scab of potatoes. It is now thought that the control is here due, not as much to change in soil acidity, as to biological agencies (see p. 53).

The application of sulphur to the soil was also found to effect a reduction of the wart disease of potato (*Synchytrium endobioticum*) by Roach, Glynne, Brierley and Crowther (*5*). The disease was not, however, entirely eradicated and it was found that its elimination was not solely a matter of increasing soil acidity. Further, as no increased efficiency could be obtained by the inoculation of the sulphur with *Thiobacillus thiooxydans*, it was concluded that the active fungicidal agent is not the hydrogen ion but is some sulphur compound other than sulphuric acid. Roach and Glynne (*6*), studying the toxic action on winter sporangia of *S. endobioticum* of various sulphur acids likely to be formed when sulphur is added to soil, found that acidified solutions of sodium thiosulphate, sodium hydrosulphite and sodium formaldehyde sulphoxylate were about ten times as toxic as sulphuric acid. In view of the generally held view that acidified solutions of these compounds are highly unstable, Roach (*7*) investigated their degree of stability and concluded that the formation of thiosulphuric acid is the probable cause of the fungicidal action of sulphur on *S. endobioticum*.

REFERENCES

(*1*) Garrett, S. D., *Root Disease Fungi*, Waltham, Mass., 1944.
(*2*) Halstead, B. D., *Spec. Bull. New Jersey agric. Exp. Sta.* 8, 1900.
(*3*) Gillespie, L. J. and Hurst, L. A., *Soil Sci.*, 1918, **6**, 219.
(*4*) Millard, W. A., *Ann. appl. Biol.*, 1923, **10**, 70.
(*5*) Roach, W. A., Glynne, M. D., Brierley, W. B. and Crowther, E. M., *Ann. appl. Biol.*, 1925, **12**, 152.
(*6*) Roach, W. A. and Glynne, M. D., *Ann. appl. Biol.*, 1928, **15**, 168.
(*7*) Roach, W. A., *J. agric. Sci.*, 1930, **20**, 74.

MECHANICAL METHODS

Of the mechanical methods of soil treatment, the most obvious when dealing with small areas of soil such as seed-beds and glasshouse soils, is to discard the infected soil, replacing it by soil free from pests or disease organisms. This method suffers from many disadvantages : the wastage caused by discarding a soil, often richer in plant foods than average farmyard manure and the necessity of supplying the fresh soil with these nutrients ; the difficulty of securing a suitable supply of uninfected soil and its storage ; the danger of reinfection of the new soil from the subsoil left in the glasshouse and from the old soil remaining on walls, pipes, tools, wheelbarrows and the workmen's boots.

The processes of cultivation such as ploughing and hoeing are frequently of great importance in the control of soil pests. Seasonal ploughing exposes the larvae and pupae of soil insects to birds and other animals or it may effect their burial. For example, as the European elm sawfly leaf-miner (*Kaliosysphinga ulmi* Sund.) pupates within one inch of the soil surface, Chrystal (*1*) recommended the removal and burial at a depth of below six inches of a thin layer of soil from beneath the tree. Theobald (*2*) after trials with various chemicals for the control of an exceptional plague of cockchafer larvae, found that rolling the affected area by means of a steam roller was effective. The practice of " Submergence ", drowning the pests by flooding the soil for a suitable length of time at the suitable period, has already been dealt with on page 34.

Of greater interest is " Crop Rotation ". The rotation of crops is the old-established device of growing a succession of crops on the same land, which brings great advantages, briefly, saving of manure, better utilization of the nutrient resources of the soil, an economical distribution of labour and finally, because continuous cropping leads to the accumulation of the particular weeds of the crop, weed control.

Considered as a method of pest control, crop rotation involves knowledge of the life history and habits of the pests to be eradicated. The pest must not be capable of over-wintering for more seasons than the period of the rotation, nor should that rotation include any crops which would serve as host plants for the organism.

These requirements may present difficulties ; for instance, *Plasmodiophora brassicae*, the organism causing clubroot of cruciferous plants, has been known to live in the soil up to six years. A six-year rotation containing no cruciferous crop would therefore be required and it would be necessary to remove all weeds able to serve as host plants.

Crop rotation, as a method of pest control, is limited to annual crops and economic conditions do not always warrant its use. The method is frequently impracticable in the intensive cultivation of market garden and glasshouse crops, where the range of crops sufficiently profitable to grow is limited. It is necessary to prevent re-contamination of the treated soil by infected manure, seed, or by drainage water from infected land.

Garrett (*3*) considered crop rotation to be the most effective method of root disease control in field crops. He concluded that a one-year break under any non-cereal crop except pasture will give adequate control of " Take-all " disease, due to *Ophiobolus graminis*. McKay (*4*) recorded the reduced loss of potato through *Verticillium albo-atrum* in a three-year rotation with grain and clover, a control associated with the short survival period of the fungus in infected plant tissue. Leukel (*5*) showed that " Purples ", a wheat disease caused by the nematode *Anguillulina tritici*, can be eradicated from infested land by growing a non-susceptible crop for two or three years. Shaw (*6*) found crop rotation the only practicable method of controlling the sugar-beet nematode, *Heterodera schachtii*, on large areas. Using crops not susceptible to the pest, a rotation of five to six years was necessary because of the persistance of the " brown cyst " stage of the eelworm. The long viability of the cysts is a serious disadvantage of the method. This may be removed if there is a successful outcome of the investigations resulting from the observation of Baunacke (*7*), that the emergence of larvae from the cysts is stimulated by root excretions from susceptible plants. Triffit (*8*) confirmed the observation for the potato strain of the eelworm (since named *H. rostochiensis*) and showed that the " hatching factor " is active only in the presence of oxygen, is non-volatile, and, to some extent, thermostable.

The existence of root excretions stimulating the emergence of encysted larvae provides new possibilities for the control of cyst-forming nematodes. It might, for example, be possible to break the resting period by stimulating the emergence of larvae which are more easily controllable than the resistant cysts. The chemical nature of the stimulant has been studied by Todd and his co-workers (*9*) who, finding it to be an acid probably containing a lactone group, have named it eclepic acid ; their most active preparations have stimulated emergence at dilutions of the order of 1 in 10^7 to 1 in 10^8. The examination of many synthetic products revealed no compound of this high activity but anhydrotetronic acid was found active at a dilution of 1 in 2000.

The possibility of adding, to the soil, chemicals which tend to neutralize the " hatching factor " was studied by Hurst and Triffit (*10*). Laboratory experiments by Smedley (*11*) indicated that the simpler chloroacetates might be of value, but the field trials of O'Brien *al et.* (*12*), of Edwards (*13*), and of Price Jones (*14*) showed that although potato yields were raised by the application of calcium chloroacetate, $Ca(CH_2Cl.COO)_2$ to eelworm-infected soil, the cyst numbers were not reduced (see also p. 273).

A modification of the principle of Crop Rotation is used to combat wireworm on newly ploughed grassland. By taking a crop susceptible to wireworm, e.g., potato, in the first year of ploughing-up, severe injury is avoided because of the presence of ample alternative food for the wireworm in the ploughed-in turf. If grown in the second year, potatoes might suffer severely as shown by Miles and Cohen (*15*) though a crop, e.g., a free-tillering wheat variety, might grow away from the wireworm attack.

REFERENCES

(*1*) Chrystal, R. N., *Agric. Gaz. Can.*, 1919, **6**, 725.
(*2*) Theobald, F. V., *J. S. E. agric. Coll., Wye*, 1927, **24**, 40.
(*3*) Garrett, S. D., *Root Disease Fungi*, Waltham, Mass., 1944.
(*4*) McKay, M. B., *J. agric. Res.*, 1926, **32**, 437.
(*5*) Leukel, R. W., *J. agric. Res.*, 1924, **27**, 925.
(*6*) Shaw, H. B., *Fmrs'. Bull. U.S. Dep. Agric.*, 772, 1916.
(*7*) Baunacke, W., *Arb. biol. Abt.* (*Anst. Reichsanst.*) *Berl.*, 1922, **11**, 185.
(*8*) Triffit, M. J., *J. Helminth.*, 1930, **8**, 19.
(*9*) Calam, C. T., Raistrick, H. and Todd, A. R., *Biochem. J.*, 1949, **45**, 513 ; Calam, C. T., Todd, A. R. and Waring, W. S., *Biochem. J.*, 1949, **45**, 520 ; Marrian, D. H., Russell, P. B., Todd, A. R. and Waring, W. S., *Biochem. J.*, 1949, **45**, 524 ; Marrian, D. H., Russell, P. B. and Todd, A. R., *Biochem. J.*, 1949, **45**, 533 ; Marrian, D. H., Russell, P. B., Todd, A. R. and Waring, W. S., *J. chem. Soc.*, 1947, p. 1365.
(*10*) Hurst, R. H. and Triffit, M. J., *J. Helminth.*, 1935, **13**, 191.
(*11*) Smedley, E. M., *J. Helminth.*, 1938, **16**, 177.
(*12*) O'Brien, D. G., Gemmel, R. R., Prentice, I. W. and Wylie, S. M., *J. Helminth.*, 1939, **17**, 41.
(*13*) Edwards, E. E., *J. Helminth.*, 1939, **17**, 51.
(*14*) Price Jones, D., *Ann. appl. Biol.*, 1947, **34**, 240.
(*15*) Miles, H. W. and Cohen, M., *Rep. ent. Field Sta., Warburton, Cheshire*, 1938, p. 8.

CHAPTER XV

TRAPS

THE successful rat-catcher, gamekeeper or angler is he who knows well the behaviour of his prey, its response to changes of external conditions. So, in pest control, the successful application of control methods is dependent on a sound knowledge of the response of the pest to external stimuli.

Higher animals have developed the faculty of controlling the response to external stimuli by memory. There is an ability to choose, the choice being ruled by previous experience and the memory of that experience. In the lower animals and plants, this power of choice is not developed and it is possible to predict the response of such an organism to tropic stimuli. The trapping of these lower organisms should be simpler than the task of the gamekeeper or rat-catcher. The rat may or may not respond to the attractive stimulus of the smell of cheese, but the moth is unable to resist the light of a candle.

But, the scientific study of behaviour among insects is juvenile and little is yet known of their tropic responses. The application of this knowledge to pest control is therefore limited, but its possibilities are indicated by the following examples.

The Aphididae or plant lice are highly specialized suctorial insects with complex life histories, the dependence of which upon external conditions has aroused much interest. *Aphis fabae* Scop.,* a common polyphagous insect is found in the autumn on a limited number of plants, notably the Spindle tree *Euonymus europoeus*. In the spring, it leaves this host to invade a wide variety of plants, such as the poppy, mangold, bean, etc. Its life history has the following sequence :

(1) The *Fundatrices* or Stem-mothers, wingless females emerging in spring from the overwintered eggs on Euonymus. These females possess that remarkable power among insects of producing young, born as miniatures of their parent. The mother is said to be " viviparous " as distinct from the more usual egg-laying or " oviparous " female insect. Further, for this act of reproduction the previous intervention of the male is unnecessary, a phenomenon known as " parthenogenesis ". The *Fundatrices* are therefore wingless, viviparous, parthenogenetic females which produce the—

(2) *Fundatrigeniae*, likewise (usually) wingless, viviparous, parthenogenetic females which live on the primary host. After the first or second generation, however, there appear the—

---

*This insect was previously known as *A. rumicis* L., a name now confined to a non-migratory form which lives on the Dock (*Rumex* spp.).

(3) *Migrantes*, which differ from their parent in being winged. At this stage, migration to the secondary host, the bean, etc., occurs. Here are produced the—

(4) *Alienicolae*, again parthenogenetic viviparous wingless females of extraordinary fecundity, the young reproducing so rapidly after birth—in nine to twenty days—that enormous numbers are produced in the summer. Successive generations of the *Alienicolae* follow until finally the—

(5) *Sexuparae* appear. These are winged forms of parthenogenetic viviparous females which migrate back to the primary host. These, in turn, give rise to the—

(6) *Sexuales*, sexually reproducing male and female insects, the latter oviparous and producing the eggs, which, overwintering upon the primary host, give rise again to the *Fundatrices*.

A most complex procedure, yet simple when compared to certain other life histories among insects. The main point of interest is that man is able experimentally to control the stages of this life history by the regulation of external conditions.

The first and less complicated feature is the influence of temperature upon the period taken in the development of the *Alienicolae* from birth to the production of the first brood. The length of this period—the Developmental period—an important factor in determining the ultimate number of insects produced, has been shown by Davidson (*1*) and by Lathrop (*2*) to vary inversely as temperature, expressed as ° F less 41.

In the same way, the appearance of the *Sexuparae* is controlled by external factors. Given favourable conditions, notably a sufficiently high temperature, the *Alienicolae* continue to be produced without the appearance of males or eggs. Marcovitch (*3*) caused the production of *Sexuparae*, not only of *A. rumicis*, but of *Capitophorus hippophaes* Koch and *A. sorbi* Kalt., in June, earlier than usual, by the exposure of the plant to a short day for about seven weeks. Marcovitch also showed that the appearance of the *Migrantes* of *A. sorbi* is governed by the increasing length of day of the spring months.

Davidson (*1*) thought that this photoperiodic effect operated via the plant, for day length would affect photosynthetic activity and the carbohydrate content of the sap whereby provoking the latent tendency of the aphids to produce winged forms. Wilson (*4*) and Kenten (*5*), however, produced winged forms even on plants exposed to long days if the insects were exposed to days of eight hours light.

A similar effect of photoperiod has been revealed in the induction of diapause, a period of dormancy which is a normal part of the life cycle of many arthropods. Way and Hopkins (*6*) reduced the percentage of pupae of the tomato moth *Diataraxia oleracea* L. entering diapause when the larvae, though exposed to long days, were fed on leaves from plants exposed to short days. The eggs of the fruit tree red spider mite *Metatetranychus ulmi* laid in summer hatch without delay whereas those laid in winter hatch only after an intervening period of diapause. Lees (*7*) showed that mites exposed to short days though feeding on long-day foliage laid "winter" eggs, whereas those exposed to long days but held on short-day plants laid "summer" eggs. The photoperiod effect is therefore exerted directly on the arthropod. The physiology of the process has been reviewed by Lees (*8*) and it is hormonal in character. The environment in some way little understood regulates the appearance or disappearance in the insect haemolymph of a chemical, a hormone, which operates the machinery controlling diapause much in the same way as insect metamorphosis is controlled by hormones. The hormone ecdysone excreted from the prothoracic gland of the silkworm and which terminates larval diapause was isolated, in small amounts by Karlson and his colleagues (*9*). Another hormone called the juvenile hormone, for its presence prevents the formation of the adult insect within the pupae, was found unexpectedly in comparative plenty in the abdomen of male Cecropia moths by Williams (*10*) who, showing that the hormone can be effective even if applied to the pupal cuticle, suggested that such hormones would be valuable insecticides, the use of which would be unlikely to select resistant strains.

REFERENCES

(*1*) Davidson, J., *Ann. appl. Biol.*, 1923, **12**, 472 ; 1929, **16**, 104.
(*2*) Lathrop, F. H., *J. agric. Res.*, 1923, **23**, 969.
(*3*) Marcovitch, S., *Science*, 1923, **58**, 537 ; *J. agric. Sci.*, 1924, **26**, 513.
(*4*) Wilson, F., *Trans. R. ent. Soc. Lond.*, 1938, **87**, 165.
(*5*) Kenten, J., *Bull. ent. Res.*, 1955, **46**, 599.
(*6*) Way, M. J. and Hopkins, B. A., *J. exp. Biol.*, 1950, **27**, 365.
(*7*) Lees, A. D., *Ann. appl. Biol.*, 1953, **40**, 449, 487.
(*8*) Lees, A. D., *The physiology of diapause in arthropods*, Cambridge Univ. Press, 1955.
(*9*) Karlson, P., *Ann. Sci. nat.*, 1956, **18**, 125.
(*10*) Williams, C. M., *Nature, Lond.*, 1956, **178**, 212.

CHEMOTROPISM

Attractants. There is ample evidence, assembled by Dethier (*1*), that chemical stimuli akin to odours are important factors directing the insect to its food plants. A classical example is due to von Frisch (*2*), who trained honey bees to select a sugar solution flavoured with a particular flower scent from among other sugar solutions. After alteration in the position of the solutions, the bees were able to find anew that flavoured with the odour to which they had been conditioned. McIndoo (*3*) showed that the steam distillate of potato foliage attracted the Colorado beetle, *Leptinotarsa decemlineata*, but he did not identify the compounds responsible for this attraction. In the case of the Mexican boll weevil, *Anthonomus grandis*, Power and Chestnut (*4*) concluded that trimethylamine was the probable attractant, and isolated this compound and ammonia from the volatile constituents of the cotton leaf. McIndoo (*5*) was doubtful whether trimethylamine had attractive properties and, indeed, whether it was possible to imitate plant odours with sufficient accuracy by chemical means. The results of Raucourt and Trouvelot (*6*) suggested that the principles of potato foliage attractant to *L. decemlineata* are not water-soluble as is trimethylamine.

Perhaps the oldest example of chemotropic traps is the bottle containing a little beer hung outside the back door to catch flies and wasps. The attractive action of beer was found by Imms and Husain (*7*) to rest not so much upon its alcohol as on its ester content. The esters of the lower fatty acids and alcohols possess the odour of certain fruits. Imms and Husain found that if, to ethyl alcohol, were added small amounts of acetic, butyric or valerianic acids, a more powerful stimulus was exercised on flies.

The correlation between attractant action and chemical constitution was studied by Cook (*8*). He showed that there is an optimum concentration for each compound, which is related to its volatility, being inversely proportional to the seventh power of the boiling point. Assessing attractant properties from the number of flies caught by the optimum concentration, he found that attractant action decreased with rise of molecular weight of the alcohols. With the esters, there was a similar decrease with rise of molecular weight, but the addition of a CH_2 group to the fatty acid radical reduced the attractiveness more than its addition to the alcohol radical. Cook did not determine the species of fly caught and, although his figures may have been affected by a fractionation of the species, they are sufficient evidence of the potential value of such studies.

In addition to the olfactory response to volatile attractants, insects have the ability to distinguish chemicals by " taste ". Taking advantage of the habit of many insects of extending their proboscis when the tarsi are moistened in sugar solution, Chadwick and Dethier (*9*), using blowflies deprived of their antenna and latella to eliminate response to odour, traced the relationship of molecular structure and chemoreception in many homologous series of simpler organic compounds.

The chemotropic action of fermenting sugars has been utilized to assist the correct timing of spray applications against codling moth *Carpocapsa pomonella*. It is assumed that the number of moths caught reflects the total number present, whereby the dates of maximum numbers are determined and the best dates for applying sprays against the newly-hatched larvae are deduced. Eyer and Rhodes (*10*), from an investigation of the chemical changes of fermenting molasses, concluded that ester formation was again the prime factor of attraction. From an extensive olfactometer study, Eyer and Medler (*11*) were unable to find any one of the fermentation products likely to be the predominant attractant though esters having a phenyl radical were consistently more effective.

The attractant properties of essential oils have been used to trap the Japanese beetle *Popillia japonica*, geraniol and eugenol being the most effective compounds. Van Leeuwen and Metzger (*12*) described a type of trap by means of which nine million beetles were destroyed by 500 traps on an area of 15 acres in 1929. The presence of traces of impurities enhances the attractivity of geraniol for which a specification was proposed by Metzger and Maines (*13*). Jones and Haller (*14*) examined geraniols meeting this specification but which differed in attractiveness.

For the eradication of the Mediterranean fruit fly *Ceratitis capitata* from Florida in 1956–7 over 50,000 traps were employed but supplies of the favoured attractant, angelica seed oil, ran out. Relief was obtained by the discovery of a synthetic attractant in the propyl ester of 6-methyl-3-*cyclo*-hexene-1-carboxylic acid (*15*).

The remarkable distances over which the males of certain moths, such as the gipsy moth *Lymantria dispar*, are guided to virgin females has been used to trap the males. Haller, Acree and Potts (*16*) considered that the attractant extracted from abdominal tips of the female moths was an alcohol which Acree (*17*) named gyptol, but in later chromatographic studies, Acree recovered three active esters of at least two alcohols, one of which is gyptol which appears to be esterified in the natural attractant. The extract of the abdominal tips is still used as the attractant in the extensive surveys carried out in infested areas of the United States, by methods surveyed by Burgess (*18*).

Trap-crops. Actual plants attractive to insect pests have usually proved the most effective lures. This fact forms the basis, not only of the cruder method of employing baits such as pieces of potato for wireworm, but of trap-crop methods, which are applicable where the polyphagous insect has a predilection for a particular plant which may, with sufficient economy, be planted with or between the rows of the crop. The trap plant, when infested, may either be collected and destroyed or be left to prevent the infection of the crop proper.

MacDougall (*19*) found that the weevil *Cryptorrhyncus lapathi* L., which attacks both alder and willow, preferred the former. To protect the osiers from the pest, he proposed the planting of alders here and there in the osier bed. The weevils were destroyed by a timely cutting away and burning of the infected branches and twigs. The preference of wireworms for wheat seedlings was used by Petherbridge (*10*) to lessen their attack on sugar-beet by the planting of wheat between the sugar-beet rows.

A marked preference by the pest for the trap crop is not always necessary. Thus to protect the main crop of swedes from the attack of the swede midge (*Contarinia nasturtii* Kieff), Taylor (*21*) proposed the planting of decoy rows on the headlands of the field. The midge lay its eggs on the first available host and the decoy rows are lifted and destroyed at the appropriate time. The main crop is thus protected from attack and grows to the stage when it can withstand the attacks of later broods of the midge.

Belyea (*22*) for the control of the white-pine weevil (*Pissodes strobi* Peck) advised the planting, with the white pine, of Scots pine which acts as a trap and reduces the infestation of the white pine. Graham (*23*) has ascribed this result to the choice by the pest of trees exposed to sunlight and suggested that the reduced infestation of the mixed stand is due to the shading, by other trees, of the white pine.

The possibility of trap cropping has been enlarged by the observation that the method may be applied to highly specific monophagous pests such as the beet eelworm *Heterodera schachtii*. The emergence of the larvae of the eelworm from its cysts is stimulated by diffusates from the roots of its host. But beet species other than the cultivated beets produce active root diffusates but in some of these species the larva is incapable of reaching maturity. By planting such wild beet species in sugar beet, Hijner (*24*) obtained a high reduction in the count of viable cysts.

Repellents. The attraction of organisms to the plant by odour is an example of positive chemotropism—the opposite, the repelling of the organism by the plant, has also received attention. Though negative chemotropism is not the basis of methods for trapping the pest, it may be dealt with here because of its relationship to attractant action.

An early application of repellents is found in the old-time preventives of plague, e.g., cinnamon and camphor. As it is now known that insects are the carriers of many of the diseases classed in those days as plagues, the success of aromatic-smelling herbs may have been due more to the repulsion of the insect rather than to a disinfectant action. Naphthalene and *p*-dichlorobenzene, the active ingredients of the familiar moth balls employed against clothes moths (Tineidae), are examples of repellents.

The need for protection, from disease-carrying insects, of service personnel in tropical theatres of the 1939–45 war prompted intensive research on repellents. This work, reviewed by Rickard Christophers (*25*) and by Dethier (*1*), placed the study of repellents on a firm scientific basis and, from it, emerged the standard service repellent consisting of a mixture of six parts of dimethyl phthalate (I), 2 parts of butopyronoxyl, better known by its trade name "Indalone" (2 : 2-dimethyl-6-carbobutoxy-2 : 3-dihydro-4-pyrone, II) and 2 parts ethylhexanediol (3-hydroxymethyl-*n*-heptan-4-ol, $CH_3.CH_2.CH_2CH(OH).CH(C_2H_5).CH_2.OH$). The success of this work encouraged extension to the protection of cattle for which purpose

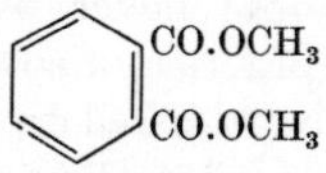

(I)

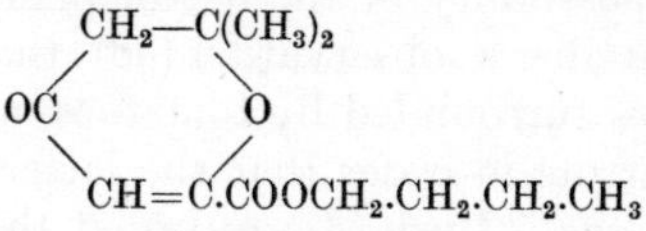

(II)

the butoxypolypropylene glycols, $C_4H_9[OCH(CH_3).CH_2]_n$, are employed (*26*). The search has revealed mosquito repellency in compounds so diverse in molecular structure (see *27*) that the interest of the chemist has been whetted. Wright (*28*) observed that many of the successful repellents absorb strongly in the far infra-red at 460 cm.$^{-1}$, an absorption band absent in non-repellent compounds of similar structure. Absorption at this wave number is associated with a particular vibration of the entire molecule and these vibrations Wright suggests excites the olfactory pigment producing a response directing the insect away from the source of odour. But water itself absorbs strongly at 457 cm.$^{-1}$ and mosquitoes are repelled by relative humidities above about 75 per cent. Wright therefore suggested that repellent substances owe their repellent action to the fact that they produce an illusion of high humidity and so evoke an avoiding reaction.

An example of attempts to protect germinating seed by means of a noxious-smelling chemical is the claim that seed treatment with turpentine or paraffin will ward off the attack of flea-beetles (*29*). Newton (*30*) tried numerous strong-smelling substances such as pyridine, turpentine and cresol derivatives, but obtained no definite indication of their repellent action. Nicotine and nicotine sulphate were more effective. On the other hand, Jenkins (*31*) found that, under favourable conditions, the treatment of turnip seed with paraffin or turpentine did reduce attack. A seed-dressing compounded of naphthalene and *p*-dichlorobenzene in kerosine solution was recommended by Walton (*32*) for flea-beetle control. Jarvis (*33*) obtained satisfactory results by sowing, between the rows of young sugar-canes, sawdust soaked with " dehydrated tar ". The powerful smell kept off the larvae of *Metaponia rubriceps* Macq. until the sets were of sufficient size to thrive despite attack.

It is an old practice to drag elderberry twigs over the germinating turnip field to drive off flea-beetles, but success may be due, not so much to repellent action or the masking of the smell of the mustard oils liberated during the germination of the turnip seed, but to the deterrent action of the dust raised in the process. Newton, however, found in his trials that whereas the entire seed of the untreated plot failed, he obtained an approximately 50 per cent " plant " on the plot treated with a steam distillate of elderberry flowers.

The use of repellents to protect foliage from leaf-eating pests is illustrated by the work of Guy (*34*) who, following up an earlier observation by Tisdale, examined the repellent properties of the thiuram sulphides. He selected thiram, $(CH_3)_2N.CS.S.S.CS.N(CH_3)_2$, for further trial and, by the addition of supplements to improve retention, devised a spray giving a high degree of protection from Japanese beetle, *Popillia japonica* (*35*).

The possibility of the use of plants themselves as repellents is suggested by Schreiber's observation (*36*) that *Pieris brassicae* never oviposits on cabbages surrounded by tomatoes, and by the old practice of sowing onions with carrots in order that the latter shall remain unattacked by carrot fly, *Psila rosae*. Bush (*37*) recorded the old belief that whiteflies are repelled from tomato houses by a few plants of Datura.

Of the use of repellents against higher animals, an example is the protection of young fruit trees from rodents by painting the stems with a sulphur-linseed oil preparation. The sulphated oil, incorrectly called a sulphonated oil, is prepared by heating linseed oil to about 270° C and slowly and carefully adding 10 per cent by weight of sulphur (*38*). Rabbits will not gnaw bark treated with this preparation which appears to be harmless to the trees.

REFERENCES

(*1*) Dethier, V. G., *Chemical insect attractants and repellents,* Blakiston Co., Philadelphia, 1947.

(*2*) Frisch, K. von, *Zool. Jb.*, 1921, iii, **38**, 449.

(*3*) McIndoo, N. E., *J. econ. Ent.*, 1926, **19**, 545.

(*4*) Power, F. B. and Chestnut, V. K., *J. Amer. chem. Soc.*, 1925, **47**, 1751.

(*5*) McIndoo, N. E., *J. agric. Res.*, 1926, **33**, 1095; see also summary of literature, *J. econ. Ent.*, 1928, **21**, 903.

(*6*) Raucourt, M. and Trouvelot, B., *C. R. Acad. Sci. Paris*, 1933, **197**, 1153.

(*7*) Imms, A. D. and Husain, M. A., *Ann. appl. Biol.*, 1920, **6**, 269.

(*8*) Cook, W. C., *J. agric. Res.*, 1926, **32**, 347.

(*9*) Chadwick, L. E. and Dethier, V. G., *J. gen. Physiol.*, 1947, **30**, 247, 255 ; 1948, **32**, 139 ; 1949, **32**, 445.
(*10*) Eyer, J. R. and Rhodes, H., *J. econ. Ent.*, 1931, **24**, 702.
(*11*) Eyer, J. R. and Medler, J. T., *J. econ. Ent.*, 1940, **33**, 933.
(*12*) Van Leeuwen, E. R. and Metzger, F. W., *Circ. U.S. Dep. Agric.* 130, 1930.
(*13*) Metzger, F. W. and Maines, W. W., *Tech. Bull. U.S. Dep. Agric.* 501, 1935.
(*14*) Jones, H. A. and Haller, H. L., *J. econ. Ent.*, 1940, **33**, 327.
(*15*) *Agric. Res. U.S. Dep. Agric.*, 1957, **6**(3), 7 ; see also Hall, S. A., Green, N. and Beroza, M., *J. agric. Food Chem.*, 1957, **5**, 663.
(*16*) Haller, H. L., Acree, F. and Potts, S. F., *J. Amer. chem. Soc.*, 1944, **66**, 1659.
(*17*) Acree, F., *J. econ. Ent.*, 1953, **46**, 313 ; 1954, **47**, 321.
(*18*) Burgess, E. D., *J. econ. Ent.*, 1950, **43**, 325.
(*19*) MacDougall, R. S., *J. Bd. Agric.*, 1911, **18**, 214.
(*20*) Petherbridge, F. R., *J. Minist. Agric.*, 1938, **45**, 23.
(*21*) Taylor, T. H., *Bull. Univ. Leeds* 82, 1912.
(*22*) Belyea, H. C., *J. For.*, 1923, **21**, 384.
(*23*) Graham, S. A., *Bull. Cornell agric. Exp. Sta.* 449, 1926.
(*24*) Hijner, J. A., *Meded. Inst. Suikerbiet. Bergen-op-Z.*, 1951, **21**, 1.
(*25*) Christophers, S. R., *J. Hyg.*, 1947, **45**, 176.
(*26*) Granett, P., Haynes, H. L., Connola, D. P., Bowery, T. G. and Barber, G. W., *J. econ. Ent.*, 1949, **42**, 281.
(*27*) King, W. V., *U.S. Dep. Agric.*, Handbook 69, 1954.
(*28*) Wright, R. H., *Canad. Ent.*, 1957, **89**, 518.
(*29*) *Adv. Leaflet, Minist. Agric.*, 109, 1946.
(*30*) Newton, H. C. F., *J.S.E. agric. Coll., Wye*, 1928, **25**, 116.
(*31*) Jenkins, J. R. W., *Welsh J. Agric.*, 1928, **4**, 334.
(*32*) Walton, C. L., *Rep. agric. hort. Res. Sta., Long Ashton*, 1935, p. 80.
(*33*) Jarvis, E., *Queensland agric. J.*, 1925, **24**, 100.
(*34*) Guy, H. C., *Bull. Delaware agric. Exp. Sta.* 206, 1937.
(*35*) Guy, H. C. and Dietz, H. F., *J. econ. Ent.*, 1939, **32**, 248.
(*36*) Schreiber, A. F., abstr. in *Rev. appl. Ent.*, 1916, A, **4**, 161.
(*37*) Bush, R., *Countryman, Idbury*, 1952, **46**, 272.
(*38*) *Bull. Virginia agric. Exp. Sta.* 126, 1932.

PHOTOTROPISM

The attraction of light for moths was employed, in 1787, by Abbé Roberjot (see *1*) for the trapping of vine moths (*Clysia* and *Polychrosis* spp.). Since that time, many applications of the method and forms of light traps have been employed against the nocturnal Lepidoptera.

The first question is why these insects, normally negatively phototropic, for their nocturnal habit would indicate a dislike of light, should be attracted at all. No reason has yet been suggested though a related phenomenon has been observed by Hewitt (*2*). The army cutworm (*Euxoa auxiliaris* Grote) is negatively phototropic, remaining in broad daylight below the soil surface ; at sunset it appears at the surface and becomes positively phototropic, moving westerly towards the sun.

The attractivity of light is dependent on its wavelength and intensity. Earlier work on the effect of colour was reviewed by Weiss (*3*) and speaking generally the shorter wavelengths of the ultraviolet to blue are more attractive than the yellows, reds and infrared. For this reason yellow and red lamps are preferred to white lamps in places where night-flying insects are a nuisance.

The efficiency of light traps as a means of crop protection was questioned by Criddle (*4*) because of the large proportion of males amongst the insects caught and because many of the females have already laid their eggs. Dewitz (*5*) found that the percentage of females attracted is dependent upon the family of Lepidoptera concerned. A similar result was obtained by Theobald (*6*), who found that the preponderance of males over females applied chiefly to the Bombycidae, Geometridae and Taeniocampae, but that large numbers of female Tortricidae, Crambidae and Tipulidae were caught. Turner (*7*) found that of the total number of moths caught during the period 8 p.m. to 10 p.m. gravid females comprised 40 per cent of the total and males 19 per cent. In the period 10 p.m. to 4 a.m. the relative number of gravid females decreased whilst that of the males and spent females increased. Williams (*8*) also found that, under English conditions, a higher proportion of females of many species of Lepidoptera were caught before midnight than after.

The second criticism advanced by Criddle was that weather conditions favourable for trapping are rare. Moonless, warm and still nights gave the highest catches in Williams's four year survey (*8*). A further objection raised by Criddle to the use of light traps was the danger of ensnaring beneficial Hymenoptera. Theobald found that such captures were few in number.

As a method of crop protection light traps have not proved effective though Herms (*9*) thought their failure under California conditions may have been due to an inadequate knowledge of insect behaviour. On the other hand light traps have found wide use for the study of insect numbers and distribution both spacially and in time. But for this purpose suction traps (see *10*) may prove more reliable.

Phototropism is probably concerned in the observation, by Folsom and Bondy (*11*), that heavy infestations of *Aphis gossypii* Glov. often follow the application of calcium arsenate to cotton. Moore (*12*) also found aphides tend to accumulate on potatoes sprayed with Bordeaux mixture. He attributed this result to the greater reflection of light from the sprayed leaves and suggested that it would be profitable to include dark colouring matter in sprays and dusts used on crops liable to aphis infestation.

REFERENCES

(*1*) Imms, A. D. and Husain, M. A., *Ann. appl. Biol.*, 1920, **6**, 269.
(*2*) Hewitt, C. G., *J. econ. Ent.*, 1927, **10**, 81.
(*3*) Weiss, H. B., *J. econ. Ent.*, 1943, **36**, 1.
(*4*) Criddle, N., *Canad. Ent.*, 1918, **50**, 73.
(*5*) Dewitz, J., *Bull. Ent. Res.*, 1912, **3**, 343.
(*6*) Theobald, F. V., *J. R. hort. Soc.*, 1926, **51**, 314.
(*7*) Turner, W. B., *J. agric. Res.*, 1920, **18**, 475.
(*8*) Williams, C. B., *Trans. R. Ent. Soc.*, 1939, **89**, 79 ; 1940, **90**, 227.
(*9*) Herms, W. B., *Hilgardia*, 1947, **17**, 359.
(*10*) Johnson, C. G. and Taylor, L. R., *Ann. appl. Biol.*, 1955, **43**, 51.
(*11*) Folsom, J. W. and Bondy, F. F., *Circ. U.S. Dep. Agric.* 116, 1930.
(*12*) Moore, J. B., *J. econ. Ent.*, 1935, **28**, 436 ; 1937, **30**, 305.

STEREOTROPISM

The tendency for insects to come to rest against a solid surface is a stereotropic response. It is related, in most cases, to the desire for shelter when the insect is inactive and is illustrated by the manner in which many insects overwinter in cracks and loose bark.

The tendency is utilized in various forms of traps, the simplest of which is the flat board employed for the capture of slugs. For the supplementary control of codling moth (*Carpocapsa pomonella*), advantage is taken of the fact that the larvae, emerging from infected fruit on the tree or ground, wander in search of suitable quarters to pass the pupal stage. Bands of dark-coloured sacking or straw are therefore tied to the tree trunk at midsummer and are periodically examined and the pupae found destroyed. Other suitable quarters such as loose bark, should be removed. Large numbers of apple blossom weevil, *Anthonomus pomorum* (L.), Curt., which also descend the tree trunk in late May and June to find shelter, were caught by Massee and Beshir (*1*) by this means, the banding material used being corrugated cardboard.

Corrugated cardboard as a banding material has also been used for codling moth traps, for wet sacking is to some extent repellent to the larvae. The cardboard band should be protected from rain by strips of waxed paper. A difficulty with paper bands is that they are often badly torn by birds and field mice. Greenslade and his co-workers (*2*) experimented with cardboard bands impregnated with various chemicals, which must be repellent to birds yet non-repellent to codling moth larvae and apple blossom weevil and must not be injurious to the tree. Solutions of certain chlorinated naphthalenes in trichloroethylene were found to be of promise, but the many factors involved created difficulties in obtaining concordant results in different seasons, and, in some, the untreated bands were the more effective traps.

Under climatic conditions when a second generation of codling moth may develop, frequent inspection of the band traps is necessary throughout midsummer. For this reason, Siegler (*3*) proposed the enclosing of the band by a wire screen of a mesh which will permit the entry of the larvae while the moths are retained. This device would also allow parasitic insects present in the larvae to escape. The impregnation of the band with a chemical toxic but non-repellent to the larvae has also been recommended. Siegler and his co-workers (*4*) found that a satisfactory material was β-naphthol in a non-volatile solvent such as lubricating oil. Steiner and Marshall (*5*) selected α-naphthylamine and β-naphthol, water-soluble derivatives being found unsatisfactory.

REFERENCES

(*1*) Massee, A. M. and Beshir, M., *J. Minist. Agric.*, 1930, **37**, 164.

(*2*) Greenslade, R. M., Massee, A. M. and Thomas, F. J. D., *Ann. Rep. E. Malling Res. Sta.* 1934, p. 180 ; Greenslade, R. M. and Massee, A. M., *Ann. Rep. E. Malling Res. Sta.* 1935, p. 177 ; Massee, A. M., Greenslade, R. M. and Brair, J. H., *Ann. Rep. E. Malling Res. Sta.* 1936, p. 232 ; Massee, A. M., Greenslade, R. M. and Duarte, A. J., *Ann. Rep. E. Malling Res. Sta.* **1937**, p. 213.

(*3*) Siegler, E. H., *J. econ. Ent.*, 1916, **9**, 517.
(*4*) Siegler, D. H., Brown, L., Ackerman, A. J. and Newcomer, E. J., *J. econ. Ent.*, 1927, **20**, 699.
(*5*) Steiner, L. F. and Marshall, G. E., *J. econ. Ent.*, 1931, **24**, 1146.

MISCELLANEOUS TRAPS

Included in this category are the many mechanical devices employed for the capture or imprisonment of the pest, in which advantage is taken not so much of the tropic responses of the pest as of some peculiarity of its habits. Thus, the baited rat trap might be regarded as dependent in action upon the chemotropic attraction of the rat by the bait ; the snare is dependent upon the habit of the pest of forming " runs " for the traversing of dense vegetation.

An example is " Grease-banding " which depends on the habit of the wingless female of Geometrid moths of crawling up the trunk of the tree to deposit its eggs on the buds and twigs. Moths of this group include the winter moth (*Operophtera brumata*), the mottled umber moth (*Hybernia defolaria* L.) and the march moth (*Anisopteryx aescularia* Schiff.), all fruit tree pests. The ascent of the female is arrested by placing a band of adhesive material round the tree trunk. The grease employed must remain sticky at winter temperatures and must be resistant to rain. Various recipes have been proposed, resin in castor-oil solution being a frequent basis, but tree-banding compositions in general use are all proprietary products (see *1*).

The tree-trunk is first scraped to remove loose bark at a point not too near the ground, for not only may dirt be splashed up and provide a passage across the grease but the winged male moths are able to carry the females in copulâ for short distances (*2*). The use of greasebands in bush plantations is not always successful because of the low position of the bands. The bands must be in position as soon as possible after leaf-fall ; leaves must be removed from the bands and the surface renewed from time to time. It is better that the grease should have no repellent action causing the moths to oviposit below the bands. If so, the area below the band should be sprayed, in early spring, with an egg-killing wash. The status of grease-banding in winter moth control has been discussed by Jary (*3*).

A second example is the " disc " method, employed against the cabbage root fly (*Hylemyia brassicae*). The fly lays its eggs just below ground level near the main root of the cabbage. To prevent oviposition near the plant, tarred felt discs are placed around the stem (4). The soil should be in a friable condition permitting a close contact with the disc. It is reported that in Holland (*5*) the employment of such discs, of which nearly one million were distributed in 1923, reduced infestation by *H. brassicae* from 90–95 per cent to 5–8 per cent.

REFERENCES

(*1*) Gliesberg, W. and Mentzel, F., *Z. PflKrankh.*, 1931, **41**, 481.
(*2*) Theobald, F. V., *J. Bd. Agric.*, 1910, **17**, 542.
(*3*) Jary, S. G., *J.S.E. agric. Coll. Wye*, 1931, **28**, 137.
(*4*) See Imms, A. D., *J. Bd. Agric.*, 1918, **25**, 59.
(*5*) Hus, P., *Rep. int. Conf. Phytopath. Holland*, 1923, p. 122.

POISON FOR USE IN TRAPS AND BAITS

As the choice of poisons for baits and traps against insects is not restricted by phytotoxic considerations, the cheaper arsenicals such as white arsenic or Paris Green were generally used.

Present day baits for soil pests such as cutworm or grasshoppers are generally based on cyclodiene insecticides such as aldrin, which was found by Weinman and Decker (*1*) to be effective against the grasshopper *Melanoplus differentialis* Thos. when applied as a suitable bait, at the low figure of 2 to 4 oz. per acre. Cereal offals or even sawdust have been used as the diluent and attractant.

Formaldehyde has been employed as a stomach poison for the control of flies. Lloyd (*2*) traced the somewhat erratic results previously obtained to the presence of formic acid and, to a lesser extent, methylamine. Formaldehyde to be used against flies should therefore be neutralized with a little lime water and should be free from the fishy odour of methylamine.

The related metaldehyde (a polymeric form of acetaldehyde, CH_3CHO) has been found especially useful against slugs (*3*), a discovery arising by accident through its employment as a solid fuel (" Meta ") by picnic parties. Miller's offals are again the favoured bait, contact with the metaldehyde causing paralysis, death following only after exposure of the slugs to low relative humidities (*4*, *5*).

Against rodents, the choice of the poison employed is regulated, apart from the factor of toxicity, by its action upon other organisms and by the results of its action upon the rodent. As there is a risk, when using poisoned baits, of domestic animals devouring the bait, a substance relatively harmless to animals other than those which are to be poisoned should be used. Secondly, it is well for sanitary reasons, especially in the case of house pests such as rats and mice, that the poison should not be immediately fatal, but that the rodent should, by the action of the poison, be forced to quit its usual haunts. To fulfil these two requirements a number of rat and mouse poisons have been suggested.

Squill. The poisonous properties of the bulbs of red squill or sea leek (*Urginia* (*Scilla*) *maritima* (L.) Baker), a liliaceous sub-tropical plant, have long been known. The precise nature of the rat-poisoning principles is unknown, but they appear to be distinct from the cardiac glucosides to which squill owes its medicinal properties. The latter compounds are present in white squill which is inferior as a rat poison. The rat-poisoning principle is water-soluble and relatively thermostable. Although destroyed by boiling with dilute acid or alkali, it can be kept for long periods without deterioration (*6*). O'Connor, Buck and Fellers (*7*) recommended, however, the product obtained by drying the bulbs at 80° C. The specificity of the action of red squill on rats and mice appears to be due to its powerful emetic action which causes vomiting by man and animals other than rodents. It has been shown to be relatively non-toxic to poultry. Particulars of suitable methods for use have been given by Munch, Silver and Horn (*8*) and by Schander and Götze (*9*).

Thallium Sulphate. In Germany, a preparation B.P. 247,249 marketed under the name " Zelio ", contained thallium sulphate as the active rat poison. It was claimed that this poison, slow in producing death, induces the rat to seek water. The vermin therefore leaves its haunts for the open where it dies. Munch and Silver (*10*) confirmed that death usually occurs on the second or third day after feeding, is due to respiratory failure, and that the minimum lethal dose, as sulphate, is 25 mg./kg. But because of the high cumulative toxicity of thallium to man, they recommended it should only be used by experienced operators against highly resistant species of rodent.

Barium Carbonate has been widely used as a rat poison in the United States. Like " Zelio " it has the advantage of being without smell and, by its action, induces the rat to seek water and die in the open.

α-Naphthyl Thiourea (ANTU). Phenyl thiourea has, to some people, an extremely bitter taste yet to others it seems tasteless. As the inability to taste it is an inherited character, and as it is regarded as non-poisonous, it is used in genetical studies. Rats used in such studies were killed, but as they avoided phenyl thiourea presumably because of its bitter taste, a search was made for a less distasteful thiourea suitable for use as a rat poison. Richter (*11*) recorded that of nine such derivatives tested, α-naphthyl thiourea (III) was readily accepted by rats to which it proved highly poisonous. The adult Norway rat is killed at doses of 6–8 mg./kg. (*12*), though sublethal doses induce a temporary tolerance (*13*). It is less toxic to other species of rat but is relatively non-poisonous to man and herbivores. Dogs are usually protected, as against red squill, by its emetic action.

$NH.CS.NH_2$ N(CH_3)$_2$

N — C

ClC CH

N = C

CH_3

(III) (IV)

Sodium Fluoroacetate. The use of sodium fluoroacetate ($FCH_2.COONa$) as a rat poison was first suggested by the National Defence Research Committee of the U.S. Office of Scientific Research and Development. It proved highly toxic to all mammals on which it was tested, dogs being killed at doses of 0.1 mg./kg. (*12*) or when fed with rats poisoned by the compound. It is therefore suitable for use as a rat poison only in the hands of experienced operators.

Potassium fluoroacetate was found by Marais (*14*) to be the toxic component of the S. African plant " Gifblaar " (*Dichapetalum cymosum*), poisonous to sheep and cattle.

Peters and his colleagues (*15*) have established that the toxicity of the fluoroacetate is due to an intervention in the Krebs tricarboxylic acid cycle, the mechanism by which the organism utilizes the energy produced by the aerobic oxidation of pyruvic acid, an end product of glycolysis. Fluorocitrate, produced from the fluoroacetate, competes with citrate as substrate for aconitase inhibiting this enzyme.

The intense physiological activity of the fluoroacetate ion and of simple derivatives such as fluoroethyl alcohol was the subject of wartime research (see *16*). Saunders and his colleagues (*17*) showed that in the series $F(CH_2)_nCOOR$, the compound is poisonous if n is an odd number but, when it is even, the compound is relatively non-toxic.

" Castrix." In searching for amoebicides, workers at I.G. Elberfeld discovered two compounds which, being too toxic for chemotherapeutic use, were developed as rat poisons. The first " Castrix ", 2-chloro-4-dimethylamino-6-methyl pyrimidine (IV), was used for the preparation of poison grains and was said to be selective in action and useful only against mice (*18*) to which the toxic dose was about 1 mg./kg. The isomeric 4-chloro-2-dimethylamino-6-methyl pyrimidine, simultaneously produced with IV in its manufacture (*19*), is inactive.

The second compound, " Muritan ", *p*-chlorophenyldiazothiourea (V), the same order of toxicity as IV, is of interest in its relationship to I as a thiourea derivative.

Dicoumarin. It has long been known that spoiled sweet clover (*Melilotus alba* Desr.) is toxic to cattle and, in 1941 Link and his colleagues (*20*) showed that the toxic principle was dicoumarin, 3 : 3′-methylenebis-(4-hydroxycoumarin), VI, for which the trade name " Dicumarol " is registered, interferes with the action of vitamin K and reduces the coagulating properties of the blood, so that trivial injury can cause a fatal haemorrhage. O'Connor (*21*) suggested its use against rodents for a daily dose of 2 mg. was fatal to rats, whereas dogs survived daily doses of 50 mg. and for therapeutic purposes, the human dose is 200–300 mg. He claimed the special virtue that, as rats treated with sub-lethal doses do not develop " bait-shyness ", pre-baiting is unnecessary. Link and his colleagues (*22*) examined the anticoagulant action of a range of 3-substituted-4-hydroxycoumarins and selected the 42nd on their list WARF42 (3-(α-acetonylbenzyl)-4-hydroxycoumarin, VII) as the most promising for development

Cl—C₆H₄—N:N.NH.CS.NH$_2$

(V)

OH OH
C C
C—CH$_2$—C
CO OC
O O

(VI)

as a rat poison. This compound is now known as **warfarin**, and not causing bait shyness, has shown exceptional promise in rodent control.

The addition of chlorine in the 4 position of the unsubstituted benzene ring of VII gives 3-(α-acetonyl-*p*-chlorobenzyl)-4-hydroxycoumarin which has been introduced in Europe as a rat poison by J. R. Geigy A.G. under the

OH
C
C—CH—
CO CH_2
O CO
CH_3

(VII)

trade name " Tomorin ". Coumachlor has been adopted as the common name of the pure compound which has the same rat-killing virtues as warfarin.

REFERENCES

(*1*) Weinman, C. J. and Decker, G. C., *J. econ. Ent.*, 1949, **42**, 135.

(*2*) Lloyd, L., *Bull. Ent. Res.*, 1920, **11**, 47.

(*3*) Gimingham, C. T. and Newton, H. C. F., *J. Minist. Agric.*, 1937, **44**, 242.

(*4*) Stringer, A., *Rep. agric. hort. Res. Sta., Bristol*, 1946, p. 87.

(*5*) Cragg, J. B. and Vincent, M. H., *Ann. appl. Biol.*, 1952, **39**, 392.

(*6*) Winton, F. R., *J. Pharm. exp. Ther.*, 1927, **31**, 123, 137.

(*7*) O'Connor, M. G., Buck, R. E. and Fellers, C. R., *Industr. engng Chem.*, 1935, **27**, 1377.

(*8*) Munch, J. C., Silver, J. and Horn, E. E., *Tech. Bull. U.S. Dep. Agric.* 134, 1929.

(*9*) Schander, R. and Götze, G., *Zbl. Bakt.*, 1930, ii, **81**, 335.

(*10*) Munch, J. C. and Silver, J., *Tech. Bull. U.S. Dep. Agric.*, 238, 1931.

(*11*) Richter, C. P., *J. Amer. med. Ass.*, 1945, **129**, 927.

(*12*) Ward, J. C., *Amer. J. publ. Hlth.*, 1946, **36**, 1427.

(*13*) Richter, C. P., *Proc. Soc. exp. Biol. N.Y.*, 1946, **63**, 364.

(*14*) Marais, J. S. C., *Onderstepoort J. vet. Sci.*, 1944, **20**, 67.

(*15*) Peters, R. A., *Proc. roy. Soc.*, 1952, **139**, B, 143.

(*16*) McCombie, H. and Saunders, B. C., *Nature, Lond.*, 1946, **158**, 382.

(*17*) Buckle, F. J., Pattison, F. L. M. and Saunders, B. C., *J. chem. Soc.*, 1949, 1471.

(*18*) Martin, H. and Shaw, H., *Brit. Intell. Obj. Sub-comm., Final Report*, 714, 1947.

(*19*) Tanner, C. C., Greaves, W. S., Orrel, W. R., Smith, N. K. and Wood, R. E. G., *Brit. Intell. Obj. Sub-comm., Final Report*, 1480, 1947.

(*20*) Campbell, H. A. and Link, K. P., *J. biol. Chem.*, 1941, **138**, 21 ; Stahmann, M. A., Huebner, C. F. and Link, K. P., *J. biol. Chem.*, 1941, **138**, 513.

(*21*) O'Connor, J. A., *Research*, 1948, **1**, 334.

(*22*) Overman, R. S., Stahmann, M. A., Huebner, C. F., Sullivan, W. R., Spero, L., Doherty, D. G., Ikawa, M., Graf, L., Roseman, S. and Link, K. P., *J. biol. Chem.*, 1944, **153**, 5.

CHAPTER XVI

THE TREATMENT OF THE CENTRES AND VECTORS OF INFECTION

THERE remain for consideration certain methods dealing more directly with the restriction of the spread of the pest, which may conveniently be classified under (I) the elimination of infection foci or centres and (II), the treatment of the carriers or vectors concerned in that spread.

THE ELIMINATION OF INFECTION CENTRES

In dealing with true parasites, it is obvious that the infected plant is the actual focus where the multiplication of the pest occurs and from which it is spread. The fundamental control method is therefore the destruction of infected plants or plant material.

General. Hand-picking is still used when labour is cheap and where the pest occurs in well-defined and clearly visible agglomerates. Examples are the collection of the colonial larvae of the Tussock moths (Lymantridae) and the destruction of the eggs where they are laid in easily-seen patches, as with certain of the Bombyx moths. Hand collection, by children, of the larvae and adults of the large and small cabbage white butterflies (*Pieris brassicae*, *P. rapae*), or of the egg clusters of the former, is a task which is encouraged by prizes given by local Gardeners' Societies. But, even with the plentiful labour sometimes found in the tropics, it is questionable whether the expenditure involved is justified. Le Pelley (*1*) in a study of hand collecting for the control of Antesia on coffee, reported adversely on the method.

The collection and destruction of diseased plant tissue is an important means of restricting spread. For the control of the brown rot of stone fruits (*Sclerotinia* spp.) the removal and burning of diseased twigs and mummified fruits is of great importance, for it is from these sources that the disease spreads in the spring. In England, the Silver Leaf Order of 1923 requires the occupier of any premises on which plum or apple trees are growing to destroy by fire on the premises, all dead wood on each plum or apple tree before the 15th July of every year, the purpose being to check the spread of the fungus *Stereum purpurem*.*

For the destruction of exposed pests it is possible to employ domestic animals as the collecting agents. Theobald (*3*) reported the value of poultry in the destruction of certain orchard pests ; he observed the extermination of the pear midge (*Contarinia pyrivora* Riley) by fowls penned under the attacked trees. Against the codling moth (*Carpocapsa pomonella*), Le Baron in 1873, suggested that pigs and sheep, allowed to run through

* See footnote on page 298.

the orchard, will eat the fallen apples before the larvae escape. Delassus (*4*) reported that the olive fly (*Dacus oleae* Gmel.) was almost eliminated, in some localities, by the persistent collection of the fallen fruit and by turning sheep and pigs into the orchards. Fulton (*5*) found that five pigs per acre controlled the apple curculio *Tachypterellus quadrigibbus* Say by clearing up the early dropped fruit. As it was only necessary to keep the pigs in the orchard for a month, the damage they caused by rooting was negligible.

The insect hibernates in one or other stage of its life cycle and usually requires some suitable shelter. If this be lacking, the pest is more exposed to its enemies, both climatic and biological, and the chance of its survival is more remote. The whole system of dormant spraying is primarily against the sheltered inactive pest. Rubbish of all sorts, which also provides a hiding-place during daytime for night-feeding pests, should be destroyed. Grass, likewise, is the home of many injurious insects during winter. Lefroy (*6*), for this reason, considered permanent grass a mistake in any commercial orchard, with the possible exception of cherry orchards. The destruction of any moss and lichen on the trunks further removes available shelter. The majority of the measures which comprise " Plant Hygiene " and " Orchard Sanitation " aim at the destruction of diseased plant tissue and at the removal of any shelter for pests.

An extreme case of shelter-requirements affording a means of control is that of the tsetse flies *Glossina palpalis* R. D. and *G. tachinoides* West., two most dangerous carriers of sleeping sickness in W. Africa. During the dry season, the breeding and survival of these tsetses are possible only along the river banks and they are confined to definite plant associations of a limited number of species of trees and shrubs. The " selective clearing " of these species along the river system renders the habitat untenable during the dry season and has resulted in the disappearance of the tsetse flies (*7*).

Eradication of the Host Plant. When no other method is possible it may be necessary to eradicate the entire crop. As an example of the method, the eradication of Mediterranean fruit fly, *Ceratitis capitata*, from Florida may be cited. This pest was discovered in Florida, in March, 1929, and within two months, was found on no fewer than 697 properties within eleven counties ; a Federal quarantine was imposed on 1st May. For its enforcement Florida was divided into three zones. In the infected zones, the area within one mile of any spot at which the fly was found, all fruit and vegetables susceptible to attack were destroyed, the planting of host plants was prohibited and all host fruits and vegetables exposed in markets and elsewhere had to be screened from attack. In the protective zones,

*To protect the cut surface, it is necessary to pay heed to the general precautions taken in pruning and to treat the wound with some protective substance. As *S. purpureum* can attack a wide variety of hosts, the eradication of all diseased material is impossible. The fungus, however, is able to enter the plant only through a wounded surface, which must therefore be protected. Brooks and Moore (*2*) found Stockholm tar, the material at one time recommended, unsatisfactory and recommended a thick white lead paint. The wound dressing should be non-injurious to bark and should leave a tough elastic film which will not crack when the wood expands or contracts. Further, it should be applicable to the we surface, a property in which the paints are deficient. Certain types of bitumen emulsion have for this reason, found favour.

the area within nine miles of the outer boundary of an infected zone, the cultivation of host fruits and vegetables was prohibited between 1st May and 1st November (with a few exceptions), the screening of host fruits and vegetables was enforced and the export of host fruits and vegetables was permitted only to states where the establishment of the fly was unlikely to occur. The third zone, the outside zone, embraced the rest of Florida. In this zone, no mature fruit, except sour limes, was allowed to remain on the trees after 15th June and the shipment of all host fruits and vegetables to states in which the establishment of the fly was considered possible, was prohibited. Woglum (*8*) recorded that, by 1st June, over half a million boxes of fruit had been destroyed and that, at the height of the eradication campaign, nearly four thousand men were employed. These rigorous measures stamped out the infestation and, on 11th November, 1930, the Federal quarantine was lifted. By the end of the year, inter-state restrictions were removed. Newell (*9*) stated that the cost of the campaign to the Federal government was over one and one-quarter million pounds. Florida remained free of the pest until it was found on 13th April, 1956. A State-Federal scheme of eradication was again begun (*10*) but whereas the earlier campaign relied almost wholly on host destruction, the development of insecticides since 1930 permitted a greater reliance on insecticidal control.

Thorough application is fundamental for success in such a campaign and, on the large scale, it is usually to be undertaken only with the stimulus of legislative action. Essentially it is dependent on the simple recognition of the pest or disease, a matter not always easy. In the case of certain virus diseases, there exist strains of host plant which, though infected, are not visibly affected. Such plants are called " carriers "—a term borrowed from analogous cases where human beings can act as infection centres of a disease without themselves suffering ill effects. The phenomenon was first observed in plants by Nishamura (*11*) with tobacco mosaic virus but naturally-occurring carriers of virus diseases of many other crop plants are now known. The existence of carrier varieties, though providing an excellent means of escaping the ill-effects of these virus diseases, introduces difficulties in their control in susceptible varieties by eradication.

Eradication of the Wild Host. In some cases, the pest is not restricted to the cultivated host and it is then necessary to remove or treat all other susceptible plants. Lefroy (*12*) gave a formidable list of wild plants which may serve as hosts for fruit pests, listing some twenty-one species which feed on hawthorn. He urged that the same control measures applied to the cultivated host should be used on the wild host.

Weed hosts also may serve as winter quarters for the pest. Pritchard and Porte (*13*) found the pathogen of Septoria leaf spot of tomato, *Septoria lycopersici* Speg., on related weeds such as common nightshade, and showed that the eradication of these removes the main means of over-wintering of the fungus. Similarly, the destruction of wild cucumbers, milkweed, pokeweed, etc., for some 50–75 yards around the field was recommended by Doolittle and Walker (*14*) to break, in winter, the continuity of the mosaic disease of cucumber.

The question of the eradication of weed hosts is complicated by the existence of physiologic races (see p. 10). Salmon gave a long list of wild plants, likely to occur near hop gardens, which can serve as hosts for the hop powdery mildew, *Sphaerotheca humuli*. The necessity of applying to these weeds the same measures as applied to the hop has disappeared, for it is known that the particular races infecting the weeds are unable to infect the hop.

Ward (*15*) and Salmon (*16*) concluded that the weed host might serve for the development of forms of pathogen with new infective properties. Salmon found that conidia of *Erysiphe graminis* from *Bromus racemosus* L. were unable directly to infect *B. commutatus* Schrad., but if transferred first to *B. hordeaceus*, the resulting conidia could infect *B. commutatus*. *B. hordeaceus* therefore was thought to serve as a bridge, affording the mildew on *B. racemosus* a passage to *B. commutatus*. Hammarlund (*17*) was unable to repeat Salmon's work and the suggestion that the infective properties of the race are modified by passage to the " bridging host " is now regarded as untenable. In many cases the " bridging host " serves for the selection of one from a mixture of physiologic races.

Eradication of " Alternate " Hosts. That barberry bushes in some way encourage the " blasting " of wheat has long been known to wheat growers, and it is recorded that in 1660, farmers around Rouen secured the enactment of a law requiring the destruction of barberry bushes in wheat areas. The authenticity of this record is in doubt (*18*) but the early colonists in America quickly adopted legislative measures for barberry eradication. In 1726, Connecticut, and in 1754, Massachusetts enforced the eradication of all barberry bushes " for it has been found by experience that the Blasting of wheat and other English grain is often occasioned by Barberry Bushes, to the great loss and damage of the inhabitants of this Province ". (*19*) Similar legislation followed in other of the eastern United States, but in England it would appear that the eradication of the barberry was thorough enough to render legislation unnecessary.

The connection between barberry rust and wheat rust was first demonstrated by Schoeler (*20*), in 1818, but the existence of heteroecism, not only with this particular rust, *Puccinia graminis*, but among various other rust fungi, was established by the researches of De Bary (*21*).

In its life history *P. graminis* has five stages : uredial, telial, sporidial, pycnial and aecial. The uredial stage, produced on susceptible wheats and grasses, is a vegetative stage and successive generations of urediospores, each with paired nuclei of opposite sex, are produced indefinitely under suitable conditions. When conditions become unfavourable the telial stage is formed and in the teliospore the paired nuclei are fused to form a diploid nucleus. Reduction division occurs when the teliospore germinates to develop a promycelium from which are produced the sporidia which can infest only certain barberry and *Mahonia* species. On this alternate host the pycnia are formed and the pycniospores which they produce are haploid. Pycniospores from different pycnia then combine, a process analogous to cross fertilization, but the nuclei of opposite sex do not at once fuse. The

" fertilized " pycnia form " cluster-cups " and in them are produced aeciospores, each with paired nuclei, which cannot infest barberry but only the appropriate cereal or grass host. The feature of the life history is that two hosts are necessary ; if one host is eradicated the pathogen will itself succumb.

In the case of stem rust of cereals (*Puccinia graminis*) the " alternate host " to be eradicated is the barberry (*Berberis* spp.). Although legal measures were taken in the older States of America, a severe outbreak of the disease in 1916 stimulated a drastic barberry eradication campaign. By 1918, concerted action was taken by the north-central and western grain-growing states of the United States and, it was estimated that by 1941, some 296 million rust-susceptible bushes had been destroyed (*22*). As a result, it is stated that the average annual loss from stem rust, which in the period 1916–20 was fifty-seven million bushels, was reduced to nine million bushels in the period 1926–30.

In certain European countries the decrease of the injury occasioned by *P. graminis* has likewise followed the diminution of numbers of the alternate host. In England, as has already been mentioned, barberry in close association with arable land is rare ; where it occurs, as in parts of Wales, the fungus is the cause of some damage (*23*). Eradication of barberry is legally enforced in Denmark and Norway. Lind (*24*) recorded that in Denmark, severe attacks of rust, prior to legal measures in 1903 enforcing eradication, used to occur every two or three years. These, however, were things of the past, though in a few places where barberry still existed the rust caused damage to the barley crop. Under certain conditions and in certain wheat-growing districts eradication of barberry is not completely successful in controlling the rust.

In the case of stem rust, the fungus is not completely dependent in all countries on the presence of barberry. Thus in Australia and South Africa, the rust is the cause of serious damage even though the barberry is rare or absent from these regions. In these cases the fungus appears to be able to survive despite the absence of the aecidial stage, and it is usually assumed that the fungus here overwinters in the uredial stage on wild grasses. Lind (*24*) traced a line dividing regions in which the urediospore cannot survive from those countries in which the urediospore is able to overwinter and in which barberry eradication is not of such prime importance.

There still remain, however, certain localities in which rust is often the cause of great loss yet where the overwintering of the urediospore is improbable. In India, for example, the urediospores cannot survive the intense heat of the plains and, in Canada, there is a long break in the wheat crop. In such cases the appearance of the disease is the result of infection windborne from the foothills of the Himalayas or from the rust-infected areas of ore southerly parts of America.

In Canada and the United States north of Texas, a careful epidemiological study is therefore necessary for the appearance, number and physiologic races of the windborne urediospores to be determined. The invasion can be severe ; Stakman (*22*) quoted an estimate that in early June 1952 the winds had carried 4000 tons of urediospores into sixteen counties of northern Oklahoma and south-central Kansas. But the need for barberry eradication is not lessened for it delays the appearance of the disease on neighbouring wheat and prevents the evolution of new physiologic races, which might be able to infest the resistant varieties by which the wheat breeder has so far kept the disease in check.

Much speculation has been provoked as to the origin of heteroecism and the reasons for the success of what, at first sight, appears such a severe handicap to the perpetuation of the species ; its survival depends upon an accidental deposition upon the right host plant twice in life. The alternate hosts of such fungi are, however, invariably members of the same plant society and, by the convenient method of overwintering thus provided, the survival of the pest is assured. Moreover, hybridization is facilitated with the consequent emergence of new physiologic races.

Heteroecism and the attendant method of control, by eradication of the alternate host, is not confined solely to pathogenic fungi. The example of *A. fabae*, in the development of which two hosts play a part, has been mentioned on p. 282. With certain aphides this migration would appear to be obligatory, with others occasional or optional ; with all migratory forms, however, dependence upon a second host plant provides a means of checking their development and spread. But better examples of heteroecism as the basis of control are to be found in medical and veterinary entomology, such as liver fluke (*Fasciola hepatica* L.) which has, as its alternate hosts, the sheep and the snail *Limnoea truncatula*. Most of the tapeworms alternate in their life history between two hosts, the pork tapeworm (*Taenia solium* Rud.) passes from the pig to human beings and is the possible cause of certain religious restrictions of the consumption of pork. Amongst other organisms there are the protozoa (*Plasmodium* spp.) responsible for malaria, the alternate host of which is the anopheline mosquito ; the Trypanosomes causing the Surra disease, the Nagana disease and the Texas fever of cattle. As however in such cases, the success of the heteroecious habit depends on dissemination of the parasites by the alternate host, these pests are considered in more detail in the discussion of " Insect Dissemination ".

REFERENCES

(*1*) Le Pelley, R. H., *Bull. ent. Res.*, 1935, **26**, 533.
(*2*) Brooks, F. T. and Moore, W. C., *J. Pomol.*, 1926, **5**, 61.
(*3*) Theobald, F. V., *J.S.E. agric. Coll. Wye*, 1923, **23**, 8.
(*4*) Delassus, —., abstr. in *Rev. appl. Ent.*, 1924, A, **12**, 186.
(*5*) Fulton, B. B., *J. agric. Res.*, 1928, **36**, 249.
(*6*) Lefroy, H. M., *J. R. hort. Soc.*, 1915, **41**, 28.
(*7*) Morris, K. R. S., *Bull. ent. Res.*, 1946, **37**, 201.
(*8*) Woglum, R. S., *Bull. Calif. Fruit Gr. Exch.* 6, 1929.
(*9*) Newell, W., *Mon. Bull. St. Plant Bd. Florida*, 1931, **15**, 49.
(*10*) Rohwer, G. G., *Bull. ent. Soc. Amer.*, 1957, **3**, (3), 38.
(*11*) Nishamura, M., *Bull. Torrey Bot. Club*, 1918, **45**, 219.
(*12*) Lefroy, H. M., *J. roy. hort. Soc.*, 1915, **41**, 28.
(*13*) Pritchard, F. J. and Porte, W. S., *Bull. U.S. Dep. Agric.*, 1288, 1924.
(*14*) Doolittle, S. P. and Walker, M. N., *J. agric. Res.*, 1925, **31**, 1.

(*15*) Ward, H. M., *Ann. Mycol. Berl.*, 1903, **1**, 132.
(*16*) Salmon, E. S., *Ann. Mycol. Berl.*, 1904, **2**, 255, 307.
(*17*) Hammarlund, C., *Hereditas*, 1925, **6**, 1.
(*18*) Fulling, E. H., *Bot. Rev.*, 1943, **9**, 483.
(*19*) *The Province Laws of Massachusetts*, 1736–1761, p. 153.
(*20*) Schoeler, Om., *Landockomninske Tidender*, 1818, 289.
(*21*) De Bary, A., *Ber. K. Preuss. Acak. Wiss. Berlin.* 1865, p. 15.
(*22*) Stakman, E. C., *Ann. appl. Biol.*, 1955, **42**, 22.
(*23*) See *Misc. Publ. Minist. Agric.*, 52, **1926**, p. 10.
(*24*) Lind, J., *Tiddsskr. Planteavl.*, 1915, **22**, 729.

THE ELIMINATION OF INFECTION VECTORS

It is possible to classify pathogens and pests, though incompletely, by the way in which they are transmitted from host to host, i.e., by what vector their spread is accomplished. The higher animals and most insects possess individual powers of locomotion which, assisted in many cases by tropic responses, secure an adequate spread. Other lower animals and plant pathogens are dependent for that distribution on accidental agencies, such as currents of air or water, and sometimes on more specialized agencies. They may, for example, invade the seed of the host and become dispersed by the process employed by that host for the dissemination of its seeds. In this way arise the seed-borne diseases, the control of which has been discussed under Seed Treatment. Further, they may be spread by the pollinating or other insect visitors of their host ; or, they may secure a sufficient spread by the agency of higher animals or by the cultural operations of man himself.

Considering in more detail the various vectors responsible for pest dissemination and dispersion, we have :

Wind Transmission. In general, dispersion by wind is uncontrollable ; Butler (*1*), however, suggested that " wind breaks " may be of use in checking the spread of wind-borne plant pathogens. He observed that, in tea gardens, those bushes to the leeward of a wind barrier suffer less from leaf diseases than those freely exposed to the wind.

In certain tea-growing districts leguminous plants are used for green manuring. As an incidental result, the tea shrub is appreciably protected from *Helopeltis theivora* Waterh., a capsid bug which is the cause of " Mosquito Blight ". Hart (*2*) concluded that *Leucaena glauca* Benth., grown as a thick hedge between the rows, was more effective for this purpose than *Tephrosia* and suggested that the green manuring crop acted as a mechanical shield against infestation from a focus of *Helopeltis*. He considered that *Leucaena*, however, may produce some substance unfavourable to the development of the pest or that its effect may be manurial.

Water Dissemination. Distribution by water becomes significant where watering or irrigation is necessary. It is especially important in glasshouse cultivation where partial sterilization of the soil is practised. Every effort is necessary, once the soil-borne pathogens have been killed by the treatment, to secure freedom from re-infection for, if accidentally introduced, the pathogen, freed from the competition of other fungi and insects, thrives at an alarming rate. Bewley and Buddin (*3*) traced severe outbreaks of the

" Damping-off " and " Buck-eye " rot of tomatoes (*Phytophthora cryptogea* Pethybr. & Laff. and *P. parasitica* Dastur) to an infected water supply. An examination of the various waters available showed that water from the mains and from deep artesian wells was relatively free, but that water from wells receiving surface-drainage contained large numbers of plant pathogens.

For the treatment of an infected water supply, Bewley suggested either filtration or sterilization by heat or chemicals. Filtration proved effective but the efficiency of the filter fell with use. Of the chemicals tested, mercuric chloride and a trade disinfectant " Chloros " proved the most toxic to fungi. Because mercuric chloride is highly poisonous, precautions are necessary to prevent human consumption of the treated water. The water was sterilized by heating to boiling point. Later, Bewley (*4*) introduced the use of Cheshunt compound (see p. 119), the solution of which is used for watering and which acts as a partial sterilizer, not only of the soil, but also of the water.

Insect Dissemination. The realization of the importance of insect vectors in the transmission of diseases arose from Waite's demonstration (*5*), in 1891, of the spread by flies and wasps of fire blight, a disease of pomaceous fruit trees due to *Bacterium amylovorum.* The diseases concerned must necessarily be confined to the smallest of parasitic organisms belonging mainly to the bacterial, protozoan and virus classes and it is on the latter group that interest has mainly centred. In only a few instances has the transmission of a virus through the true seed been established and, apart from the vegetative reproduction of diseased plants, the chief agency spreading the virus from plant to plant has proved to be insects. The insect transmission of virus has served as the spearhead of research on the virus diseases and has a rich literature surveyed by Storey (*6*) and Bawden (*7*).

Insects may be mechanical disease vectors as in fire blight mentioned above, when the insect serves merely as a means of conveyance though it may also provide, in the wounds it produces, a point of entry for the pathogen. But in other cases, and particularly with the viruses, the successful distribution of the disease can only be achieved by a limited number of insect species. Curly top of sugar-beet is disseminated only by the leaf-hopper *Eutettix tenellus* (*8*) ; aster yellows only by the leaf-hopper *Cicadula sexnotata* Fall. Storey (*9*) drew attention to a specificity between viruses grouped according to symptoms and the insect families ; thus the mosaic viruses are transmitted mainly by aphides, the yellows by leaf-hoppers. This correlation may be due to the manner in which the insect accomplishes inoculation for Dykstra and Whitaker (*10*) showed that, among even the aphides, certain *Myzus* spp. readily transmitted leaf roll of potato whereas *Macrosiphum solanifolii* Ashm. generally failed to do so : *Myzus* spp. habitually fed in the phloem whereas *M. solanifolii* often fed in the vascular tissues. But other evidence points to a more specific relationship between vector and virus. It is now established that, in many cases, a latent period is necessary before the insect, after feeding on a diseased plant, can become infective to a healthy plant. An early example is due to Kunkel (*11*) who found a period of ten days in the transmission of aster yellows by *C. sexnotata*. Moreover it is a general rule that those vectors which show this latent period remain infective for periods much longer than those vectors which are able to transmit the disease immediately after feeding on an infected plant. Rand and Pierce (*12*) suggested that the latent period was required for the virus to multiply in the vector to a point when an infective dose can be emitted. Other investigators, for instance, Bawden (*7*, p. 92) considered that the latent period is merely in time taken from the virus to reach the salivary glands of the vector. But many virologists, for instance Maramorosch (*13*), now consider that the latent period, the persistence of infectivity and the specific relationship of vector and virus are evidence that the insect here plays a part more extensive than that of vector, approaching that of alternate host. Indeed, Maramorosch considered convincing the evidence, first produced by Kunkel (*14*), of multiplication of the virus in the arthropod vector though he offered no explanation of why the insect can supply the nucleoprotein required for the multiplication of the virus and yet remain healthy and of apparently unimpaired fecundity.

It is certain, however, that the insect plays an important part in the overwintering of the virus. The carrying-over of the curly top virus was studied by Carsner and Stahl (*15*). As the disease can only be transmitted by *E. tenellus* and as they showed that the insects reared from eggs are non-virulent, these workers concluded that the appearance of the disease in spring is due to the presence of insects, viruliferous as a result of feeding upon the preceding year's crop, upon susceptible weeds or on old beets growing in the neighbourhood. It is significant that *E. tenellus* is one of the few jassids which overwinter as adults.

The control of these arthropod-borne pathogens rests, therefore, upon two main processes, the destruction of infected plant material and the inhibition of the effectiveness of the vector. An illustration is the so-called deterioration of the potato. The falling-off of yield and the general unsatisfactory results following the practice of continually growing home-grown " seed " in certain localities, was given several explanations. One hypothesis suggested that the potato, a plant of cold climates, deteriorated through an over-ripeness of the seed when grown under warm conditions, but Brown and Blackman (*16*) showed that neither early lifting nor the shading of the plant lifted for seed affected productivity, provided virus diseases were absent. Alternatively, it was suggested that continued asexual propagation was responsible for deterioration; yet the saffron crocus, because of its irregular chromosome complement, is sterile and the cultivated variety has survived for nearly four thousand years despite continued asexual propagation (*17*). The theory now generally accepted is that deterioration is the result of the accumulation of virus diseases. The control of deterioration therefore becomes an efficient eradication of those plants showing symptoms of virus infection and, as far as practicable, the control of the insects responsible for the spread of these diseases. Brown and Blackman (*16*) found that rogueing was effective in restoring the cropping vigour of a deteriorated potato crop. But rogueing alone was shown by Doncaster and Gregory (*18*) to be ineffective in the potato fields of southern England. They regarded the clearing of the land, prior to planting, of all potato " volunteers " as a more important control measure. The use of insecticides to reduce the vector population appears to be a feasible control measure. Early tests by Austin and Martin (*19*), who added contact insecticides to modified Bordeaux mixture used for the control of *Phytophthora infestans* in potatoes, revealed that the yield from " seed " saved from potatoes sprayed in two previous seasons was significantly greater than that of " seed " from potatoes unsprayed in those seasons. Doncaster and Gregory found, however, that fumigation with nicotine, although reducing the aphid population, had little effect on the spread of leaf roll and rugose mosaic, presumably because the winged migrants had already transmitted a substantial amount of the viruses. The more extended trials of Broadbent and his colleagues (*20*) showed that the timely use of both contact and systemic aphicides stopped the spread of leaf roll virus though was less effective against virus Y.

But experience has shown that in districts where for climatic or other reasons the insect vectors are scarce, the rate of deterioration is slow. For this reason, potato growers in the south of England in general prefer Scotch seed, the lowland farmers of N. Wales secure their seed from farms at a higher elevation. Maldwyn Davies and Whitehead (*21*) have shown, by an epidemiological study of the aphides of the potato crop, the relative importance of factors such as the production of winged aphides, the extent of their migration and its dependence on wind and other weather conditions, and the development of the aphis population following colonization, in determining the spread of virus diseases. Their work has provided a scientific basis for the selection of areas suitable for seed potato production.

Cultural and Accessory Vectors. The processes of cultivation may themselves be responsible for the dispersion of pests. A surprising number of agencies have been found, not only amongst the implements and materials used, but also in connection with methods of transport and marketing. International exchange of plant products and nursery stock is an outstanding example. Although in this case the plants themselves are the carriers, other factors such as methods of transport, packing, etc., are sometimes involved.

General cleanliness of implements is always desirable, but its value in avoiding the spread of disease is often overlooked. In glasshouse cultivation, especially during and after soil sterilization, the spades, wheelbarrows and forks, contaminated with untreated soil, may easily bring about a re-infection. As regards machinery, Leukel (*22*) established that an infected threshing machine was causing the spread of the " Purples " of wheat (*Anguillulina tritici*) among farms of a certain threshing ring. In horticulture, the pruning knife may prove a dangerous carrier, especially of the bacterial and virus diseases. Attention is to be directed to the spread of disease by marketing infected products. The use of returnable baskets, in the marketing of fruit, is objectionable because of the risk of their carrying disease to other orchards and nurseries (*23*).

Glasshouses may be infected unwittingly by nursery workers carrying disease organisms upon their boots, clothing or hands. Bewley (*24*) recommended that assistants working in diseased glasshouses should not enter houses where healthy plants were growing until their hands had been washed and their clothes either stoved or exposed to direct sunlight. An extreme case arises through the high infectivity of the virus causing tobacco mosaic. Valleau and Johnson (*25*) found that the drying of tobacco at 165° F for forty minutes did not reduce the infectivity of diseased tobacco, and it has been suggested that the mosaic virus carried on the fingers of cigarette smokers may prove dangerous to other susceptible plants such as tomato.

Farmyard manure has been found responsible for the introduction and spread of pests. Bewley showed that straw manure may reasonably be regarded with suspicion and, in one instance (*26*), traced the introduction of cucumber anthracnose (*Colletotrichum lagenarium* (Passr.) Ell. & Hals.) to this source. The alarming rate of spread of flag smut of wheat (*Urocystis tritici* Koern.) in New South Wales was attributed by Clayton (*27*) to its presence in straw manure. As the spores of the fungus are not killed by passage through the alimentary canal, he suggested the substitution of oat for wheat in the horses' ration. McKay and Pool (*28*), in an investigation of the transmission of the leaf spot of beet (*Cercospora beticola*), found that the fungus was apparently destroyed by passage through the alimentary canal. The greatest danger of infection was in the uneaten portions of the beet which become mixed with the manure. As the organism is also destroyed by heat, these workers recommended the ensilage of the green beet tops for the pathogen is unable to survive the temperature of this process.

By far the most important vector of disease is the plant itself, for it is almost exclusively upon the living plant including cuttings, bulbs, fruits, and seeds that diseases and pests are carried over long distances. Instances of an epiphytotic following the importation of infected plants into a region previously free from the scourge are innumerable and serious for reasons which have been discussed under " Biological Control ". Thus the steady improvement in transport facilities was held by Jensen (*29*) to be a not altogether unmixed blessing. He correlated the appearance in Europe, about the year 1840, of potato blight (*Phytophthora infestans*) with the introduction of the speedier steamship for transoceanic traffic. The fungus, a native of the northern Andes, is killed by exposure for even short intervals to a temperature not many degrees above that of a normal summer in temperate climates. In the days before steam navigation, infected tubers would thus be sterilized during the period taken by the sailing ship in crossing the tropics, either in the passage to North America or to Europe. The introduction of the steamboat enabled a crossing of the equatorial zone in a time insufficient to ensure a complete destruction of the hyphae in infected tubers, and the fungus was thus able to survive the transportation from its native home. Jensen's hypothesis has been criticized by Reddick (*30*).

Intercontinental air transport provides a dangerous vehicle as shown by the 2442 interceptions of insects and plant diseases made at United States air-ports in 1945. Cooley (*31*) suggested that, in addition to inspection at ports of entry, periodic pest and disease surveys in those foreign countries from which plants or unprocessed plant products are imported, and continuous surveys in the vicinity of the ports of entry to detect the survival of new pests or diseases, are necessary precautions.

Unrestricted import of plant material is undoubtedly to be regarded as a menace, and to safeguard against the importation of infected plants legislation is the only practical measure. In Egypt, for example, by Law 1 of 1916, there is a specific prohibition of importation of cotton plants and seeds, cotton ginned and unginned, and of cotton wool. Alternatively there may be inspection of all or certain plant material, together with, in some cases, fumigation or quarantine and the destruction of infected material. By the Government Order of 1917, no plant other than fruit, vegetable or sugar-cane, could be imported into British India unless sterilized with hydrocyanic acid gas at one of the specified ports. In the United States, the Federal Quarantine Act of 1912 provides for the quarantine of imported plants and the destruction of all material suspected of infection.

To safeguard against the spread of pests between neighbouring countries, international as opposed to internal legislation is generally necessary. If the passage of the pest from one country to the next involves no sea voyage, it would be necessary first to restrict the points of entry in order to permit the satisfactory inspection of the whole of the imported material. In 1881, an International Phylloxera Convention of a number of European countries was held which resulted in the establishment of measures restricting the movement of vines from one country to another—measures which have done much to prevent the spread of the pest beyond certain well-marked zones.

Considered as a method of crop protection, legislative restriction of the importation of plant products involves economic and political consequences. Although the method may be applied for reasons which rest on a firm scientific basis, it may lead to retaliatory measures by exporting countries or states economically affected by the restrictions. Further, the interference with the normal flow of trade may lead to resentment by purchasers with resultant changes in demand. It is impossible therefore to pursue the discussion of the economic aspects, which have been critically reviewed in a report from the University of California (*32*).

REFERENCES

(*1*) Butler, E. J., *Fungi and Diseases of Plants*, Calcutta, 1918, p. 109.

(*2*) Hart, S. J. G., abstr. in *Rev. appl. Ent.*, 1924, A, **12**, 10.

(*3*) Bewley, W. F. and Buddin, W., *Ann. appl. Biol.*, 1921, **8**, 10.

(*4*) Bewley, W. F., *Rep. exp. Res. Sta., Cheshunt*, 1921, p. 38.

(*5*) Waite, M. B., *Proc. Amer. Assoc. Adv. Sci.*, 1891, p. 315.

(*6*) Storey, H. H., *Bot. Rev.*, 1939, **5**, 240.

(*7*) Bawden, F. C., *Plant Viruses and Virus Diseases*, 3rd Ed., Waltham, Mass., 1950.

(*8*) Ball, E. D., *Bull. U.S. Dep. Agric. Bur. Ent.*, 66, 1909, iv, p. 33.

(9) Storey, H. H., *2nd Cong. Int. Path. C. R.* 11, 471, 1934.
(10) Dykstra, T. P. and Whitaker, W. C., *J. agric. Res.*, 1938, **57**, 319.
(11) Kunkel, L. O., *Amer. J. Bot.*, 1926, **13**, 646.
(12) Rand, F. V. and Pierce, W. D., *Phytopathology*, 1920, **10**, 189.
(13) Maramorosch, K., *Advanc. Virus Res.*, 1955, **3**, 221.
(14) Kunkel, L. O., *J. econ. Ent.*, 1938, **31**, 20.
(15) Carsner, E. and Stahl, C. F., *J. agric. Res.*, 1918, **14**, 393 ; 1924, **28**, 297.
(16) Brown, W. and Blackman, V. H., *Ann. appl. Biol.*, 1930, **17**, 1.
(17) For discussion see Bijhouwer, A. P. C., *J. Pomol.*, 1931, **9**, 122.
(18) Doncaster, J. P. and Gregory, P. H., *The Spread of Virus Diseases in the Potato Crop*, London, H.M. Stat. Off., 1948.
(19) Austin, M. D. and Martin, H., *J.S.E. agric. Coll. Wye*, 1933, **32**, 49.
(20) Broadbent, L., Burt, P. E. and Heathcote, G. D., *Ann. appl. Biol.*, 1956, **44**, 256.
(21) Davies, W. M., *Bull. ent. Res.*, 1932, **23**, 535 ; *Ann. appl. Biol.*, 1934, **21**, 283 ; 1935, **22**, 106 ; 1936, **23**, 401 ; Davies, W. M. and Whitehead, T., *Ann. appl. Biol.*, 1935, **22**, 549 ; 1938, **25**, 122 ; Whitehead, T., Currie, J. F. and Davies, W. M., *Ann. appl. Biol.*, 1932, **19**, 529.
(22) Leukel, R. W., *J. agric. Res.*, 1924, **27**, 925.
(23) Report on Fruit Marketing in England and Wales, *Econ. Ser. Min. Agric.*, 15, 1927, p. 27.
(24) Bewley, W. F., *Diseases of Glasshouse Plants*, London, 1923, p. 31.
(25) Valleau, W. D. and Johnson, E. M., *Phytopathology*, 1927, **17**, 513.
(26) Bewley, W. F., *Rep. exp. Res. Sta. Cheshunt*, 1921, p. 32.
(27) Clayton, E. S., *Agric. Gaz. N.S.W.*, 1925, **36**, 860.
(28) McKay, M. B. and Pool, V. W., *Phytopathology*, 1918, **8**, 119.
(29) Jensen, J. L., *Mem. Soc. nat. Agric. France*, 1887, **131**, 31.
(30) Reddick, D., *Phytopathology*, 1928, **18**, 483.
(31) Cooley, C. E., *J. econ. Ent.*, 1947, **40**, 129.
(32) Smith, H. S. *et al.*, *Bull. Calif. agric. Exp. Sta.*, 553, 1933.

AUTHOR INDEX

A

Aamodt, O. S., 30, 32
Abel, A. L., 242
Abbott, W. S., 169, 175
Ackerman, A. J., 292
Acree, F., 196, 197, 286, 289
Adams, J. F., 123, 127, 128
Adams, R., 183
Adrian, E. D., 221, 225
Ainsworth, M., 68
Alamercery, J., 241
Aldridge, W. N., 223, 226
Allan, J. M., 253
Allen, R. F., 16, 23
Allen, T. C., 157, 159, 226
Allen, W. W., 240, 241
Ambrose, A. M., 182
Amos, A., 128
Anderson, H. W., 138, 139
Anderson, J. A., 19, 23
Andreae, S. R., 240, 241
Andreae, W. A., 240, 241
Andreasen, A. H. M., 64, 68
Andrews, W. H., 154
Angell, H. R., 23, 253
Angus, T. A., 56, 57
Appel, O., 16, 23, 140, 145, 267, 268
Aragão, H. B., 58, 59
Archer, W. A., 197
Arens, K., 14, 22
Armitage, H. M., 48, 51
Armstrong, G., 216
Armstrong, S. F., 31, 33
Arnaudi, C., 24
Arnold, E. L., 258
Arnold, M. H. M., 159
Arnstein, H. R. V., 54, 56
Arthur, J. C., 259
Arthur, J. M., 173, 175, 201, 202
Atsumi, K., 182
Auclair, J. L., 26
Audoynaud, A., 119, 122
Audus, L. J., 236, 238
Auerbach, C., 9
Augustinsson, K. B., 225
Austin, M. D., 76, 77, 162, 163, 190–2, 194, 306, 309
Avens, A. W., 154, 190, 194, 199, 200
Avery, A. G., 9
Avery, S., 147, 148, 154

B

Back, E. A., 27, 28, 37, 41, 253
Bain, S. M., 125, 126
Baird, W., 72, 76
Baker, A. W., 182
Baker, F. E., 157, 159
Ball, E. D., 308
Balls, A. K., 226
Ballu, T., 68
Balsom, E. W., 210, 215
Barber, C. A., 12
Barber, D. R., 268
Barber, G. W., 289
Barker, B. T. P., 107, 108, 111, 124–6, 142, 145
Barker, C. H., 199, 200
Barnes, M. M., 229, 231
Barnes, R. A., 18, 23
Barr, T., 76
Barratt, R. W., 129, 133
Barrons, K. C., 243
Bartell, F. E., 70, 72
Barth, M., 124–6
Barthel, W. F., 168, 175
Bartlett, P. B., 215
Bassett, H., 93, 96
Bassi, A., 52, 55
Bateman, E. W., 64, 68
Batt, R. F., 76
Baumann, G., 267
Baunacke, W., 280, 281
Bawden, F. C., 22, 24, 304, 305, 308
Beach, S. A., 112, 115
Beacher, J. H., 211, 216
Beament, J. W. L., 230, 231
Beans, H. T., 148, 154
Beaumont, A., 39, 41
Beck, S. D., 28
Beckley, V. A., 51
Bedford, Duke of, 75, 77, 122, 193, 235
Bell, E. P., 243, 244
Bell, J. M., 116, 122
Belyea, H. C., 286, 289
Ben-Amotz, Y., 72, 80, 193
Bennett, F. T., 142, 145
Bennett, S. H., 23, 199, 200, 202
Beran, F., 190, 192, 194
Bergeson, G. B., 273, 277
Bergmann, F., 102, 105
Bergold, G. H., 58, 59
Berkeley, M. J., 4, 6
Berlese, A., 47, 51
Beroza, M., 172, 175, 289
Berry, W. E., 114, 115, 128
Beshir, M., 291
Bessey, C. E., 148
Bewley, W. F., 31, 33, 40, 41, 122, 142, 145, 269, 276, 303, 304, 307–9
Bickel, H., 179, 181, 182
Bidstrup, P. L., 200
Biffen, R. H., 7, 9
Bigelow, C., 163
Bijhouwer, A. P. C., 309
Bird, F. T., 59
Blackith, R. E., 174, 175

Blackman, G. E., 233, 235, 239
Blackman, V. H., 305, 309
Blakeslee, A. F., 9
Blakeslee, E. B., 272, 277
Blauvelt, W. E., 228, 231
Blijdorp, P. A., 198, 200
Blin, H., 196, 197
Blinn, R. C., 215
Bliss, C. I., 91, 94, 96, 199, 200
Bliss, D. E., 55, 56
Blodgett, F. M., 66, 68
Blondeau, R., 244
Bluestone, H., 212, 216
Boam, J. J., 176–9, 181, 182
Bobilioff, W., 110, 112
Bodenheimer, F. S., 51
Bodenstein, O. F., 176
Bodnár, J., 257, 258, 263
Bohonos, N., 138, 139
Bolle, J., 58, 59
Bolley, H. L., 233, 235, 256, 258, 259, 263, 265
Bondy, F. F., 290
Bonner, J., 234, 235
Bonns, W. L., 128
Bonrath, W., 260, 261, 265
Booer, J. R., 262–5, 274, 277
Borash, A. J., 175
Borchers, F., 151, 154
Borei, H., 156
Börner, C., 271, 277
Böttcher, F. K., 171, 175
Bottomley, W., 165, 166
Boursnell, J. C., 223, 226
Bousquet, E. W., 200, 202
Bovien, P., 155, 156
Bovington, H. H. S., 201, 202
Bowery, T. G., 289
Boyce, A. M., 199, 200, 248, 249
Boyd, A. E. W., 234
Boyd, O. C., 122, 123
Boyle, L. W., 257, 258
Bracey, P., 173, 175
Bradbury, D., 241
Bradbury, F. R., 216
Bradsher, C. K., 133
Brair, J. H., 291
Branas, J., 123, 126
Brazzel, J. H., 225
Brenchley, W. E., 32, 33
Brett, C. C., 263, 265
Brian, P. W., 53–6, 139, 140, 146
Bridges, P. M., 175
Brierley, W. B., 13, 23, 278, 279
Briggs, F. N., 257–9
Briscoe, M., 76
Brittain, R. W., 64, 68
Brittain, W. H., 277
Broadbent, L., 26, 28, 306, 309
Broadfoot, W. C., 53, 55
Brodaty, E., 180, 182
Brook, M., 143, 145
Brooks, F. T., 17, 23, 298, 302
Brown, A. J., 257, 258
Brown, A. W. A., 62, 68, 166, 167
Brown, F. C., 133
Brown, G. T., 193
Brown, H. D., 205, 206, 215
Brown, L., 292
Brown, W., 14, 15, 18, 22, 23, 143, 145, 305, 309
Bruce, W. N., 208, 215
Brunn, L. K., 159
Buc, H. E., 90
Bucha, H. C., 241, 242
Buchanan, J. B., 56
Buck, R. E., 293, 296
Buckle, F. J., 296
Buckley, T. A., 178, 182
Buckner, A. J., 216
Buddin, W., 277, 303, 308
Bulger, J. W., 157, 159, 252
Bunbury, H. M., 81
Burchfield, H. P., 79, 80, 89, 90, 92, 96, 118, 122, 136, 137
Burdette, R. C., 189, 193
Burgen, A. S. V., 223, 226
Burger, O. F., 52, 55
Burgess, E. D., 286, 289
Burk, D., 102, 105
Burke, E., 152–4
Burrill, T., 5
Burt, P. E., 309
Busbey, R. L., 252
Bush, R., 288, 289
Bushland, R. C., 49, 51, 225
Busvine, J. R., 90, 204, 210, 214–6, 253
Butenandt, A., 177, 182
Butler, E. J., 21, 24, 33, 36, 303, 308
Butler, M. L., 144, 146
Butler, O., 116, 120, 122, 127, 128, 249, 250
Buttress, F. A., 256
Byrde, R. J. W., 136, 137

C

Cahn, R. S., 176–182
Cairns, H., 263, 265
Calaby, J. H., 59
Calam, C. T., 281
Calhoun, S. L., 234
Cameron, F. K., 149, 154
Campau, E. J., 68
Campbell, F. L., 87, 89, 90, 150, 152, 154, 155, 159, 166, 167, 182
Campbell, H. A., 296
Campbell, I. G. M., 168, 175
Campbell, R. E., 277
Carbone, D., 21, 24
Carleton, W. M., 64, 68
Carman, G. E., 203, 205, 215
Carroll, J., 194
Carruth, L. A., 157, 159
Carsner, E., 304, 309
Carter, R. H., 203, 215
Carter, W., 273, 277
Caryl, C. R., 76
Casanges, A. H., 252
Casida, J. E., 221, 225, 226
Chadwick, L. E., 207, 215, 223, 275, 285, 289

Challenger, F., 108, 111
Chamberlain, R. W., 174, 176
Chamberlin, J. C., 76–8, 80, 188, 189, 193
Chandler, W. A., 123, 145
Chapman, P. J., 152, 155, 190, 194, 199, 200
Chapman, R. K., 225, 226
Chatt, J., 130, 133
Chefurka, W., 225
Chester, F. D., 119, 122
Chester, K. S., 21, 24, 255
Chestnut, V. K., 284, 288
Chisholm, R. D., 273, 277
Chomette, A., 267
Christensen, C. M., 12, 13
Christensen, J. J., 12, 13
Christophers, S. R., 287, 289
Chrystal, R. N., 279, 281
Clark, B. S., 144, 146
Clark, D., 159
Clark, E. L., 218, 225
Clark, E. P., 177, 178, 182, 183
Clark, G. L., 76
Clark, J. F., 124–6
Clark, R. V., 23
Clayton, E. E., 248, 249, 252, 263, 265
Clayton, E. S., 307, 309
Cleveland, C. R., 190, 193
Clifford, A. T., 149, 154
Clowes, G. H. A., 199, 200
Cohen, M., 281
Collie, B., 72, 76
Collins, D. L., 68
Collison, P., 160
Comes, O., 17, 23
Conners, I. L., 13
Connola, D. P., 289
Cook, A. H., 54, 56
Cook, F. C., 149–152, 154, 253
Cook, M. T., 19, 23
Cook, W. C., 35, 36, 285, 288
Cooke, A. R., 242
Cooley, C. E., 308, 309
Cooley, J. S., 128
Cooley, R. A., 154
Coon, B. F., 201, 202
Coons, G. H., 12, 18, 21, 23, 24
Cooper, W. F., 70, 72, 159
Coquillett, D. W., 245, 248
Corey, R. A., 221, 225
Corteggiani, E., 222, 225
Cotter, G. J., 225
Cotton, R. T., 253
Cox, C. E., 131, 135, 136
Crafts, A. S., 233, 235, 237, 238, 241, 244
Cragg, J. B., 296
Craig, L. C., 165, 167
Craigie, J. H., 11, 13, 16, 23
Crane, M. B., 28
Cranham, J. E., 231
Crauford-Benson, H. J., 86, 88, 89
Cressman, A. W., 189, 193
Crichton-Browne, J., 156
Cricks, W. P., 76
Criddle, N., 290
Cristol, S. J., 204, 212, 215, 216
Crombie, L., 168, 175, 197
Crosier, W., 265
Cross, E. W., 12
Cross, G. L., 24
Crossman, S. S., 186, 193
Crow, J. F., 207, 215
Crowdy, S. H., 139, 140
Crowther, E. M., 278, 279
Croxall, H. E., 265
Crump, L. M., 275, 277
Cullman, V., 175
Cummings, O. K., 225
Cunliffe, N., 42, 43
Cunningham, H. S., 17, 23, 136, 137, 267
Cupples, H. L., 70, 72
Currie, J. F., 309
Currier, H. B., 241, 244
Curtis, K. M., 14, 23
Curtis, P. J., 139
Cutkomp, L., 171, 175
Cutler, D. W., 275, 277

D

Daams, J., 241
Dagley, S., 97, 104
Dahm, P. A., 175, 214, 216
Daines, R. H., 264, 265
Dantony, E., 75, 76
Danysz, J., 57
Darbishire, F. V., 270, 276
Darkis, F. R., 253
Darley, M. M., 125, 126
Darlington, C. D., 9
Darnell Smith, G. P., 257, 258
Darpoux, H., 39, 41
David, E., 75–7, 121
David, W. A. L., 173, 175
Davidow, B., 214, 216
Davidson, J., 26, 28, 283, 284
Davidson, W. M., 196, 197
Davies, D. W., 264, 265
Davies, E. C. H., 76
Davies, R. G., 192, 194, 229, 231
Davies, W. M., 50, 51, 306, 309
Davis, R. J., 34, 36
Davison, A. N., 223, 226
Dawsey, L. H., 189, 193
Dawson, H. M., 122
Dearborn, F. E., 150, 154
DeBach, P., 49, 51
de Bary, A., 4, 6, 16, 23, 300, 303
de Castella, F., 110, 112
Decker, G. C., 43, 216, 293, 296
DeEds, F., 159
Deichmann-Gruebler, W., 202
Delage, B., 123, 126
Delassus, —, 298, 302
DeLong, D. M., 125, 126, 155, 156
Demolon, A., 276
den Boer, P. J., 231
Dennis, R. W. G., 256
De Ong, E. R., 76–8, 80, 121, 123, 161, 163, 187–9, 193, 248, 249, 274, 277
Deroin, J., 225

DeRose, H. R., 239
Desalbres, L., 210, 216
Dethier, V. G., 26, 28, 284, 285, 287–9
Dewitz, J., 290
d'Hérelle, F. H., 56, 57
Dicke, R. J., 157, 159
Dickson, J. G., 16, 23, 36
Dietvorst, F. C., 229, 231
Dietz, H. F., 200, 202, 289
Diggle, W. M., 224, 226
Dillon Weston, W. A. R., 16, 23, 41, 257, 258, 262, 263, 265
Dills, L. E., 195, 196
Dimond, A. E., 129–131, 133
Dixon, S. E., 182
Dodd, A. P., 51
Doherty, D. G., 296
Doncaster, J. P., 306, 309
Doolittle, S. P., 19, 20, 24, 299, 302
Doran, W. L., 14, 23, 109, 111
Dorman, S. C., 225
Dorogin, G., 141, 145
Doull, J., 225
Dowson, W. J., 37, 41, 43
Drake, C. J., 43
Drake, N. L., 215
Dresden, D., 206, 207, 215
Driggers, B. F., 161, 163
Duarte, A. J., 291
Dubacquié, L., 124, 126
DuBois, K. P., 219, 225
Duddington, C. L., 55, 56
Dudgeon, G. C., 46
Dufrénoy, J., 19, 23
Duggar, B. M., 128
Dulac, J., 123, 126
Dumas, J. B. A., 271, 277
Dunbar, C. O., 120, 123
Dunez, A., 276
Duncanson, L. A., 130, 133
Dunegan, J. C., 139
Dunlap, V. C., 36
Dunn, E., 277
Dunnam, E. W., 243
Dupire, A., 209, 216
Durkee, A. B., 126
Durrant, R. G., 93, 96
Durrell, L. W., 14, 22
Dusey, F., 112, 115
Dustan, A. G., 52, 55
Dutton, W. C., 113, 115, 200
Dworak, M., 20, 24
Dyer, H. A., 154
Dykstra, T. P., 304, 309

E

Eagleson, C., 172
Eaker, C. M., 215
Eaton, J. K., 229, 231
Ebeling, W., 72, 188, 189, 193, 194
Edgecombe, A. E., 21, 24
Edwards, E. E., 268, 277
Eldred, D. N., 249
Elliott, M., 170, 171, 175
Elliott, M. I., 203, 215
Emerson, O. H., 53, 55
Emerson, R. L., 138, 139
Emmens, C. W., 90, 96
English, L. L., 72, 189, 193
Ennis, W. B., 239
Ephraim, F., 119, 122
Erdtman, H., 18, 23
Eriksson, J., 11, 13
Erkerson, S. H., 16, 23
Esdorn, I., 262, 263, 265
Essig, E. O., 272, 277
Evans, A. C., 26, 28, 70–2, 75, 76, 78, 80, 89
Eyer, J. R., 285, 289
Eyre, J. V., 108, 111, 115

F

Fajans, E., 77–80, 193
Fales, J. H., 176
Fallscheer, H., 79, 80
Fargher, R. G., 142, 145
Farkas, A., 144, 146
Farley, A. J., 107, 111
Farsted, C. W., 25, 28
Fawcett, C. H., 238
Fawcett, H. S., 52, 55
Fay, R. W., 216
Feichtmeir, E. F., 109, 111
Feils, G., 252
Feldberg, W., 225
Felix, E. L., 136, 137, 257
Fellers, C. R., 293, 296
Fellows, H., 36
Fennessy, B. F., 59
Ferguson, J., 97, 98, 104, 206, 253
Fernald, C. H., 148, 154
Feytaud, J., 161, 163
Fildes, P., 102, 105
Filinger, G. A., 50, 51
Filmer, R. S., 161, 163
Fink, D. E., 150, 154, 159
Finn, T. P., 238
Finney, D. J., 91–6
Fischer, E., 101, 104
Fischer, G. W., 13
Fisher, R. W., 223, 225, 226, 228
Fisher, W. B., 62, 68
Fiske, W. F., 46, 51
Fitzgibbon, M., 265, 266
Fitzhugh, O. G., 225
Fitz-James, P., 56, 57
Fleck, E. E., 203, 215
Flegg, P. B., 271, 277
Fleming, A., 54, 56
Fleming, W. E., 157, 159, 271, 277
Flenner, A. L., 133
Flint, W. P., 43
Fogg, G. E., 70, 72, 235
Foister, C. E., 36, 41, 263, 265
Folkers, K., 139
Folsom, J. W., 290
Fontaine, T. D., 19, 24
Foote, M. W., 137

Forch, C., 97, 104
Ford, J. H., 139
Forsyth, W., 106, 111, 112
Fowler, H. D., 17, 23
Fox Wilson, G., 48, 51, 251
Foy, N. R., 254, 255
Fraenkel, G. S., 14, 22, 26–8
Franklin, H. J., 34, 36
Fransen, J. J., 180, 182
Frawley, J. P., 219, 225
Frear, D. E. H., 190, 194, 200, 202
Frederichs, K., 152, 154
Frederick, W. J., 37, 41
Frew, J. G. H., 31, 33, 42, 53
Friend, H., 77
Frisch, K. von, 284, 288
Frohberger, P. E., 266, 267
Fryer, J. C. F., 13
Fryer, P. J., 100, 112
Fuchs, W. J., 12
Führer, H., 97, 104
Fujitani, J., 167, 174
Fukuto, T. R., 224, 226, 228
Fulling, E. H., 303
Fulmek, L., 151, 154
Fulton, B. B., 155, 156, 298, 302
Fulton, H. R., 15, 23
Fulton, R. A., 159
Furr, J. R., 110, 112

G

Gadd, C. H., 31, 33
Gaddum, J. H., 91, 96
Gage, J. C., 224, 226
Gahan, J. B., 215
Galloway, L. D., 142, 145
Garbowski, L., 141, 145
Gardner, M. W., 17, 23
Garrett, S. D., 32, 33, 53–6, 279–281
Gasser, R., 135, 220, 225, 228, 229, 231
Gassner, G., 30, 31, 33, 249, 250, 262, 263, 265
Gast, A., 240, 241
Gastine, G., 76, 77, 119, 122
Gatai, K., 204
Gatterdam, P. E., 225
Gauhe, A., 24
Gäumann, E., 19, 20, 22, 23
Gauss, W., 267
Gavaudan, P., 207, 215
Gayon, U., 116, 122, 124, 126
Gemmel, A. R., 281
George, S. W., 177, 182
Gerber, N. N., 18, 23
Gersdorff, W. A., 170–2, 175, 182, 197
Getzin, L. W., 225
Geuther, T., 228, 229
Giannotti, O., 214, 216
Gibbs, W., 72, 76
Gifford, C. M., 273, 277
Gillam, A. E., 168, 175
Gillander, H. E., 192, 194
Gillespie, L. J., 278, 279
Gillette, C. P., 149, 154
Gillham, E. M., 88, 89, 196, 197
Gilligan, G. M., 120, 123
Gilman, J. C., 259
Gilpatrick, J. D., 277
Gimingham, C. T., 107, 111, 124–6, 155, 156, 166, 176, 181, 182, 194–6, 200, 296,
Ginsburg, J.M., 153, 155, 161, 163
Girard, A., 270, 276
Glasgow, H., 274, 277
Glasgow, R. D., 68
Gleisberg, W., 292
Glover, L. C., 69, 72, 163, 196, 225
Glover, L. H., 161, 163
Glover, P. E., 40, 41
Glynne, M. D., 34, 36, 109, 111, 278, 279
Gnadinger, C. B., 167, 170, 174
Goble, G. J., 199, 200
Godfrey, G. H., 277
Goette, M. B., 216
Goetze, G., 57, 190, 194
Goskøyr, J., 131
Goldsworthy, M. C., 90, 120, 123, 129, 133
Good, R. J., 70, 72
Gooden, E. L., 182
Goodey, T., 51
Goodhue, L. D., 68, 179, 182
Goodwin, W., 107, 111, 115, 141, 142, 145, 149, 153–5
Gordon, H. T., 207, 208, 215
Gottlieb, D., 14, 22
Götze, G., see Goetze, G.
Gough, G. C., 10, 12
Gough, H. C., 88, 90, 248, 249, 271, 277
Grace, N. H., 237, 238, 265
Graf, J. E., 38, 41
Graf, L., 296
Graham, J. J. T., 162, 163
Graham, S. A., 162, 163, 186, 193, 286, 289
Grainger, J., 39, 41
Granett, P., 289
Grantham, R. J., 133
Graves, A. H., 15, 23
Gray, G. P., 187, 193, 247, 249
Greaney, F. J., 259
Greaves, W. S., 145, 215, 234, 265, 296
Green, A. A., 87, 89, 173, 175
Green, E. L., 90, 120, 123, 129, 133, 188–190, 193, 194
Green, J. R., 187, 193
Green, N., 168, 169, 171, 174, 175, 289
Greenberg, H., 175
Greenslade, R. M., 28, 71, 72, 226, 291
Greeves, T. N., 263, 265
Gregory, P. H., 306, 309
Greig Smith, R., 275, 277
Grevillius, A. Y., 26, 28
Griffin, E. L., 189, 193
Griffiths, J. T., 155, 156
Griffiths, M. A., 257, 258
Griswold, G. H., 251
Grob, H., 228
Gross, P. M., 253
Grove, J. F., 139, 201, 202
Groves, K., 79, 80
Grubb, N. H., 20, 24, 127 ,128

Grummitt, O., 229, 231
Guba, E. F., 249, 250
Guest, H. R., 175
Gulland, J. M., 174
Gunn, D. L., 14, 22
Gunther, F. A., 180, 182, 203, 205, 206, 215
Guthrie, F. E., 210, 216
Guozdenović, F., 141, 145
Gustafson, C., 201, 202
Guterman, C. E. F., 258
Guy, H. G., 90, 158, 159, 288, 289
Gysin, H., 135, 226–8, 241

H

Haag, H. B., 182
Haas, A. R. C., 250
Haber, F., 64, 68, 247
Hagan, E. C., 225
Hailer, E., 258, 259
Haldane, J. B. S., 101, 104
Hall, S. A., 289
Hall, W. C., 241
Hall, W. E., 225
Haller, H. L. (J.), 159, 172, 175, 177–180, 182, 196, 197, 202, 215, 285, 286, 289
Halstead, B. D., 142, 145, 278, 279
Hamilton, C. C., 190, 194
Hamilton, J. M., 88, 90, 110–2, 115, 140, 145
Hamm, P. C., 243, 244
Hammerlund, C., 300, 303
Hamner, C. L., 235, 238, 240, 241
Hampe, P., 67, 68
Hanna, A. D., 49, 51
Hanna, W. F., 11, 13
Hannay, C. L., 56, 57
Hanriot, M., 178, 182
Hansberry, (T.) R., 161–3, 165, 167, 177, 179, 182, 196, 197
Hansch, C., 238
Hansen, B., 235, 238
Hansen, H. N., 12, 13
Hansen, J. W., 200
Hanson, K. R., 183
Harden, A., 23
Harder, H., 169, 175
Harkins, W. D., 73, 76
Harman, J. W., 80
Harper, S. H., 168, 169, 175, 178, 196, 197
Harris, H. H., 157, 159
Harris, L. E., 234, 235
Harris, W. D., 230, 231
Harrison, A., 175, 176
Harrison, C. M., 207, 215
Harry, J. B., 121, 123, 140, 145
Hart, S. J. G., 303, 308
Hartley, C., 6
Hartley, G. S., 73, 76, 224, 226
Hartmann, M., 74, 76
Hartzell, A., 70, 72, 114, 115, 162, 163, 171, 173, 175, 197, 201, 202, 250, 251
Harvey, C. C., 15, 23
Harvey, R. B., 23
Harvill, E. K., 173, 175, 201, 202
Haseman, L., 25, 27
Hassebrauk, K., 30, 33
Hassel, O., 209, 216
Hatfield, W. C., 23
Hatton, R. G., 28
Hawkes, J. G., 8, 9
Hawkins, L. A., 23
Hayes, R. A., 215
Haynes, E. P., 147
Haynes, H. L., 175, 227, 228, 289
Hazelton, L. W., 225
Headlee, T. J., 161, 163, 164, 191, 194
Heal, R. E., 215
Heald, F. D., 257, 258
Heath, D. F., 226
Heath, G. D., 64, 65, 68
Heathcote, G. D., 309
Heatherington, W., 51
Heberlein, C., 118, 122
Hedenburg, O. F., 173, 175
Hedrick, U. P., 127, 128
Heiberg, B. C., 19, 24
Heimpel, A. M., 57
Hein, R. E., 175
Heinemann, H., 68
Helson, V. A., 244
Hemmi, H., 97, 104, 215
Hemming, H. G., 53–5, 139, 146
Hemstreet, C., 21, 24
Henderson, M. R., 176, 181
Henderson, V. E., 277
Henderson Smith, J., 22, 24, 90, 96
Henglein, A., 225
Hennig, E., 254, 255
Hensill, G. S., 80
Hepburn, J. R. I., 118, 122
Herman, F. A., 152, 155
Herms, W. B., 290
Herrick, G. W., 251
Heuberger, J. W., 129, 133
Hewitt, C. G., 289, 290
Hewlett, P. S., 95, 96
Hey, G. L., 198, 200, 202
Hickman, C. J., 121, 123
Hicks, C. S., 165, 166
Hietala, P. K., 20, 24
Higgins, B. B., 15, 23
Higgins, J. C., 271, 277
Higgons, D. J., 231
Hijner, J. A., 287, 289
Hilchey, J. D., 89, 90
Hilgendorff, G., 147, 154
Hill, A. V., 253
Hill, D. L., 223, 225
Hiltner, L., 256, 260, 265, 272, 275, 277
Hilton, W., 182
Hinshelwood, C. N., 97, 104
Hirst, E. L., 17, 23
Hitchcock, A. E., 234, 235
Hobson, R. P., 169, 175
Hochstein, P. E., 135, 136
Hodson, W. E. H., 268
Hoffhine, C. E., 139
Hoffmann, O. L., 238, 239, 241
Hoffman-Bang, N., 68
Holland, E. B., 120, 123, 148, 154, 249, 250

Holland, E. G., 225
Holloway, J. K., 55
Hollrung, M., 256
Holly, K., 235
Holmes, H. L., 159
Holt, J. J. H., 96, 104
Holton, C. S., 267
Hood, C. E., 153, 155
Hooker, J. D., 176, 181
Hopf, H. S., 222, 225
Hopkins, B. A., 283, 284
Hopkins, D. E., 49, 51
Hopkins, D. P., 36
Hopkins, F. G., 108,111
Hopton, C. U., 174
Horn, E. E., 293, 296
Horne, A. S., 18, 23
Horsfall, J. G., 109, 111, 120, 123, 125–9, 133, 134, 144–6, 257, 258
Horton, E., 115, 141, 145
Hoskins, W. M., 71, 72, 77, 78, 80, 113, 115, 193, 207, 208, 215, 216
Howard, A., 33, 35, 36, 42, 43
Howard, F. L., 140, 145
Howard, L. O., 46, 51, 249
Howard Jones, G., 259
Howe, M. F., 14, 22
Howerton, P. W., 216
Howlett, F. S., 114, 115, 127, 128
Hoyle, G., 222, 225
Hoyt, L. F., 162, 163
Hsu, C. T., 118, 122
Hubanks, P. E., 215
Huebner, C. F., 296
Hueck, H. J., 231
Hueper, W. C., 202
Hughes, L. C., 17, 23
Hugill, J. A. C., 159
Huisman, H. O., 139, 229, 231
Hummer, R. W., 181, 182, 230, 231, 243
Hunt, N. R., 273, 276, 277
Hurd, A. M., 17, 23
Hurst, L. A., 278, 279
Hurst, R. H., 281
Hus, P., 292
Husain, M. A., 285, 288–290
Hutchinson, H. B., 275, 277
Hutson, J. M., 159
Hysop, G. R., 234, 235

I

Ichikawa, S. T., 273, 277
Ikawa, M., 159, 296
Imms, A. D., 250, 288–290, 292
Incho, H. H., 175
Ingle, L., 211, 216
Ingram, J. M. A., 238
Irving, G. W., 19, 20, 24
Isenbeck, K., 12
Ishikawa, T., 177, 182
Ivens, G. W., 239
Ivy, E. E., 220, 235
Iwanowski, D., 4

J

Jackson, C. H. N., 41
Jackson, M. L., 66, 68
Jacobs, S. E., 56, 57
Jacobson, M., 196, 197
Jaeger, A. O., 73
Jaeger-Draafsel, E., 231
James, H. C., 51
Jameson, H. R., 277
Janaki Ammal, E. K., 9
Janes, R. J., 68
Jang, R., 226
Jansen, E. F., 223, 236
Janssen, M. J., 133
Jaquiss, D. B., 183
Jarvis, E., 288, 289
Jary, S. G., 76, 77, 162, 163, 190–2, 194, 292
Jeffery, R. N., 165, 167
Jeger, O., 157, 159
Jenkins, J. R. W., 288, 289
Jenkins, R. R., 249, 250
Jensen, J. L., 256, 267, 268, 307, 309
Jeppson, L. R., 229, 231
Jepson, F. P., 31, 33
Jernakoff, M., 90
Jerrel, E. A., 57
Johns, I. B., 159
Johnson, A. C., 252
Johnson, B., 28
Johnson, C. G., 290
Johnson, E. M., 307, 309
Johnson, F. A., 157, 159
Johnson, G. A., 225
Johnson, G. V., 152
Johnson, H. W., 263, 265
Johnson, J., 16, 23, 26, 276, 278
Johnson, J. C., 258, 259
Johnson, J. P., 277
Johnson, J. R., 53, 56
Johnson, S. W., 119, 122
Johnson, W. G., 249
Johnstone, C., 65, 68
Jones, E. S., 34, 36
Jones, H. A., 176, 180–2, 285, 289
Jones, J. K. N., 17, 23
Jones, L. K., 88, 90
Jones, L. R., 35, 36
Jones, M. A., 197
Jones, R. G., 139
Jones, R. L., 239
Judenko, E., 51
Julien, J. B., 23

K

Kaars Sijpesteijn, A., 130–3
Kagi, H., 74, 76
Kagy, J. F., 198–200
Kalajev, A., 24
Kariyone, T., 177, 182
Karlson, P., 284
Kearns, C. W., 204, 208, 211, 212, 215, 216
Kearns, H. G. H., 23, 62, 71, 72, 76, 84, 89, 113, 115, 163, 187, 192–4, 198, 200–2, 268

Keen, B. A., 235
Keil, H. L., 140, 145
Keilin, D., 249
Keitt, G. W., 88, 90
Kellerman, W. A., 256, 259, 260, 265
Kelley, V. W., 188, 193
Kelly, R. B., 157, 159
Kelsall, A., 121, 123, 152, 155
Kelsey, J. A., 142, 145
Kemp, H. J., 25
Kenaga, E. E., 181, 182, 230, 231, 252
Kendrick, J. B., 17, 23, 259
Kennedy, J. S., 68
Kent, N. L., 32, 33
Kent, W. G., 186, 193
Kenten, J., 283, 284
Kern, J. G., 81
Ketelaar, J. A. A., 223, 226
Kienholz, J. R., 139
Kilby, B. A., 225
Kilgore, B. W., 147, 154
Killip, E. P., 181
Kilmer, G. W., 215
King, C. G., 194
King, H. E., 9
King, L. J., 236, 238
King, W. V., 215, 289
Kingscote, A. A., 182
Kirby, A. H. M., 150, 154, 231
Kirkpatrick, A. F., 247, 249
Kirkpatrick, T. W., 40, 41, 45, 46, 51
Kittleson, A. R., 135
Klein, W. H., 240, 241
Klöpping, H. L., 130, 133
Klotz, L. J., 18, 23
Knight, H., 76–8, 80, 188–190, 193, 194, 247, 249
Knight, T. A., 9, 12
Knipling, E. F., 49, 51
Knülsi, E., 241
Koch, A., 274, 277
Koch, R., 4
Koebele, A., 46, 50
Koechlin, H., 215
Kokoski, F. J., 122, 123
Kolbezen, M. J., 226, 228
Kollros, J. J., 215, 216, 225
Koolhaas, D. R., 178, 182
Koopman, H., 241
Koopmans, M. J., 135
Kornfeld, E. C., 138, 139
Korsmo, E., 232, 234
Kovács, A., 14, 22
Krahl, M. E., 199, 200
Kramer, J. A., 238
Kramer, O., 145
Krijgsman, B. J., 206, 215
Kroemer, K., 127, 128
Krueger, G. von, 221, 225
Krukoff, B. A., 176, 181
Kuehl, F. A., 139
Kuenen, D. J., 231
Kühn, J., 255, 256
Kuhn, R., 14, 24, 27, 28
Kükenthal, H., 217
Kunkel, L. O., 268, 304, 305, 309

L

Labatut, R., 210, 216
Lacey, M. S., 54, 56
Lackman, W. H., 244
Ladell, W. R. S., 247, 249
LaDue, J. P., 215
LaForge, F. B., 168, 169, 171, 172, 177, 182
Lambert, E. B., 38, 41
Lamberton, J. A., 183
Lambrech, J. A., 228, 238
Lane, W. J., 226
Lane, W. R., 63, 68
Lange, W., 221, 225
Langenbuch, R., 208, 216
Langmuir, I., 73, 76
Lapham, M. H., 33, 36
LaPidus, J. B., 28
Larrimer, W. H., 43
Last, F. T., 30, 32
Lathrop, F. H., 283, 284
Latus, M., 241
Läuger, P., 180, 182, 203, 204, 215
Laurent, E., 30, 32
Lawes, J. B., 30
Lawrence, W. J. C., 276, 278
Leach, B. E., 139
Leach, B. R., 271, 277
Leach, J. G., 18, 20, 23, 24, 263, 265
Leach, L. D., 35, 36
Lear, B., 252, 277
Le Baron, W., 147, 154, 297
Lebediantzeff, A., 270, 276
Le Conte, J. L., 61, 68
Lee, C. S., 196, 197
Lee, H. A., 111
Lees, A. D., 283, 284
Lees, A. H., 12, 13, 27, 28, 142, 145, 193
Lefroy, H. M., 157, 159, 251, 298, 299, 302
Le Goupil, —., 252
Lenz, W., 177, 182
Leonard, C. S., 154
Leonian, L. H., 13
Leopold, A. C., 240, 241
Le Pelley, R. H., 25, 28, 297, 302
Leslie, P. H., 57
Leszczenko, P., 141, 145
Levan, A., 9
Leukel, R. W., 255, 257, 258, 280, 281, 306, 309
Lewallen, L. L., 208, 216
Lewis, S. E., 200, 225, 252
Lhoste, J., 267
Lidov, R. E., 212, 213, 216
Liebig, J. von, 30, 32
Liener, I. E., 27, 28
Lies, T., 202
Lilley, C. H., 68
Lind, J., 301, 303
Lindgren, D. L., 215, 253
Lindquist, D. A., 214, 216
Lineweaver, H., 102, 105

Linford, M. B., 55, 56
Link, A. de S., 24
Link, G. K. K., 21, 24
Link, K. P., 16, 23, 159, 295, 296
Lipke, H., 27, 28
Lipmann, F., 199, 200
Little, J. E., 136, 137
Little, V. A., 182
Little, W. C., 30, 32
Llewellyn, M., 226
Lloyd, L., 249, 251, 293, 296
Locke, S. B., 140, 145
Lockley, R. M., 59
Lodeman, E. G., 167, 174, 186, 193, 194, 196
Loeffler, L., 57
Loh, T. C., 141, 145
London, F., 100, 104
Loomis, W. F., 199, 200
Loos, C. A., 43
Lord, K. A., 197, 222, 225
Lorenz, W., 221, 225
Loveday, P. M., 216
Lovett, A. L., 153, 155
Löw, I., 19, 24, 27, 28
Lowe, V. H., 112, 115
Lowenstein, O., 87, 90
Lowry, P. R., 69, 72, 163, 196
Lubatti, O. F., 252
Ludwig, R. A., 15, 23, 132, 133
Luers, H., 215
Lukens, R. J., 135, 136
Lundbäck, S. V., 154
Luther, E. E., 149, 154
Lutman, B. F., 127, 128

M

Ma, R., 24
Maan, W. J., 13
MacDaniels, L. H., 110, 112
MacDougall, R. S., 286, 289
MacGill, E. I., 33, 36
Mach, E., 107, 111
Machacek, J. E., 53, 55, 259
Mack, G. L., 71, 72, 140, 145
Mackie, W. W., 257–9
Maclachlan, G. A., 241
MacLeod, G. F., 66–8
MacMillan, J., 139
Mader, E. O., 122, 123, 128
Maeda, S., 228
Magerlein, B., 215
Magie, R. O., 128
Maines, W. W., 285, 289
Malelu, J. S., 259
Mallman, W. L., 21, 24
Mally, C. W., 246, 249
Maltais, J. B., 26, 28
Mamelle, J., 274, 277
Mangini, —., 107, 111
Mann, H. D., 215
Manns, T. F., 129, 133
Manske, R. H. F., 159
Marais, J. S. C., 294, 296
March, R. B., 208, 212, 216, 223, 225, 226
Maramorosch, K., 305, 309
Marcovitch, S., 155, 156, 283, 284
Margot, A., 135
Marke, D. J. B., 65, 68
Markwood, L. N., 163, 164, 166
Marlatt, C. L., 46, 51
Marmoy, C. J., 265, 266
Marrian, D. H., 281
Marryat, D. C. E., 16, 23
Marsh, P. B., 144, 146
Marsh, R. W., 23, 41, 71, 72, 76, 113, 115, 121, 123, 125, 126, 128, 133, 159, 163, 187, 193
Marshall, G. E., 291, 292
Marshall, J., 77–80, 154, 155
Marshall, R. P., 273, 276, 277
Martin, D. F., 225
Martin, H(enri), 180, 182, 203, 204, 215
Martin, H(ubert), 70–2, 75–80, 84, 89, 107, 108, 111, 113–7, 120, 121, 204, 215, 225, 232, 235, 265, 296, 306, 309
Martin, J. P., 111
Martin, J. T., 92, 96, 178, 179, 182
Martin, W. H., 128
Mason, B. J., 70, 72
Massee, A. M., 25, 28, 51, 268, 291
Massee, G., 15, 23
Masson, E., 118, 122
Matsubara, H., 174, 176
Matsui, M., 169, 175
Matthews, A., 272, 276, 277
Mattson, A. M., 221, 225
Mattson, E. L., 225
Maxon, M. G., 226
May, C., 20, 24, 114, 115, 127, 128
May, E., 151, 154
McAllister, L. C., 158, 159
McAlpine, D., 128, 258, 259
McAtee, W. L., 45, 46
McBeth, C. W., 273, 277
McCall, G. L., 199, 200
McCallan, S. E. A., 88, 90, 99, 104, 108–111, 114, 115, 117, 122, 124–6, 133, 134, 140, 141, 145
McCarter, W. S. W., 68
McCartney, W., 182
McClintock, J. A., 62, 68
McColloch, J. W., 25, 27
McCombie, H., 225, 296
McDaniel, A. S., 111
McDonnell, C. C., 160, 162, 163
McFarland, R. H., 175
McGovran, E. R., 197
McGowan, J. C., 54, 56, 144, 146
McGregor, J. K., 182
McGregor, W. S., 225
McIndoo, N. E., 149, 150–2, 154, 161–3, 183, 196, 197, 284, 288
McIntosh, A. H., 88, 89
McKay, M. B., 23, 280, 281
McKay, R., 122, 123
McKee, R. K., 143, 145
McKenzie, A., 23
McKinlay, K. S., 231

McKinney, H. H., 34, 36, 42, 43
McKinney, K. B., 28
McLean, R., 253
McMartin, A., 265
McNew, G. L., 79, 80, 92, 96, 136, 137
McWhorter, F. P., 142, 145
Means, O. W., 222, 225
Medler, J. T., 285, 289
Mehrotra, K. N., 222, 225
Meijer, T. M., see Meyer, T. M.
Melander, A. L., 190, 194
Melander, L. W., 16, 23
Melhus, I. E., 259
Meltzer, J., 229, 231
Mendes, C., 47, 51
Menschikov, G., 164, 166
Mentzel, F., 292
Mentzer, C., 239
Menusan, H., 195, 196
Menzel, K. C., 128
Meredith, C. H., 54, 56
Merrell, D. J., 207, 215
Métalnikov, S., 56, 57
Métalnikov, S. S., 56, 57
Metcalf, R. L., 171, 175, 205, 207–9, 212, 215, 216, 223–6, 228, 229, 231
Metcalfe, T. P., 239
Metchnikoff, E., 52, 55, 56
Metzger, F. J., 246, 249
Metzger, F. W., 285, 289
Meyen, F. J. F., 4, 6
Meyer, A., 121, 123
Meyer, H., 33
Meyer, H. H., 97, 104
Meyer, K. H., 97, 104, 215
Meyer, T. M., 178, 179, 182
Miles, H. W., 281
Millard, W. A., 53, 55, 278, 279
Millardet, A., 5, 60, 75, 77, 115, 116, 122, 124, 126
Miller, A. C., 174, 176
Miller, C. S., 241
Miller, D., 277
Miller, D. M., 241
Miller, E. C., 104, 105
Miller, J. A., 104, 105
Miller, J. G., 66, 68
Miller, L. P., 104, 108, 109, 111, 134
Miller, P. R., 39, 41
Miller, P. W., 268
Millerd, A., 234, 235
Mills, J. E., 155, 156
Mills, W. D., 39, 41
Minarik, C. E., 241
Minshall, W. H., 242, 244
Misaka, K., 161, 163
Missiroli, A., 215
Mitchell, J. G., 234
Mitlin, N., 170–2, 175, 197
Miyajima, S., 182
Miyoshi, M., 15, 23
Moilliet, J. L., 72, 76
Molinas, E., 272, 277
Molho, D 239
Molz, E., 33, 36, 272, 277
Mond, R. L., 118, 122
Monroe, C. M., 134
Monteith, J., 35, 36
Montgomery, H. B. S., 87, 89, 127, 128, 140, 145
Moore, J. B., 290
Moore, M. H., 87, 88, 127, 128, 140, 145
Moore, R. H., 197
Moore, W., 66, 68, 96, 104, 157, 159, 162, 163, 186, 193, 248–50, 253
Moore, W. C., 12, 17, 23, 257, 258, 298, 302
Moorefield, H. (H.), 208, 216, 228
Mori, A., 107, 111
Morris, H. E., 152–4, 187, 193
Morris, H. M., 87, 89, 176, 182, 200
Morris, K. R. S., 302
Morris, L. E., 142, 145
Morris, W. T., 139
Morrison, F. O., 88, 90
Morse, W. J., 141, 145
Morstatt, H., 6
Moznette, G. F., 153, 155
Mueller, G. F., 153, 155
Mueller, G. P., 215
Muir, R. M., 238
Mulholland, T. C. P., 139
Muller, H. J., 9
Müller, H., 33, 36
Müller, P., 180, 182, 203, 204, 206, 215
Mullins, L. J., 206, 210, 215
Munck, J. C., 293, 294, 296
Munger, F., 159
Munson, R. G., 159
Munson, S. C., 150, 154
Murtwyler, C., 174
Murphy, D. F., 200, 202
Murphy, P. A., 122, 123
Murray, D. R. P., 101, 104, 190, 194
Murray, R. L., 81
Muskett, A. E., 88, 90, 122, 123, 128, 263, 265
Muth, F., 107, 111
Myers, J. G., 45, 46
Myers, K., 59
Mylius, A., 215

N

Nachmansohn, D., 102, 105
Nagai, K., 177, 182
Nattrass, R. M., 145
Nealon, E. J., 162, 163
Neatby, K. W., 11, 13
Neifert, I. E., 253
Neilson-Jones, W., 54, 56
Nelson, C, I., 20, 24
Nelson, F. C., 87, 90
Nelson, H. D., 249
Nelson, O. A., 166
Newcomer, E. J., 292
Newell, J., 276, 278
Newhall, A. G., 252, 258, 273, 276, 277
Newman, F. C., 168, 175

Newman, M. S., 215
Newton, H. C. F., 287–9, 296
Newton, R., 19, 23
Nielsen, L. W., 140, 145
Nienow, I., 138, 139
Nikitin, A. A., 120, 123
Nishamura, M., 299, 302
Nitsche, G., 151, 154
Nixon, M. W., 276
Nobbe, F., 257, 258
Nord, F. F., 109, 111
Norman, A. G., 238
Norton, L. B., 152, 154, 155, 162, 163, 165–7, 179, 182
Nostitz, A. von, 272, 277
Nottle, R. A., 166
Nougaret, R. L., 33, 36
Nutman, P. S., 235
Nuttall, W. H., 70, 72, 159
Nutting, M. D. F., 226

O

Oaks, A., 241
Oberlin, C., 270, 276
O'Brien, D. G., 281
O'Brien, M., 39, 41
O'Brien, R. D., 222, 224–6, 228
O'Colla, P., 211, 216
O'Connor, J. A., 296
O'Connor, M. G., 293, 296
O'Donnell, F. G., 273, 276, 277
Oettingen, W. F. von, 201, 202, 252
Ogilvie, L., 255, 265
O'Kane, W. C., 69, 72, 87, 90, 151, 154, 162, 163, 195, 196, 253, 271, 277
Oliver, J., 76
O'Meara, P., 150, 153–5
Omerod, E. A., 183
Ono, M., 182
Orchard, O. B., 180, 182
Orékhov, A., 164, 166
Orrell, W. R., 145, 215, 265, 296
Otto, H. D., 205, 206, 215
Overley, F. L., 187–9, 191, 193
Overman, R. S., 296
Overton, E., 97, 104
Owen, J. H., 19, 23, 24
Owen, O., 247, 249
Owings, J. F., 241
Oxford, A. E., 139
Oxley, T., 176, 181

P

Pachéco, H., 239
Page, A. B. P., 174, 175, 252
Paine, S. G., 33
Painter, R. H., 12, 13
Pallansch, M. J., 28
Palmer, R. C., 77
Parfentjev, I. A., 153, 155
Parke, T. V., 139
Parker, J. R., 34, 36
Parker, T., 250, 251
Parker, W. B., 80, 183
Parker, W. L., 211, 216
Parker-Rhodes, A. F., 91–4, 96, 109, 111, 117, 129, 133
Parkin, E. A., 87, 89, 173, 175
Parnell, F. R., 9
Parrott, P. J., 112, 115
Parson, H. E., 263, 265
Pastac, I., 142, 145, 233, 235
Pasteur, L., 4, 17, 23
Patkaniane, A., 145
Patrigeon, G., 119, 122
Patten, A. J., 150, 153-5
Patterson, N. A., 51, 159
Pattison, F. L. M., 296
Patton, R. L., 199, 200
Pauling, L., 104
Payne, D. J. H., 200
Pearce, G. W., 149, 152, 154, 155, 190, 194, 225
Pearce, T. J. P., 76
Peck, R. L., 139
Peet, C. H., 200, 202
Pellegrini, J. P., 176
Peltier, G. L., 37, 38, 41
Pemberton, C. E., 27, 28, 37, 41
Penman, H. L., 270, 276
Peoples, S. A., 244
Pepper, B. B., 157, 159, 161, 163
Perkins, R. C. L., 51
Perry, A. S., 208, 216
Perry, J. W., 72, 76
Persing, C. O., 200
Petch, T., 52, 55
Peters, B. G., 273, 277
Peters, G., 64, 68
Peters, R. A., 294, 296
Petersen, S., 267
Peterson, P. D., 90, 114
Petherbridge, F. R., 41, 123, 183, 277, 286, 289
Pethybridge, G. H., 257, 258
Phillips, G. L., 249
Phipers, R. F., 177–180, 182
Pickard, J. P., 204, 215
Pickering, S. U., 75, 77, 82, 84, 116–8, 120, 122, 124, 126, 149, 154, 186, 193, 232, 235, 257, 276, 277
Pickett, A. D., 49, 51, 159
Pickles, A., 32, 33
Pictet, A., 164, 166
Pielou, D. P., 215
Pierce, L., 140, 145
Pierce, W. D., 305, 309
Pierpont, R. L., 201, 202
Pilat, M., 155, 156
Pinckard, J. A., 253
Pirie, H., 97, 98, 104, 253
Plackett, R. L., 95, 96
Plank, H. K., 196, 197
Platt, A. W., 25, 28
Platz, G. A., 14, 22
Pollacci, E., 107, 111
Pollard, A. G., 271, 277
Pool, V. W., 23

Poole, J. B., 24
Popenoe, C. H., 194–6
Porte, W. S., 299, 302
Portecorvo, G., 12, 13
Portele, K., 111
Portsmouth, G. B., 43
Posnette, A. F., 48, 51
Posnjak, E., 116, 122
Potter, C., 87–9, 171, 175, 196, 197, 222, 225
Potter, M. C., 21, 24
Potts, S. F., 64, 66, 68, 286, 289
Poussel, H., 207, 215
Powell, A. R., 111
Powell, D., 121, 123
Power, F. B., 283, 288
Powers, G. E., 191, 194
Pozefsky, A., 176
Pramer, D., 139, 140
Pratt, F. S., 247, 249
Prelog, V., 157, 159
Prentice, I. W., 281
Prescott, J. A., 270, 276
Preston, N. C., 274, 277
Prévost, B., 115, 122, 123, 126, 255, 256
Price Jones, D., 281
Priess, H., 180, 182
Prill, E. A., 175
Pritchard, F. J., 299, 302
Probert, M. E., 142, 145
Purdy, L. H., 267

Q

Quaintance, A. L., 186, 193
Quastel, J. H., 234, 235
Quarterman, K. D., 216
Quale, H. J., 246–250

R

Rabaté, E., 233, 235
Rache, J., 216
Radeleff, R. D., 215
Rader, W. E., 134
Radomski, J. L., 214, 216
Raistrick, H., 138, 139, 281
Raleigh, W. P., 122
Ramsay, J. A., 222, 225
Ramsbottom, J. K., 268
Ramsey, G. B., 19, 24
Rand, F. V., 305, 309
Rasmussen, I. M., 215
Rasmussen, N. H., 68
Ratcliffe, F. N., 59
Rathbun-Gravett, A., 6.
Raucourt, M., 209, 216, 284, 288
Rawlins, W. A., 128
Ray, B. R., 70, 72
Rayner, M. C., 54, 56
Raynor, R. N., 235
Read, W. H., 65, 68, 180, 182
Read, W. J., 57
Ready, D., 240, 241
Reckendorfer, P., 111, 124, 126
Reddick, D., 307, 309
Reed, G. M., 43
Reed, J. C., 148, 154
Refai, F. J., 25, 27
Reiber, H. G., 235, 244
Reid, J. H., 251
Reid, W. J., 125, 126
Reimer, F. C., 10, 13
Remer, W., 31, 33
Rennerfelt, E., 18, 23
Reynolds, E. S., 19, 24
Rey-Pailhade, J. de, 108, 111
Rhodes, E. O., 194
Rhodes, H., 285, 289
Ricaud, J., 116, 122
Rich, S., 70, 80, 134, 144–6
Richards, A. G., 171, 175
Richardson, C. H., 159, 161, 163–7, 189, 193, 198–200
Richardson, H. H., 89, 90, 252
Richardson, L. T., 55, 56, 266, 267
Richardson, R. E., 81
Richter, C. P., 294, 296
Richter, L., 275, 277
Riehm, E., 260, 261, 265, 267, 268
Riemschneider, R., 205, 206, 215
Ripley, L. B., 155, 156
Ripper, W. E., 49, 51, 68, 200, 226, 231
Roach, W. A., 20, 24, 25, 28, 109, 111, 278, 279
Roan, C. C., 228
Roark, R. C., 177, 181, 196, 197, 253
Robbins, W. W., 235
Roberts, A. W. R., 96, 104
Roberts, H. A., 235
Roberts, J. W., 140, 145
Robertson, A., 177, 178, 182, 183
Robertson, A. G., 41
Robertson, J., 73, 76, 106, 107, 111
Robinson, D. B., 23
Robinson, R. H., 75, 76, 148, 149, 154
Robson, J. M., 9
Roche, J. N., 194
Rodier, W., 46, 49
Roeder, K. (D.), 166, 167, 207, 215, 222, 225
Roeder, S., 166, 167
Roemer, T., 12
Rogers, B. J., 241
Rogers, E. F., 215
Rogoff, W. M., 212, 216
Rohwer, G. G., 302
Roseman, S., 296
Rotschy, A., 164, 166
Rudolfs, W., 163, 164
Ruhland, W., 124, 126
Rumm, C., 123, 126
Rupprecht, G., 107, 111
Rusby, G. L., 178, 182
Russell, E. J., 270, 275–8
Russell, E. W., 235
Russell, J., 252
Russell, P. B., 281
Ruston, D. F., 9
Ruzicka, L., 167, 168, 170, 174

S

Safro, V. I., 114, 115
Sajó, K., 187, 193
Salaman, R. N., 21, 22, 24
Saleh, A. M., 17, 23
Salmon, E. S., 20, 24, 30, 77, 86, 89, 108, 109, 111, 113, 115, 141, 142, 195, 196, 258, 259, 300, 303
Salmon, S. C., 25
Salzberg, P. L., 200, 202
Sampson, K., 264, 265
Samuels, C. D., 188, 193
Sanders, G. E., 114, 115, 121, 123
Sanders, H. J., 169, 175
Sanford, G. B., 53, 55
Sankowsky, N. A., 90
Sardiña, J. R., 21, 24
Sass, J. E., 264, 265
Sasser, E. R., 6
Sauchelli, V., 111
Saunders, B. C., 225, 295, 296
Saunders, C. B., 254, 255
Saunders, D. H., 166, 167
Savage, W. G., 57
Savige, W. E., 183
Savit, J., 215, 216, 225
Scales, A. L., 225
Schaffer, P. S., 159
Schaffnit, E., 32, 33
Schallek, W., 166, 167
Schander, R., 57, 124, 126–8
Schanderl, K., 127, 128
Schechter, M. S., 169, 175, 196, 197
Schechtman, J., 118, 122
Schenck, S. L., 216
Schmid, H., 179, 181, 182
Schmidt, E. W., 145
Schneider, A., 215
Schoeler, Om., 300, 303
Schoene, D. L., 239, 241
Schoene, W. J., 149, 154
Schrader, G., 216–9, 221, 223, 225, 252
Schreiber, A. F., 197, 288, 289
Schroeter, J., 11, 12
Schulthuss, H. H., 255, 256
Schulze, C., 276, 277
Schuster, J., 16, 23
Schwaebel, —., 272, 277
Schwartz, A. M., 72, 76
Sciarini, L. J., 109, 111
Scott, W. M., 112, 115
Seibt, S., 174
Seiferle, E. J., 157, 159
Selmi, F., 107, 111
Sempio, C., 107, 109, 111
Serfaty, A., 222, 225
Seshadri, T. R., 179, 182
Sessions, A. C., 120, 123, 125, 126, 128
Sestini, F., 107, 111
Sexton, W. A., 235, 238, 239
Shafer, G. D., 155, 156, 186, 193, 247, 249
Sharvelle, E. G., 136, 137, 267
Shaw, H., 127, 128, 140, 145, 159, 161, 163, 192, 194, 198, 200, 202, 225, 265, 296
Shaw, H. B., 280, 281
Shaw, W. S., 238, 239
Shelford, V. E., 38, 41
Shepard, H. H., 6, 161, 163, 165, 167, 253
Shill, A. C., 248, 250
Shimada, M., 182
Siang, W. N., 267
Sicard, L., 116–8, 122
Siegler, E. H., 159, 194–6, 291, 292
Sievers, A. F., 183, 196, 197
Silver, J., 293, 294, 296
Simonart, P., 139
Simpson, A. C., 119, 200
Sinclair, J., 267, 268
Sirrine, F. A., 112, 115
Sisler, H. D., 131, 135, 136
Skerrett, E. J., 205, 215
Slade, R. E., 216, 235, 238
Slater, E. C., 200
Small, T., 40, 41
Smallman, B. N., 222, 225
Smedley, E. M., 281
Smieton, M. J., 143, 145
Smissman, E. E., 27, 28
Smith, A. C., 176, 181
Smith, A. E., 238, 239
Smith, C. L., 215
Smith, C. M., 153, 155
Smith, C. R., 164–7
Smith, H. S., 6, 12, 13, 48, 51, 309
Smith, H. V., 156
Smith, H. W., 253
Smith, J. N., 200
Smith, K. M., 58, 59
Smith, L. E., 159
Smith, L. M., 66–8
Smith, M. A., 119, 129, 133
Smith, M. C., 156
Smith, N. K., 145, 215, 265, 296
Smith, O. J., 90
Smith, R. E., 12, 13, 163
Smith, R. H., 68, 77, 78, 80, 186, 193
Smith, T. O., 120, 122
Smithies, R. H., 200
Snapp, O. I., 190, 191, 194
Soenen, A., 144, 146
Soloway, S. B., 168, 175, 216
Somers, E., 78, 80
Sosnovsky, G., 134, 136
Sousa, A. A., 175
Southern, B. L., 257, 258
Späth, E., 165, 166
Spencer, E. Y., 224, 226
Spero, L., 296
Speroni, G., 208, 216
Speyer, E. R., 48, 51, 157, 159, 180, 182, 247, 249–251
Speyer, W., 192, 194
Speziale, A. J., 243, 244
Spillane, J. T., 225
Spinks, G. T., 30, 32
Sprague, R., 144, 146
Sproston, T. J., 137
Spuler, A., 153, 155, 187–191, 193, 194

Stahl, C. F., 305, 309
Stahmann, M. A., 18, 19, 23, 24, 226, 296
Stakman, E. C., 11–13, 30, 32, 301, 303
Stammers, F. M. G., 64
Standen, H., 216
Standen, J. H., 80
Staniland, L. N., 25, 28, 39, 41, 191, 192, 194, 268
Stanley, W. W., 64, 68, 156
Stansbury, H. A., 175
Stapley, J. H., 277
Starkey, R. L., 275, 277
Staudermann, W., 86, 89, 144, 145
Staudinger, H., 167–170, 174
Stedman, E., 226, 228
Steer, W., 51, 86, 89, 161, 163, 192, 194, 198, 200, 202
Steiner, L. F., 291, 292
Steiner, P., 152, 154
Steinhaus, E. A., 52, 55–7
Stellwaag, F., 72
Sternburg, J., 208, 215, 216
Stevens, N. E., 3, 6
Stevenson, H. A., 231
Stewart, A. W., 81
Stewart, W. D., 80
Stocken, L. A., 150, 154
Stoddard, E. M., 133
Stoker, R. I., 197
Storey, H. H., 304, 308, 309
Störmer, K., 275, 277
Storrs, E. E., 89, 90
Stoughton, R. H., 36, 41
Stover, R. H., 34, 36
Streeter, L. R., 110, 111, 122, 123, 148, 154, 163, 164
Strickland, A. H., 48, 51
Stringer, A., 25, 28, 89, 180, 182, 215, 276
Stubbs, J., 139, 140
Stubbings, W. W., 76
Stützer, A., 265
Subramaniam, T. S., 182
Sudhoff, R. W., 81
Suit, R. F., 128
Sullivan, W. N., 159, 166, 167, 172, 175, 182, 197
Sullivan, W. R., 296
Sumerford, W. T., 208, 215
Swain, A. F., 52, 55, 249
Swanson, C. P., 238
Swanson, C. R., 238, 239
Sweetman, H. L., 38, 41, 45, 46, 155, 156
Swezey, O. H., 51
Swingle, D. B., 152–4
Swingle, H. S., 151, 154, 190, 191, 194
Swingle, M. C., 88, 90, 171, 175
Swingle, W. T., 123, 124, 126, 256, 259, 260, 265
Synerholm, M. E., 175, 237, 238
Szembel, S. J., 141, 145
Szeöke, E., 14, 22
Szyszkowski, B. von, 97, 104

T

Taber, W. C., 116, 122
Taff, A. W., 169, 175
Taillade, M., 107, 111
Takahashi, W. N., 143, 145
Takei, S., 177, 182
Talbert, T. J., 113, 115
Tanner, C. C., 145, 215, 265, 296
Tartar, H. V., 148, 154
Tattersfield, F., 86, 89, 90, 155, 156, 166, 169, 171, 175–9, 181, 182, 194–8, 200
Taubenhaus, J. J., 19, 23
Tauber, O. E., 155, 156
Taylor, C. B., 53, 55
Taylor, J., 65, 68
Taylor, J. W., 257, 258
Taylor, L. R., 290
Taylor, R. E., 16, 23
Taylor, T. H., 286, 289
Taylor, T. H. C., 47, 51
Tehon, L. R., 38, 41
Templeman, W. G., 234, 235, 239, 265, 266
Terényi, A., 257, 258, 263, 265
ter Horst, W. P., 136, 137
Thatcher, R. W., 110, 111, 147, 154, 163, 164
Theobald, F. V., 279, 281, 290, 292, 297, 302
Thiem, H., 183, 271, 277
Thomas, E. L., 253
Thomas, F. J. D., 277, 291
Thomas, I., 183
Thompson, H. E., 238, 241
Thompson, H. V., 59
Thompson, R., 278
Thompson, R. H. S., 150, 154
Thompson, W. R., 45, 46
Thomson, W., 85
Thomson, W. E. F., 41
Thorn, G. D., 132, 133
Thorne, G., 51
Thornton, H. G., 32, 33, 235
Thornton, N. C., 36
Thorold Rogers, M. A., 139
Thorpe, W. H., 12, 13
Thorsteinson, A. J., 26, 28
Thung, T. H., 22, 24
Thurston, H. W., 123, 145
Tilemans, E. M. J., 128
Tilford, P. E., 127, 128
Tims, E. C., 19, 24
Tischler, N., 182, 243, 244
Tisdale, W. B., 34, 36
Tisdale, W. H., 16, 23, 133, 257, 258
Tobias, J. M., 206, 215, 216, 222, 225
Todd, A. R., 280, 281
Todd, C. W., 241, 242
Tolba, M. K., 17, 23
Tomlinson, J. R., 176
Tomkins, R. G., 89, 90
Toms, B. A., 68
Tonkin, W. H., 253
Townsend, A. A., 66, 68
Trappmann, W., 151, 154
Traube, I., 96, 97, 104

Trevan, J. W., 91, 96
Triffit, M. J., 280, 281
Trischmann, H., 19, 24
Trotter, —., 46, 51
Trouvelot, B., 284, 288
Truffaut, G., 142, 145, 233, 235, 271
Tso, T. C., 165, 167
Tucker, R. P., 107, 111, 187, 193
Tudor, P., 68
Tukey, H. B., 235, 238
Tull, J., 255, 256
Tunell, G., 116, 122
Turner, N., 166, 167
Turner, P. E., 32, 33
Turner, W. B., 290
Turrell, F. M., 110, 112
Tutin, F., 84, 171, 175, 191, 192, 194
Twarog, B. M., 222, 225
Tydeman, H. M., 28

U

Udey, E. C., 128
Uhlenbroek, J. H., 135, 241
Ullyett, G. C., 49, 51
Underhill, J. C., 207, 215
Unwin, C. H., 133
Uppal, B. N., 97, 104, 259
Urbschat, E., 266, 267
Uvarov, B. P., 36, 41, 151, 154

V

Valleau, W. D., 23, 307, 309
van Asperen, K., 210, 216
van Bavel, C. H. M., 270, 276
van der Kerk, G. J. M., 130, 132, 133
van der Linden, T., 209
van der Meulen, P. A., 152, 155
van der Veen, R., 229, 231, 242
Van Everdingen, E., 39, 41
van Leeuwen, E. R., 152, 155, 158, 159, 285, 289
van Overbeek, J., 244
van Poeteren, N., 45, 46
van Slogteren, D. H. M., 24
van Slogteren, E., 21, 24
van Slyke, L. L., 154
Varadarajan, S., 179, 182
Vassiliev, I. V., 52, 55
Vasudeva, R. S., 18, 23
Venanzi, L. M., 130, 133
Verguin, J., 37, 41
Vermorel, V., 75, 76, 116
Verschaffelt, E., 26, 28
Vincent, M. H., 296
Vincent, W. B., 163
Virtanen, A. I., 20, 24
Vlitos, A. J., 236, 238
Voegtlin, C., 150, 154
Voelkel, H., 152, 154
Vogelbach, C., 212, 216
Vogt, E., 107, 111, 227
Volch, W. H., 113–5, 149, 156
Volk, A., 32, 33

W

Wachs, H., 173, 175
Waeffler, R., 135
Wagner-Jauregg, T., 202
Wain, R. L., 76, 117, 121, 123, 125, 126, 180, 182, 200, 204, 215, 237, 238
Waite, M. B., 13, 113, 115, 141, 145, 163, 304, 308
Waksman, S. A., 137, 139, 275, 277
Walker, E., 126, 264, 265
Walker, G. L., 90
Walker, H. W., 155, 156
Walker, J. C., 18, 19, 23, 24, 35, 36
Walker, M. M., 41
Walker, M. N., 299, 302
Walker, T. W., 278
Wallace, E., 113–5
Wallace, T., 107, 111, 114, 115
Wallen, V. R., 138, 139
Walton, C. L., 191, 192, 194, 268, 288, 289
Walton, R. C., 113, 115
Wampler, E. L., 71, 72
Warburg, J. W., 200
Warburg, O., 247, 249
Ward, J. C., 296
Ward, H. M., 11, 13, 300, 303
Wardlaw, C. W., 34, 36
Ware, W. M., 20, 24, 142, 145
Waring, W. S., 281
Waterhouse, W. L., 11, 13
Waters, H. A., 121, 123
Waters, W. A., 64, 66, 68
Watzl, O., 190, 194
Way, A. M., 140
Way, M. J., 197, 283, 284
Waygood, E. R., 240, 241
Webb, E. C., 223, 225, 226
Webster, R. L., 154, 155
Weed, A., 173, 175
Weed, R. M., 104, 108, 111, 133
Weiant, E. A., 207, 215
Weigel, C. A., 249
Weindling, R., 53, 55
Weinman, C. J., 216, 293, 296
Weiss, H. B., 289, 290
Wellman, F. L., 35, 36
Wellman, R. H., 88, 90, 123, 133, 134, 145
Wells, R. W., 215
Welsh, J. H., 207, 215
Werotte, L., 144, 146
Wesenberg, G., 260
Wessels, J. S. C., 242
West, B., 134
West, T. F., 168, 169, 175
Westgate, W. A., 69, 72, 163, 196
Whaley, F. R., 123, 145
Whaley, W. M., 183
Wheatley, W., 215
Whetstone, R. R., 134, 225
Whetzel, H. H., 141, 145
Whiffen, A. J., 138, 139
Whitaker, W. C., 304, 309
Whitcomb, W. D., 251
White, D. A., 166

White, R. P., 110, 111
White, R. W., 130, 133, 224, 226, 241
Whitehead, C., 183
Whitehead, T., 306, 309
Whitfield, F. G. S., 68
Whittaker, V. P., 102, 105
Whittingham, D. J., 159
Wiant, J. S., 19, 24
Wickert, J. N., 76
Wiehe, P. O., 265
Wiersma, C. A. G., 166, 167
Wiesmann, R., 207, 208, 227, 228
Wiesner, K., 159
Wigglesworth, V. B., 161, 163, 171, 175, 207, 215
Wightman, F., 238
Wilbrink, G., 268
Wilcox, H., 24
Wilcoxon, F(rank), 70, 72, 88, 90, 108–111, 114, 115, 117, 122, 124–6, 140, 145, 153, 155, 162, 163, 201, 202, 250, 251
Wilcoxon, F(redericka), 196, 197
Wild, H., 206, 215
Wilkes, B. G., 76
Wilkins, A., 192, 194
Wilkinson, E. H., 117, 121, 123, 125, 126
Willaman, J. J., 16, 23, 248–250
Williams, C. B., 33, 36, 290
Williams, C. M., 284
Williams, R. T., 200
Wilson, A. R., 234
Wilson, C. S., 173–5
Wilson, F., 283, 284
Wilson, H. F., 66, 68, 150, 154
Wilson, I. B., 102, 105
Wilson, R. A., 139
Wiltshire, S. P., 107, 111, 187
Wingard, S. A., 20, 21
Winteringham, F. P. W., 172, 174, 175, 208, 209, 216, 252
Winton, F. R., 296
Witman, E. D., 123
Wöber, A., 116, 117, 122, 140, 145
Woglum, R. S., 246, 248–250, 299, 302
Woke, P. A., 108, 182
Wolf, F. A., 253
Wolf, F. T., 134
Wood, J. I., 3, 6
Wood, R. E. G., 145, 215, 265, 296
Woodcock, D., 136, 137, 205, 215
Woodfin, J. C., 145
Woodman, R. M., 69, 70, 72, 75, 77, 78, 80, 82, 84
Woodward, R. C., 277
Woolley, D. W., 137
Wormald, H., 20, 24, 258, 259
Wormald, L. K., 115
Worrall, L., 7, 9
Worsley, R. R. de G., 177, 182
Wright, J. M., 54, 56, 139, 140
Wright, R. H., 287, 289
Wüthrich, E., 140, 145
Wyckoff, R. W. G., 58, 59
Wylie, S. M., 281

Y

Yadoff, O., 67, 68
Yamafugi, K., 58, 59
Yamamoto, R., 167, 174
Yap, F., 55, 56
Yarnold, G. D., 70, 72
Yarwood, C. E., 112
Yeager, J. F., 150, 154
Yersin, H., 267
Yoshihara, I., 59
Yothers, W. W., 186, 193
Youden, W. J., 250, 251
Young, H. C., 109, 111, 113, 115
Young, H. D., 160, 163
Young, P. A., 187, 193
Young, T. R., 55

Z

Zajic, E., 165, 166
Zeid, M. M. I., 172, 175
Zimmerman, P. W., 234, 235, 237, 238
Zopf, W. F., 55, 56
Zukel, J. W., 158, 159, 230, 231
Zundel, G. L., 257, 258

SUBJECT INDEX

A

Acetanilide, as fungicide, 142
3-(α-Acetonylbenzyl)-4-hydroxycoumarin, see warfarin
3-(α-Acetonyl-*p*-chlorobenzyl)-4-hydroxycoumarin, see coumachlor
Acetylcholine, enzymic hydrolysis of, 101 ; role in insects, 222 ; role in organophosphate poisoning, 221
Acidity of cell sap, in relation to disease resistance, 17
Acidity of cell sap, in relation to spray damage, 128
Acidity of soil and incidence of disease, 278
Acquired immunity, in plants, 21
Acridiidae, bacteria as control agents against, 56
" Acti-dione ", 138
Activators, as spray and dust components, 67
Acyrthosiphum pisi, nature of host resistance to, 26
Adherence, see tenacity
Adhesives, see stickers
Advancing contact angle, 70
Aedes aegypti, action of pyrethrins on, 173
— use of hydrocarbon oils against, 191
Aegerita webberi, as biological control agent, 52
Aerosol method of spray distribution, 65
" Aerosol OT ", 73
Affinin, as insecticide, 196
African pyrethrum, 167
Agglutination, test, for differentiation of fungal strains, 21
Agricere, 275
" Agri-mycin ", as fungicide, 137
" Agrosan G ", 261
Ahasversus advena, as test organism in bioassay, 58
" Alanap-1 ", 239 ; " Alanap-2 ", 239 ; " Alanap-3 ", 239
Alcohols, attractant properties of, 285 ; as fungicides, 97
Alcohols, sulphated, as spreaders, 73
Aldrin, 212
Alfalfa caterpillar, see *Colias philodice eurytheme*
Alfalfa fatigue, 269
Alienicolae, 283
Aliphatic thiocyanates, as insecticides, 200
Alkoxyethylmercuric compounds, as fungicides, 262
Alkyl sulphates, as spreaders, 73
Alkyl 4 : 4 : 6-trimethyl tetrahydropyrimidines, as fungicides, 134
Allethrin, 169
Allethrolone, 169
Allyl alcohol, as weedkiller, 243
Allyl mercuric chloride, 264
Allyl phenols, thiocyanoalkyl esters of, as insecticides, 201
Allyl sulphides, as factors in resistance to fungal attack, 19
Alternaria solani, use of calcium arsenate against, 141 ; use of griseofulvin against, 139 ; use in bioassay, 92
Alternate host, 300
Aluminium naphthenate, use in oil sprays, 190
Amaroids, as insecticides, 183
Amides, organic, as insecticides, 196
Amino acids, as components of spore excretions, 124 ; in relation to aphid attack, 26
N-(2-Aminoethyl)stearamide, as fungicide, 134
2-Amino-4-nitro-6-methylphenol, insecticidal properties of, 199
Aminotriazines, as weedkillers, 240
3-Amino-1 : 2 : 4-triazole, as defoliant, 240 ; as weedkiller, 240
Amiton, as acaricide, 218

" Amizol ", 240
Ammonium bicarbonate, 119
— carbonate, chemistry of, 119 ; use in copper sprays, 119
— diamminochromium tetrathiocyanate, as insecticide, 157
— dinitrocresylate, as weedkiller, 234
— reineckate, as insecticide, 157
— sulphamate, as weedkiller, 233
— sulphate, as weedkiller, 233
— thiocyanate, phytotoxic properties of, 158
Amphipathy, 73
Amur cork tree, insecticidal fruits of, 196
Anabasine, as insecticide, 164
Anabasis aphylla, insecticidal properties of, 164
Anacyclin, as insecticide, 197
Anacyclus spp., as insecticides, 167, 197
Angelica seed oil, as insect attractant, 286
Anguillulina dipsaci, as cause of soil sickness, 269 ; hot-water treatment against, 268
— *tritici*, control by seed selection, 255 ; crop rotation against, 280 ; flotation methods of seed treatment, 255 ; transmission by cultural operations, 306
Anhydrotetronic acid, use as hatching factor, 280
Anilides, as fungicides, 142
Aniline point, 190
Animal oils, see glyceride oils
Anionic detergents, 74
Anisoplia austriaca, biological control of, 52
Anisopteryx aescularia, use of greasebands against, 292
Annona spp., as insecticides, 196
Antagonistic action, amongst fungi, 53 ; in mixture of toxicants, 94
Antesia spp., control by hand collection, 297
Anthonomus grandis, attraction by trimethylamine, 284 ; use of silicofluorides against, 155
— *pomorum*, use of banding traps against, 291
Anthracene oils, as insecticides, 185, 191
Anti-auxins, 240
Antibiotics, as biological control agents, 53 ; as fungicides, 137
Antibody formation in plants, 21
Anticholinesterases, 102, 222
Anti-dusts, use in seed disinfectants, 265
Antigen-antibody specificity, 100
" Antinonnin ", 158, 198
Antioxidants, use with pyrethrum extracts, 169 ; use with thiodiphenylamine, 158
Ants, in relation to insect attack, 48 ; repellents for, 48
Antu, as rat poison, 294
Aonidiella aurantii, action of hydrocarbon oils on, 189 ; use of hydrogen cyanide against, 247
Aphelenchoides ritzema-bosi, hot-water treatment against, 268
Aphelinus mali, as biological control agent, 47
Aphididae, as virus vectors, 304 ; attractant action of spray residues on, 290 ; influence of external factors on life-history of, 283 ; nature of resistance to, 25 ; use of DNC against, 198
Aphis fabae, life history of, 282 ; ovicidal properties of cresols on, 199 ; see also *A. rumicis*
— *gossypii*, attraction to white surface, 290
— *pomi*, action of tar oils on, 191 ; use of organic thiocyanates against, 201
— *rumicis*, action of fatty acids on, 195 ; action of hydrocarbon oils on, 189 ; action of nicotine derivatives on, 165 ; action of pyrethrins on, 171 ; action of rotenoids on, 179 ; as test organism in bioassay, 86 ; effects of nutrition on reproduction of, 26 ; use of organic thiocyanates against, 200
— *sorbi*, influence of environmental factors on, 283
Aphycus lounsburyi, as biological control agent, 48
Apple, blossom weevil, see *Anthonomus pomorum* ; codling moth, see *Carpocapsa pomonella* ; collar blight, see *Bacterium amylovorum* ; crown gall, see *Bacterium tumefaciens* ; scab, see *Venturia inaequalis* ; sucker, see *Psyllia mali* ; woolly aphis, see *Eriosoma lanigerum*
" Aramite ", 230
Archips argyrospila, hydrocarbon oils as ovicides against, 190
Ardea ibis, as biological control agent, 45
Area of spread, as index of spreading properties,

Armillaria mellea, control by fumigation, 55
Aromatic hydrocarbons, 184 ; ovicidal properties of, 191 ; phytotoxic properties of, 187
Arsenates, metallic, relative toxicities of, 151
Arsenical compounds, action on insects, 150 ; as insecticides 147 ; correctives for use with, 152 ; deterrent action of, 151 ; effects of sub-lethal doses of, 152 ; fungicidal properties of, 141 ; phytotoxic properties of, 152 ; supplements for use with, 153
Arsenious oxide, as insecticide, 149
Arsenites, metallic, relative toxicities of, 151
Aryloxyalkylcarboxylates, as weedkillers, 235
Asarinin, as pyrethrum synergist, 172
Ascochyta pisi, control by seed selection, 254
Aspergillus niger, action of dithiocarbamates on, 131
Asphaltogenic acids, phytotoxic properties of, 187
Aspidiotus perniciosus, use of hydrocarbon oils against, 186, 191
Atomisation, as method of spray application, 63
Attractant action and chemical constitution, 285
Attractants, use of, 284
Auramines, as fungicides, 142
Availability, 67 ; of arsenicals in relation to insecticidal action, 151 ; of copper compounds in relation to fungicidal action, 126
Azobenzene, as acaricide, 228 ; as fumigant, 245
Azurin, 119

B

Bacillus cereus var. *mycoides*, metabolic activities of, 236
— *salutaris*, as biological control agent, 56
— *sotto*, as biological control agent, 57
— *thuringiensis*, as biological control agent, 56
— *typhi murium*, use against rodents, 57
Bacteria, as biological control agents, 56 ; in relation to soil sterilization, 275
Bacterium amylovorum, insect transmission of, 304 ; use of resistant stocks against, 10 ; use of streptomycin against, 128
— *carotovorum*, antibody formation, 21
— *lathyri*, potash manuring for control of, 31
— *pruni*, use of zinc sulphate–lime against, 140
— *tumefaciens*, stock-scion influence on resistance to, 20
Bait crops, 285
BAL, 150
" Balling " of dusts, 63, 67
Banana, Panama wilt, see *Fusarium oxysporum* f. *cubense*
Bandarine, 75
Banding traps, 291
Barbasco, as insecticide, 176
Barberry, eradication of, 300
Barium carbonate, as rat poison, 294
— polysulphide, as fungicide, 112
— silicofluoride, as insecticide, 155
— tetrasulphide, as fungicide, 112
Barley, covered smut, see *Ustilago hordei* ; gout fly, see *Chlorops taeniopus* ; loose smut, see *Ustilago nuda* ; stripe, see *Helminthosporium gramineum*
" Bayer 17147 ", 220
Bean, resistance to aphid attack, 25
Beaumont's rules, 39
Beauveria bassiana, as pathogen of silkworm, 52
Beet, Leafhopper, see *Eutettix tenellus*
Beneficial fungi, spread of, 52
Beneficial insects, 46 *et seq.*
Bentonite, use with copper fungicides, 120 ; use with sulphur, 106
Benzene, as fumigant, 253
Benzene hexachloride, see BHC
Benzoquinones, as fungicides, 136
2(3)-Benzoxazolinone, as factor in disease resistance, 20
2-Benzoylbenzoic acids, as weedkillers, 239
Benzyl benzoate, as acaricide, 228

Beta-oxidation, 237
BHC, as insecticide, 209 *et seq.*; as soil fumigant, 274 ; comparative insecticidal properties of isomers of, 209 ; off-flavours produced by, 210
Bimodal probit curves, 130
Bindex, 75
Bioassay, principles of, 86 *et sq.*
Biochemical lesions, 96
Biologic races, see physiologic races
Biological assay, see bioassay
Biological control, 44 *et seq.*
Birds, as biological control agents, 44, 45 ; repellents for use against, 291 ; status as pests, 46
Bis(*p*-chlorophenoxy)methane, as acaricide, 229
Bis *OO*-diethylphosphorothionic anhydride, see TEPP
Bis(2-hydroxy-5-chlorophenyl)sulphide, as fungicide, 145
1 : 3-Bis(3-hydroxy-2 : 2 : 2-trichloroethyl)urea, as weedkiller, 242
2 : 3-Bismercaptopropanol, 150
Bisphenols, as fungicides, 144
Bis-*NNN'N'*-tetramethylphosphorodiamidic anhydride, see schradan
Bitumen emulsions, as wound dressings, 298
Black currant, big bud of, see *Eriophyes ribis*
"—Leaf 40 ", 160
— scale, see *Saissetia oleae*
" Bladan ", 216
Bluestone, see copper sulphate
Boiling point and toxicity, 96, 186, 190
Bombycidae, use of light traps against, 190
Borax-casein, as spreader, 75
Bordeaux mixture, action on fungus, 115 *et seq.*; alkaline, 117 ; as emulsifier, 82, 192 ; as fungicide, 123 *et seq.*; chemistry of, 116 ; equal lime, 117 ; excess lime, 127 ; instant, 116 ; neutral, 117 ; phytocidal action of, 126 ; weathering of, 124
Bordorite, 120
*iso*Bornyl thiocyanoacetate, as insecticide, 201
Boron, influence on host-parasite relationship, 32
" Bottom ventilation ", 29, 40
Botrytis allii, host resistance to, 18, 19
Botrytis cinerea, heterocaryosis in, 12 ; penetration by, 15
— spp., use of griseofulvin against, 139 ; use of pentachloronitrobenzene against, 143
" Bouisol ", 120
Breaking of emulsions, 82
" Breeze ", 73
Brevicoryne brassicae, effect of nutrition on reproduction of, 26
Bridging hosts, 11, 300
Brining of seed, 255
British gum, as sticker, 80
B(ritish) P(atent) 241,568 (anti-dust), 265
B.P. 247,249 (thallium sulphate), 294
B.P. 251,330 (magnesium arsenate), 150
B.P. 350,642 (dispersible salicylanilide), 81
B.P. 392,556 (copper pastes), 81, 120
B.P. 401,707 (nicotine), 162
B.P. 425,295 (DNC), 235
B.P. 439,435 (detergents), 74
B.P. 488,428–9 (anti-oxidants), 158
B.P. 547,871 (DDT), 202
B.P. 592,788 (pyrotechnic mixtures), 65
B.P. 598,927 (glyodin), 134
2-Bromo-4 : 6-dimethylphenylnitramine, as weedkiller, 239
Bromomethane, 252
Brown rot, see *Sclerotinia laxa*
Brown tail moth, see *Euproctis chrysorrhoea*
Build-up of spray residue, 77
Burgundy, mixture, as fungicide, 118 *et seq.*, chemistry of, 118
Butopyronoxyl, as insect repellent, 287
Butoxypolypropylene glycols, as insect repellents, 287

β-Butoxy-β′-thiocyanodiethyl ether, as insecticide, 200
Butter-yellow, carcinogenetic properties of, 104
*iso*Butylamides, as insecticides, 196
n-Butyl " carbitol " thiocyanate, as insecticide, 200, 202
N-*iso*Butyl-2 : 6 : 8-decatrienamide, as insecticide, 196
N-*iso*Butyl hendecenamide, as pyrethrum synergist, 173
N-*iso*Butyl-2 : 4 : 8 : 10-tetraenamide, as insecticide, 197
2-(*p*-*tert.*-Butylphenoxy)-*iso*propyl 2′-chloroethyl sulphite, as acaricide, 230

C

Cabbage, aphis, see *Brevicoryne brassicae* ; butterflies, see *Pieris* spp.; yellows, see *Fusarium conglutinans*
Cactoblastis cactorum, as biological control agent, 50
Cadmium salts, as fungicides, 140
Calandra granaria, action of chlorinated ethylenes on, 99 ; action of tar oils on, 190 ; as test organism in bioassay, 86 ; toxicity of organic vapours to, 98
Calcium arsenates, action of lime on, 152 ; as insecticides, 148 ; fungicidal properties of, 141; phytotoxic properties, 152
— arsenite, as insecticide, 148
— caseinate, see lime casein
— chloroacetate, use against eelworms, 281
— cuprites, in aged Bordeaux mixture, 118
— cuprocyanide, as factor in cyanide damage, 249
— cyanide, as fumigant, 246 ; use against soil pests, 274
— hydroxide, as constituent of Bordeaux mixture, 116 ; as corrective for arsenical sprays, 152 ; interaction with lead arsenate, 152
— polysulphides, see lime sulphur
— silicofluoride, as insecticide, 155
— thiocarbonate, as soil sterilizing agent, 271
Callitroga hominivorax control by irradiation, 49
Calomel, see mercurous chloride
Calonectria graminicola, use of mercuric chloride against, 260
Calosoma sycophanta, as biological control agent, 46
Campbells Patent Sulphur Vaporizer, 107
Camphor, as repellent, 287
Capitophorus hippophaes, influence of external factors on life-history of, 283
Capric acid, as insecticide, 194
Captan, as fungicide, 134 ; mode of action of, 135
Carbamates, as insecticides, 226 ; as weedkillers, 238
Carbanilates, as weedkillers, 238
Carbazole derivatives, as insecticides, 159
Carbolic acid, see cresylic acid
Carbolineums, 187 ; as insecticides, 187 ; as soil sterilizing agents, 272
Carbon dioxide, as factor in arsenical damage, 152 ; as stimulant for spore germination, 14 ; role in fungicidal action of copper derivatives, 124, 257
Carbon disulphide, as soil sterilizing agent, 271
— tetrachloride, use in fumigation, 253
" Carboxide ", 253
Carboxylase, inhibition by quinones, 137
p-Carboxyphenyl trichloromethylthiolsulphonate, as fungicide, 135
Carpocapsa pomonella, as factor in brown rot attack, 15 ; biological control of, 45, 50 ; use of bait traps against, 285 ; use of band traps against, 291 ; use of pigs and sheep against, 297 ; use of ryania against, 50
" Carriers " of virus diseases, 299
Carrot fly, see *Psila rosae*
Caseinates, as spreaders, 75
Castor oil, as fungicide, 142
" Castrix ", 295
" Catalytic sulphur ", 114
Catechol, as factor in disease resistance, 19
Cationic detergents, 74
Centipedes, as biological control agents, 51
Cephus cinctus, nature of resistance to, 25

Ceratitis capitata, climate and distribution of, 37 ; eradication of, 298 ; role of essential oils in attack by, 27 ; use of attractants against, 285
Ceratostomella paradoxa, see *Endoconidiophora paradoxa*
Cercospora beticola, stomatal length and resistance to, 14 ; transmission by diseased beet, 307
— *melonis*, immunity from, 10
" Cerenox ", 266
" Ceres " powder, 256, 259
" Ceresan ", European, 261 ; U.S.A. 261
"— M ", 261
"— New improved ", 261
"— Universal Trochenbeize ", 261
Cerium salts, as fungicides, 140
Cetylpyridinium chloride, as spreader, 78
Cetyltrimethylammonium bromide, as spreader, 74
Cevadine, as insecticide, 157
Cevine, as insecticide, 157
Chelating agents, toxicity of, 103
" Chemically-toxic " substances, 99
Chemotropism, as basis of control measures, 284 ; as factor in host selection, 26 ; role in fungal infection, 15
Cheshunt compound, as fungicide, 119 ; as soil sterilizing agent, 304 ; as water sterilizer, 304
Chinese yam bean, as insecticide, 177
Cholinesterases, inhibition by insecticides, 101, 221 *et seq.*
Chloranil, 136, 266
Chlorbenside, as acaricide, 230
Chlordane, as insecticide, 211
Chlordene, 211
Chlorethylenes, as fumigants, 253 ; as soil sterilizing agents, 272
Chlorfenson, as acaricide, 230
Chlorinated camphene, 211
Chlorinated naphthalenes, as bird repellents, 291
Chlorinated terpenes, as insecticides, 210
Chloroacetates, use against eelworms, 281
" Chlorbenzilate ", 229
p-Chlorobenzyl *p*-chlorophenyl sulphide, see chlorbenside
p-Chlorobenzyl *p*-fluorophenyl sulphide, see fluorbenzide
2-Chloro-4 : 6-bis(ethylamino)-*s*-triazine, as weedkiller, 240
α-Chloro-*NN*-diallylacetamide, as weedkiller, 243
7-Chloro-4 : 6-dimethoxycoumaran-3-one-2-*spiro*-1′-(2′-methoxy-6′-methyl-*cyclo*hex-2′-en-4′-one), 139
2-Chloro-4-dimethylamino-6-methyl pyrimidine, as rat poison, 295
1-Chloro-2 : 2-di(*p*-chlorophenyl)ethane, 204
2-Chloro-1 : 1-di(*p*-chlorophenyl)-2-methylpropane, 205
Chlorodinitrobenzenes, as soil fumigants, 272
1-Chloro-2 : 4-dinitronaphthalene, as fungicide, 144
2-Chloroethyl p-chlorobenzenesulphonate, as acaricide, 230
2-Chloro-3-hydroxy-1 : 4-naphthoquinone, as fungicide, 136
4-Chloro-2-methylphenoxyacetic acid, see MCPA
6-Chloro-2-methylphenoxyacetic acid, as component of MCPA, 235
p-Chlorophenyl benzenesulphonate, see fenson
p-Chlorophenyl *p*-chlorobenzenesulphonate, see chlorfenson
p-Chlorophenyldiazothiourea, as rat poison, 295
3-(*p*-Chlorophenyl)-1 : 1-dimethylurea, see monuron
N-*p*-Chlorophenyl-*N′N′*-dimethylurea, see monuron
Chlorophenyl dimethylurea trichloroacetate, as weedkiller, 243
3-(*p*-Chlorophenyl)-5-methyl rhodanine, as fungicide, 132 ; as nematicide, 132
p-Chlorophenyl phenyl sulphone, as acaricide, 229
1-(*p*-Chlorophenyl)-1-phenyl-2 : 2 : 2-trichloroethane, 204, 206
Chlorops taeniopus, early sowing for control of, 42 ; use of phosphatic manure against, 31
" Chloros ", for water disinfection, 304
Chlorphenolmercury, 260
Chlorpicrin, as soil fumigant, 273
" Chlorthion ", 218
Chromaphis juglandicola, biological control of, 52
Chromium derivatives, as fungicides, 140 ; as insecticides, 157

Chrysanthemic acid, 168
Chrysanthemum carneum, as insecticide, 167
— *cineraefolium*, as insecticide, 167
— *roseum*, as insecticide, 167
Chymotrypsin, action of DFP on, 223
Cicadula sexnotata, as vector of aster yellows, 304
Cinerins, as insecticides, 168
Cinerolone, 168
Cinnamon, as repellent, 287
CIPC, as weedkiller, 238
" Citrofume ", 246
Citrus canker, see *Xanthomonas citri* ; cottony cushion scale, see *Icerya purchasi* ; mealy bug, see *Pseudococcus citri* ; purple scale, see *Lepidosaphes bechii* ; scab, see *Spaceloma fawcetti*
Cladosporium fulvum, control by cultural methods, 40 ; use of chloronitrobenzene against, 143 ; use of salicylanilide against, 142
Climatic conditions and degree of attack, 36 *et seq.*
Clysia spp., use of carbazole derivatives against, 159 ; use of light traps against, 289
Coal tar oils, see hydrocarbon oils
Cocarboxylase, inhibition by captan, 135
Coccobacillus acridiorum, as biological control agent, 56
Coconut fatty acids, as insecticides, 195
Codling moth, see *Carpocapsa pomonella*
Coffee mealy bugs, see *Pseudococcus citri* and *P. lilacinus* ; rust, see *Hemileia vastatrix*
Colchicine, 9
Colias philodice eurytheme, biological control of, 56
Colletotrichum circinans, host resistance to 19
— *falcatum*, influence of soil conditions on attack by, 33
— *lagenarium*, transmission by manure, 307
— *lindemuthianum*, nature of host resistance to, 20
— *lini*. use of thiram against, 266
Colloidal sulphurs, 81
Colorado beetle, see *Leptinotarsa decemlineata*
Competitive inhibition of enzymes, 102
Complement fixation test, to differentiate fungal strains, 21
Compound G 4, 144
Concentrate spraying, 63
Coniothyrium spp., resistance to, in raspberry, 13
Contact angles, as indices of wetting and spreading properties, 70 *et seq.*
Contact insecticides, 61 ; comparison of, 86
Contarinia nasturtii, control by trap cropping, 286
— *pyrivora*, use of poultry against, 297
Conventional spraying, 120
" Coposil ", 120
Copper acetates, as fungicides, 120
— acetoarsenite, as insecticide, 148 ; see also Paris green
— ammonium silicates, as fungicides, 120
— arsenates, basic, as fungicides, 121
— arsenite, as insecticide, 148
— bicarbonate, 124
— carbonates, basic, as fungicides, 118 ; as seed disinfectants, 257
— chlorides, basic, see copper oxychloride
— 3 : 5-di*iso*propyl salicylate, as fungicide, 121
— dusts, 121
— fungicides, 115 *et seq.*; action on fungus, 123 *et seq.*; phytotoxic action of, 126
— hydroxide, as fungicide, 116
— lime dusts, 121
— naphthenates, as fungicides, 121
— nature of fungicidal action of, 123 *et seq.*
— oxychloride, as fungicide, 120
— resinate, as fungicide, 121
— sulphate, action on charlock, 233 ; action on fungus spores, 123 ; as seed disinfectant, 255 ; as weedkiller, 233 ; monohydrated, 257 ; pentahydrated, 116 ; reaction with ammonium hydroxide, 119 ; reaction with lime, 116 ; reaction with sodium carbonate, 118 ; " snow ", 116
— sulphates, basic, 117

Copper trioxysulphate, 116
— zeolites, as fungicides, 120
— zinc chromates, as fungicides, 121, 141
— see also cuprous and cupric salts
Correctives, as spray components, 152
Corrosive sublimate, see mercuric chloride
Corrugated paper, use in insect traps, 291
Corticium solani, use of mercuric chloride against, 263
Cotton, black-arm disease, see *Xanthomonas malvacearum* ; jassid, see *Empoasca facialis* ; worm, see *Prodenia litura*
Cotton, varieties resistant to jassid, 7
Cottonseed oil, as spreader, 76
Coumachlor, as rat poison, 295
Coverage, as factor in spray performance, 67
Cracca spp., see *Tephrosia* spp.
" Crag Herbicide 2 ", 242
Crambidae, use of light traps against, 290
Cranberries, flooding of, as control measure, 34
Creaming of emulsions, 82
Creosote, see hydrocarbon oils
" Cresatin ", 144
m-Cresol acetate, as fungicide, 144
Cresols, as ovicides, 198 ; as repellents, 288 ; as soil insecticides, 272
Cresylic acid, 272 ; as soil insecticide, 272
Cresylmercuric cyanide, as seed disinfectant, 260
Crop rotation, as control measure, 232, 279
Crotonaldehyde, as fungicide, 144
Crown-gall, see *Bacterium tumefaciens*
Cryolite, as insecticide, 155
Cryptoloemus montrouzieri, as biological control agent, 48, 50
Cryptorrhynchus lapathi, control by trap crops, 286
Cubé, as insecticide, 176
Cucumber Anthracnose, see *Colletotrichum lagenarium* ; blotch, see *Cercospora melonis*
— mosaic, control by eradication of weed hosts, 299
— use of resistant stocks, 10
Cultivation, as control measure, 279
Cumulative action, of copper, 123
Cumulative frequency curve, 90
Cupram, 119
Cuprammonium carbonate, as fungicide, 119
— sprays, 118 *et seq.*
— sulphate, as fungicide, 119
Cupric chloride, as fungicide, 121
— dimethyl dithiocarbamate, as fungicide, 131
— dinitrocresylate, as fungicide, 121
— hydroxide, as component of Bordeaux mixture, 117
— oxide, as fungicide, 120
— oxinate, as fungicide, 121
— phosphate, as fungicide, 120, 121
— phthalate, as fungicide, 121, 125
— sebacate, as fungicide, 121, 125
— silicate, as fungicide, 120
— sulphanilate, as fungicide, 121
— sulphide, as fungicide, 121
Cuprous cyanide, as insecticide, 157 ; use in poison baits, 157
— oxide, as fungicide, 120 ; for seed treatment, 257
— thiocyanate, as insecticide, 157
Currant, big bud see *Eriophyes ribis*
Cuscuta spp., control by seed selection, 254
Cuticle thickness and resistance to disease, 16
" Cut-off ", in toxicity of homologues, 98, 195
Cutworms, effects of climatic factors on, 35
Cyanide fumigation, see hydrocyanic acid
" Cyanogas ", 246
Cyclethrin, as insecticide, 169

Cyclethrolone, 169
Cyclodiene insecticides, 211 *et seq.*; use in poison baits, 293
Cycloheximide, as fungicide, 54, 138

D

2,4-D, as weedkiller, 234 *et seq.*
Dactylella ellipsospora, as biological control agent, 55
Dactylopius spp., as biological control agents, 50
Dacus oleae, collection of fallen fruit as control measure against, 298
Dalapon, as weedkiller, 242
Damping off, see *Pythium ultimum*
David's powder, 121
DCU, as weedkiller, 242
DDD, as insecticide, 204
" D-D Soil Fumigant ", 273
DDT, 103, 202 *et seq.*; action on mites, 49; as fumigant, 245; as insecticide, 206; dehydrochlorination of, 204; detoxication of, 208; insect resistance to, 207; isomers of, 202, 204; isosters of, 204
DDT carbinol, as synergist for DDT, 208
DDT-dehydrochlorinase, 208
Defoliation, as control measure, 40, 232
Degree of fineness, see particle size
Degulia spp., see *Derris* spp.
Deguelin, 178
Dehydrorotenone, 180
Delphinium spp. as insecticides, 196
Demeton, as insecticide, 217
—— O, 218
—— O methyl, 218
—— S, 218
—— S methyl, 218
Demissidine, 27
Demissin, 27
Derrin, 177
Derris spp. as insecticides, 176; piscicidal properties of, 180
Derris resin, chemistry of, 177 *et seq.*; deterrent action of, 180; evaluation of, 180; solvents for, 180
Deterioration of asexually-produced plants, 305
Developmental period, 283
Devil's shoestring, as insecticide, 177
Dextrine, as sticker, 80
DFP, physiological activity of, 221
Dialkyldithiocarbamates, as fungicides, 129
Diamondback moth, see *Plutella maculipennis*
1 : 1-Dianisyl neopentane, as insecticide, 205
Diapause in insects, 283
Diataraxia oleraceae, photoperiodic effects on, 283
Diazinon, 220
Dibasic lead arsenate, 148
Dibenzothiazine, see thiodiphenylamine
1 : 2-Dibromo-3-chloropropane, as soil fumigant, 273
Dicalcium hydrogen arsenate, see calcium arsenate
Dichapetalum cymosum, toxic components of, 294
Dichlone, as fungicide, 136; particle size effects, 92
Dichloral urea, as weedkiller, 242
4 : 4-Dichloroazobenzene, as acaricide, 229
p-Dichlorobenzene, as fumigant, 253; as repellent, 287; as soil fumigant, 272
o-Dichlorobenzene hexachloride, as insect repellent, 209
1 : 1-Dichloro-2 : 2-di-(*p*-chlorophenyl)ethane, see DDD
1 : 1-Dichloro-2 : 2-di-(*p*-chlorophenyl)ethylene, 203
Dichlorodifluoromethane, use in aerosols, 65
1 : 8-Dichloro-3 : 6-dinitrocarbazole, as insecticide, 159
Dichlorodiphenyltrichloroethane, 202
ββ'-Dichloroethyl ether, as fumigant, 253

ββ′-Dichloroethyl sulphide, as mutagen, 8
Dichloroethylene, as soil fumigant, 272
2 : 3-Dichloro-1 : 4-naphthoquinone, see dichlone
1 : 1-Dichloro-1-nitroethane, as fumigant, 253
2 : 4-Dichlorophenol, as weedkiller, 237
2 : 4-Dichlorophenoxyacetic acid, see 2,4-D
2 : 4-Dichlorophenoxyalkylcarboxylic acids, β-oxidation of, 237
Di-(*p*-chlorophenyl)chloromethane, as synergist for DDT, 208
3-(3 : 4-Dichlorophenyl)-1 : 1-dimethylurea, see diuron
Di-(*p*-chlorophenyl)ethane, 204 ; as synergist for DDT, 208
1 : 1-(Dichlorophenyl)ethanol, as acaricide, 229 ; as synergist for DDT, 208
Di-(*p*-chlorophenyl)ethynyl carbinol, as synergist for DDT, 208
3-(3 : 4-Dichlorophenyl)-1-methyl-1-*n*-butylurea, see neburon
(Dichlorophenyl)methyl carbinol, 229
1 : 1-Di-*p*-chlorophenyl-2-nitrobutane, as insecticide, 205
1 : 1-Di-*p*-chlorophenyl-2-nitroethane, as insecticide, 204, 205
4 : 4′-Dichlorophenyl sulphone, as acaricide, 229 ; as stomach poison, 204
1 : 1-Dichlorophenyl-2 : 2 : 2-trichloroethanol, 228
1 : 2-Dichloropropane, as soil fumigant, 273
1 : 3-Dichloropropene, as soil fumigant, 273
2 : 2-Dichloropropionic acid, see dalapon
Dichloropropylene, as soil fumigant, 273
Dicoumarin, as rat poison, 295
" Dicumarol ", 295
Di*cyclo*hexylamine salt of dinex, as ovicide, 199
Dieldrin, 212
Diethylaminoethyloleylamine hydrochloride, as spreader, 74
OO-Diethyl *O*-3-chloro-4-methylcoumarinyl phosphorothioate, 219
OO-Diethyl *S*-*p*-chlorophenylthiomethyl phosphorodithioate, 219
Diethyl 2-chlorovinyl phosphate, 221
OO-Diethyl *O*-2 : 4-dichlorophenyl phosphorothioate, 219
OO-Diethyl *S*-2-diethylaminoethyl phosphorothiolate, see Amiton
OO-Diethyl *S*-2-ethylthioethyl phosphorodithioate, 218
OO-Diethyl *S*-2-ethylthiomethyl phosphorodithioate, 128
OO-Diethyl *O*-2-ethylthioethyl phosphorothionate, see Demeton-O
OO-Diethyl *S*-2-ethylthioethyl phosphorothionate, see Demeton-S
OO-Diethyl *O*-(2-*iso*propyl-4-methyl-6-pyrimidyl)phosphorothioate, 220
OO-Diethyl *O*-4-methyl-7-coumarinyl phosphorothioate, 219
Diethyl *p*-nitrophenyl phosphate, 224
OO-Diethyl *O*-*p*-nitrophenyl phosphorothioate, see parathion
Diethyl phenyl phosphates, as cholinesterase inhibitors, 223
Diethyl phosphorofluoridate, 221
Differential hosts, 11
Differential wetting, as reason for selective weedkillers, 234
Diffusivity, 270
Dihydrorotenone, 180
1 : 2-Dihydroxypyridazine-3 : 6-dione, 239
3 : 5-Dihydroxy-*trans*-stilbene, as fungicide, 18
" Dilan ", 205
2 : 3-Dimercaptopropanol, as antidote to arsenicals, 150
" Dimetan ", as insecticide, 227
Dimethylaminoazobenzene, carcinogenetic properties of, 104
Dimethylaniline, as fungicide, 143
Dimethyl 1-carbomethoxy-1-propen-2-yl phosphate, as insecticide, 220
2 : 2-Dimethyl-6-carbobutoxy-2 : 3-dihydro-4-pyrone, see butopyronoxyl
OO-Dimethyl *O*-2-chloro-4-nitrophenyl phosphorothioate, 218
OO-Dimethyl *O*-3-chloro-4-nitrophenyl phosphorothioate, 218
Dimethyl 2 : 2-dichlorovinyl phosphate, 221
OO-Dimethyl *S*-(1 : 2-di(ethoxycarbonyl)ethyl phosphorodithioate, see malathion
5 : 5-Dimethyldihydroxyresorcinol diethylcarbamate, as insect repellent, 226
5 : 5-Dimethyldihydroxyresorcinol dimethylcarbamate, see " Dimetan "
OO-Dimethyl *O*-2-ethylthioethyl phosphorothioate, see Demeton-O-methyl
OO-Dimethyl *S*-2-ethylthioethyl phosphorothioate, see Demeton-S-methyl
Dimethyl 1-methylvinyl phosphate, 221
2 : 3-Dimethyl-1 : 4-naphthoquinone, as fungicide, 136

OO-Dimethyl *O*-*p*-nitrophenyl phosphorothioate, see methyl parathion
OO-Dimethyl *S*-4-oxobenzotriazine-3-methyl phosphorodithioate, 220
β-[2-(3 : 5-Dimethyl-2-oxy*cyclo*hexyl)-2-hydroxyethyl]glutarimide, as fungicide, 138
2 : 4-Dimethylphenoxyacetic acid, as weedkiller, 235
Dimethyl phthalate, as insect repellent, 287
Dimethyl sulphate, solubility of hydrocarbon oils in, 184, 192
Dimethyl 2 : 2 : 2-trichloro-1-hydroxyethyl phosphonate, 221
OO-Dimethyl *O*-2 : 4 : 5-trichlorophenyl phosphorothioate, see korlan
Dinaphthylmethane disulphonate, as dispersing agent, 81.
Dinex, as insecticide, 199
2 : 4-Dinitro-*o*-*sec*.butylphenol, see dinoseb
Dinitrocaprylphenyl crotonate, as fungicide, 144
Dinitro-*o*-cresol, see DNC
Dinitrocresylates, as insecticides, 198 ; as ovicides, 198, as weedkillers, 233
Dinitro*cyclo*hexylphenol, as ovicide 199
Dinitrophenols, as insecticides, 198 ; as weedkillers, 233
2 : 4-Dinitro-6-(2-octyl)phenyl crotonate, as acaricide, 144 ; as fungicide, 144
2 : 4-Dinitro-1-thiocyanobenzene, as fungicide, 144
Dinoseb, as weedkiller, 234
Diphenyl, as fungicide, 144
Diphenyl sulphone, as acaricide, 229
Diplodia natalensis, onion resistance to, 19
Diplumbic hydrogen arsenate, see lead arsenate
Dipping tests of wetting properties, 69
Di-*iso*propyl fluorophosphonate, 221
Diprion hercyniae, use of virus against, 58
DIPS, 121
" Dipterex ", 221
Dipyridyls, insecticidal properties of, 164
Direct fungicides, comparison of, 86
Disease resistance, acidity of cell sap and, 17 ; effect of fertilizers on, 30 *et seq.* ; hereditable character of, 7 ; mechanical strength of cell wall and, 14, 15 ; plant toxins and, 18, 20 ; protein specificity and, 18 ; specificity of food requirements and, 17
Disodium ethylene bisdithiocarbamate, see nabam
Dispersing agents, 80
" Dispersols ", 74
Dissemination of pathogens, by cultural operations, 306 ; by implements, 306 ; by insects, 304 ; by transport, 307 ; by water, 303 ; by wind, 303
Distillate washes, see hydrocarbon oils
" Disyston ", 218
Dithiocarbamates, as fungicides, 128 *et seq.*
Dithiocarbamyl compounds, effect on yeasts, 131
Diuron, as weedkiller, 241
DMC, 229
DNBP, 234
DNC, activation of 234 ; as insecticide, 158, 198 ; as weedkiller, 233
DNOC, 233
DNOCHP, see dinex
Dodder spp., removal of, from seed samples, 254 ; status as pests, 2
Dodecyl nicotinium bromide, as insecticide, 162
Dodecyl sodium sulphate, as spreader, 73
Dodecyl thiocyanate, as insecticide, 200
Dominant genes, 8
" Dorlone ", 273
Dormant washes, 61, 187
Dosage, calculation of, in fumigation, 245
Dose metameter, 90
Dosis curativa, 261
Dosis tolerata, 261
Dossier machine, 254
Dow ET-57, 219
Drag-sheets, use in fumigation, 64
Drained salts, 250
" Dreft ", 73
Drosophila melanogaster, as test organism in bioassay, 88

Dry-mix Sulphur Lime, as fungicide, 107
Duboisia hopwoodii insecticidal alkaloids of, 165
Dusting v. spraying, 63 *et seq.*
" Dusting tendency ", 64
" Dutch rules ", 39
" Dutox ", 155
Dyestuffs, as fungicides, 142

E

Eau Celeste, 119 ; modified, 119
Eau Grison, 112
Ecdysone, as insect hormone, 284
Ecliptic acid, as eelworm hatching factor, 280
Ecological islands, 45
Ectophytic fungi, 13 ; use of direct fungicides against, 60
ED50, 91
Elafrosin, 147
Elateridae, trap crops against, 286
Elderberry, repellent properties of, 288
Electrostatic factors in spraying and dusting, 66
Electrostatic machines for dust application, 67
Elliptone, 178
Emerald green, 147
Empoasca facialis, cotton varieties resistant to, 7
Emulsifiable concentrates, 84
Emulsification, 81
— degree of, and toxicity, 142, 189
Emulsifying agents, 81 *et seq.*
Emulsions, breaking of, 82 ; creaming of, 82 ; inversion of, 83 ; preferential retention in, 77 ; preparation of, 83 ; properties of, 82 ; stability of, 82 ; stock, 83 ; types of, 83
Encarsia spp., as beneficial insects, 48
Endoconidiophora paradoxa, use of organomercury compounds against, 263
Endophytic fungi, 13 ; use of protective fungicides against, 60
" Endothal ", 243
3 : 6-Endoxohexahydrophthalates, as weedkillers, 243
Endrin, 213
Entomogenous fungi, as biological control agents, 52
Entomophthora spp., as biological control agents, 52
Enzyme-substrate specificity, 18, 100
Enzymes, inhibition of, as biochemical lesion, 101
Epidemiology, as basis of control measures, 38
Epilachna varivestis, meteorological conditions and distribution of, 38
" EPN ", 221
Epoxidation, as factor in action of cyclodiene insecticides, 214
Eradication, of alternate host, 330 *et seq.*; of host, 298 *et seq.*
Erbon, 243
Erigeron affinis, as insecticide, 196
Eriophyes ribis, nature of resistance to, 27
Eriosoma lanigerum, biological control of, 47 ; nature of resistance to, 25
Eserine, 226
Essential metabolites, 102
Essential oils, as factors of plant resistance, 27
Esters, as attractants, 285
Ethyl acetate, as fumigant, 253
— 4′4′-dichlorobenzilate, as acaricide, 229
— hexanediol, as insect repellent, 287
O-Ethyl *O*-*p*-nitrophenyl phenylphosphonothioate, 221
O-Ethyl phenylcarbamate, as weedkiller, 238
Ethylene dibromide, as fumigant, 253 ; as soil fumigant, 273
— dichloride, as fumigant, 253
— diisothiocyanate, as fungicide, 132
— oxide, as fumigant, 253
— thiuram monosulphide, as fungicide, 132
Ethylmercuric chloride, as seed disinfectant, 261

Ethylmercuric hydroxide, reactions of, 263
— phosphate, as seed disinfectant, 261
N-(Ethylmercuri)-*p*-toluenesulphonamide, as seed disinfectant, 261
Eucalyptus, as repellent, 196
Eugenol, as attractant, 285
Euproctes chrysorrhoea, host selection by, 26
European corn borer, see *Pyrausta nubilalis*
European pine sawfly, see *Neodiprion sertifer*
European spruce sawfly, see *Diprion hercyniae*
Eutettix tenellus, as vector of " curly top ", 38, 304 ; climatic conditions and distribution of, 38
Euxoa auxiliaris, phototropism of, 289
Exanthema of citrus, 249
Excess-lime Bordeaux mixture, 127
Exobasidium vexans, control by pruning, 43
Eye-spotted bud moth, see *Spilonota ocella*

F

Fagaramide, as pyrethrum synergist, 172
Fasciola hepatica, heteroecism of, 302
Fatty acids, as attractants, 285 ; as insecticides, 194 ; phytotoxic properties of, 195
— alcohols, sulphated, as spreaders, 73
Fenchyl thiocyanoacetate, as insecticide, 201
Fenson, as acaricide, 230
Fenuron, as weedkiller, 240
Ferbam, as fungicide, 129
Ferrous sulphate, as corrective for polysulphide sprays, 113 ; as " marker ", 113 ; as weed-killer, 233
Ferric dimethyldithiocarbamate, see ferbam
" Ferrox sulphur ", 106
Fertilizers, influence on degree of attack, 30 *et seq.*
Field performance, factors affecting, 67, 85
Fillers, for seed treatment dusts, 265
Film coverage of spray deposit, 78
Film method of bioassay, 87
Fire Blight, see *Bacterium amylovorum*
Fish glue, as sticker, 75
Fish oil, as sticker, 153
" Fixed " nicotines, as insecticides, 161
Flax, wilt, see *Fusarium lini*, seed treatment of, 266
Flea-beetles, use of repellents against, 288
Flooding of soil, as control measure, 34
Flotation sulphurs, 106
Flour paste, as sticker, 80
Flour sulphur, 106
Flower of sulphur, 106
Fluorbenside, as acaricide, 230
Fluorides, insecticidal properties of, 155 ; toxicity to higher animals, 156
Fluoroacetates, toxic properties of, 294
Fluted scale, see *Icerya purchasi*
" Fly-free " dates, 42
Fog appliances, 63
Fogs, insecticidal, 63
" Folosan ", 143
Forecasting of epidemics, 39
Forficula auricularia, use of fluorides against, 155
Formaldehyde, action on seed grain, 258 ; action on fungal spore, 258 ; as seed disinfectant, 258 ; as soil sterilizing agent, 273 ; use in baits, 293
Formalin, 273
" Freezing ", in fumigation, 246
French green, 147
Fritfly, see *Oscinella frit*
Fruit tree red spider, see *Metatetranychus ulmi*
Frukusgrün, 147

Fuchsine, as fungicide, 143
" Fumazone ", 273
Fumigation, 64, 245 *et seq.*; of nursery stock, 252
Fundatrices, 282
Fundatrigeniae, 282
Fungal antagonism, 53
Fungicides, comparison of efficiencies of direct, 86; comparison of efficiencies of protective, 86, 88; direct, 60; eradicative, 60; functions of, 60; importance of timing of application of protective, 60; protective, 60
Fungi, as biological control agents, 52 *et seq.*; mechanism of infection by, 13 *et seq.*
Furethrin, as insecticide, 169
Furethrolone, 169
Fusarium caeruleum, use of organomercury compounds against, 263; use of tecnazene against, 143
— *conglutinans*, effect of soil moisture on parasitism of, 34; effect of soil temperature on attack by, 35; nature of host resistance to, 19
— *culmorum*, pH of cell sap and resistance to, 17
— *lini*, as cause of soil sickness, 269; nature of host resistance to, 16, 19; protein specificity and resistance to, 20
— *lycopersici*, nature of host resistance to, 19
— *nivale*, nature of host resistance to, 20; use of mercuric chloride against, 260
— *oxysporum*, influence of fertilizers on attack by, 32; pH of cell sap and resistance to, 17
— — f. *cubense*, biological control of, 54; control by flooding, 34
— — f. *pisi*, pathogenicity of, 12

G

Gamma irradiation, as control measure, 49
" Gardinols ", 73
Gelatine, as spreader, 75
Geometridae, use of grease bands against, 292; use of light traps against, 290
Geotropism, 14
Geraniol, as insect attractant, 285
German pyrethrum, 167
Germination injury, by copper sulphate, 257; by formaldehyde, 258; by organomercury compounds, 264
" Germisan ", 260
" Gesarol ", 202
Gibberella saubinetii, resistance to, 16
Gifblaar, see *Dichapetalum cymosum*
Gipsy moth, see *Lymantria dispar*
" Gix ", 203
Gliotoxin, as fungicide, 53, 137
Globulins and disease resistance, 20
Glossina spp., control by selective clearing, 40, 298
Glucocheriolin, as factor in host selection, 26
Glucosides, see glycosides
Glutathione, as factor in fungicidal action of sulphur, 108
Glyceride oils, as fungicides, 142; as insecticides, 194; as spreaders, 76; as stickers, 79; use with lead arsenate, 153
Glyceryl oleate, use in oil sprays, 190
Glycosides, as factors in disease resistance, 19; as factor in insect resistance, 26
Glyodin, as emulsifier, 74; as fungicide, 133
Glyoxalidines, as fungicides, 133
Gnomonia veneta, forewarning by, 39
Gooseberry, American mildew, see *Sphaerotheca mors-uvae*; sawfly, see *Pteronidia ribesii*
Goulac, 75
Gout fly, see *Chlorops taeniopus*
Grafting, use in resistant varieties, 10
Grain weevil, see *Calandra granaria*
Grape vine, anthracnose, see *Gnomonia veneta*; black rot, see *Guignardia bidwellii*; downy mildew, see *Plasmopara viticola*; moths, see *Clysia* and *Polychrosis* spp.; phylloxera, see *Phylloxera vitifoliae*
Gray flotation sulphur, 106
Grease banding, as control measure, 292

Green American hellebore, see *Veratrum viride*
Green manuring, as control measure, 53, 279
Green oils, 185
— sulphur, 106
Griseofulvin, as fungicide, 54, 138
" Guesarol ", 202
Guignardia bidwellii, resistance of grapes to, 14
" Guthion ", 220
" Gypsine ", 148
Gyptol, as insect attractant, 286

H

Haber's rule, 64, 247
Haiari, as insecticide, 176
Half-white oils, 185
Haptotropism, 15
" Hatching factor ", 280
Heap method of seed treatment, 256
Heat, as soil sterilizing agent, 270 ; use for seed treatment, 267
Helicopters, for spray application, 63
Heliopsis spp., as insecticides, 197
Hellebore, as insecticide, 156
Helminthosporium avenae, action of organomercury compounds on, 262
— *gramineum*, action of organomercury compounds on, 262 ; control by seed selection, 255
— *sativum*, effect of formaldehyde on attack by, 259 ; phytotoxins produced by, 15
Helopeltis theivora, wind breaks for control of, 303
Hemileia vastatrix, climatic conditions and prevalence of, 37
2-Hendecyl-2-oxazoline, as fungicide, 134
HEOD, 213
Heptachlor, 212
2-Heptadecyl-2-imidazoline, see glyodin
Herbicides, 232 *et seq.*
Hessian fly, see *Phytophaga destructor*
Heterocaryosis, 12
Heterodera marioni, as cause of soil sickness, 269 ; biological control of, 55
Heterodera rostochiensis, use of soil fumigants against, 273
— *schachtii*, control by trap cropping, 286 ; crop rotation against, 280 ; status of parasitic nemas in control of, 51 ; stimulation by host plant diffusates, 280, 286
Heteroecism, 300
Hexachlorobenzene, as seed protectant, 267
1 : 2 : 3 : 4 : 7 : 7-Hexachlorobi*cyclo*[2 : 2 : 1]hepten-5 : 6-bisoxymethylene sulphite, 215
Hexachlorodi*cylco*pentadiene, 211
1 : 2 : 3 : 4 : 10 : 10-Hexachloro-6 : 7-epoxy-1 : 4 : 4a : 5 : 6 : 7 : 8 : 8a-octahydro-1 : 4-5 : 8-dimethanonaphthalene, 213
1 : 2 : 3 : 4 : 10 : 10-Hexachloro-1 : 4 : 4a : 5 : 8 : 8a-hexahydro-1 : 4-5 : 8-dimethanonaphthalene, 213
Hexachloro*cyclo*hexane, see BHC
Hexachlorotetra*cyclo*dodecadiene, 212
Hexaethyl tetraphosphate, 217
Hexahydro-1 : 3 : 6-thiadiazepine-2 : 7-dione, as fungicide, 132
2-*cyclo*Hexyl-4 : 6-dinitrophenol, see dinex
HHDN, 213
Hippodamia convergens, action of hydrogen cyanide on, 247
Histidine, effect on fungitoxicity of dithiocarbamates, 131
Hochberg-LaMer generator, 63
Homologous series, 96 ; relationships between toxicities of members of an, 96 *et seq.*, 195
Hop, aphis, see *Phorodon humuli* ; powdery mildew, see *Sphaerotheca humuli*
Hoplocampa testudinea, use of nicotine against, 161 ; use of quassia against, 183
" Horlin ", 271
Host selection, by insects, 26
Hot-pressed naphthalene, 250
Hot-water method of seed treatment, 267
Hybernia defolaria, use of grease bands against, 292

Hydrated lime, see calcium hydroxide
Hydrocarbon oils, as carriers in sprays, 63 ; as fungicides, 141 ; as insecticides, 183 *et seq.* ; as ovicides, 190 *et seq.*; as spreaders, 76 ; as stickers, 79 ; as weedkillers, 244 ; classification of, 183 ; phytotoxic properties of, 187 ; Waterman analysis of, 190
Hydrochloric acid, as factor in cyanide damage, 248
Hydrocyanic acid, action on insect, 247 ; action on plant, 248 ; as fumigant, 245 *et seq.*; liquid, 246 ; resistance of insects to, 248 ; use for soil treatment, 274
Hydrogen cyanide, see hydrocyanic acid
— ion concentration, of cell sap in relation to disease resistance, 17 ; of insect gut, in relation to the toxicity of arsenicals, 151 ; of soil, in relation to disease intensity, 278
— sulphide, as factor in fungicidal action of sulphur, 93, 108 ; as factor in phytotoxic action of polysulphides, 113 ; as fungicide, 108
Hydrotropism, 14
15-Hydroxydeguelin, 178
15-Hydroxyrotenone, 178
3-Hydroxymethyl-*n*-heptan-4-ol, see ethylhexane diol
m-Hydroxyphenyl trimethylammonium methyl sulphate dimethylcarbamate, 226
6-Hydroxy-3(2H)pyridazinone, 239
8-Hydroxyquinoline, see oxine
Hydroxythiazoles, as insecticides, 201
Hyperparasitism, 44
Hypersensitivity, as factor of disease resistance, 16 ; as factor in insect attack, 27
Hypoderma spp., use of organophosphates against, 219

I

Icerya purchasi, biological control of, 46 ; encouragement by DDT sprays, 49 ; fumigation against, 245
" Igepals ", as spreaders, 74
" Igepons ", as spreaders, 73
2-Imidazolines, as fungicides, 134
3-(α-Iminoethyl)-5-methyltetronic acid, as weedkiller, 240
Immunity, see disease resistance
Immunization of plants, 21
Immunological methods, use in virus diagnosis, 21
Impatiens balsamina, fungicides from, 136
Implements, as disease vectors, 306
Importation, restriction of, as control measure, 308
" Indalone ", 287
Independent action of toxicants, 94
Index of variation, 91
3-Indoleacetic acid, as plant hormone, 237
Infection centres, elimination of, 297 *et seq.*
Infection vectors, elimination of, 303 *et seq.*
Inherent toxicity, 67
Initial deposit, 77
Initial retention, as factor in protective efficiency, 67
Insect injury and fungus attack, 15
Insecticides, contact, 61 ; direct, 61 ; functions of, 61 ; protective, 61 ; stomach, 61, *et seq.* ; systemic, 60
Insectivorous birds, as biological control agents, 44, 45
Insects, dissemination of disease by, 304 *et seq.*
Instant Bordeaux, 116
Instant Dip, 263
Inversion, of emulsions, 83 ; of suspensions, 78
Inverted pan method of steam sterilization, 270
IPC, see propham
Iron sulphate, see ferrous sulphate
Irradiation, as control measure, 49
Irrigation, as control measure, 34
" Iscothan ", 144
Isodrin, 213
" Isolan ", 227
Isoquassin, 183
Isosesamin, as pyrethrum synergist, 172

Isosteric compounds, 204

J

Jamaica quassia, 183
Japanese beetle, see *Popillia japonica*
" Jaundice " of silkworms, 58
Johnson's mixture, 119
Joint action of toxicants, 94
Junonia coenia, bacterial pathogens of, 57
Juvenile hormone, 284

K

Kakothrips robustus, influence of soil conditions on attack by, 33
Kaliosysphinga ulmi, control by cultivation, 279
" Karathane ", 144
Kerosine, see hydrocarbon oils
Kinesis, 14
" Knock-down ", 170
" Knock-out " point, 89
" Kolodust ", 106
" Kolofog ", 106
Korlan, 219

L

Laboratory trials, see bioassay
Lantana camara, biological control of, 50
Larkspur, insecticidal properties of, 196
Latent virus, 58
Lauric acid, as insecticide, 195
Lauryl thiocyanate, as insecticide, 200
LD 50, 91
Lead arsenate, acid, 148 ; basic, 148
— arsenates, action of hard water on, 153 ; action of lime on, 152 ; action of lime sulphur on, 113, 141 ; action of soaps on, 153 ; as fungicides, 141 ; as insecticides, 148 *et seq.*; foliage damage by, 152 ; versus calcium arsenates, 148
— chromate, as insecticide, 157
Leaf excretions, and aphid resistance, 25 ; as factors in phytotoxic action of arsenicals, 153 ; in relation to fungal attack, 14 ; in relation to the action of copper fungicides, 125
Lecanium corni, use of tar oils against, 190
Lecinthinase, in relation to pathogenicity, 57
Lepidoptera, use of light traps against, 289
Lepidosaphes beckii, biological control of, 32
— *ulmi*, action of hydrocarbon oils on eggs of, 186 ; biological control of, 49
Leptinotarsa decemlineata, migration of, 1 ; mode of attraction to host plant, 284 ; nature of resistance to, 26 ; tolerance to DDT, 208
Leptosphaesia herpotrichoides, use of sulphuric acid against, 233
" Lethane 60 ", 200
" Leytosan ", 261
Lichens, use of dinitrocresols against, 198
Light traps, 289 *et seq.*
Lignin pitch, 75
Lime, as corrective for arsenical sprays, 152 ; phytotoxic action of, 127 ; status and degree of attack, 32
— casein, as spreader, 75 ; use with arsenicals, 153
— sulphur, as fungicide, 112 *et seq.* ; chemistry of, 112 ; defoliation by, 114 ; dry, 112 ; interaction with lead arsenate, 113, 141 ; phytotoxic action of, 113
Linamarine, as factor of disease resistance, 19
Lindane, action on insect, 210 ; as insecticide, 209 *et seq.* ; as fumigant, 245 ; as soil fumigant, 274
Lipoid solubility and toxic action, 97, 126, 264
Liquidusters, 64
Lithium salts and disease resistance, 32

Lithobius forficatus, as biological control agent, 51
Liver of sulphur, as fungicide, 112 ; as seed disinfectant, 256
Lonchocarpus spp., as insecticides, 176 *et seq.*
London forces, 100, 205
London purple, as insecticide, 148
" Loro ", 200
Lousewort, insecticidal properties of, 196
Lubricating oils, see hydrocarbon oils
Lycopersicin, as factor in disease resistance, 19
Lycopersicum esculentum, insecticidal properties of, 196
Lycophotia margaritosa, effect of climatic factors on, 35
Lygaeus kalmii, ovicidal action of cresols on, 199
Lygus pabulinus, action of glyceride oils on, 194 ; ovicidal properties of hydrocarbon oils on, 190
Lymantria dispar, biological control of, 46 ; chemotropic responses of, 286 ; use of lead arsenate against, 148 ; use of virus against, 58
Lymantria monacha, use of virus against, 58
Lymantridae, control by hand picking, 297

M

Macrosiphum rosae, action of fatty acids on, 195
— *solanifolii,* as virus vector, 304
Magnesium arsenate, as insecticide, 150
Malaccol, 178
Malachite, 118
— Green, as fungicide, 143 ; use with sulphur, 106
Malacosoma americana, action of arsenicals on, 150
Malaoxon, 224
Malathion, 73, 220
Maleic hydrazide, as weedkiller, 239
Malic acid, as component of spore excretions, 124
Mamelia brassicae, action of rotenoids on, 180
Mammea americana, as insecticide, 196
Mandibulate insects, 24
Maneb, as fungicide, 129
" Manganar ", 150
Manganase arsenates, as insecticides, 150
Manganous ethylene bisdithiocarbamate, see maneb
Mangold, leaf spot, see *Uromyces betae*
Manuring, as control measure, 30
March moth, see *Anisopteryx aescularia*
" Marker ", use in sprays, 113
Maximum initial retention, 77
MCPA, as weedkiller, 234, 235
Mechanics of toxic action, 96 *et seq.*
Median effective dose, 91
Median lethal dose, 91
Mediterranean fruit fly, see *Ceratitis capitata*
Melanoplus differentialis, use of poison baits against, 293
Melilotus alba, toxic components of, 295
Mendelian laws of heredity, 7
Mercurated hydrocarbons, as seed disinfectants, 261
Mercuric bromide, action on fungus spore, 263
— chloride, action on fungus spore, 263 ; as fungicide, 140 ; as seed disinfectant, 260 ; as soil sterilizing agent, 274 ; as water disinfectant, 304
— cyanide, as seed disinfectant, 262
Mercurous chloride, use against diptera, 274
Mercury, as fungicide, 264, 274
— , organic compounds of, see organomercury compounds
" Mersolates ", 73
" Meta ", 293
Metaldehyde, use against slugs, 293
Metamorphosis in insects, hormonal control of, 284
Metanicotine, as insecticide, 165

Metanopia rubriceps, use of repellents against, 288
Metarrhizium anisopliae, as biological control agent, 52
" Metasystox ", 218
Metatetranychus ulmi, action of diphenyl sulphones on, 229 ; encouragement by tar oil washes, 49 ; hydrocarbon oils, as ovocides against, 190 ; photoperiodic effect on, 283
Meteorological conditions and degree of attack, 38
Meteorological forecasting of epidemics, 29
Methanesulphonyl fluoride, as fumigant, 252
6-Methoxy-2(3)-benzoxazolinone, as factor in insect resistance, 27
Methoxychlor, as insecticide, 203
Methoxyethylmercuricarbide, as seed disinfectant, 262
Methoxyethylmercuric chloride, as seed disinfectant, 261
— hydroxide, reactions of, 263
— silicate, as seed disinfectant, 261
2-Methoxy-1 : 4-naphthoquinone, as fungicide, 136
Methyl bromide, as fumigant, 252 ; as soil fumigant, 273
Methyl celluloses, as dispersing agents, 81
2-Methyl-4-chlorophenoxyacetic acid, see MCPA
γ-(2-Methyl-4-chlorophenoxy)butyric acid, as weed killer, 237
2-Methyl-4 : 6-dinitrophenol, see DNC
Methyl-3-*cyclo*hexene-1-carboxylic acid, propyl ester, as attractant, 286
Methyl *iso*thiocyanate, as fumigant, 132
2-Methylmercapto-1 : 4-naphthoquinone, as fungicide, 136
Methylmercuric dicyandiamide, as seed disinfectant, 261
Methyl parathion, as insecticide, 218
2-(1-Methyl-*n*-propyl)-4 : 6-dinitrophenol, see dinoseb
3-(1-Methyl-2-pyrrolidyl)-pyridine, see nicotine
N-Methylsulphonyl-*N*-trichloromethylthio-4-chloroaniline, as fungicide, 135
Methyl triphenyl phosphonium derivatives, as insecticides, 158
2 : 2′-Methylene bis(p-chlorophenol), as fungicide, 144
3 : 3′-Methylene bis(4-hydroxycoumarin), see dicoumarin
Methylmercuric iodide, 262
Mexican bean bettle, see *Epilachna varivestis*
" MGK 264 ", 173
Micronutrients, influence on fungal attack, 32
Microtus agrostis, birds as control agents of, 45
Migrantes, 283
Millet, grain smut, see *Sphacelotheca sorghi*
Mineral oils, see hydrocarbon oils
Miscible oils, 84
" Mitin FF ", 203
Mitis green, 147
Molasses, as supplements for Bordeaux mixture, 120
Molecular shape and toxicity, 181
Monochlorobenzene hexachloride, as insect repellent, 209
Monochus spp., possible role as control agents, 51
Monosulphide sulphur, 112
Monuron, as weedkiller, 241
Mortality data, interpretation of, 90 *et seq.*
Moss, use of nitrophenols against, 198
Mottled umber moth, see *Hybernia defolaria*
Multistage parasitism, 47
" Muritan ", 295
Musarin, 54
Musca domestica, action of cyclodiene insecticides on, 214 ; action of hydrocarbon oils on, 186 ; action of pyrethrins on, 170 ; use of traps against, 285 ; use of veratrine alkaloids against, 157
Muscardine disease of silkworm, 52
Mussel scale, see *Lepidosaphes ulmi*
Mustard oils, as factors in host selection, 26
Mutagens, use in plant breeding, 8
Mutation, 8
Mycosphaerella pinodes, control by seed selection, 254
" Mylone ", 133

Myxomatosis, use for rabbit control, 58
Myzus persicae, food finding by, 26

N

Nabam, as fungicide, 129 ; oxidation products of, 132
" Nacconols ", 74
Naphthalene, action on plant, 250 ; as fumigant, 250 ; as repellent, 287 ; grade 16, 250
Naphthenates, use in oil sprays, 190
Naphthols, for treatment of band traps, 291
Naphthoquinones, as fungicides, 136
α-Naphthylacetic acid, as weedkiller, 234, 237
α-Naphthylamine, use in band traps, 291
1-Naphthyl *N*-methylcarbamate, as insecticide, 227
N-1-Naphthylphthalamic acid, as weedkiller, 239
N-1-Naphthylphthalimide, as weedkiller, 239
α-Naphthyl thiourea, see ANTU
Narcosis, 97
Neburon, as weedkiller, 242
Nectoria diploae, as biological control agent, 52
Nematodes, as biological control agents, 51
" Nemagon ", 273
" Neocid ", 202
Neodiprion sertifer, biological control of, 58
Neonicotine, as insecticide, 164
Neoquassin, 183
Neostigmine, 226
" Neotran ", 229
Nicotiana spp., insecticidal alkaloids of, 164
Nicotine, action on insect, 161 ; as fumigant, 161, 251 ; as insecticide, 160 *et seq.* ; as ovicide, 161 ; as repellent, 288 ; dusts, 163 ; " fixed ", 161 ; toxicity in relation to structure, 164
— peat, 161
— soaps, 162
— sulphate, 160 ; as stomach poison, 161 ; as repellent, 288
— tannate, as stomach poison, 161
Nicotinium oleate, as insecticide, 162
Nicotinium salts, as insecticides, 161
Nicotyrine, as insecticide, 165
Nicouline, 177
" Nirit ", 144
" Nirosan ", 159
" Nirosit ", 159
Nitramines, as weedkillers, 239
Nitrobenzenes, as soil sterilizing agents, 272
Nitrogenous manures, influence on degree of attack, 30 *et seq.*
3-Nitro-4-hydroxybenzoic acid, as weedkiller, 240
" Nomersan ", 266
Nonachlor, 212
Non-competitive enzyme inhibition, 102
Non-ionic detergents, 74
Normal equivalent deviation, 91
Nornicotine, as insecticide, 164
" NPD ", 217
Nutrition of host plant—relation to attack, 25 *et seq.*

O

Oat, fritfly, see *Oscinella frit* ; loose smut, see *Ustilago avenae*
Octachlor, 211
Octachlorodi*cyclo*pentadiene, isomers of, 211
1 : 2 : 4 : 5 : 6 : 7 : 10 : 10-Octachloro-4 : 7 : 8 : 9-tetrahydro-4 : 7-*endo*methylene-indane, 211
Oecophylla longinoda, control of, 49
Oedema, 188
Oil-flocculated suspensions, 78

Oils, see glyceride oils and hydrocarbon oils
Olefines, see hydrocarbon oils
OMPA, 217
Oncopeltus fasciatus action of chloro*cyclo*pentadienes on, 212
Onion, neck rot, see *Botrytis allii* ; smudge, see *Colletotrichum circinans* ; smut, see *Urocystis cepulae*
Oospora pustulans, use of dips against, 263
Operophthera brumata, cresols, as ovicides against, 192 ; use of grease bands against, 292
Ophiobolus graminis, biological control of, 53 ; effect of sulphuric acid on, 233 ; soil conditions and parasitism of, 34 ; use of crop rotation against, 280
Optical activity, in relation to host specificity, 17 ; in relation to toxicity, 165, 170
Opuntia spp., biological control of, 50
Orchard sanitation, 298
Organomercury compounds, action on seed grain, 264 ; action on fungus spore, 264 ; as seed disinfectants, 260 *et seq.* ; chemistry of, 262 ; structure in relation to toxicity, 261
Organophosphorus compounds, as acaricides, 218 ; as anticholinesterases, 221 ; as insecticides, 216 *et seq.* ; as nematicides, 219 ; as ovicides, 222
Oryctolagus cuniculus, use of virus against, 58
Oryzaephilus surimanensis, as test organism in bioassay, 86
Osage orange, see *Toxylon pomiferum*
Oscinella frit, early sowing for control of, 42 ; parasitism of, by nematodes, 51
Osmium compounds, as fungicides, 140
Overton-Meyer theory, 97
Ovex, as acaricide, 230
Ovicides, 61 ; bioassay of, 86
Oxidation-reduction potential, and disease resistance, 17
Oxidative phosphorylation, inhibition by nitrophenols, 199
Oxine, as fungicide, 121
Oxytetracycline, 137
Oystershell scab, see *Lepidosaphes ulmi*

P

Pachyrrhizone, as insecticide, 179
Pachyrrhizus erosus, insecticidal properties of, 177
" Panogen ", 261
Paradichlorobenzene, see *p*-Dichlorobenzene
Paraffin, see hydrocarbon oils
" Paraffinicity ", as factor in insecticidal action of hydrocarbon oils, 190
Paraoxon, 224
Parathion, 218
Paris green, as insecticide, 147 ; use in baits, 293
— purple, 148
Parthenogenesis, 282
Partial sterilization of soil, 269 *et seq.* ; effects on plant growth, 274 ; effects on soil microflora, 275
Particle size, in relation to application, 64 ; in relation to tenacity, 79 ; in relation to toxicity, 92, 109
Particulate sulphur, 107
Partition coefficient and toxicity, 97, 195
Pasteurization of soil, 270
Patents, see B.P., Swiss P., U.S.P.
Pea, aphid, see *Acyrthosiphum pisum* ; thrips, see *Kakothrips robustus*
Peach, aphid, see *Myzus persicae* ; tree borer, see *Sanninoidea exitosa* ; yellows, use of warm water treatment against, 268
Pear, midge, see *Contarinia pyrivora* ; use of resistant stocks, 10
Pectinases and pathogenicity, 18
Pediculoides ventricosus, as biological control agent, 47
Pellicularia filamentora, use of formaldehyde against, 259
Pellitorine, 167, 197
Pellitory root, 167
Pemphigus betae, irrigation as control measure against, 34
Penetrating properties, 71
Penicillin, 54
Penicillium griseofulvin, fungal antibiotics from, 138

Pentachloronitrobenzene, as fungicide, 143
Pentathionic acid, as factor in fungicidal action of sulphur, 109
PEPS, as sticker, 80
Perchloromethyl mercaptan, as toxophore, 135
" Perenox ", 120
Periodic system and fungicidal properties of elements, 140
Periplaneta americana, action of arsenicals on, 150 ; relative toxicities of cyclodiene insecticides to, 214
Peronospora tabacina, use of benzene against, 253 ; use of *p*-dichlorobenzene against, 253
" Perozid ", 140
Petroleum oils, see hydrocarbon oils
Phaedon cochleariae, action of pyrethroids on, 171
Phellodendron amurense, insecticidal properties of, 196
Phenols, as factors of disease resistance, 18 ; use for soil treatment, 272
Phenothiazine, see thiodiphenylamine
3-Phenyl-1 : 1-dimethylurea, see fenuron
Phenylmercuric acetate, as seed disinfectant, 261
Phenylmercuric chloride, as fungicide, 140
Phenylmercuric triethanolamine lactate, as fungicide, 140
Phenylmercury urea, as seed disinfectant, 261
1-Phenyl-3-methyl-5-pyrazolyl dimethylcarbamate, as insecticide, 227
Phenylureas, as weedkillers, 241
Philothion, 108
Phoma betae, use of seed protectants against, 266
Phorodon humuli, " stripping " for control of, 40 ; use of quassia against, 183
" Phosdrin ", 220
Phosphates, organic, see organophosphates
Phosphatic manures, influence on degree of attack, 30
Phosphine A.C.R., 143
Phosphonium derivatives, as insecticides, 158
Photoperiodic effects on life histories of insects and mites, 283
Phototropism, 14 ; as basis of control measures, 289
Phthalamates, as weedkillers, 238
Phylloxera vitifoliae, flooding for control of, 34 ; soil type and attack by, 33 ; use of carbon disulphide against, 271 ; use of resistant root-stocks against, 10
" Physically-toxic " substances, 99
Physiologic races, 10
Physostigmine, 226
Phytophaga destructor, " fly-free " dates, 42 ; varietal resistance to, 25
Phytophthora colocasiae, action of alcohols on, 97
— *cryptogea*, water transmission of, 304
— *infestans*, control by haulm destruction, 232 ; effect of fertilizers on, 30 ; forecasting of epidemics of, 39 ; hot air treatment against, 267 ; resistance to, 14
— *parasitica*, water transmission of, 304
Picraena excelsa, as insecticide, 183
Picrasmin, as insecticide, 183
Pieris spp., control by hand-picking, 297 ; host selection by, 26 ; use of repellents against, 288
Pine oils, as spreaders, 76
Pinosylvin, as factor in resistance to fungal attack, 18
Piperidinium reineckate, as insecticide, 158
3-(2-Piperidyl)-pyridine, see neonicotine
Piperine, as pyrethrum synergist, 173
Piperonal, 172, 173
Piperonyl butoxide, as pyrethrum synergist, 170
Piperonyl cyclonene, as DDT synergist, 209 ; as pyrethrum synergist, 173
Pissodes strobi, trap cropping against, 286
Plant breeding, 7 *et seq.*
Plant growth substances, use in seed treatment, 265
— hygiene, 298
— toxins and disease resistance, 18
Plasmodiophora brassicae, control by crop rotation, 280 ; control by liming, 278 ; use of chloronitrobenzenes against, 143 ; use of mercuric chloride against, 274

Plasmopara viticola, action of copper sulphate on, 128 ; as test organism in bioassay, 86 ; foretelling attack by, 39 ; use of Bordeaux mixture against, 115; use of dinitrothiocyanobenzene against, 144
Plesiocoris rugicollis, action of tar oils on, 192
Pleurotropis parvulus, as biological control agent, 47
Plum, brown rot, see *Sclerotinia laxa* ; silver leaf, see *Sterum purpureum.*
Plutella maculipennis, host selection by, 26 ; use of insecticides in biological control of, 49
Poabius bilabiatus, as biological control agent, 51
Poison baits, 293
Polar groups, 73
Pollen, effect of sulphur on, 110
Polychrosis botrana, action of nicotine on, 161 ; use of carbazole derivatives against, 159 ; use of light traps against, 289
Polyethylene polysulphides, as stickers, 80
Polyglycerol esters, as spreaders, 74
Polyhedral diseases, 58
Polyploidy, 8, 9
Polypropylene glycol esters of 2, 4-D, as weedkillers, 235
Polyspora lini, use of thiram against, 266
Polysulphide sulphur, as factor in the fungicidal action of sulphur, 109 ; fungicidal properties of, 113
Polysulphides, as fungicides, 112 *et seq.* ; as insecticides, 112 ; as seed disinfectants, 259 ; chemistry of, 112
Polyvinyl acetate, as sticker, 80
Popillia japonica, deterrent effect of arsenicals on, 151 ; traps for use against, 285 ; use of repellents against, 288
Pot fumigation, 246
Potassic fertilizers, effect on degree of attack, 30
Potassium cyanide, as fumigant, 246 ; use for soil treatment, 274
— permanganate, as fungicide, 141
— polysulphide, see liver of sulphur
— sulphocarbonate, use for soil treatment, 271
— thiocarbonate, use for soil treatment, 271
Potato, beetle, see *Leptinotarsa decemlineata* ; blight, see *Phytophthora infestans* ; brown scab, see *Streptomyces scabies* ; deterioration of, 305 ; dry rot, see *Fusarium caeruleum* ; leaf hopper, see *Empoasca mali* ; leaf roll, vectors of, 306 ; Rhizoctonia disease, see *Pellicularia filamentora* ; skin spot, see *Oospora pustulans* ; wart, see *Synchytrium endobioticum*
" Potosan ", 219
Powdery mildews, physiologic races of, 10
Precipitin test, for differentiation of physiologic races, 21
Preferential retention, 77
Presoakage method of seed disinfection, 207
Prickly pear, see *Opuntia* spp.
Pristiphora erichsonii, bacterial pathogens of, 57
Probit, use in interpretation of mortality data, 90 *et seq.*
Prodenia eridania, action of rotenoids on, 180
Prodenia litura, biological control of, 46 ; control by defoliation, 232
Promecotheca reichei, biological control of, 47
Propham, as weedkiller, 238
*iso*Propyl carbanilate, see propham
*iso*Propyl 3-chlorophenylcarbamate, as weedkiller, 238
*iso*Propyl 2 : 4-dichlorophenoxyacetate, as weedkiller, 235
n-Propyl isome, 173
*iso*Propyl-3-methyl-5-pyrazolyl dimethylcarbamate, as insecticide, 227
2-*n*-Propyl-4-methyl-5-pyrazolyl-(6)-dimethylcarbamate, as insecticide, 227
*iso*Propyl *N*-phenylcarbamate, see propham
Protective colloids, use in spray materials, 80, 81
— fungicides, 60 ; comparison of, 86
— insecticides, 61
"— stupefaction ", 247
Protein specificity and disease resistance, 20
Proteolytic enzymes, as factors in disease resistance, 18
Protocatechnic acid, as factor in disease resistance, 19

Protoparce quinquemaculata, action of nicotine on, 162
Protozoa, as factor in soil sickness, 275
Pruning, as control measure, 40, 43 ; transmission of disease by, 306
Pseudococcus citri, biological control of, 48 ; control by pruning, 40
— *lilacinus*, biological control of, 48 ; control by banding, 48
— *njalensis*, biological control of, 48
Pseudomonas medicaginis, control by seed selection, 254
Psila rosae, date of planting and attack by, 42 ; use of repellents against, 288
Psyllia mali, action of tar oils on, 192 ; biological control of, 52 ; cresols as ovicides against, 192 ; use of organic thiocyanates against, 201
Puccinia glumarum, influence of nutrient conditions on attack by, 30 ; influence of soil conditions on attack by, 33 ; nature of resistance to, 7, 16
Puccinia graminis, control by barberry eradication, 300 ; epidemiology of, 38 ; influence of manuring on attack by, 30, 31 ; physiologic races of, 11 ; resistance to in relation to cell sap acidity, 17 ; use of cycloheximide against, 138 ; use of early maturing varieties against, 43
" Puratised N5D ", 140
" Pyramat ", 227
Pyrausta nubilalis, resistance to, 27 ; use of ryania against, 157
Pyrethric acid, 168
Pyrethrin I, 167
Pyrethrin II, 167
Pyrethroids, 169 ; action on insects, 171
Pyrethrol, 167
Pyrethrolone, 167
Pyrethrone, 167
Pyrethrum spp., as insecticides, 167, *et seq.*
Pyrethrum synergists, 172 *et seq.*
Pyridine, as insecticide, 164 ; as repellent, 287
" Pyrolan ", 227
Pyrotechnic smoke generators, 63
Pyrrol, as insecticide, 164
3-(2-Pyrrolidyl)-pyridine, 164
Pythium ultimum, biological control of, 55 ; use of cuprous oxide against, 257 ; use of thiram against, 55, 266

Q

Quantal response, 90
Quarantine, as control measure, 308
Quassia amara, 183
Quassia, as insecticide, 183
Quassiins, 183
Quaternary ammonium derivatives, as spreaders, 74
Quebracho-fixed nicotine, 161
Quick-breaking emulsions, 83
Quillaja saponaria, 75
Quinone oxime benzoylhydrazine, as seed protectant, 266
Quinones, as fungicides, 136, 266

R

Rabbits, Rodier system of control, 46 ; use of repellents against, 288 ; use of virus against, 58
Radix pyrethri, 167
" Randox ", 243
Rape oil, as insecticide, 194
Rat poisons, 293 *et seq.*
Rats, use of bacteria against, 57
Receding contact angle, 70
Recessive genes, 8
Red oils, 185
— scale, see *Aonidiella aurantii*
— squill, as rat poison, 293
Reinecke's salt, as insecticide, 157
Repellents, use as control measure, 287

Resistant varieties, production of, 7 *et seq.* ; use of, 9 *et seq.*
Rhodanines, as fungicides, 132, 133
Rhodnius prolixus, action of nicotine on, 161
Robigalia, 4
Rodent repellents, 288
Rodier system of rabbit control, 46
Rodolia cardinalis, use as biological control agent, 46
Roh-ten, 177
Root diffusates, as factor in eelworm attack, 2801 biological activity of, 232
— knot, see *Heterodera marioni*
Rose, powdery mildew, see *Spaerotheca pannosa*
Rosemary, as repellent, 196
Rota method of sulphur application, 107
Rotenoids, biosynthesis of, 179 ; chemistry of, 177 *et seq.* ; insecticidal properties of, 179 *et seq.*
Rotenone, 177
Rotenonic acid, 179
Rubber accelerators, as fungicides, 128
" Run-off ", 78
Rust fungi, physiologic races, 10
Ryania, insecticidal properties of, 157
Ryania speciosa, 157
Ryanodine, as insecticide, 157
Rye, Fusarium disease, see *Calonectria graminicola*

S

Sabadilla, as insecticide, 157
Sabadillin, as insecticide, 157
Safety cyanide packages, 246
St. Urbansgrün, 147
Saissetia oleae, biological control of, 48
Salicylanilide, as fungicide, 142
Salicylic acid, as seed disinfectant, 256
Salmonella enteritidis, use against rats, 57
Salmonella typhimurium, as biological control agent, 57
San José scale, see *Aspidiotus perniciosus*
" Santomerse ", 74
Sanninoidea exitosa, use of *p*-dichlorobenzene against, 272
Sap, acidity of, in relation to disease resistance, 17 ; in relation to spray damage, 128
" Sapamines ", 74
Sapindus utilis, 75
Saponins, as spreaders, 75
Saturated hydrocarbons, 184
" Savon pyrèthre ", 169
Scabrin, insecticidal properties of, 197
Scheele's green, 148
Schistocerca spp., use of bacteria against, 56
Schoenocaulon spp., as insecticides, 157
Schradan, as insecticide, 217 ; metabolism of, 224
Schweinfurtergrün, 147
Sclerotinia fructigena, wound entry by, 15
— *laxa*, control by hand-picking, 297 ; resistance of stone fruit to, 14 ; susceptibility to and toughness of cuticle, 16
— *sclerotiorum*, use of methyl bromide against, 273
— *trifoliorum*, as factor of clover sickness, 269
Sclerotium rolfsii, phytotoxins produced by, 15
Scutigerella immaculata, biological control of, 51
Sea Leek, as rat poison, 293
Secondary deposit, 77
Seed cleaning, 254
— disinfectants, 255 *et seq.*
— protectants, 254
— selection, as control measure, 254
— steeps, 255
— treatment with repellents, 287

Selection of resistant varieties, 7
Selective clearing, for tsetse fly control, 40, 298
Selenia tetralunaria, phenols as ovicides against, 198
" Semesan ", 260
Septoria lycopersici, control by weed eradication, 299
Sereh disease of sugar cane, resistance to, 10 ; use of warm water treatment against, 268
Serological methods, of differentiating physiologic races, 21; use in plant pathology, 21
Sesame oil, as pyrethrum synergist, 172
Sesamin, 172
Sesamolin, as pyrethrum synergist, 172
" Sesin ", 236
Sesone, as weedkiller, 236
" Sevin ", 227
Sexuparae, 283
Shading action of spray residues, 127
" Shirlan ", 142
Silesiagrün, 147
Silica, role in plant resistance, 25
Silicofluorides, as insecticides, 155
Silver leaf, see *Stereum purpureum*
— salts, as fungicides, 140
Silvex, as weedkiller, 236
" Simazin ", 241
Similar joint action, 94
Sinalbin, as factor in host selection, 26
Sinapsis arvensis, control by weedkillers, 233
Sinigrin, as factor in host selection, 26
" Sinox ", 233
Skim milk, as spreader, 75
Slugs, traps for, 291
Slurry methods of seed treatment, 261
Smokes, insecticidal, 63, 64
Soap bark, as spreader, 75
Soaps, as emulsifiers, 82 ; as fungicides, 142 ; as insecticides, 194 ; as spreaders, 73
Soda Bordeaux, 118
— process of fumigation, 248
Sodium alkyl sulphates, as spreaders, 73
— aluminofluoride, as insecticide, 155
— arsenate, as fungicide, 141
— arsenite, as fungicide, 141
— carbonate, as fungicide, 141
— carboxymethylcellulose, as dispersing agent, 81
— chloride, for seed treatment, 255
— cyanide, as fumigant, 246 ; -magnesium sulphate fumigant, 247 ; -sodium bicarbonate fumigant, 247
— 2 : 4-dichlorophenoxyethyl benzoate, as weedkiller, 236
— 2 : 4-dichlorophenoxyethyl sulphate, as weedkiller, 236
— 2 : 2-dichloropropionate, as weedkiller, 242
— 3 : 5-dinitro-*o*-cresylate, as weedkiller, 233
— dinonyl sulphosuccinate, as spreader, 78
— dithionite, as fungicide, 93
— dodecyl benzenesulphonate, as spreader, 74
— dodecyl sulphate as spreader, 73
— fluoride, as insecticide, 155
— fluoroacetate, as rat poison, 294
— Sodium fluosilicate, as insecticide, 155
— hydroxide, as fungicide, 141
— laurate, as insecticide, 195
— methyldithiocarbamate, as fungicide, 131
— *N*-1-naphthylphthalamate, as weedkiller, 239
— oleate, as insecticide, 195
— silicofluoride, as insecticide, 155
— sulphide, as fungicide, 141
— tetrathionate, as fungicide, 93
— trichloroacetate, as weedkiller, 242

Soil acidity, and incidence of disease, 278 ; effect of ammonium sulphate on, 233
— antiseptic, 271
— bacteria, effects of partial sterilization on, 275
— conditions, and degree of attack, 33 *et seq.*
— diffusion in, 270
— fumigant, 271
— fungi, effects of partial sterilization on, 275
— protozoa, effects of partial sterilization on, 275
— sickness, 269
— temperature and incidence of disease, 35, *et seq.*
— treatment, 269 *et seq.*
Solanidine, as factor in resistance to Colorado beetle, 27
Solanum chacoense, resistance to Colorado beetle, 27
— *demissium*, resistance to Colorado beetle, 27
— *polyadenium*, resistance to aphids, 25
Solar Distillate, 186
" Solbar ", 112
" Soluble " caseins, as spreaders, 75
Soufre noir, 106
Sowing, modification of time of, as control measure, 41 *et seq.*
Soybean, nature of insect resistance of, 27
Soybean flour, as sticker, 80
Soyin, 27
" Spans ", as spreaders, 74
" Sparklet " bulb application, 65
Specific surface activity, 96
Specificity of food requirements and disease resistance, 18
" Spergon ", 266
Sphaceloma fawcetti, climatic conditions and prevalence of, 37
Sphacelotheca sorghi, use of sulphur against, 259
Sphaerostilbe aurantiicola, as biological control agent, 52
Sphaerotheca humuli, action of alkali on, 141 ; action of sulphur on, 108 ; nature of resistance to, 20 ; use of petroleum oils against, 142 ; use of salicylanilide against, 142
— *mors-uvae*, use of kerosine emulsion against, 142 ; use of sodium arsenate against, 141 ; use of sodium carbonate against, 141 ; use of sulphur against, 110
— *pannosa*, use of hydrocarbon oils against, 142
Spilonota ocellana, biological control of, 50
Spongospora subterranae, use of mercuric chloride against, 263
Spore excretions, as factor in action of copper fungicides, 124
" Sporeine ", 56
Spray deposit, physiological effects on plants, 127
Spray injury, by arsenicals, 152 ; by copper fungicides, 126 ; by hydrocarbon oils, 187 ; by lime sulphur, 113 ; types of, 110
— load, see spray residue
— residue, effects on beneficial fungi, 52 ; factors affecting amount of, 67
Spraying, 67 *et seq.*
Spreaders, 67 *et seq.*
Spreading coefficient, 70
— properties, assessment of, 69
Squill, as rat poison, 293
Stavesacre, as insecticide, 196
Stemphylium sarcinaeforme, action of sulphur on, 93 ; use in bioassay, 93 ; use of dithiocarbamates against, 129
Stereoisomers, differential toxicity of, 103, 165, 170 ; role in disease resistance, 17
Stereotropism, 15 ; as basis for trapping, 291
Stereum purpureum, orchard sanitation against, 297 ; resistance to and gum formation, 17
Stickers, 67, 77 *et seq.*
Stimulation of plants, by hydrocyanic acid, 249 ; by organomercury compounds, 264 ; by soil antiseptics, 274 ; by sulphur, 110
Stock emulsions, 83
Stock solutions, 116
Stock, influence on resistance of scion, 20
" Stoddard Solvent ", as weedkiller, 244
Stomach poisons, 147 *et seq.*
Strawberry, tarsonemid mite, see *Tarsonemus fragariae*

Streptomycin, as fungicide, 137
Streptomyces griseus, fungal antibiotics from, 137, 138
— *scabies*, control by green manuring, 53 ; control by increasing soil acidity, 278 ; soil temperature and attack by, 35 ; use of formaldehyde against, 259 ; use of mercuric chloride against, 263 ; use of organomercury compounds against, 263
Stripping, as control measure, 40
" Strobane " 211
Structural complementariness, 100, 205, 206
Structural topography, 205, 206
Suberization and plant resistance, 16
Submergence, as control measure, 34, 279
Substrate inhibition, 101
Suction traps for insects, 290
Suctorial insects, 25
Sugar beet, curly top, transmission by leaf hoppers, 304 ; leaf spot, see *Cercospora beticola* ; nematode, see *Heterodera schachtii* ; root aphis, see *Pemphigus betae*
— cane, blight, see *Tomaspis saccharina*, pineapple disease, see *Endoconidiophora paradoxa* ; red rot, see *Colletotrichum falcatum* ; Sereh disease, varieties resistant to, 10
Sulfotep, as fumigant, 217, 245 ; as insecticide, 217
Sulphamic acid, as weedkiller, 233
Sulphated alcohols, as spreaders, 73
Sulphated linseed oil, as rodent repellent, 288
" Sulphenone ", 229
Sulphite cellulose lye, see sulphite lye
— lye, as emulsiffer, 83 ; as protective colloid, 81 ; as spreader, 74
" Sulphoergethan ", 271
Sulphonated fatty acids, as spreaders, 73
"— Lorol ", 73
" Sulphoxide ", 173
Sulphoxylic acid, as factor in fungicidal action of sulphur, 94
Sulphur, as fumigant, 107 ; as fungicide, 106 *et seq.* ; as soil fungicide, 278 ; as seed disinfectant, 259 ; bacteria, see *Thiobacillus thiooxydans* ; " balling " of, 106 ; defoliation by, 110 ; flotation, 106 ; green, 106 ; hydrolysis of, 93 ; mode of action, as fungicide, 93 ; phytotoxic properties of, 110 ; wettable, 107
Sulphur dioxide, as factor in fungicidal action of sulphur, 109
Sulphur nitride, as insecticide, 159
Sulphuric acid, as factor in fungicidal action of sulphur, 109, 278 ; as weedkiller, 233
Sulphur-shy varieties, 110
Sulphur-sun scald, 110
Sulphuryl fluoride, as fumigant, 252
Sumatra-type derris, 176
Sumatrol, 178
Surface activity and molecular structure, 72 ; and spreading properties, 72 *et seq.*
— adsorption, 72
— tension, as factor of spreading efficiency, 70
Surinam quassia, 183
Swede, midge, see *Contarinia nasturtii*
Swiss patent, 236,227 (nitroalkanes), 204
Synchytrium endobioticum, nature of resistance to, 20 ; soil moisture and attack by, 34 ; use of sulphur against, 278 ; varieties resistant to, 9
Synergism, 94 ; between nitrocresols and hydrocarbon oils, 199 ; pyrethrum synergists, 172 *et seq.*
Synthetic detergents, as spreaders, 73
Systemic fungicides, 60, 138
Systemic insecticides, 60, 217
" Systox ", 218

T

2, 4, 5-T, as weedkiller, 235
2-(2, 4, 5-TP), see silvex
Tachypterellus quadrigibbus, use of pigs against, 298
Taenia solium, heteroecism of, 302
Taeniocampae, use of light traps against, 290
Tank-mix method of emulsification, 83

Tar acids, 185
—, as wound dressing, 186
— bases, 185
— distillates, 185
— oils, see hydrocarbon oils,
Tarsonemus fragariae, warm-water treatment against, 268
Taxis, 14
TCA, 242
TDE, 203
Tea, blister blight, see *Exobasidium vexans* ; mosquito blight, see *Helopeltis theivora* ; shot hole borer, see *Xyleborus fornicatus*
Tecnazene, as fungicide, 143
" Tedion ", 229
" Teepol ", 73
" Telone ", 273
Tenacity, 67, 85 ; factors affecting, 77
Tenebrio molitor, action of pyrethroids on, 171
Tent fumigation, 245
TEPP, 217 ; action as ovicide, 222
Tephrosia spp., as insecticides, 176 *et seq.* ; use as wind break 303
Tephrosin, 178
" Tergitols ", 73
Termites, late sowing to avoid attack of, 42
Terpenes and plant resistance, 17
" Terramycin ", as fungicide, 137
Tetrachloro-*p*-benzoquinone, see chloranil
2 : 4 : 5 : 4′-Tetrachlorodiphenyl sulphone, as acaricide, 229
Tetrachlorodiphenylethane, 203
Tetrachloroethane, as fumigant, 251
Tetrachloroethylene, as soil fumigant, 273
2 : 3 : 4 : 5-Tetrachloronitrobenzene, as fungicide, 143
2 : 3 : 5 : 6-Tetrachloronitrobenzene, see tecnazene
Tetraethyl dithionopyrophosphate, see sulfotep
Tetraethyl pyrophosphate, as insecticide, 217
2 : 3′ : 4 : 5′-Tetrahydroxystilbene, as fungicide, 18
Tetramethylthiuram disulphide, see thiram
Tetramethylthiuram oxide, 130
1 : 3 : 6 : 8-Tetranitrocarbazole, as insecticide, 159
Tetra-*n*-propyl dithionopyrophosphate, as insecticide, 217
Tetranychus bimaculatus, see *T. urticae*
— *opuntiae*, as biological control agent, 80
— *telarius*, see *T. urticae*
— *urticae*, action of diphenyl sulphone on, 230 ; action of phenyl benzenesulphonate on, 230 ; use of flour paste against, 80 ; use of naphthalene against, 250
Thallium, as rat poison, 294
" Thanite ", 201
Thermodynamic activity, 97
Thigmotropism, 15
" Thimet ", 218
Thioarsenates, as fungicides, 141
Thiobacillus thiooxydans, use with sulphur for soil treatment, 278
Thiocarbamyl carbamyl sulphide, as fungicide, 130
Thiocyanates, organic, as insecticides, 200
Thiocyanoacetates, as insecticides, 201
Thiocyanoalkyl esters of allyl phenols, as insecticides, 201
2-Thiocyanoethyl laurate, as insecticide, 200
α-Thiocyanoketones, as insecticides, 201
Thiocyanopropyl-phenyl ether, as insecticide, 201
" Thiodan ", 214
2-Thio-3 : 5-dimethyl tetrahydro-1 : 3 : 5-thiadiazine, as fungicide, 133 ; as weedkiller, 241
Thiodiphenylamine, as insecticide, 158
Thiophosphates, metabolism of, 224
Thiosulphuric acid, as factor in the fungicidal action of sulphur, 93, 279
Thioureide ion, as factor in the fungicidal action of dithiocarbamates, 130
Thiram, as fungicide, 55, 129 ; as repellent, 287 ; as seed protectant, 266

Thiuram sulphides, as fungicides, 129 ; as repellents, 288
Thixotropy, use in paste products, 81
Thrips tabaci, soil conditions and attack by, 33
Thuja plicata, resistance to fungal attack, 18
Thujaplicins, as factors in disease resistance, 18
" Thylox " sulphur, 106
Tilletia caries, control by basic copper carbonate, 257 ; control by basic copper sulphate, 257 ; control by copper sulphate, 257 ; control by organomercury compounds, 262 ; date of sowing and attack by, 42
— *contraversa*, control by hexachlorobenzene, 267
— *foetida*, see *T. caries*
Timbo, 176
Time of sinking tests of penetrating properties, 71
Time of sowing, adjustment of, as control measure, 41 *et seq.*
Tineidae, use of repellents against, 287
" Tinocine D ", 162
Tip burn of potato, use of Bordeaux mixture against, 127
Tipulidae, use of light traps against, 290
Titaniagrün, 147
Tobacco, blue mould, see *Peronospora tabaci* ; extracts, as insecticides, 160
Tobacco mosaic, acquired immunity from, 21 ; transmission by cultural operations, 307 ; use of malachite green against, 143
— " shreds ", use in fumigation, 251
Todd generator, 63
Tolylmercuric acetate, as seed disinfectant, 261
Tomaspis saccharina, lime status and attack by, 32
Tomatidine, as factor in disease resistance, 19
Tomatine, as factor in disease resistance, 19 ; as factor in insect resistance, 27
Tomato, bacterial spot, see *Xanthomonas vesicateria* ; damping off, see *Phytophthora cryptogea* ; insecticidal properties of, 196 ; leaf mould, see *Cladosporium fulvum* ; moth, see *Diataraxia oleraceae* ; Septoria leaf spot, see *Septoria lycopersici* ; streak, effect of potash fertilizers on, 31 ; stripe, see streak ; Verticillium wilt, see *Verticillium albo-atrum*
" Tomorin ", 296
Tortricidae, use of light traps against, 290
Toxaphene, as insecticide, 211
Toxic action, mechanics of, 96 *et seq.*
Toxicarol, chemistry of, 177 ; insecticidal properties of, 179
Toxicity, evaluation of, 85 *et seq.*
Toxylon pomiferum, resistance factor of, 18
T.P. arsenate, 148
Trap crops, 286 *et seq.*
Traps, as control agents, 282 *et seq.*
Traube's rule, 97
Tree-banding compositions, see grease bands
Trialeurodes vaporariorum, biological control of, 48 ; use of tetrachloroethane against, 251
s-Triazole derivatives, as weedkillers, 240
Tribolium castaneum, as test organism in bioassay, 86
Trichloroacetic acid, as weedkiller, 242
2 : 6 : 7-Trichlorocamphene, as insecticide, 210
1 : 1 : 1-Trichloro-2 : 2-di-(*p*-chlorophenyl)ethane, see DDT
1 : 1 : 1-Trichloro-2 : 2-di-(*p*-chlorophenyl)ethylene, 203
1 : 1 : 1-Trichloro-2 : 2-di-(*p*-fluorophenyl)ethane, 203
1 : 1 : 1-Trichloro-2 : 2-di-(*p*-hydroxyphenyl)ethane, 205
1 : 1 : 1-Trichloro-2 : 2-di-(*p*-methoxyphenyl)ethane, see methoxychlor
2 : 3 : 3-Trichloro-1 : 1-diphenyl propylenes, as insecticides, 206
2 : 4 : 5-Trichlorodiphenyl sulphone, as acaricide, 229
1 : 1 : 1-Trichloro-2 : 2-di-(*p*-tolyl)ethane, 203
Trichloroethylene, as soil fumigant, 273 ; use in cyanide fumigation, 247
2 : 2 : 2-Trichloro-1-hydroxyethyl dimethyl phosphonate, as insecticide, 221
Trichloromethanesulphenyl chloride, see perchloromethyl mercaptan
Trichloro-*S*-methoxymethane thiol, as fungicide, 135
N-Trichloromethylmercapto-4-*cyclo*hexene-1 : 2-dicarboximide, see captan
Trichloronitromethane, as soil fumigant, 273
2 : 4 : 5-Trichlorophenoxyacetic acid, see 2, 4, 5-T
2-(2 : 4 : 5-Trichlorophenoxy)ethyl 2 : 2-dichloropropionate, as weedkiller, 243

2 : 4 : 5-Trichlorophenoxypropionic acid, see silvex
1 : 2 : 4-Trichloro-3 : 5-dinitrobenzene, as fungicide, 143
1 : 3 : 5-Trichloro-2 : 4 : 6-trinitrobenzene, as fungicide, 143
Trichoderma lignorum, see *T. viride*
— *viride*, as biological control agent, 53
Trihedralized configuration, 205
Trimethylamine, as insect attractant, 284
Trimethylene dithiocyanate, as insecticide, 201
Triosephosphate dehydrogenase, inhibition by dithiocarbamates, 131
Triphenyl phosphine, as insecticide, 158
Triplumbic arsenate, 148
Tris(2 : 4-dichlorophenoxyethyl)phosphite, as weedkiller, 236
" Trithion ", 219
" Tritons ", as spreaders, 74
Tropisms, 14
" Tropotox ", 237
Tsetse flies, see *Glossina* spp.
Tuba-root, 176
Tubatoxin, 177
Turgidity, as factor in disease resistance, 16
Turnip, white rot, see *Bacterium carotovorum*
Turntable methods of bioassay, 87
Turpentine oils, 187 ; as repellent, 287
Tussoch moths, see Lymantridae
" Tweens ", as spreaders, 74
Two-solution method of emulsification, 83
Tylenchinema oscinellae, as possible biological control agent, 51

U

Uncinula necator, use of sulphur against, 106
Undecylic acid, as insecticide, 195
Undrained salts, 250
U(nited) S(tates) P(atent) 1,254,908 (dry lime sulphur), 112
U.S.P. 1,344,018 (magnesium arsenate), 150
U.S.P. 1,420,978 (magnesium arsenate), 150
U.S.P. 1,466,983 (magnesium arsenate), 150
U.S.P. 1,550,650 (sulphur), 106
U.S.P. 1,880,404 (nitro-derivatives), 198
U.S.P. 1,931,367 (insecticidal product), 155
U.S.P. 1,954,171 (copper product), 120
U.S.P. 1,958,102 (copper product), 120
U.S.P. 2,004,788 (copper product), 120
U.S.P. 2,028,091 (spreaders), 73
U.S.P. 2,037,090 (sulphur product), 106
U.S.P. 2,051,910 (copper product), 120
U.S.P. 2,061,185 (sulphur product), 106
U.S.P. 2,062,911 (Reinecke's salt), 157
U.S.P. 2,069,568 (sulphur product), 106
U.S.P. 2,069,710 (sulphur product), 106
U.S.P. 2,098,257 (sulphur product), 114
U.S.P. 2,101,645 (sulphur nitride), 159
U.S.P. 2,107,058 (nicotine-peat), 161
U.S.P. 2,152,236 (nicotine derivatives), 161
U.S.P. 2,157,861 (copper zeolites), 123
U.S.P. 2,202,145 (sesame oil), 172
U.S.P. 2,209,184 (organic thiocyanates), 201
U.S.P. 2,217,611 (organic thiocyanates), 201
U.S.P. 2,220,521 (organic thiocyanates), 201
U.S.P. 2,277,744 (sulphamic acid), 233
U.S.P. 2,326,350 (pyrethrum synergists), 173
U.S.P. 2,348,949 (sabadilla), 157
U.S.P. 2,349,771 (chloranil), 266
U.S.P. 2,390,911 (sabadilla), 157
U.S.P. 2,516,186 (" Dilan "), 205

U.S.P. 2,565,471 (toxaphene), 211
U.S.P. 2,576,081 (endoxohexahydrophthalic acid), 244
U.S.P. 2,576,666 (cyclodiene insecticides), 216
U.S.P. 2,635,977 (cyclodiene insecticides), 216
U.S.P. 2,642,354 (dalapon), 243
U.S.P. 2,670,282 (aminotriazole), 241
U.S.P. 2,695,258 (pyrotechnic mixtures), 68
U.S.P. 2,717,851 (cyclodiene insecticides), 216
U.S.P. 2,761,806 (organophosphorus nematicide), 219
Unsaturated hydrocarbons, 184
Unsulphonated residue, 184
Uraniagrün, 147
Urginia maritima, as rat poison, 293
Urocystis cepulae, temperature and attack by, 35
— *tritici*, straw as vector of, 307
Uromyces betae, influence of potash fertilizers on attack by, 30
" Urox ", 243
" Uspulun ", 260
" Uspulun Trockenbeize ", 260
Ustilago avenae, action of basic copper carbonate on, 257 ; control by organomercury compounds, 262 ; effect of soil moisture on attack by, 34
— *hordei*, use of basic copper carbonate against, 257 ; use of sulphur against, 259
— *kolleri*, use of copper carbonate against, 257
— *levis*, see *U. kolleri*
— *nuda*, control by seed selection, 254 ; hot-water treatment against, 267 ; use of organomercury compounds against, 260
— *tritici*, hot-water treatment against, 267 ; use of organomercury compounds against, 260

V

Variability, in probit analysis, 92
Vedalia, see *Rodolia cardinalis*
Vegetable oils, see glyceride oils
Venturia inaequalis, as test organism in bioassay, 88 ; forecasting of correct time of spray application against, 39 ; leaf hairiness as resistance factor, 14 ; use of bis(2-hydroxy-5-chlorophenyl) sulphide against, 145 ; use of lime sulphur against, 112 ; use of organomercury compounds against, 140
Veratridine, as insecticide, 157
Veratrine, as insecticide, 157
Veratrum album, as insecticide, 156
— *viride*, as insecticide, 157
Verdigris, as fungicide, 120
Vert de Montpelier, 120
Verticillium albo-atrum, cultural control of, 40, 280 ; effect of manuring on attack by, 31
Vigour tolerance, 208
Violet 5BO, as fungicide, 143
Virus diseases, 3, 57 ; acquired immunity from, 21 ; eradication of, 299 ; use of insecticide against vectors of, 306 ; warm-water treatment against, 268
Viscosity, as factor of spray penetration, 71 ; as factor of spray retention, 78 ; in relation to insecticidal properties of hydrocarbon oils, 189
Volatility, in relation to toxicity, 253
" Volck ", 186

W

Warfarin, as rat poison, 295
Warm-water treatment, as control measure, 268
Washing soda, see sodium carbonate
Water supply, disinfection of, 304
— transmission of disease organisms, 303
Waterman analysis of hydrocarbon oils, 190
Weather conditions and degree of attack, 36 *et seq.*
" Weedazol ", 240
Weedkillers, 232 *et seq.*
Weeds, as crop pests, 2 ; biological control of, 50 ; chemicals for use against, 232 *et seq.*

Western red cedar, see *Thuja plicata*
Wettable sulphurs, 106
Wetting properties, assessment of, 69
Wheat, black rust see *Puccinia graminis* ; bunt, see *Tilletia caries* ; flag smut, see *Urocystis tritici* ; loose smut, see *Ustilago tritici* ; purples, see *Anguillulina tritici* ; smut, see *Tilletia caries* ; stem sawfly, see *Cephus cinctus* ; take-all disease, see *Ophiobolus graminis* ; yellow rust, see *Puccinia glumarum*
White arsenic, use in poison baits, 293
— haiari, as insecticide, 176
— hellebore, as insecticide, 156
— oils, 185
— squill, as rat poison, 293
Whizzed naphthalene, 250
Wind transmission, 303
— breaks, as control measure, 303
Wind-blown sulphur, 106
Wipfelkrankheit, 58
Wireworms, see Elateridae
Woolly aphis, see *Eriosoma lanigerum*
Wound dressings, 298

X

Xanthates, as fungicides, 130
Xanthomonas citri, climate and attack by, 37
— *malvacearum*, atmospheric conditions and attack by, 36 ; oxidation-reduction potential and resistance to, 17
— *stewarti*, use of streptomycin against, 138
— *vesicatoria*, cell sap acidity and resistance to, 17
X-irradiation, as control measure, 49
Xyleborus fornicatus, influence of nitrogenous manuring on damage by, 31

Z

" Zelio ", 294
Zinc ammonium silicates, as fungicides, 120
— arsenite, as insecticide, 149
— dimethyldithiocarbamate, see ziram
— ethylene bisdithiocarbamate, see zineb
— fluoroarsenate, as insecticide, 149
Zinc oxide, for seed treatment, 258
Zinc sulphate-lime, as fungicide, 140
Zineb, as fungicide, 129
Ziram, as fungicide, 129

[illegible]